Molecular Biology: Sequencing Techniques and Applications

Molecular Biology: Sequencing Techniques and Applications

Editor: Gildroy Swan

R CALLISTO REFERENCE

www.callistoreference.com

Callisto Reference,
118-35 Queens Blvd., Suite 400,
Forest Hills, NY 11375, USA

Visit us on the World Wide Web at:
www.callistoreference.com

ISBN: 978-1-63239-811-6 (Hardback)

Cataloging-in-publication Data

Molecular biology : sequencing techniques and applications / edited by Gildroy Swan.
   p. cm.
Includes bibliographical references and index.
ISBN 978-1-63239-811-6
1. Molecular biology. 2. Molecular biology--Technique. 3. Molecular genetics. 4. Genetics--Technique. I. Swan, Gildroy.
QH430 .M65 2017
572.8--dc23

Table of Contents

Preface...VII

Chapter 1 **The Cytochrome P450 Superfamily Complement (CYPome) in the Annelid**
Capitella teleta...1
Chris A. Dejong, Joanna Y. Wilson

Chapter 2 **RADIA: RNA and DNA Integrated Analysis for Somatic Mutation Detection**.................15
Amie J. Radenbaugh, Singer Ma, Adam Ewing, Joshua M. Stuart, Eric A. Collisson,
Jingchun Zhu, David Haussler

Chapter 3 **The Putative *Leishmania* Telomerase RNA (*Leish*TER) Undergoes**
***Trans*-Splicing and Contains a Conserved Template Sequence** 26
Elton J. R. Vasconcelos, Vinícius S. Nunes, Marcelo S. da Silva, Marcela Segatto,
Peter J. Myler, Maria Isabel N. Cano

Chapter 4 **Exome-Wide Somatic Microsatellite Variation is Altered in Cells with DNA**
Repair Deficiencies..40
Zalman Vaksman, Natalie C. Fonville, Hongseok Tae, Harold R. Garner

Chapter 5 **Sequencing, Annotation and Analysis of the Syrian Hamster**
(*Mesocricetus auratus*) Transcriptome..54
Nicolas Tchitchek, David Safronetz, Angela L. Rasmussen, Craig Martens,
Kimmo Virtaneva, Stephen F. Porcella, Heinz Feldmann, Hideki Ebihara,
Michael G. Katze

Chapter 6 **Novel Biogenic Aggregation of Moss Gemmae on a Disappearing African Glacier**...........65
Jun Uetake, Sota Tanaka, Kosuke Hara, Yukiko Tanabe, Denis Samyn,
Hideaki Motoyama, Satoshi Imura, Shiro Kohshima

Chapter 7 **Whole-Genome Sequencing of the World's Oldest People** 75
Hinco J. Gierman, Kristen Fortney, Jared C. Roach, Natalie S. Coles, Hong Li,
Gustavo Glusman, Glenn J. Markov, Justin D. Smith, Leroy Hood,
L. Stephen Coles, Stuart K. Kim

Chapter 8 **Early Chordate Origin of the Vertebrate Integrin αI Domains** 85
Bhanupratap Singh Chouhan, Jarmo Käpylä, Konstantin Denessiouk,
Alexander Denesyuk, Jyrki Heino, Mark S. Johnson

Chapter 9 **Deep Sequencing of HIV-1 near Full-Length Proviral Genomes Identifies High**
Rates of BF1 Recombinants Including Two Novel Circulating Recombinant
Forms (CRF) 70_BF1 and a Disseminating 71_BF1 among Blood Donors in
Pernambuco, Brazil..101
Rodrigo Pessôa, Jaqueline Tomoko Watanabe, Paula Calabria, Alvina Clara Felix,
Paula Loureiro, Ester C. Sabino, Michael P. Busch, Sabri S. Sanabani

Chapter 10   **Genomic Characterization and Phylogenetic Position of Two New Species
in *Rhabdoviridae* Infecting the Parasitic Copepod, Salmon Louse
(*Lepeophtheirus salmonis*)**...114
Arnfinn Lodden Økland, Are Nylund, Aina-Cathrine Øvergård, Steffen Blindheim,
Kuninori Watanabe, Sindre Grotmol, Carl-Erik Arnesen, Heidrun Plarre

Chapter 11   **Detection Theory in Identification of RNA-DNA Sequence Differences
Using RNA-Sequencing**..130
Jonathan M. Toung, Nicholas Lahens, John B. Hogenesch, Gregory Grant

Chapter 12   **Zinc-Finger Nuclease Knockout of Dual-Specificity Protein Phosphatase-5
Enhances the Myogenic Response and Autoregulation of Cerebral Blood Flow
in FHH.1^{BN} Rats**..142
Fan Fan, Aron M. Geurts, Mallikarjuna R. Pabbidi, Stanley V. Smith,
David R. Harder, Howard Jacob, Richard J. Roman

Chapter 13   **De novo Assembly of the Grass Carp *Ctenopharyngodon idella* Transcriptome
to Identify miRNA Targets Associated with Motile Aeromonad Septicemia**....................152
Xiaoyan Xu, Yubang Shen, Jianjun Fu, Liqun Lu, Jiale Li

Chapter 14   **Exome Sequencing Identifies Three Novel Candidate Genes Implicated in
Intellectual Disability**..162
Zehra Agha, Zafar Iqbal, Maleeha Azam, Humaira Ayub, Lisenka E. L. M. Vissers,
Christian Gilissen, Syeda Hafiza Benish Ali, Moeen Riaz, Joris A. Veltman,
Rolph Pfundt, Hans van Bokhoven, Raheel Qamar

Chapter 15   **High Diversity and Low Specificity of Chaetothyrialean Fungi in Carton
Galleries in a Neotropical Ant–Plant Association**..173
Maximilian Nepel, Hermann Voglmayr, Jürg Schö nenberger, Veronika E. Mayer

Chapter 16   **On the Importance of the Distance Measures Used to Train and Test
Knowledge-Based Potentials for Proteins**...183
Martin Carlsen, Patrice Koehl, Peter Røgen

Chapter 17   **An Odorant-Binding Protein Is Abundantly Expressed in the Nose and in the
Seminal Fluid of the Rabbit**..201
Rosa Mastrogiacomo, Chiara D´Ambrosio, Alberto Niccolini, Andrea Serra,
Angelo Gazzano, Andrea Scaloni, Paolo Pelosi

Chapter 18   **Migratory Birds Reinforce Local Circulation of Avian Influenza Viruses**...................212
Josanne H. Verhagen, Jacintha G. B. van Dijk, Oanh Vuong, Theo Bestebroer,
Pascal Lexmond, Marcel Klaassen, Ron A. M. Fouchier

Permissions

List of Contributors

Index

Preface

Molecular biology studies the biological activity of cells that occurs between various forms of biomolecules. Some examples of biomolecules are carbohydrates, lipids, nucleic acids and peptides. This book on molecular biology seeks to enumerate the cellular function of biomolecules and its application to various fields. Most of the topics introduced in this book cover new techniques and applications of molecular biology. The various sub-fields of molecular biology along with technological progress that have future implications are also glanced at. It will provide comprehensive knowledge to the readers. This book would be helpful for students and researchers associated with molecular modeling, genomics and pharmaceuticals. For all those who are interested in molecular biology, this book can prove to be an essential guide.

The information shared in this book is based on empirical researches made by veterans in this field of study. The elaborative information provided in this book will help the readers further their scope of knowledge leading to advancements in this field.

Finally, I would like to thank my fellow researchers who gave constructive feedback and my family members who supported me at every step of my research.

Editor

The Cytochrome P450 Superfamily Complement (CYPome) in the Annelid *Capitella teleta*

Chris A. Dejong, Joanna Y. Wilson*

Department of Biology, McMaster University, Hamilton, Ontario, Canada

Abstract

The Cytochrome P450 super family (CYP) is responsible for a wide range of functions in metazoans, having roles in both exogenous and endogenous substrate metabolism. Annelids are known to metabolize polycyclic aromatic hydrocarbons (PAHs) and produce estrogen. CYPs are postulated to be key enzymes in these processes in annelids. In this study, the CYP complement (CYPome) of the annelid *Capitella teleta* has been robustly identified and annotated with the genome assembly available. Phylogenetic analyses were performed to understand the evolutionary relationships between CYPs in *C. teleta* and other species. Predictions of which CYPs are potentially involved in both PAH metabolism and steroidogensis were made based on phylogeny. Annotation of 84 full length and 12 partial CYP sequences predicted a total of 96 functional CYPs in *C. teleta*. A further 13 CYP fragments were found but these may be pseudogenes. The *C. teleta* CYPome contained 24 novel CYP families and seven novel CYP subfamilies within existing families. A phylogenetic analysis identified that the *C. teleta* sequences were found in 9 of the 11 metazoan CYP clans. Two CYPs, CYP3071A1 and CYP3072A1, did not cluster with any metazoan CYP clans. We found xenobiotic response elements (XREs) upstream of *C. teleta* CYPs related to vertebrate CYP1 (CYP3060A1, CYP3061A1) and from families with reported transcriptional upregulation in response to PAH exposure (CYP4, CYP331). *C. teleta* had a CYP51A1 with ~65% identity to vertebrate CYP51A1 sequences and has been predicted to have lanosterol 14 α-demethylase activity. CYP376A1, CYP3068A1, CYP3069A1, and CYP3070A1 were the most appropriate candidates for steroidogenesis genes based on their phylogeny and warrant further analyses, though no specific aromatase (estrogen synthesis) candidates were found. Presence of XREs upstream of *C. teleta* CYPs may indicate a functional aryl hydrocarbon receptor in *C. teleta* and candidate CYPs for studies of PAH metabolism.

Editor: John A. Craft, Glasgow Caledonian University, United Kingdom

Funding: Natural Sciences and Engineering Research Council of Canada Discovery and Accelerator Program (grant #2011R00152; http://www.nserc-crsng.gc.ca/index_eng.asp) and Ontario Ministry of Research and Innovation Early Researcher Award (ER10-07-197; http://www.ontario.ca/ministry-research-innovation). The funders had no role in study design, data collection and analysis, decision to publish, or preparation of the manuscript.

Competing Interests: The authors have declared that no competing interests exist.

* Email: joanna.wilson@mcmaster.ca

Introduction

The Cytochrome P450 (CYP) superfamily of protein enzymes are found in all domains of life [1,2]. CYPs catalyze a monooxygenase reaction [3] of compounds that fall into two general categories: exogenous (i.e. xenobiotics) and endogenous (e.g. steroids and lipids) substrates. CYPs are involved in both the synthesis and catabolism of important biological signaling molecules. CYPs involved in metabolism of endogenous substrates typically act on a small number of very similar, structurally related molecules. CYPs responsible for metabolism of xenobiotics generally have more flexible active sites to allow them to act on a wider array of substrates.

All newly identified CYPs are named by the Cytochrome P450 nomenclature committee, using standard conventions for this gene superfamily. CYPs are named by amino acid sequence identity; genes with 40% and 55% identify are placed in the same family and subfamily, respectively [4]. CYPs are named by family and subfamily using a numeral and letter, respectively. The specific gene is given a number, by order of discovery [4]. For example

CYP19A1 is in family 19, subfamily A and has a gene number of 1.

Since the early 2000's there have been several studies focused on the CYP genome complements (CYPomes) in metazoans, with studies completed on vertebrates [5–7], hemichordates [8], insects [9], crustaceans [10], and Cnidaria [11]. Many more CYPomes have been partially completed and unpublished CYPomes have been made available on the Cytochrome P450 webpage [12]. The smallest number of genes in a metazoan CYPome was found in the sponge *Amphimedon queenslandica* (35 CYP genes) and the largest metazoan CYPome identified so far included ~235 genes in the lancelet, *Branchiostoma floridae* [13]. Vertebrate genomes typically contain 57–102 CYP genes [14].

Vertebrate steroidogenesis is well understood; the specific genes and proteins and the substrates and intermediates involved have been identified. CYPs and the hydroxysteroid dehydrogenases (HSDs) are the primary enzymes responsible for vertebrate steroidogenesis. The first step in the steroid pathway is the long-chain cleavage of cholesterol to pregnenolone via CYP11A [15]. The production of estradiol (18 carbon) from lanosterol (30

Table 1. Subset of highly conserved motifs across the *Capitella teleta* CYPome.

CYP	K-helix		Meander Coil	
	AA	EXXR	AA	FDPER
CYP10B1	341	ETFR	393	FKPER
CYP20A1	338	ESLR	390	FDPER
CYP26D1	348	EVLR	400	FDPDR
CYP3052A1	362	ELLR	415	FEPER
CYP3052A10	365	ELLR	418	FQPER
CYP3052A2	361	ELLR	414	FEPER
CYP3052A3	361	ELLR	414	FEPER
CYP3052A4	349	ELLR	378	..PER
CYP3062A1	359	EVYR	412	FNPDN
CYP3062A2	372	EVYR	425	FNPNR
CYP3063A1	358	EIMR	411	FNPDR
CYP3064A1	335	EVLR	388	FNPSR
CYP3065A1	379	ETYR	432	FRPER
CYP3065A2	336	ETYR	389	FRPER
CYP3065A3	336	ETYR	389	FRPER
CYP3065A4	377	ECYR	430	FKPER
CYP3065B1	375	ETFR	428	FKPER
CYP3066C1	380	ESLR	432	FNPKR
CYP3067A1	344	ESFR	400	FKYDR
CYP3068A1	336	EMLR	388	FDPYR
CYP3069A1	312	ETLR	364	FNPDQ
CYP3070A1	366	ETLR	418	FNPDR
CYP331A2	363	ETLR	418	FEPER
CYP371B1	403	EALR	455	FIPER
CYP372A1	346	ESFR	391	FIPER
CYP39B1	347	ESIR	398	FKPDR
CYP44C1	359	EGFR	411	FIPER
CYP4AT1	356	ESLR	409	YDPER
CYP4BK4	304	ESMR	356	FRPDR
CYP4EE1	399	ESLR	452	YNPER
CYP4V25	367	ETLR	419	FIPDR
CYP51A1	364	ETLR	416	FNPDR

Two motifs (K-helix, and meander coil) are represented in an aligned format to show conservation across the *C. teleta* CYPs. Bolded letters represent conserved residues. AA is the amino acid number where the motif begins in each gene. The expected motif sequence is given in each heading for comparison. The glutamic acid and arginine residues in the meander coil are conserved across the entire CYPome.

carbon) is a six to eight enzymatic step process and involves CYPs from families 11, 17, 19, 21 [15]. The sex steroids are one of the end products of steroidogensis. CYP19A has the aromatase function, which is responsible for estrogen production from androgen precursors. The CYP19A gene has only been found in chordates, though is predicted to have more ancestral origins [16].

Capitella teleta is a polychaete annelid found in marine environments along the Pacific and Atlantic shores around the continental United States, Japan and the Mediterranean [17]. There has been an interest in determining the identify and function of CYPs in *C. teleta*, primarily focused on deciphering their ability to metabolize xenobiotics and polycyclic aromatic

hydrocarbons (PAHs; see [18,19]). This stems from research that found *C. teleta* to be the most opportunistic invertebrate after a 1969 oil spill in Massachusetts [20] and a concentration dependent increase in CYP-dependent activity with exposure to PAHs [21] in *Capitella* spp, More recently, differences in tolerance to PAHs and capacity for PAH metabolism amongst *Capitella* species have been investigated [22,23]. Two CYPs in *C. teleta*, CYP331A1 (a novel family) and CYP4AT1, have been identified and their expression was increased in response to various PAHs, suggesting a possible role of these CYPs in PAH metabolism [18].

Invertebrate endocrine systems have been much less studied that their vertebrate counterparts. Yet, data show that multiple

Table 2. Subset of less conserved motifs across the *Capitella teleta* CYPome.

CYP	I-helix			Heme Loop	
	AA	[A/G]GX[D/E]T[T/S]		AA	PFXXGXRXCXG
CYP10B1	284	GAVETT		415	PFGHGARMCIG
CYP20A1	281	AGFHTT		410	PFGFGKRKCLG
CYP26D1	289	AGYETT		421	PFGSGSRSCAG
CYP3052A1	307	AGTATT		440	PFGAGPRVCLG
CYP3052A2	307	AGTATT		439	PFGAGPRVCMG
CYP3052A3	308	GGTATT		439	PFGAGPRVCLG
CYP3052A4	295	AGTSTT		401	PFGAGPRVCLG
CYP3062A2	319	AGTESM		447	PFGAGMRRCPG
CYP3063A1	301	AGTETS		437	PFGAGKRKCIG
CYP3064A1	281	GVSDGS		410	PFSTGQRSCVG
CYP3065A1	322	DSLDTL		452	PFGVGPRSCPG
CYP3065A2	279	DSLDTL		409	PFGVGPRSCPG
CYP3065A3	279	DSLDTL		409	PFGVGPRSCVG
CYP3065A4	320	DALDSL		450	PFGLGPRACAG
CYP3066C1	323	SGHSTV		452	PFGMGPRSCIG
CYP3067A1	286	....NT		424	AFGS...LCPG
CYP3068A1	278	ASQETL		411	PFGAGNRTCVG
CYP3069A1	264	GAQETL		383	PFGGGAHACVG
CYP3070A1	309	AGQETT		436	PFSLGQRSCLG
CYP331A2	306	AGYDTT		438	PFGAGPRNCIG
CYP371B1	346	GAVDTT		476	PFGFGARSCIG
CYP372A1	281	AGIDST		414	PFGYGPRMCIG
CYP372B1	280	PNIEIEDRST		421	PFSHGLRACPG
CYP39B1	282	ASLANA		419	PFGGGRFQCPG
CYP44C1	301	DGMITT		432	PFSCGPRMCPG
CYP4AT1	299	EGHDTT		429	PFSAGPRNCIG
CYP4EE1	340	EGHDTT		472	PFSAGPRNCIG
CYP4V25	308	EGHDTT		439	PFSAGLRNCIG
CYP51A1	306	AGQHTS		437	PFGAGRHRCIG

Two motifs (I-helix, and heme loop) are represented in an aligned format to show conservation across the *C. teleta* CYPs. Bolded letters represent conserved residues. AA is the amino acid number where the motif begins in each gene. The expected motif sequence is given in each heading for comparison. The cysteine residue in the heme loop are conserved across the entire CYPome. Note the lack of conservation in CYP372B1 (I-helix) and CYP3067A1 (I-helix and heme loop).

endocrine active agents, sometimes including steroids typical in vertebrates, are present in invertebrate lineages [24]. Annelids are one group of invertebrates thought to produce and utilize the vertebrate sex steroid estradiol. *C. teleta* and *Platynereis dumerilii*, another marine annelid, had estrogen receptors (ERs) that responded to exogenous estrogen and regulated downstream gene expression, the first species with this function identified outside the vertebrates [25]. The annelid *Nereis virens* had detectable aromatase activity, likely occurring in the gut epithelium [26]. Despite having detectable aromatase activity, the protein responsible for this function remains unknown. CYPome studies in annelid species may provide clues to the evolution of the steroidogenesis pathway in metazoa and whether annelid invertebrates utilize the same enzymes for *de novo* sex steroid production.

The objective of this study was to annotate the *C. teleta* CYPome. The *C. teleta* CYPome is the first detailed analysis of a lophotrochozoan CYPome, providing important information on CYP content and evolution in an understudied metazoan superphyla. This study examines the potential role of the various CYPs in exogenous metabolism, particularly PAHs, and hypothesizes which CYPs may have a role in *C. teleta* steroidogenesis.

Results

Eighty-four full length CYPs were identified and annotated from the *C. teleta* assembly (v1); the entire list of CYPs, their genomic location, size, and nomenclature are provided in Table S1. There were twelve partial CYP sequences identified that aligned well with existing ESTs but could not be completed based on the current assembly (Table S2). There were thirteen partial

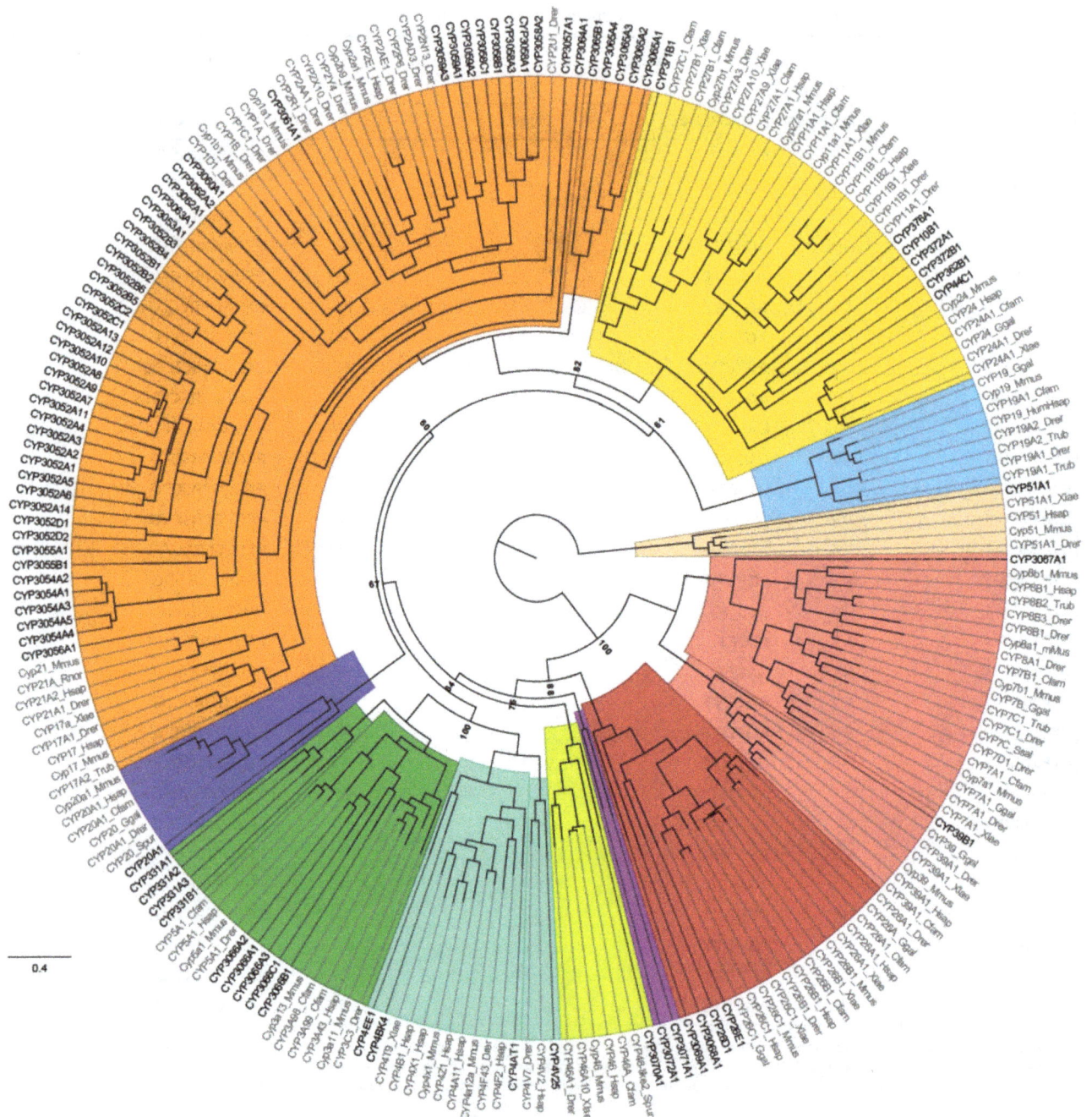

Figure 1. Phylogenetic tree of Cytochrome P450s in metazoa. The tree was completed on RaxML using non-parametric bootstrapping with a gamma distribution. The tree was rooted with CYP51. The black names are the *Capitella teleta* sequences. The tree is colour coded by clan: clan 2 orange, clan 3 dark green, clan 4 teal, clan 7 salmon, clan 19 light blue, clan 20 dark blue, clan 26 red, clan 46 lime green, clan 51 beige, mitochondrial clan yellow, and the two sequences that do not fit into a clan (CYP3071A1 and CYP3072A1) are purple.

CYP sequences identified that lacked any EST support (Table S3); whether these were genes or pseudogenes remains unclear. Based on the names assigned by the cytochrome P450 nomenclature committee, the predicted *C. teleta* CYPs were found in 9 of the 11 known metazoan CYP clans [13] and predicted 24 novel CYP families and 7 novel CYP subfamilies.

All of the full length CYPs contained at least some signature CYP motifs (Tables 1 and 2; Tables S4 and S5). The I-helix motif [A/G]GX[D/E]T[T/S] [27]; had conservation of at least three of the six amino acids in all but thirteen *C. telata* CYPs

(CYP3065A1–4, CYP3065B1, CYP3066A1–3, CYP3066C1, CYP3067A1, CYP372B1, and CYP39B1). The remaining CYPs had obvious sequence homology, with a majority of the conservation at the ends in the I-helix motif, even though this is the most poorly conserved motif of the four examined. The K-helix motif was fully conserved across all of the *C. teleta* sequences with no exceptions to the E-X-X-R consensus sequence (Table 1) [27]. The meander coil was conserved across all of the annotated sequences, although CYP372B1 and CYP4EE1 has substitutions for the first two amino acids in the motif (Table 1). Lastly, the

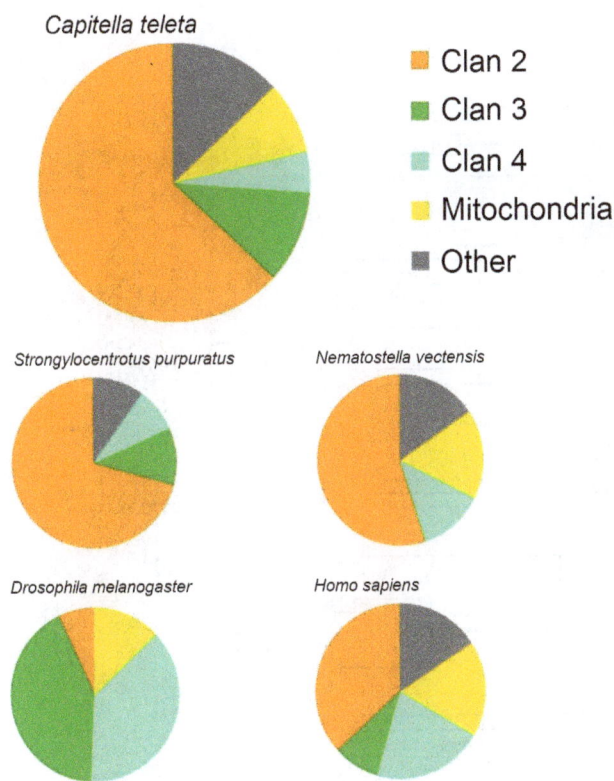

Figure 2. Distribution of the major Cytochrome P450 clans in five different species. *Capitella teleta*, *Strongylocentrotus purpuratus* [8], *Nematostella vectensis* [11], *Drosophila melanogaster* [9], and *Homo sapiens* [5] are compared.

cysteine residue in the heme binding loop is highly conserved, with very few exceptions [28], and this residue was present in all of the *C. teleta* sequences (Table 2). There was clear homology in the heme loop motif across all of the *C. teleta* sequences except for a gap in the motif in CYP3067A1. Interestingly, CYP3067A1 had a gap in both the heme loop and I-helix motifs (Table 2).

The phylogenetic relationships among the genes of the *C. teleta* CYPome is shown in Figure 1 and the distribution of these genes in the major clans (clans 2, 3, 4, and mitochondrial) shown in Figure 2. A majority of the *C. telata* CYPs were in clan 2, accounting for ~60% of the CYPome (Figure 2); Six of these genes were the most basal sequences within this clan (Figure 1). 33 genes were clustered as a distinct sister group to the CYP1 and CYP2 genes, without a single vertebrate sequence (Figure 2). Five sequences clustered with the vertebrate CYP1s and eight sequences were clearly clustered within the CYP2s. As expected, CYP2U1 was the most basal of the CYP2 genes (Figure 2). There were no *C. telata* sequences that clustered with the CYP17 or CYP21 sequences.

Clan 3 and 4 contained nine and four CYPs, respectively, while six genes were from the mitochondrial clan (Figure 2). A single *C. telata* sequence was found to cluster with CYP4V (CYP4V25), CYP7s/CYP306s (CYP3067A1), CYP11A (CYP376A1), CYP20 (CYP20A1), CYP27s (CYP371B1), CYP39 (CYP39B1), CYP46 (CYP3070A1), CYP51 (CYP51A1). Interestingly, there were a small numbers of genes (4 sequences) that clustered with CYP26s (Figure 1).

Figures 3, 4, and 5 shows phylogenies for clans 2 (Figure 3), 3 and 4 (Figure 4), and the mitochondrial clan (Figure 5). Invertebrate sequences were added to those sequences included in Figure 1 to help resolve and increase bootstrap support for internal branching arrangements within each clan (Figures 3–5). The addition of invertebrate sequences to the larger phylogeny interfered negatively with tree construction, producing a phylogeny with less robust bootstrap values. In the clan 2 phylogeny (Figure 3), sequences from *C. elegans* and *D. pulex* were added to the analysis; the *C. elegans* sequences clustered with the CYP2s. The CYP3058 family, clustered closest with the *C. elegans* sequences. The large cluster of clan 2 *C. teleta* CYPs remained on their own, as in the large phylogeny (Figure 1), basal to the rest of the clan 2 sequences. Clan 3 and 4 were sister clans (Figure 1) and were included together on the same clan phylogeny (Figure 4) with added sequences from *C. elegans*, *H. robusta* and *D. pulex*. The additional *C. elegans* sequences clustered closest with the CYP331 family, which were basal in clan 3 in the large phylogeny (Figure 1). The *D. pulex* sequences clearly clustered with the CYP4Vs, including the *C. telata* CYP4V25 (Figure 4). In the mitochondrial clan phylogeny (Figure 5), CYP10B1, CYP362B1, CYP44C1, CYP372A1, and CYP372B1, clustered with CYP36 from *D. pulex* and CYP44 from *C. elegans*. CYP371B1 clustered with *H. robusta* CYP371A1.

Table 3 provides the upstream XREs of *C. teleta* genes from CYP families CYP331 and CYP4. The CYP1-like genes, CYP3060A1 and CYP3061A1, were also examined. These CYP genes were either closely related to vertebrate CYP1s (CYP3060A1, and 3061A1) or genes that were upregulated in response to PAH exposure (CYP331 and CYP4 [18]. CYP331A1 had three XREs within 10 kb of the start site, the remaining CYP331A genes had no XREs. CYP331B1, CYP3060A1, CYP3061A1, and CYP4AT1 each had one XRE 10 kb upstream. Multiple XREs were found upstream of CYP4V25 (two) and CYP4BK4 (four). Only CYP4EE1, of the *C. teleta* CYP4s, had zero XREs upstream of the start site. There was a p-value of $5.96e-05$ and q-value of 1 for each of the sites, this was calculated using overall genome base frequencies.

Discussion

CYPome Annotation

Annotation of CYPomes can be challenging when working in species that are distantly related to those with a defined CYPome, because the searches are based on homology to known, yet distant sequences: *C. teleta* is the first lophotrochozoan to have its CYPome annotated and vertebrate sequences were primarily used in our initial searches. These reference vertebrate sequences were well curated, with very high confidence in their annotation, including exon boundaries, making any manual corrections from the PASA output for *C. teleta* more reliable. Annotations of *C. teleta* were additionally verified using *C. elegans* and *D. pulex* sequences for unique hits and no significant regions were found that the vertebrate sequences missed. Overall, our analysis predicted eighty-four full length CYPs, and identified twelve partial CYP sequences that aligned well with existing ESTs, and thirteen partial CYP sequences that lacked any EST support. Our analysis of the *C. teleta* CYPome has identified 24 novel families and 7 novel subfamilies. CYP26 contained two new subfamilies and CYP4, CYP10, CYP39, CYP44, CYP352 each had one new subfamily in *C. teleta*. The CYPomes of non-chordate phyla often contain novel CYP families [12,14]: *C. elegans* contained 14 unique families [12] and the *D. melanogaster* CYPome contained 24 families with most families unique to arthropods [9].

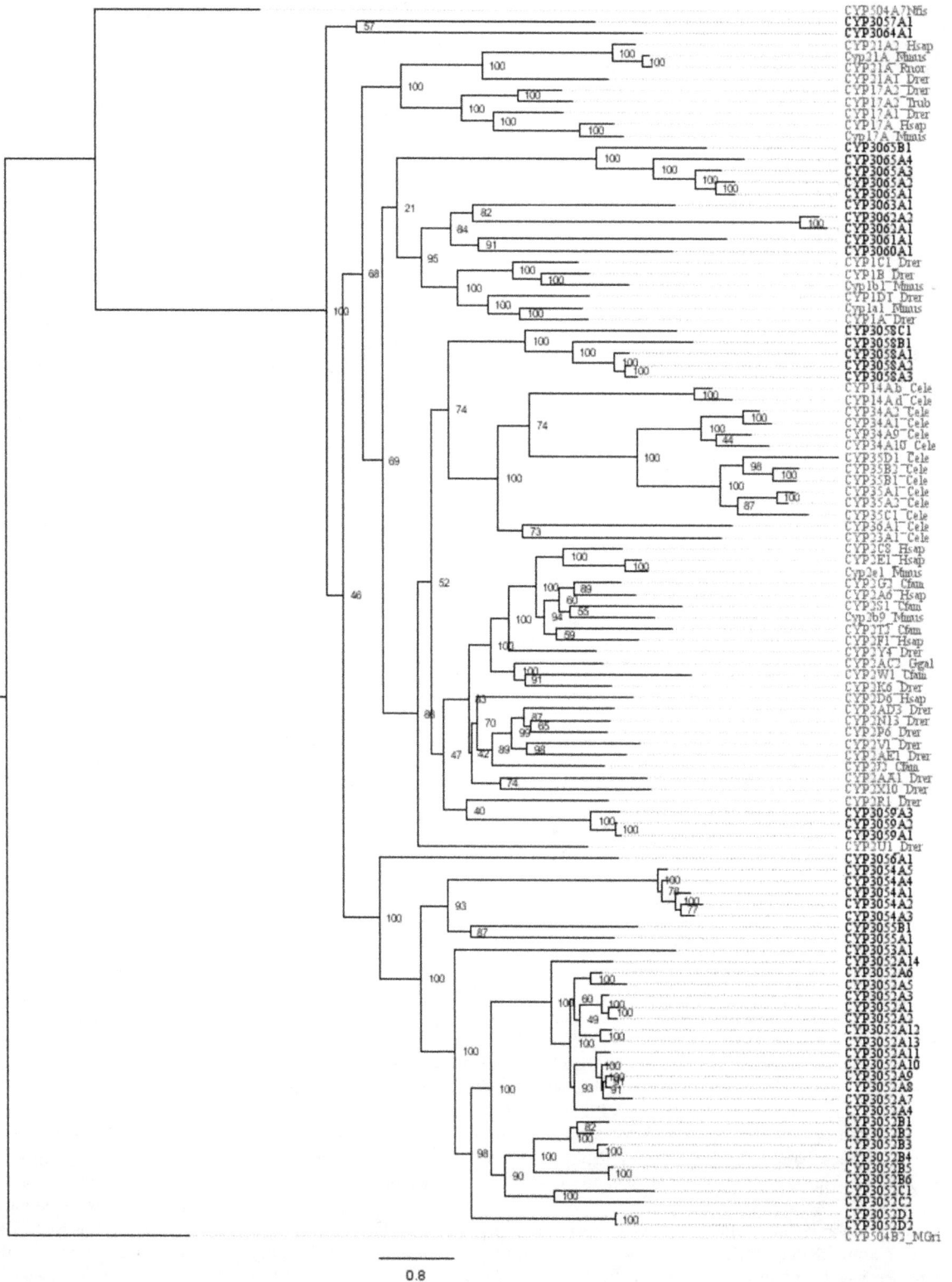

Figure 3. Phylogeny of Cytochrome P450 clan 2. The sequences are identical to those in Figure 1 with added invertebrate sequences to increase internal node resolution. The tree was completed on RaxML using non-parametric bootstrapping with a gamma distribution. *C. teleta* sequences are in black, all other sequences are in gray. The phylogeny was rooted using fungal CYP86s.

During manual annotation it was important to ensure that genes had a length of ~1500 bp, the average length for a CYP. Start (ATG) and stop (TAA/TAG/TGA) codons were noted and always present, as well as appropriate splice signals (GT/AG) [29] at intron/exon boundaries. The numbers of exons were not well conserved between related CYPs in other species. Exon number was taken into consideration between related sequences within *C. teleta* during searches for missing exons where EST data was lacking for annotation.

All of the fully annotated *C. teleta* CYPs had EST support covering all or almost all of the gene. A notable exception was CYP3052C1, which was missing EST data for exons one and two. Homology searches in the expected upstream and downstream regions were able to identify the missing exons. There were 12 incompletely annotated CYPs with EST support and these were presumed to be functional, full length genes though they could not be resolved with the existing genome assembly. Thus, the total number of CYPs identified in *C. telata* was 96, which fall into the range predicted by Nelson and colleagues [13] and is comparable to the 50–100 genes found in vertebrate CYPomes [5–7], 120 genes in the sea urchin *S. purpuratus* [8], 75 genes in the crustacean *Daphnia pulex* [10], 83 genes in the insect *D. melanogaster* [9] and 82 genes in the sea anemone *N. vectensis* [11]. *C. teleta* had an average number of CYPs for a metazoan CYPome.

There were 13 gene fragments that may be pseudogenes. These fragments lacked EST support or had identifiable early stop codons. The number of possible pseudogenes per functional gene (.14) is higher than noted in other species: *Daphnia pulex* has.04 [10], *C. elegans* has.1 [12] and *D. melanogaster* has.08 [9] pseudogene per gene. It is possible that a small number of these fragments were functional genes. One CYP on scaffold 342 had EST support but had an in frame stop codon in the first exon.

To provide support for the annotation process, the identified CYPs were examined for conserved CYP motifs. The heme binding region starts around amino acid 430 and has a well conserved motif of PFXXGXXRXCXG (the 'X' represents a non conserved amino acid); the cysteine, until recently, was considered the only absolutely conserved amino acid in all known CYPs, although exceptions have been documented [28]. There are three other well conserved motifs: portions of the I-helix, [A/G]GX[E/D]T[T/S], located around amino acid 300; K-helix, EXXR, located around amino acid 360; and an area known as the 'meander coil', FDPER, located around amino acid 410 [30]. The K-helix motif is incredibly well conserved in CYPs, with only a handful of known exceptions to the two conserved amino acids [31]. These motifs are important when analyzing potential CYPs, if one or more of these regions are missing, or out of place, it is likely that the gene was constructed incorrectly, a pseudogene or not a CYP at all.

The high conservation in the motifs (Tables 1 and 2) was expected and supports our annotation of these genes as CYPs. The least conserved domain is the I-helix, and our findings in *C. teleta* support this; CYP3065A1–4, CYP3065B1, CYP3066A1–3, CYP3066C1, CYP3067A1, and CYP372B1, CYP39B1 all have lower conservation in the I-helix. The gaps in CYP3067A1 and the insertion in CYP372B1 are peculiar. Whether these genes are fully functional may be questioned, yet, there is EST support to show that they are expressed.

The *C. teleta* CYPome phylogenetic analysis (Figure 1) contains almost exclusively vertebrate sequences, along with the *C. telata* sequences we identified. The arrangement of the clans was consistent with previous work, down to the family level of the known sequences [6,7]. It was difficult to add any sequences outside of vertebrates because of their divergence from vertebrate and *C. teleta* sequences and the lack of sequences that would help provide definitive phylogenetic relationships. When *Drosophila melanogaster* sequences were added to the phylogeny the bootstrap support was very weak, especially in clan 2–4 where most *D. melanogaster* sequences were added, and this is likely due to the evolutionary distance between vertebrates, insects and annelids. The *D. melanogaster* CYPome paper [9] provides a prime example of the difficulty in creating phylogenies between vertebrates and invertebrate CYP sequences. There were major branches (i.e. those that separate clans) with less than 10% bootstrap support [9]. The more recent *D. pulex* CYPome [10] had much better support at the clan level (support beyond the clan level was not provided), which was due to increased saturation in arthropod CYPs from available insect CYP sequences and basal chordate CYPs. As genome sequences become available from a wider array of species across the major metazoan phyla, the evolutionary distance between CYPomes will be reduced and help improve the phylogenetic analyses.

The clan phylogenies (Figures 3–5) have additional invertebrate sequences to help resolve nodes within the major clans found in the *C. teleta* CYPome. The phylogenies are rooted using plant and fungi CYPs, these sequences were the closest CYPs to the clans that were being rooted on the CYP webpage [12]. Using closely related clans as an outgroup, such as clan 46 for the clan 3 and 4 phylogeny, did not provide robust bootstrap support. The clan 2 phylogeny (Figure 3) included sequences from *C. elegans*, the clan 3 and 4 phylogeny (Figure 4) included *D. pulex*, *D. magna*, and *C. elegans*. sequences, and the mitochondrial phylogeny (Figure 5) incorporated sequences from *D. pulex*, *Caenorhabditis spp.*, and *H. robusta*. The addition of these sequences in the clan phylogenies increased the support for the internal nodes (data not shown). The sequence similarity data and phylogenetic analyses have provided information to infer the placement of the *C. teleta* CYPs into clans and assign nomenclature.

Clan Distribution

C. teleta possesses CYPs from all the metazoan clans except for 19 and 74. Clan 19 has not been found outside chordates [32] and clan 74 had not been found outside anemone and placozoa [13], although it has been recently found in amphioxus [16]. *C. teleta* is the first protostome analyzed to have representation in clan 46. CYP46 is the only clan 46 CYP gene in vertebrates and functions as a cholesterol 24-hydroxylase in the brain [33]. Since *C. teleta* CYP3070A1 had only 35% identity with human CYP46A1, it is difficult to predict whether the function is conserved in the *C. telata* ortholog. *In silico* molecular docking or 3D modeling of the protein may help support or refute the possibility that cholesterol is a substrate of CYP3070A1.

There are 53 full length clan 2 CYPs in *C. teleta*, representing ~60% of the total CYPome (Figure 2). This is the second largest in relative size for known CYPomes and is smaller only to *S. purpuratus* (~70%, Figure 2). Insects generally have only 5.5–10% of their CYPs in clan 2 [10]. The function of many insect

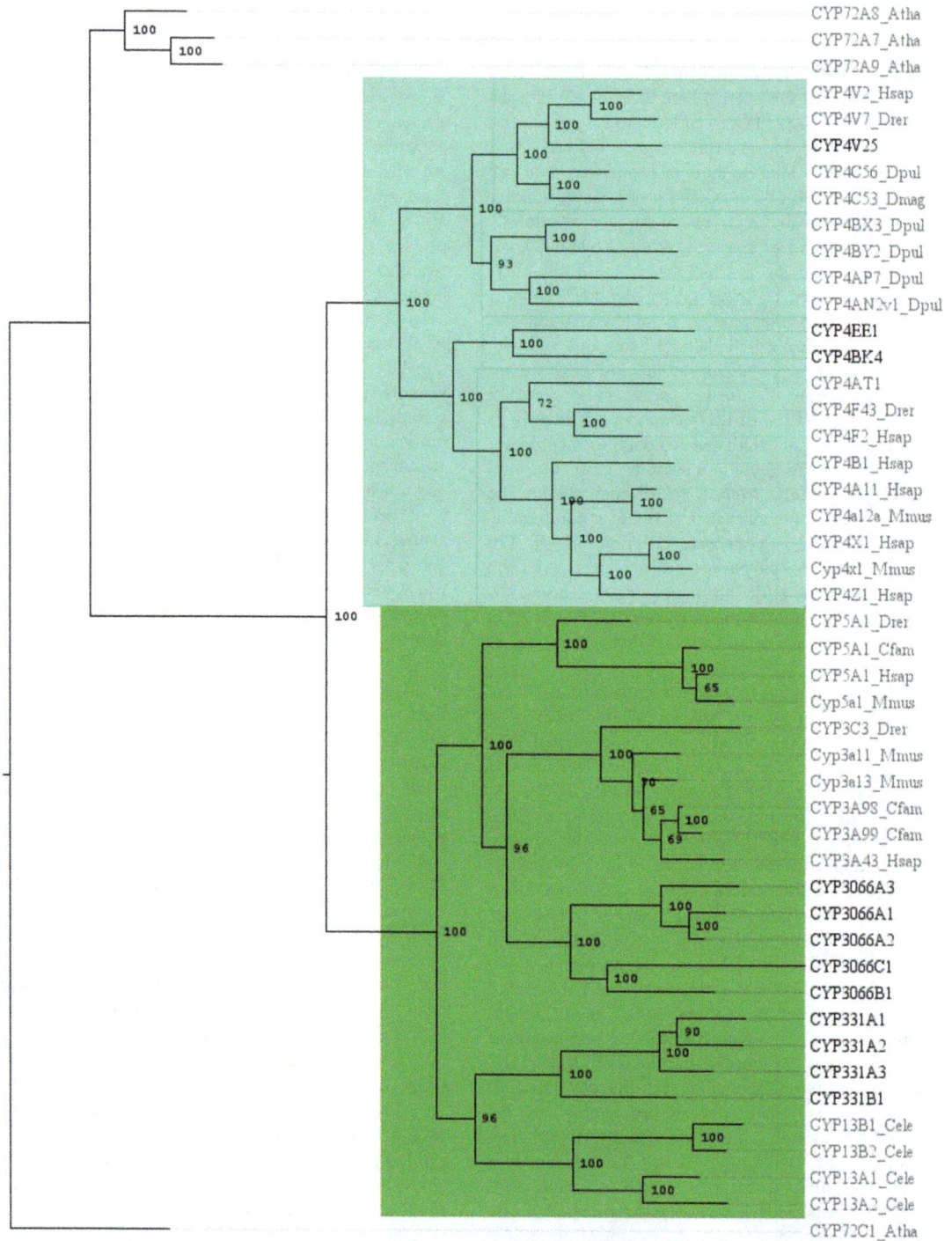

Figure 4. Phylogeny of Cytochrome P450 clan 3 and 4. The sequences are identical to those in Figure 1 with added invertebrate sequences to increase internal node resolution. The tree was completed on RaxML using non-parametric bootstrapping with a gamma distribution. *C. teleta* sequences are in black, all other sequences are in gray. The phylogeny was rooted using fungal *Arabidopsis thaliana* CYP72s.

clan 2 CYPs are unknown but some are known for their role in ecdysone synthesis [10]. Clan 2 CYPs are much more important for metabolism of exogenous compounds in mammals [34].

All of the *C. telata* clan 2 CYPs were located in novel CYP families; indeed the *C. telata* clan 2 sequences had 14 novel CYP families made up from 20 subfamilies. It has been postulated that a large number of CYPs related to families involved in exogenous metabolism (i.e. families 1–4) may suggest evolutionary pressure towards diverse function [11]. The largest family was CYP3052 with 24 sequences; these sequences made up the majority of the large standalone cluster of 33 *C. teleta* CYPs on the phylogenetic tree (Figure 1). If this family of *C. teleta* CYPs follows the trend of other large CYP families, namely families CYP1–4, then these proteins may be involved in xenobiotic metabolism.

There were five novel families with a single sequence each (CYP3060–3063 and CYP3065) that were CYP1-like. There were fewer CYP1-like genes in *C. telata* than were found in *S. purpuratus* (11) but similar to what is typical (3–4 CYP1 genes) in vertebrates [8,12]. Of the two families that grouped with CYP2s, CYP3058 clustered more closely with *C. elegans* sequences than vertebrate CYP2 sequences in the clan 2 phylogeny (Figure 3). The other family, CYP3059, clustered with vertebrate CYP2R, although the bootstrap support in the clan phylogeny was quite low (Figure 3) suggesting the placement of this family is uncertain with respect to the vertebrate CYP2 families. The function of the *C. elegans* CYPs are unknown but vertebrate CYP2s are well known for their role in xenobiotic metabolism [34]. CYP3057A1 and CYP3064A1 were basal in this clan and had high divergence from the remaining sequences.

The clan 3 phylogeny had sequences from across all metazoan phyla. Clan 3 contains families CYP3 and CYP5 in vertebrates, but is represented by different families in invertebrates such as families CYP6 and CYP9 [35]. Mammalian CYP3s are known to have very flexible active sites that can accommodate structurally diverse substrates. CYP3A4 is the most important enzyme involved in drug metabolism in humans but other CYP3s are also important in metabolism of endogenous and exogenous compounds [33]. Clan 3 CYPs are involved in both endogenous and exogenous metabolism in arthropods [10]. *C. teleta* had two clan 3 families with a total of nine CYPs; both families were novel. *N. vectensis* had 20 clan 3 CYPs [11], *S. purpuratus* had 10 [8], and *D. melanogaster* has an expanded clan 3 with 36 CYPs [9]. Mammals appear to have a much smaller number of clan 3 genes than many invertebrate species; humans have just five clan 3 sequences from a single subfamily [5].

The *C. teleta* clan 3 sequences included CYP331A1, which had been previously described [18]. CYP331A1 had increased expression from exposure to benzo[α]pyrene (BaP) and fluoranthene, two PAHs [18]. The CYP331 family has been expanded in this annotation with two more *CYP331A* genes and the *CYP331B1* gene.

The CYP4 family was expanded in *D. melanogaster* (32) and other insects [9], but was relatively limited in *N. vectensis* (3) [11]. There were five clan 4 CYPs in *C. teleta* and all were from the CYP4 family. CYP4V25 was an ortholog to CYP4Vs yet was below the 55% sequence identity threshold used during standard nomenclature. All of the top BLAST hits for CYP4V25 were CYP4Vs from various species (data not shown). Furthermore, CYP4Vs have been found in molluscs and crustaceans (DR

Nelson, personal communication). Collectively, this information supports the placement of this sequence into the CYP4V family despite the low sequence identity to other gene members. Little is known of CYP4 function outside vertebrates. CYP4C has a role in juvenile hormone synthesis in the cockroach *Blaberus discoidalis* [36]. In vertebrates, CYP4s primarily metabolize endogenous compounds, specifically fatty acids, although they do metabolize some exogenous pharmaceuticals (e.g. erythromycin) [37]. Yet, even in mammals the function of CYP4V is unknown [33].

Like the CYP4Vs, the function of CYP20A1 remains unclear in vertebrates. The *C. teleta* CYP20A1 is ~40% identical to other CYP20A1s but is a clear ortholog (Figure 1) with no other closely related sequences. CYP20A1 has been documented in invertebrates such as *S. purpuratus* and *H. robusta* [13]. It is interesting that CYP20A1 has unknown function yet has such clear homology between annelids and vertebrates.

CYP10 has been identified in molluscs and has been suggested as the only family in the mitochondrial clan in molluscs [38]. Since orthologs have now been identified in two major phyla, the *C. telata* CYP10 may suggest that CYP10 is present in all lophotrochozoans. Interestingly, CYP10 was not the only mitochondrial CYP in *C. telata*. CYP44 was placed in the mitochondrial clan; a CYP44 homolog has also been found in *C. elegans* [12], roundworms and molluscs [13]. Thus, CYP10 and CYP44 may be expected mitochondrial CYPs in lophotrochozoans.

PAH and xenobiotic metabolism

C. teleta has been long known to metabolize PAHs in sediment, and it has been suggested that CYPs were responsible [39,40]. BaP was metabolized, likely by CYPs, in another annelid, *Nereis virens* [40]. There is conflicting data on whether CYPs are transcriptionally upregulated after PAH exposure in *N. virens* (primarily tested with BaP exposure), with some studies reporting a 2-fold increase in CYPs and others reporting no change (reviewed by [40]). In *C. teleta*, two CYPs (CYP4AT1 and CYP331A1) had a 1.9–2.6 fold increased expression with exposure to some PAHs, including BaP [18]. CYP4AT1 had 1.25 to 1.9 fold increase in gene expression after exposure to PAH contaminated sediments [18]. Interestingly, there were 3 genes found for *C. teleta* in the CYP331 family, two of which were in the same (CYP331A) subfamiliy.

An important factor to consider is the presence of CYP co-enzymes in these reactions such as cytochrome P450 reductase and cytochrome b5 [41]. The current putative JGI *C. teleta* transcriptome assembly predicts the presence of cytochrome P450 reducatase (estExt_Genewise1Plus.C_990037) and cytochrome b5 (estExt_Genewise1.C_2000018), which shows the presence of coenzymes necessary for functional CYP activity in this species.

Vertebrates, including mammals, mediate metabolism of PAHs through CYP1A and CYP1 gene expression is increased with exposure to PAHs (reviewed in [43]), through transcriptional activation via the aryl hydrocarbon receptor (AHR) pathway [44]. In mammals, the CYP2B and CYP2C subfamilies are also important for PAH metabolism [42] but these subfamilies are not present in all vertebrates. The AHR is activated by planar PAHs and halogenated aromatic hydrocarbons; TCDD is the ligand with the highest affinity for this receptor in many species [45]. AHRs transcriptionally regulate a battery of genes through interaction with a specific sequence, the xenobiotic or dioxin

(C)

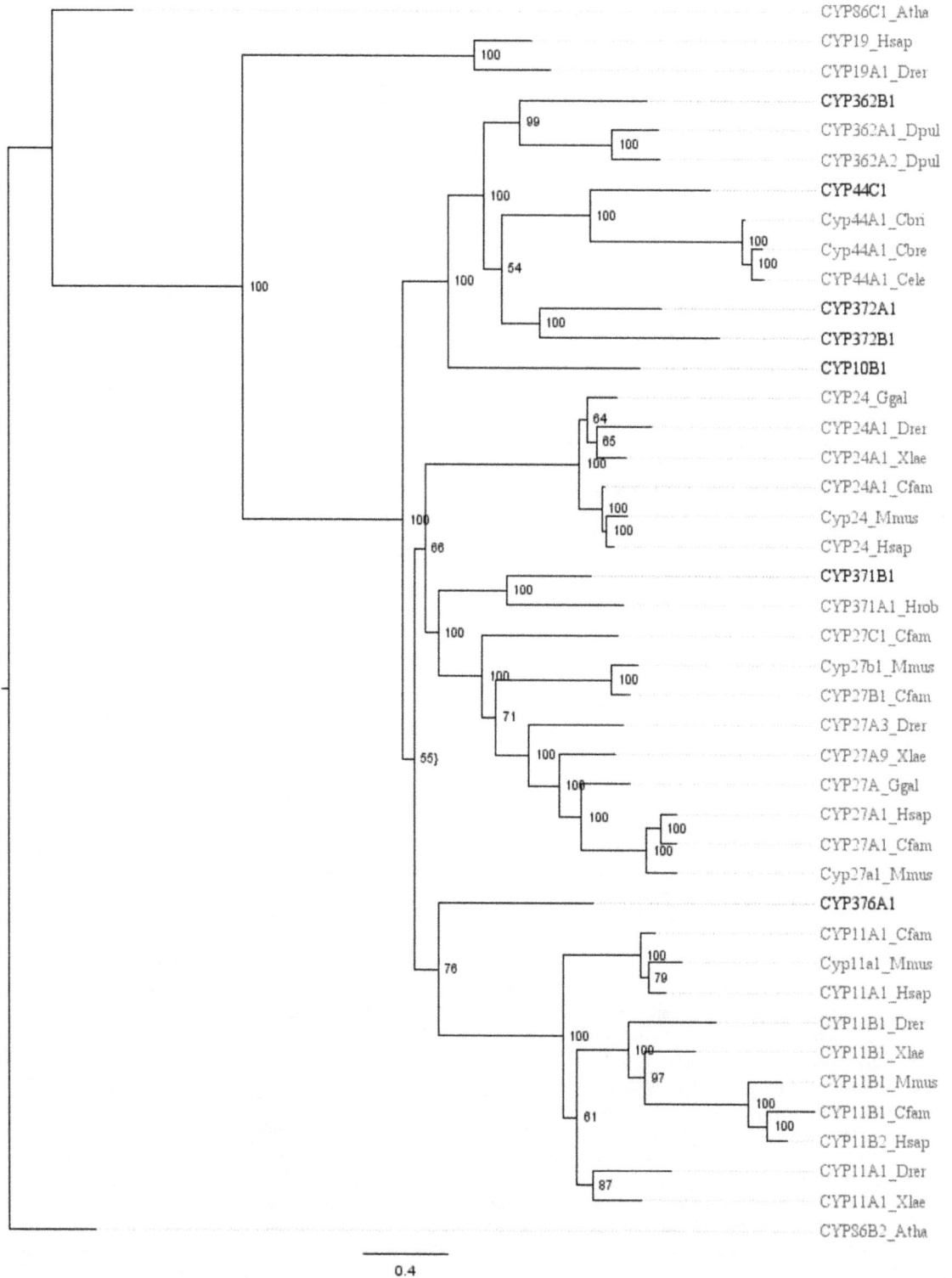

CYP86C1_Atha
CYP19_Hsap
CYP19A1_Drer
CYP362B1
CYP362A1_Dpul
CYP362A2_Dpul
CYP44C1
Cyp44A1_Cbri
Cyp44A1_Cbre
CYP44A1_Cele
CYP372A1
CYP372B1
CYP10B1
CYP24_Ggal
CYP24A1_Drer
CYP24A1_Xlae
CYP24A1_Cfam
Cyp24_Mmus
CYP24_Hsap
CYP371B1
CYP371A1_Hrob
CYP27C1_Cfam
Cyp27b1_Mmus
CYP27B1_Cfam
CYP27A3_Drer
CYP27A9_Xlae
CYP27A_Ggal
CYP27A1_Hsap
CYP27A1_Cfam
Cyp27a1_Mmus
CYP376A1
CYP11A1_Cfam
Cyp11a1_Mmus
CYP11A1_Hsap
CYP11B1_Drer
CYP11B1_Xlae
CYP11B1_Mmus
CYP11B1_Cfam
CYP11B2_Hsap
CYP11A1_Drer
CYP11A1_Xlae
CYP86B2_Atha

0.4

Figure 5. Phylogeny of Cytochrome P450 mitochondrial clan. The sequences are identical to those in Figure 1 with added invertebrate sequences to increase internal node resolution. The tree was completed on RaxML using non-parametric bootstrapping with a gamma distribution. *C. teleta* sequences are in black, all other sequences are in gray. The phylogeny was rooted using CYP19s and *Arabidopsis thaliana* CYP86s.

response element (XRE or DRE) [46]. Many AHR ligands, including PAHs, are also substrates for CYP1 enzymes [45]. AHRs are present in invertebrates and the amino acid sequence of the DNA binding domain is similar to that found in vertebrates. Indeed, AHRs from *Drosophila* [47], *C. elegans* [48], and *Mya arenaria* [49] are capable of binding with the mammalian XRE sequence. Therefore, we examined the upstream region of the CYP1-like (CYP3060A1 CYP3061A1), CYP331 and CYP4AT genes in *C. telata* to determine if XREs were present (Table 3). CYP331A1 had three XREs within 10 kb of the start site, but CYP4AT1 had only one. The difference in the number of XREs between these two CYPs may explain the difference in expression during BaP exposure, since there is a relationship between the number of XREs and the relative upregulation of the gene [50]. The CYP1-like *C. teleta* sequences had one XRE and CYP4BK4 had four XREs in the 10 kb upstream region. Should the AHR have a role in regulating gene transcription in *C. teleta* after exposure to PAHs, we would predict that CYP4BK4 would have the greatest transcriptional response. Considering the structural link between AHR ligands and CYP1 substrates in vertebrates, we might speculate that CYP4BK4 be a primary candidate gene for studies of PAH metabolism in this species. Future PAH exposure studies in *C. teleta* will shed light on to role of the AHR and XREs in *C. teleta* and the potential role these CYPs may play in PAH metabolism.

Steroidogenic CYPs

CYP51A1 enzymes are responsible for lanosterol-14-alpha-demethylation; the conversion of lanosterol into cholesterol [51]. Cholesterol is the precursor to steroids and this function is expected in all species with endogenous steroid production. The next step in vertebrate steroidogenesis is cholesterol-side-chain-cleavage, which is completed by CYP11A1 in vertebrates and converts cholesterol to pregnenolone [15]. There was one *C. teleta* CYP (CYP376A1) that clustered with the CYP11 family in the phylogenetic tree (Figure 1) and is the best candidate for cholesterol side-chain-cleavage function in *C. teleta*. CYP11B

functions in the synthesis of cortisol and coticosterone [15,52], which are not expected in annelids since these molecules have not been found in amphioxus, *Ciona intestinalis*, or sea urchins [53].

CYP17A1 functions as a 17-alpha-hydroxylase, which is responsible for converting pregnenolone into DHEA. The production of DHEA is the next step in steroidogenesis after side-chain-cleavage and before the production of androgens [15]. Since there is no *C. telata* CYP that clusters with the CYP17 genes from vertebrates, it is difficult to predict which CYP is likely to complete this function at this time. There were many clan 2 CYPs identified but whether 17-alpha-hydroxylase activity is mediated by one of them is unclear. Analyzing the single copy clan 2 CYPs (e.g. CYP3057A1 and CYP3064A1) would be an appropriate place to begin the search for a 17-alpha-hydroxylase enzyme.

Detectable estrogen production has been documented in annelids [26], yet there was no CYP19 identified in the *C. teleta* CYPome. This is not surprising, as a CYP19 has not been identified outside of chordates and sea anemone had no CYP19 [11], in spite of endogenous estrogen production [54]. It has been postulated that another CYP has the aromatase function outside of chordates [11]. There are many CYPs identified in *C. teleta*, the most promising candidates genes for steroidogenic functions are the single copy CYPs from clan 2, CYP376A1 from the mitochondrial clan, CYP3068A1 or CYP3069A1 from clan 26 and CYP3070A1 from clan 46. All of these CYPs should be further examined by *in silico* methods for their potential ability to bind the intermediates of the steroidogenic pathway.

Conclusion

Capitella teleta has an interesting complement of CYPs. CYPs were found in nine of the eleven metazoan CYP clans. There were a total of 24 novel CYP families; careful study will be required to determine their function. The annotation of the *C. teleta* CYPome will make annotating other lophotrochozoan CYPomes easier. With additional annelid and other lophotrochozoan CYP sequences, we will better understand which of the novel CYP families and subfamilies discovered here are specific to annelids. *C.*

Table 3. Xenobiotic response elements upstream of *C. teleta* CYPs.

CYP	Number of XREs 10 kb upstream
CYP331A1*	3
CYP331A2	0
CYP331A3	0
CYP331B1	1
CYP4V25	2
CYP4AT1*	1
CYP4BK4	4
CYP4EE1	0
CYP3060A1	1
CYP3061A1	1

Each gene was searched for the consensus xenobiotic response element (XRE) sequence TNGCGTG [39], 10 kb upstream of the start site. Genes were chosen based on homology to vertebrate CYP1s and genes from clan 3 and 4. Genes were searched in the 10 kb upstream region. The asterisks mark the genes which were transcriptionally upregulated with exposure to the PAHs benzo[α]pyrene, 3-methylcholanthrene, or fluoranthene [18]. p-value of 5.96e−05 and q-value of 1 for all sites.

teleta is known to survive well in polluted environments and two CYP genes CYP331A1 and CYP4AT1 were known to be transcriptionally regulated by PAHs [18]. Indeed, several more closely related homologs were identified in this study. XRE sequences were found upstream in several of these genes suggesting that CYP331A1, CYP331B1, several CYP4s and the CYP1-like CYP3060A1 and CYP3061A1 genes may be in the AHR gene battery. Empirical testing will be needed to demonstrate this and explore their possible role in PAH metabolism. Functional hypotheses were raised for several of the *C. teleta* CYPs. CYP51A1 is very likely to catalyze the production of cholesterol, due to a ~65% amino acid identity and clear orthology to other CYP51 sequences. Yet, the steroidogenic pathway was not completely identified. Cholesterol side-chain-cleavage has been hypothesized as the function of CYP376A1. Still, there are no obvious candidates for 17α-hydroxylase and aromatase enzymes, which are carried out by CYP17A and CYP19A, respectively, in vertebrates. Considering that *C. teleta* produces *de novo* estradiol, these reactions are likely undertaken by other CYPs. Future studies on invertebrate steroidogenesis should focus on the CYPs with low copy number and phylogenetic positions close to vertebrate steroidogenic CYPs shown in this study. *In silico* protein folding and docking studies may provide important clues to narrow the number of candidates genes for steroidogenic CYPs and direct future functional studies.

Methods

The *C. teleta* genome used for this study was version 1 of the assembly (Joint Genome Institute, University of California); the genome assembly had approximately 7.9x coverage with 21,042 scaffolds with a total size of 333.7 Mb. The EST database (National Center for Biotechnology Information, July 2012) had approximately 130,000 reads. The other sequences used in phylogenetic analyses were retrieved from the Cytochrome P450 web-page [12]. Many vertebrate sequences were used in the analyses, although there was a focus on *Danio rerio, Mus musculus,* and *Homo sapiens*; species which have had rigorous annotation of their CYPome [7,12,33]. A select number of CYP sequences from invertebrates were included: *Haliotis diversicolor* [12], *Crassostrea gigas* (NCBI), *Daphnia pulex* [10] and *Helobdella robusta* (JGI). For some phylogenies, *Caenorhabditis spp.* sequences were added [12].

Gene annotation

The *Capitella teleta* EST database was assembled with PASA (r2012-06-25) [55], to align and extend the ESTs to each other and to align them to the *C. teleta* genome. Homology searching of ESTs was performed using all CYPs from human and zebrafish using tBLASTn (v2.2.27) [56]. Hits were compiled and the regions hit were autonomously counted via a custom Perl script. This approach allowed for many CYPs to be used as inputs for homology searches. Since CYPs can have <15% sequence identity from each other in their amino acid sequences, the use of a wide variety of CYPs during homology searching maximizes the number of unique hits. The hit regions were checked against the PASA outputs, overlaps collected, and approximate gene regions predicted. The putative gene regions were compared to previously annotated CYPs and gene boundaries were adjusted using Artemis (v14.0.0) [57] according to homology. Since there are no closely related species with their CYPome analyzed or a large EST/refseq database, exact exon boundaries were difficult to annotate with very high certainty. The exon boundaries were examined for appropriate splice signals [29] and to ensure that the boundaries

were located appropriately to the reading frame. When there were large gaps in a gene, FASTA (36.3.5e) [58] searches against genome scaffolds were completed to find these missing regions, rather than BLAST, because FASTA has increased sensitivity.

Once annotated, CYPs were compared to the automated gene calls in September 2012 on JGI. There were no CYPs on JGI that were not found by the above method. The manual annotation made for more appropriate splice sites, with the JGI annotations at times leaving out segments or entire exons.

Phylogenetic Analyses of CYP sequences

Alignments were created in MUSCLE (v3.8.31) [59] and manually refined in Mesquite (v2.75) [60] at the amino acid level. The N- and C-termini of CYPs are more divergent and were hard masked from further analysis. ZORRO (r2011-12-01) [61], a soft masking tool, was used on the remainder of the alignment. The phylogenetic analysis was conducted using a total of 220 sequences on RAxML (v7.4.2) [62] with 100 bootstraps using the slower algorithm (-b) with a gamma distribution. The clan 2 phylogeny had additional *C. elegans* sequences, the clan 3 and 4 phylogeny included from *Daphnia pulex, Daphnia magna* and *C. elegans* sequences, and the mitochondrial phylogeny incorporated sequences from *D. pulex, Caenorhabditis spp., and H. robusta.* The maximum likelihood analyses were based on the VT substitution model with fixed base frequencies (phylogeny of all clans), MTMAM substitution model with fixed base frequencies (clan 2 phylogeny), LG substitution model with empirical base frequencies (clan 3 and 4 phylogeny), or JTT substitution model with empirical base frequencies (mitochondrial clan phylogeny). The appropriate models were determined by ProtTest (v3.2) [63]. To root the phylogenetic trees, CYPs outside the clans were chosen; the clan 2 phylogeny used CYP family 504 genes from fungus (*Magnaporthe grisea* and *Nassarius fischeri*), the clan 3 and 4 phylogeny used CYP72s from *Arabidopsis thaliana*, and the mitochondrial clan phylogeny used CYP86s from *A. thaliana* and vertebrate CYP19s. These roots were selected based on closely related out-groups from David Nelson's "singlefam tree" on the Cytochrome P450 webpage [12].

All predicted *C. teleta* CYP genes were named by the cytochrome P450 nomenclature committee using the sequences provided, synteny data available and the phylogenetic trees generated in this study. STRAP (r2013-02-26) [64] was used for the motif work and Figtree (v1.4.0) was used to generate the figures of the phylogenetic trees.

Searches for the xenobiotic response element (XRE, TNGCGTG) [65] in the 10 kb upstream region of the predicted start site in each gene of families *CYP331, CYP4* and *CYP3061* used the MEME suite (v4.9.1) [66].

Supporting Information

Table S1 Cytochrome P450 superfamily complement in *Capitella teleta*. Temporary names were based off the scaffold they were found on. Length is in amino acids. CYPs were named by the CYP nomenclature committee. There are a total of 84 full length CYPs. Only complete CYPs are listed.

Table S2 Incomplete cytochrome P450s in *Capitella teleta*. Temporary names are based off the scaffold they were found on. The listed CYPs are not full length and are missing exons but have EST support.

Table S3 Cytochrome P450 fragments in *Capitella teleta*. Temporary names are based off the scaffold they were found on. None of these fragments have EST support, except for p_342, suggesting they may be pseudogenes. P_342 had an early stop codon and is a pseudogene.

Table S4 Highly conserved motifs across the *Capitella teleta* CYPome. Two motifs (K-helix, and meander coil) are represented in an aligned format to show conservation across the *C. teleta* CYPs. Bolded letters represent conserved residues. AA is the amino acid number where the motif begins in each gene. The expected motif sequence is given in each heading for comparison. The glutamic acid and arginine residues in the meander coil are conserved across the entire CYPome.

Table S5 Less conserved motifs across the *Capitella teleta* CYPome. Two motifs (I-helix, and heme loop) are

represented in an aligned format to show conservation across the *C. teleta* CYPs. Bolded letters represent conserved residues. AA is the amino acid number where the motif begins in each gene. The expected motif sequence is given in each heading for comparison. The cysteine residue in the heme loop are conserved across the entire CYPome. Note the lack of conservation in CYP372B1 (I-helix) and CYP3067A1 (I-helix and heme loop).

Acknowledgments

We thank Jed Goldstone and David Nelson for review of the annotations and nomenclature of the sequences.

Author Contributions

Conceived and designed the experiments: JYW CAD. Performed the experiments: CAD. Analyzed the data: CAD. Contributed reagents/materials/analysis tools: CAD. Wrote the paper: JYW CAD.

References

1. Nelson DR (2011) Progress in tracing the evolutionary paths of cytochrome P450. BBA Proteins Proteom 1814: 14–18.
2. Nelson DR (1999) Cytochrome P450 and the individuality of species. Arch Biochem Biophys 369: 1–10.
3. Nebert DW, Gonzalez FJ (1987) P450 genes: Structure, evolution, and regulation. Annu Rev Biochem 56: 945–993.
4. Nelson DR, Koymans L, Kamataki T, Stegeman JJ, Feyereisen R, et al. (1996) P450 superfamily: Update on new sequences, gene mapping, accession numbers and nomenclature. Pharmacogenet Genomics 6: 1–42.
5. Lewis DFV (2004) 57 varieties: The human cytochromes P450. Pharmacogenomics 5: 305–318.
6. Nelson DR (2003) Comparison of P450s from human and fugu: 420 million years of vertebrate P450 evolution. Arch Biochem Biophys 409: 18–24.
7. Goldstone J, McArthur A, Kubota A, Zanette J, Parente T, et al. (2010) Identification and developmental expression of the full complement of cytochrome P450 genes in zebrafish. BMC Genomics 11: 643
8. Goldstone J, Hamdoun A, Cole B, Howard-Ashby M, Nebert D, et al. (2006) The chemical defensome: Environmental sensing and response genes in the *Strongylocentrotus purpuratus* genome. Dev Biol 300: 366–384.
9. Tijet N, Helvig C, Feyereisen R (2001) The cytochrome P450 gene superfamily in drosophila melanogaster: Annotation, intron-exon organization and phylogeny. Gene 262: 189–198.
10. Baldwin W, Marko P, Nelson D (2009) The cytochrome P450 (CYP) gene superfamily in *Daphnia pulex*. BMC Genomics 10: 169–181.
11. Goldstone JV (2008) Environmental sensing and response genes in cnidaria: The chemical defensome in the sea anemone *Nematostella vectensis*. Cell Biol Toxicol 24: 483–502.
12. Nelson DR (2009) The cytochrome P450 homepage. Hum Genomics 4: 59–65.
13. Nelson DR, Goldstone JV, Stegeman JJ (2013) The cytochrome P450 genesis locus: The origin and evolution of animal cytochrome P450s. Philos Trans R Soc B Biol Sci 368: 20120474.
14. Nelson DR (2011) Progress in tracing the evolutionary paths of cytochrome P450. BBA Proteins proteom 1814: 14–18.
15. Baker ME (2011) Origin and diversification of steroids: Co-evolution of enzymes and nuclear receptors. Mol Cell Endocrinol 334: 14–20.
16. Callard GV, Tarrant AM, Novillo A, Yacci P, Ciaccia L, et al. (2011) Evolutionary origins of the estrogen signaling system: Insights from amphioxus. J Steroid Biochem Mol Biol 127: 176–188.
17. Blake JA, Grassle JP, Eckelbarger KJ (2009) *Capitella teleta*, a new species designation for the opportunistic and experimental capitella sp. I, with a review of the literature for confirmed records. Zoosymposia 2: 25–53.
18. Li B, Bisgaard HC, Forbes VE (2004) Identification and expression of two novel cytochrome P450 genes, belonging to \CYP4\ and a new CYP331 family, in the polychaete *Capitella capitata* sp.I. Biochem Biophys Res Commun 325: 510.
19. Selck H, Palmqvist A, Forbes VE (2003) Biotransformation of dissolved and sediment-bound fluoranthene in the polychaete, capitella sp. I. Environmental toxicology and chemistry 22: 2364–2374.
20. Sanders HL, Grassle JF, Hampson GR, Morse LS, Garner-Price S, et al. (1980) Anatomy of an oil spill: Long-term effects from the grounding of the barge florida off West Falmouth, Massachusetts. J Marine Res 38: 265–380.
21. Lee RF, Singer SC (1980) Detoxifying enzymes system in marine polychaetes: Increases in activity after exposure to aromatic hydrocarbons. Rapp P-v Reun Cons Int Explor Mer 179: 29–32.
22. Linke-Gamenick I, Forbes VE, Mendez N (2000) Effects of chronic fluoranthene exposure on sibling species of capitella with different development modes. Mar Ecol Prog Ser 203: 191–203.
23. Bach L, Palmqvist A, Rasmussen LJ, Forbes VE (2005) Differences in PAH tolerance between capitella species: Underlying biochemical mechanisms. Aquat toxicol 74: 307–319.
24. Janer G, Porte C (2007) Sex steroids and potential mechanisms of non-genomic endocrine disruption in invertebrates. Ecotoxicol 16: 145–160.
25. Keay J, Thornton JW (2009) Hormone-activated estrogen receptors in annelid invertebrates: Implications for evolution and endocrine disruption. Endocrinol 150: 1731–1738.
26. Garciaa-Alonso J, Rebscher N (2005) Estradiol signalling in *Nereis virens* reproduction. Invertebr Reprod Dev 48: 95–100.
27. Werck-Reichhart D, Feyereisen R (2000) Cytochromes P450: A success story. Genome Biol 1: 1–9
28. Sezutsu H, Le Goff G, Feyereisen R (2013) Origins of P450 diversity. Philos Trans R Soc B Biol Sci 368: 20120428.
29. Mount SM (1982) A catalogue of splice junction sequences. Nucleic Acids Rese 10: 459–472.
30. Sezutsu H, Le Goff G, Feyereisen R (2013) Origins of P450 diversity. Philos Trans R Soc B Biol Sci 368: 20120428.
31. Rupasinghe S, Schuler MA, Kagawa N, Yuan H, Lei L, et al. (2006) The cytochrome P450 gene family CYP157 does not contain EXXR in the K-helix reducing the absolute conserved P450 residues to a single cysteine. FEBS Letters 580: 6338.
32. Reitzel AM, Tarrant AM (2010) Correlated evolution of androgen receptor and aromatase revisited. Mol Biol Evol 27: 2211–2215.
33. Nebert DW, Wikvall K, Miller WL (2013) Human cytochromes P450 in health and disease. Philosl Trans R Soc B Biol Sci 368: 20120431.
34. Nebert DW, Russell DW (2002) Clinical importance of the cytochromes P450. The Lancet 360: 1155–1162.
35. Verslycke T, Goldstone JV, Stegeman JJ (2006) Isolation and phylogeny of novel cytochrome P450 genes from tunicates (ciona spp.): A CYP3 line in early deuterostomes? Mol Phylogenet Evol 40: 760.
36. Bradfield JY, Lee YH, Keeley LL (1991) Cytochrome P450 family 4 in a cockroach: Molecular cloning and regulation by regulation by hypertrehalosemic hormone. Proc Natl Acad Sci U S A 88: 4558–4562.
37. Kalsotra A, Turman CM, Kikuta Y, Strobel HW (2004) Expression and characterization of human cytochrome P450 4F11: Putative role in the metabolism of therapeutic drugs and eicosanoids. Toxicol Appl Pharmacol 199: 295–304.
38. Nelson DR (1998) Metazoan cytochrome P450 evolution. Comp Biochem Physiol C PharmacolToxicol Endocrinol 121: 15–22.
39. Gardner WS, Lee RF, Tenore KR, Smith LW (1979) Degradation of selected polycyclic aromatic hydrocarbons in coastal sediments: Importance of microbes and polychaete worms. Water Air Soil Pollut 11: 339–347.
40. Lee RF (1998) Annelid cytochrome P-450. Comp Biochem Physiol C Pharmacol Toxicol Endocrinol 121: 173–179.
41. Porter TD (2002) The roles of cytochrome b5 in cytochrome P450 reactions. J Biochem Mol Toxicol 16: 311–316.
42. Rendic S, Carlo FJD (1997) Human cytochrome P450 enzymes: A status report summarizing their reactions, substrates, inducers, and inhibitors. Drug Metab Rev 29: 413–580.
43. Oost Rvd, Beyer J, Vermeulen NPE (2003) Fish bioaccumulation and biomarkers in environmental risk assessment: A review. Environ Toxicol Pharmacol 13: 57.
44. Hahn ME, Woodin BR, Stegeman JJ, Tillitt DE (1998) Aryl hydrocarbon receptor function in early vertebrates: Inducibility of cytochrome P450 1A in agnathan and elasmobranch fish. Comp Biochem Physiol C Pharmacol Toxicol Endocrinol 120: 67–75.

45. Hahn M (2002) Aryl hydrocarbon receptors: Diversity and evolution. Chem Biol Interact 141: 131–160.

46. Denison MS, Fisher J, Whitlock J (1988) The DNA recognition site for the dioxin-ah receptor complex. nucleotide sequence and functional analysis. J Biol Chem 263: 17221–17224.

47. Kozu S, Tajiri R, Tsuji T, Michiue T, Saigo K, Kojima T (2006) Temporal regulation of late expression of bar homeobox genes during drosophila leg development by spineless, a homolog of the mammalian dioxin receptor. Dev Biol 294: 497.

48. Powell-Coffman JA, Bradfield CA, Wood WB (1998) *Caenorhabditis elegans* orthologs of the aryl hydrocarbon receptor and its heterodimerization partner the aryl hydrocarbon receptor nuclear translocator. Proc Nat Acad Sci 95: 2844–2849.

49. Butler RA, Kelley ML, Powell WH, Hahn ME, Van Beneden RJ (2001) An aryl hydrocarbon receptor (AHR) homologue from the soft-shell clam, mya arenaria: Evidence that invertebrate AHR homologues lack 2, 3, 7, 8-tetrachlorodibenzo-p-dioxin and beta-naphthoflavone binding. Gene 278: 223–234.

50. Rushmore TH, Pickett C (1990) Transcriptional regulation of the rat glutathione S-transferase ya subunit gene. characterization of a xenobiotic-responsive element controlling inducible expression by phenolic antioxidants. J Biol Chem 265: 14648–14653.

51. Lamb DC, Kelly DE, Kelly SL (1998) Molecular diversity of sterol 14alpha-demethylase substrates in plants, fungi and humans. FEBS Lett 425: 263–265.

52. Baker ME (2011) Insights from the structure of estrogen receptor into the evolution of estrogens: Implications for endocrine disruption. Biochem Pharmacol 82: 1–8.

53. Holland L, Albalat R, Azumi K, Benito-Gutierrez E, Blow M, et al. (2008) The amphioxus genome illuminates vertebrate origins and cephalochordate biology. Genome Res 18: 1100–1111.

54. Twan W, Hwang J, Chang C (2003) Sex steroids in scleractinian coral, *Euphyllia ancora*: Implication in mass spawning. Biol Reprod 68: 2255–2260.

55. Haas BJ, Delcher AL, Mount SM, Wortman JR, Smith Jr RK, et al. (2003) Improving the arabidopsis genome annotation using maximal transcript alignment assemblies. Nucleic Acids Res 31: 5654–5666.

56. Altschul SF, Gish W, Miller W, Myers EW, Lipman DJ (1990) Basic local alignment search tool. J Mol Biol 215: 403–410.

57. Rutherford K, Parkhill J, Crook J, Horsnell T, Rice P, et al. (2000) Artemis: Sequence visualization and annotation. Bioinform 16: 944–945.

58. Pearson W, Lipman D (1988) Improved tools for biological sequence comparison. Proc Natl Acad Sci U S A 85: 2444–2448.

59. Edgar RC (2004) MUSCLE: Multiple sequence alignment with high accuracy and high throughput. Nucleic Acids Res 32: 1792–1797.

60. Maddison W, Maddison D (2011) Mesquite: A modular system for evolutionary analysis. Available: http://mesquiteproject.org. Accessed 2013 Oct 3.

61. Wu M, Chatterji S, Eisen JA (2012) Accounting for alignment uncertainty in phylogenomics. PLoS One 7: e30288.

62. Stamatakis A (2006) RAxML-VI-HPC: Maximum likelihood-based phylogenetic analyses with thousands of taxa and mixed models. Bioinform 22: 2688–2690.

63. Abascal F, Zardoya R, Posada D (2005) ProtTest: Selection of best-fit models of protein evolution. Bioinform 21: 2104–2105.

64. Gille C, Frommel C (2001) STRAP: Editor for STRuctural alignments of proteins. Bioinform 17: 377–378.

65. Sun YV, Boverhof DR, Burgoon LD, Fielden MR, Zacharewski TR (2004) Comparative analysis of dioxin response elements in human, mouse and rat genomic sequences. Nucleic Acids Res 32: 4512–4523.

66. Bailey TL, Elkan C (1994) Fitting a mixture model by expectation maximization to discover motifs in bipolymers. Proc Int Conf Intell Syst Mol Biol 2: 28–36.

RADIA: RNA and DNA Integrated Analysis for Somatic Mutation Detection

Amie J. Radenbaugh[1]*, Singer Ma[1], Adam Ewing[1], Joshua M. Stuart[1], Eric A. Collisson[2], Jingchun Zhu[1], David Haussler[1,3]*

1 University of California Santa Cruz Genomics Institute, Department of Biomolecular Engineering, University of California Santa Cruz, Santa Cruz, California, United States of America, 2 Division of Hematology/Oncology, University of California San Francisco, San Francisco, California, United States of America, 3 Howard Hughes Medical Institute, Chevy Chase, Maryland, United States of America

Abstract

The detection of somatic single nucleotide variants is a crucial component to the characterization of the cancer genome. Mutation calling algorithms thus far have focused on comparing the normal and tumor genomes from the same individual. In recent years, it has become routine for projects like The Cancer Genome Atlas (TCGA) to also sequence the tumor RNA. Here we present RADIA (RNA and DNA Integrated Analysis), a novel computational method combining the patient-matched normal and tumor DNA with the tumor RNA to detect somatic mutations. The inclusion of the RNA increases the power to detect somatic mutations, especially at low DNA allelic frequencies. By integrating an individual's DNA and RNA, we are able to detect mutations that would otherwise be missed by traditional algorithms that examine only the DNA. We demonstrate high sensitivity (84%) and very high precision (98% and 99%) for RADIA in patient data from endometrial carcinoma and lung adenocarcinoma from TCGA. Mutations with both high DNA and RNA read support have the highest validation rate of over 99%. We also introduce a simulation package that spikes in artificial mutations to patient data, rather than simulating sequencing data from a reference genome. We evaluate sensitivity on the simulation data and demonstrate our ability to rescue back mutations at low DNA allelic frequencies by including the RNA. Finally, we highlight mutations in important cancer genes that were rescued due to the incorporation of the RNA.

Editor: Srikumar P. Chellappan, H. Lee Moffitt Cancer Center & Research Institute, United States of America

Funding: This work was supported by the National Cancer Institute [U24CA143858 to AJR, SM, AE, and JZ; R01CA180778 to JMS; R01CA194003 to EAC; and U24CA180951 to JZ]; the National Human Genome Research Institute [U01ES017154 to AJR]; a gift from Edward Schulak [to AE]; and the Howard Hughes Medical Institute [to DH]. The results published here are in whole or part based upon data generated by the TCGA Research Network: http://cancergenome.nih.gov/, a project of the National Cancer Institute. The remaining funders had no role in study design, data collection and analysis, decision to publish, or preparation of the manuscript.

Competing Interests: The authors have declared that no competing interests exist.

* Email: aradenba@soe.ucsc.edu (AJR); haussler@soe.ucsc.edu (DH)

Introduction

Much of our current understanding of cancer has come from investigating how normal cells are transformed into cancerous cells through the stepwise acquisition of somatic genomic abnormalities. These events include point mutations, insertions and deletions (INDELs), chromosomal rearrangements, and changes to the copy number of segments of DNA. Transforming a normal human cell into a malignant, immortal cancer cell line requires an estimated five to seven genetic alterations in key genes and pathways [1,2]. Not surprisingly, much research has been devoted to determining how cancer cells are able to acquire their abilities through the accumulation of somatic mutations.

The Cancer Genome Atlas (TCGA) project has produced exome-wide data from thousands of tumors and patient-matched normal tissues. With the development of RNA Sequencing (RNA-Seq) [3], TCGA began providing an additional high-throughput tumor sequence dataset. These three datasets consisting of tumor and patient-matched normal DNA and tumor RNA have become a new standard in cancer genomics. RNA-Seq enables one to investigate the consequences of genomic changes in the RNA transcripts they encode to better characterize 1) germline variants, 2) somatic mutations, and 3) variants in the RNA that are not found in the DNA that could be the result of RNA editing [4].

Over the next few years, many more whole-genome and exome-capture DNA and RNA-Seq BAM (the binary version of Sequence Alignment/Map [5]) files will become available. TCGA has collected over 10,000 tissue samples from more than 20 types of cancer. There is a clear need for an efficient method for the combined analysis of patient-matched tumor DNA, normal DNA, and tumor RNA. Here we present a method called RADIA to identify and characterize alterations in cancer using DNA and RNA obtained by high-throughput sequencing data.

Somatic mutation calling is traditionally performed on patient-matched pairs of tumor and normal genomes/exomes [6–11]. The ability to accurately detect somatic mutations is hindered by both biological and technical artifacts that make it difficult to obtain both high sensitivity and high specificity. Different mutation calling algorithms often disagree about putative mutations in the same source data, and frequently have discernible systematic differences due to the trade-off between sensitivity and specificity [12]. This is especially true for somatic mutations with low variant allele frequencies (VAFs). By creating an algorithm that utilizes both DNA and RNA, we have increased the power to detect somatic mutations, especially at low variant allele frequencies.

RADIA combines patient-matched tumor and normal DNA with the tumor RNA to detect somatic mutations. The DNA Only Method (DOM) (Figure 1) uses just the tumor/normal pairs of DNA (ignoring the RNA), while the Triple BAM Method (TBM) (Figure 1) uses all three datasets from the same patient to detect somatic mutations. The mutations from the TBM are further categorized into two sub-groups: RNA Confirmation and RNA Rescue mutations (Figure S1). RNA Confirmation mutations are those that are made by both the DOM and the TBM due to the strong variant read support in both the DNA and RNA. RNA Rescue mutations are those that had very little DNA support, hence not called by the DOM, but strong RNA support, and thus called by the TBM. RNA Rescue mutations are typically missed by traditional methods that only interrogate the DNA.

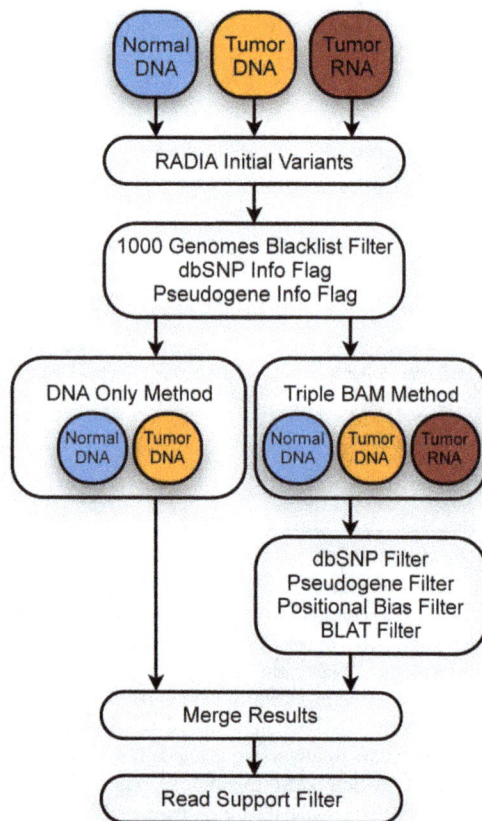

Figure 1. Overview of the RADIA work-flow for identifying somatic mutations. The normal DNA, tumor DNA, and tumor RNA BAMs are processed in parallel and initial low-level variants are identified. The variants are filtered by the DNA Only Method using the pairs of normal and tumor DNA and by the Triple BAM Method using all three datasets. The mutations from the two methods are merged and output in VCF format.

We have applied RADIA to data derived from over 3,300 patients representing 15 different cancer types from TCGA (Table S1). Overall, the RNA Rescue mutations that are made possible by the incorporation of the RNA-Seq data provide a two to seven percent increase in somatic mutations compared to the DOM (Table S1). Many of these mutations were new discoveries that were not previously found by other mutation calling algorithms in TCGA. Of these new discoveries, some mutations were found in well-known cancer genes that were heavily mutated in a specific cohort. We also find mutations in new samples where the same gene has already been identified as harboring mutations in other samples from the cohort. When these RNA Rescue mutations are added to the DNA Only mutations, these genes achieve a statistically significant overall mutation rate for the cohort.

Here we specifically focus on results from 177 endometrial carcinoma [13] and 230 lung adenocarcinoma [14] patients from TCGA. To demonstrate the increase in sensitivity from including the tumor RNA-Seq dataset, we simulated mutations by spiking them into the tumor DNA and tumor RNA of a breast cancer patient using bamsurgeon (https://github.com/adamewing/bamsurgeon). We also evaluated sensitivity and precision on the endometrial carcinoma and lung adenocarcinoma data using validation data that was generated by TCGA. We highlight RNA Rescue mutations found by the TBM in tumor suppressor genes such as *TP53*, *STK11*, and *CDKN2A* in lung adenocarcinoma.

Methods

RADIA operates on two or more BAM files, producing somatic mutation calls through a series of steps outlined in Figure 1. Each step in this process is described in detail, beginning with the initial selection of sites for further processing and ending with a description of filters used to eliminate false positives while maintaining true positives.

2.1 Variant Detection with RADIA

RADIA is typically run on three BAM [5] files consisting of a pair of patient-matched tumor and normal genomes and a tumor transcriptome and outputs germline (inherited) variants and somatic Single Nucleotide Variants (SNVs). Here we focus specifically on the detection of somatic SNVs with RADIA. The DOM is run on the pairs of tumor and matched-normal DNA while the TBM is applied to the DNA and RNA triplets. After the DOM and TBM specific filters, the results are merged and run through a final read support filter (Figure 1). If RNA-Seq data is not available, RADIA can utilize paired tumor and normal DNA genomes using the DOM to detect germline variants and somatic SNVs.

Internally, RADIA uses the samtools [5] mpileup command (version 0.1.18) to examine the pileups of bases in each sample in parallel. A heuristic algorithm determines the existence and type of variant at any given position based on the user-configurable minimum thresholds for overall depth, variant depth, Base Alignment Quality (BAQ) [15], and mapping quality. Initially, RADIA requires a minimum overall depth of four bases, minimum variant depth of two bases, minimum phred BAQ of 10, and minimum phred mapping quality of 10. These initial calls are lenient in coverage and provide a good baseline set of calls for further filtering.

RADIA scans pileups of reads across the reference genome and outputs variants in Variant Call Format (VCF) (https://github.com/samtools/hts-specs). For each position, summary information such as the overall depth, allele specific depth and frequency, average BAQ base quality, average mapping quality, and the fraction of reads on the plus strand are calculated for both the DNA and RNA. All of this information is used during the filtering process.

2.2 Variant Filtering

After the initial variants are detected, a number of filters are applied to remove false positive variants that result from biological and technical artifacts. Each filter is described here in detail.

2.2.1 Filtering Around INDELs. Many current mutation calling algorithms have a pre-processing step to account for misaligned reads around INDELs. This realignment step is computationally expensive and relies on accurately predicting the location of INDELs which itself is not a trivial problem. Base Alignment Quality (BAQ) is an alternative option for dealing with alignment ambiguity around INDELs. It calculates the probability that a base has been misaligned and returns the minimum of the original base quality and the base alignment quality. BAQ is run by default when executing a samtools mpileup command and has been shown to improve SNP calling accuracy [15]. We use the extended version of BAQ (option –E) that is activated by default in the latest version of samtools (0.1.19) for increased sensitivity and slightly lower specificity [5].

2.2.2 1000 Genomes Blacklist Filter. The 1000 Genomes Project coined the term "accessible genome" to be the part of the reference genome that is reliable for accurate variant calling after removing ambiguous or highly repetitive regions [16]. Since the reference genome is incomplete, repetitive in places, and does not represent human genetic variation comprehensively, reads often get mapped incorrectly in locations outside the accessible genome (inaccessible sites), leading to false positive variant calls. Over 97% of inaccessible sites are due to high copy repeats or segmental duplications. In the pilot, the 1000 Genomes Project determined that 85% of the reference sequence and 93% of the coding region was accessible. Due to longer read lengths (75–100 bp) and improvements to both paired end protocols and sequence alignment algorithms, the accessible genome increased in Phase I to 94% of the reference and 98% of the coding region [17]. We filter variants that are not in the accessible genome using the Phase I mapping quality and depth blacklists (ftp://ftp-trace.ncbi.nih.gov/1000genomes/ftp/phase1/analysis_results/supporting/accessible_genome_masks/).

2.2.3 Strand-Bias Filter. It has recently been shown that variant allele reads that occur exclusively on one strand are largely associated with false positives [8]. In order to account for this technical artifact, we filter based on the variant allele strand bias. If we have at least four total reads supporting the variant allele, then we apply the strand bias filter if more than 90% of the reads are on the forward strand or more than 90% are on the reverse strand.

2.2.4 Filtering by mpileup Support. RADIA can be executed on patient-matched pairs of tumor and normal DNA samples using the DOM to identify germline variants and somatic mutations. We first compare the matched normal DNA to the human reference genome. We require the normal DNA to pass the mpileup support filters described in Table 1 for all germline variants.

If no germline variant is found, we compare the tumor DNA to the matched normal DNA and the reference genome to search for somatic mutations. We require the normal DNA and tumor DNA to pass the mpileup support filters shown in Table 1 for all somatic variants. To ensure that we have the power to detect a possible germline variant at this site, we require that the germline DNA depth is 10 or more.

We use the Triple BAM Method to augment our somatic mutation calls using both the pairs of DNA and the RNA-Seq data. The normal DNA, tumor DNA, and tumor RNA must pass the mpileup support filters shown in Table 2 for all somatic mutations. We require at least one read with a minimum BAQ phred score of 15 in the tumor DNA. To rule out possible germline variants, we again require that the normal DNA depth is 10 or more. In addition, we filter out calls that overlap with common SNPs that are not flagged as clinically relevant and found in at least one percent of the samples in dbSNP [18]. We downloaded this subset of dbSNP from the "Common SNPs" track on the UCSC human genome browser [19,20]. We found that many false positive variants overlapped with earlier versions of dbSNP. These variants were due to technical artifacts and were removed from subsequent versions of dbSNP [21]. Therefore, we filter out all variants that overlap with dbSNP versions 130, 132 or 135 (ftp://ftp.ncbi.nih.gov/snp/). The TBM calls are subjected to further filtering procedures as shown in Figure 1 and described below.

2.2.5 Pseudogene Filter. We noticed that many of our TBM mutations overlapped with predicted pseudogenes. Although expressed pseudogenes have recently been reported to be significant contributors to the transcriptional landscape and shown to play a role in cancer progression [22], mutations that overlap with predicted pseudogenes have a high false positive rate. Sequence similarity of pseudogene copies to their parent genes leads to uncertainty in alignment within these regions. Because of these technical artifacts, we remove TBM mutations that overlap with pseudogenes annotated in GENCODE by the ENCODE project (version 19) [23] and predicted by RetroFinder (version 5) [23,24]. We downloaded the pseudogene annotations from the following tracks on the UCSC human genome browser [19,25]: Gene Annotations from ENCODE/GENCODE and Retroposed Genes. The predicted pseudogenes occupy 1.5% of the total genome.

2.2.6 Highly Variable Genes Filter. We remove TBM mutations that overlap with families of genes that have high sequence similarity. Some examples of these gene families are Human Leukocyte Antigens (HLAs), Ribosomal Proteins (RPLs), and immunoglobulins. While mutations in these genes may exist, special processing would be needed to distinguish them from false positive calls due to misaligned reads. We annotate the mutations using SnpEff [26] and filter out the following five gene families: RPLs, RP11s, HLAs, IGHVs and IGHCs.

2.2.7 Positional Bias Filter. False positive calls are associated with misaligned reads where the alternative allele is consistently within a certain distance from the start or end of the read. The positional bias filter is applied when 95% or more of the reads that have an alternative allele are such that the alternate allele falls in the first third or last third of the read.

2.2.8 BLAT Filter. We observed multiple instances where RNA-Seq reads appeared to be incorrectly mapped due to the added difficulties in aligning RNA-Seq data, such as dealing with hard to identify splice junctions and multiple gene isoforms. To guarantee that the RNA-Seq reads that support a variant do not map better to another location in the genome, we created a BLAT filter. All of the RNA-Seq reads that support a variant are extracted from the BAM file and aligned to the human genome using BLAT [27]. If the read maps to another location with a better score, the read is rejected. After using BLAT on each read, we again require that there are at least four valid reads that support the variant and that 10% or more of the reads support the variant.

2.2.9 Read Support Filter. We merge the calls from the DOM and the TBM and apply one final filter. We require that each somatic mutation be supported by at least four "perfect" reads. We define a perfect read as follows:

Table 1. DNA Only Method mpileup Support Filters.

Filter	Germline	Somatic	
	Normal DNA	Normal DNA	Tumor DNA
Min Total Depth	10	10	10
Min Alt. Depth	4	NA	4
Min Alt. Percent	10%	NA	10%
Min Avg. Alt. BAQ	20	NA	20
Max Alt. Strand Bias	90%	NA	90%
Max Alt. Percent	NA	2%	NA
Max Other Percent	2%	2%	2%

The germline variants and somatic mutations from the DOM are filtered according to the parameters described here. The minimum average alternative read BAQ filter uses the phred scale. The maximum other percent restricts the percentage of reads that are allowed to support an additional alternative allele.

1. Minimum mapping quality of read is 10
2. Minimum base quality of alternative allele in read is 10
3. Minimum base qualities of the five bases up- and down-stream of the alternative allele are 10
4. Read is properly paired
5. Read has fewer than four mismatches across its entirety when compared to the reference
6. Read does not require an insertion or deletion to be mapped

After determining the number of perfect reads that support the reference and the alternative at a coordinate, we re-apply the strand bias filter to guarantee that no more than 90% of the total perfect reads are from one strand.

Results

We evaluate the sensitivity of RADIA using simulation data that was generated from patient data. We also measure the sensitivity and precision of RADIA using patient and validation data generated by TCGA. All patients in this study provided written informed consent to genomic studies in accordance with local Institutional Review Boards (Table S2) and the policies and guidelines outlined by the Ethics, Law and Policy Group from TCGA. All patient data is anonymous and was originally collected for routine therapeutic purposes.

3.1 Sensitivity on Simulation Data

In order to evaluate sensitivity and demonstrate the increase in power from including the RNA-Seq data, we simulated somatic mutations starting from patient data. We spiked mutations into a pair of breast cancer tumor DNA and tumor RNA samples using bamsurgeon (https://github.com/adamewing/bamsurgeon), a tool we developed to generate simulation data that closely mimics actual experimental data from high-throughput sequencing datasets. Bamsurgeon first determines the loci that have an appropriate DNA and RNA depth to spike in mutations. It then extracts the reads at the loci, adjusts the VAF according to the user-defined VAF distribution, and then re-maps the reads (Figure S2). This simulation strategy is more sophisticated than simply generating simulated reads from a reference genome, as it retains the biological and technical artifacts that are inherently present in next generation sequencing data. We performed two spike-in experiments: one varying the DNA VAF while holding the RNA VAF constant, and one varying the RNA VAF while holding the DNA VAF to 10% or less.

3.1.1 Sensitivity on Variable DNA-Constant RNA Simulation Data. To evaluate the sensitivity of RADIA, we spiked in 1,594 mutations to the tumor DNA sequence with a variant allele frequency ranging from 1–50% and to the tumor RNA sequence at a constant frequency of 25%. The overall sensitivity rate averaged across all VAFs is 85% consisting of 1,351 out of 1,594 spiked in mutations (Figure 2A). Of the 243 calls that

Table 2. Triple BAM mpileup Support Filters.

Filter	Somatic		
	Normal DNA	Tumor DNA	Tumor RNA
Min Total Depth	10	1	10
Min Alt. Depth	NA	1	4
Min Alt. Percent	NA	NA	10%
Min Avg. Alt. BAQ	NA	15	15
Max Alt. Strand Bias	NA	90%	90%
Max Alt. Percent	10%	NA	NA
Max Other Percent	10%	10%	2%

The somatic mutations from the TBM are filtered according to the parameters shown here.

were filtered out, over 50% are removed because they failed to meet the minimum variant allele frequency, more than 20% land in blacklist regions that the method ignores, and nearly 20% are discarded due to the BLAT filter. The number of mutations that are rejected by the full list of filters can be found in Figure S3.

3.1.2 Sensitivity on Low Frequency DNA-Variable RNA Simulation Data. To demonstrate the ability of the TBM to rescue calls at low DNA VAFs, we spiked in 1,761 mutations to the tumor RNA sequence with a variant allele frequency ranging from 1–50% and to the tumor DNA sequence at a frequency of 10% or less. Most of the mutations by the DOM are filtered out due to the low allelic frequency in the DNA (Figure S4). For the mutations

that have sufficient read support in the RNA, these low DNA VAFs are rescued back (Figure 2B).

3.2 Precision and Sensitivity on Patient Data

We made somatic mutation calls on 177 non-hypermutated TCGA endometrial carcinoma samples [13]. All 177 tumor and matched normal whole exome sequencing and RNA-Seq alignments in BAM [5] format were downloaded from TCGA at the Cancer Genomics Hub (CGHub, https://cghub.ucsc.edu, Table S2). The exomes were sequenced using the Illumina Genome Analyzer II, and the paired-end sequencing reads were aligned by

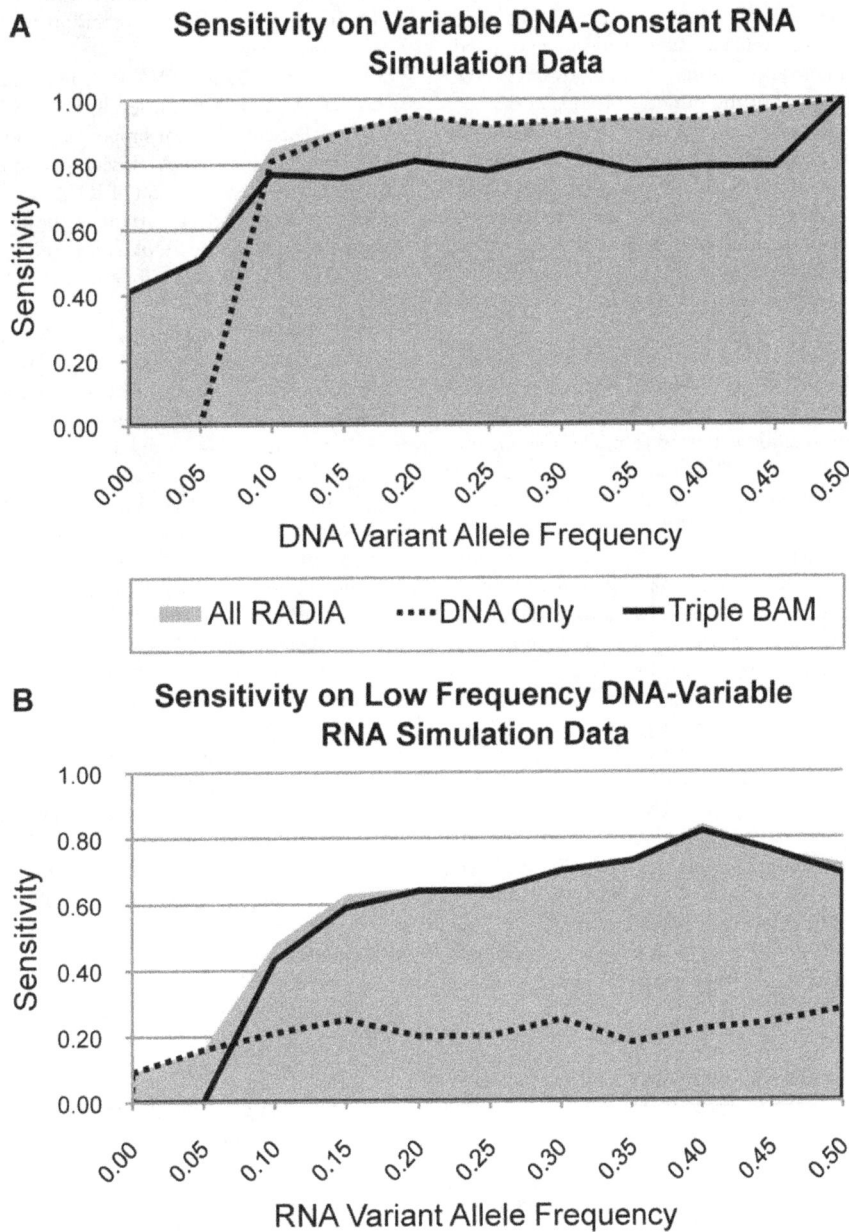

Figure 2. Sensitivity of RADIA on simulation data. Artificial mutations were spiked into the tumor DNA and RNA BAM files of a breast cancer patient using bamsurgeon. (A) Mutations were spiked into the DNA at variant allele frequencies distributed from 1–50% and into the RNA at a constant 25%. The overall sensitivity of RADIA was 85%. RNA Rescue calls from the Triple BAM method detected the mutations that had a DNA VAF less than 10%. (B) Mutations were spiked into the DNA at 10% or less and into the RNA distributed from 1–50%. Most of the DOM mutations are filtered due to the low DNA allelic frequency. The mutations that have adequate RNA read support are rescued back at these low DNA allelic frequencies.

BWA [28]. The RNA was sequenced using the Illumina Genome Analyzer II, and the single-end sequencing reads were aligned by MapSplice (V2) [29].

3.2.1 RADIA Precision on Endometrial Carcinoma Patient Data. For the study on endometrial carcinoma by TCGA [13], mutations were submitted by three independent TCGA Genomic Data Analysis Centers (GDACs). These mutations were merged and targeted for custom recapture and resequencing using new cDNA libraries from the tumor and normal DNA samples [13]. We downloaded the validation BAMs containing the results of the hybrid capture and resequencing of targeted mutations from CGHub (https://cghub.ucsc.edu, Table S2). We utilized the identical validation criteria used by the TCGA Endometrial Analysis Working Group to validate the somatic mutations detected by RADIA [13]. For each somatic mutation, we queried the patient-matched tumor and normal validation data. We required at least 10 reads in both the tumor and normal data in order to determine if a call validated, otherwise we classified it as ambiguous. If the variant was present at low levels in both datasets, we also classified it as ambiguous. Otherwise, we determined whether a mutation validated as germline/LOH, somatic, or neither according to Table 3. In addition, any RNA Rescue call in the "Not Validated" group that overlapped with a COSMIC somatic mutation that was confirmed in another study was considered as validated.

We made a total of 27,900 somatic mutation calls over 177 endometrial samples, of which the DOM and TBM made 27,390 and 6,325 calls respectively. Of the 6,325 TBM calls, there were 5,815 RNA Confirmation mutations that were made by both the DOM and TBM signifying high DNA and RNA support, and importantly, a total of 510 RNA Rescue mutations that were missed by the DOM.

Using the validation strategy described above, we demonstrate that the overall precision for RADIA is 98% (Figure 3A). Due to lack of coverage or uncertainty in the tumor and normal validation BAMs, a total of 1,825 calls were considered to be ambiguous. Of the remaining 26,075 mutations called by RADIA, 25,520 validated as somatic, 271 validated as germline/LOH variants and 284 did not validate. The precision of calls made by the DOM and the TBM was 98% and 98.5% respectively. For the RNA Confirmation mutations made by both the DOM and the TBM, the precision was 99.3%. There were 510 RNA Rescue mutations made only by the TBM, and even though most of these calls were not targeted for validation, the precision was 74%. For the 510 RNA Rescue calls, 251 were classified as ambiguous, 6 validated as Germline/LOH, and 61 did not validate. Of the remaining 192 RNA Rescue mutations that validated, 178 (93%) were verified using the validation BAMs and 14 (7%) were confirmed as somatic mutations in COSMIC.

We next examined the precision of the DOM with varying RNA-Seq reads supporting the variant allele as well as the precision of RNA Rescue mutations with differing levels of DNA supporting reads. Sixty-two percent of the DOM mutations were covered by reads in the RNA-Seq data, and 29% had at least 10 RNA-Seq reads covering the mutation. Nearly half (44%) had at least one RNA read supporting the DNA variant allele, while 25% of the DOM mutations had at least four supporting RNA reads. The precision of the DOM is lowest (92%) with no RNA-Seq support, increases to 95% with weak RNA-Seq support (at least one but less than five supporting reads), and increases to 99.3% for RNA Confirmation mutations. Overall, mutations that are detected by the DOM validate above 92%, regardless of the RNA-Seq support, and the precision increases as the RNA-Seq support increases.

On the other hand, RNA Rescue mutations weakly supported by the DNA validate at low levels. For RNA Rescue mutations, we require at least one variant supporting read in the DNA in order to distinguish between RNA Rescue mutations and possible RNA editing events. The precision of RNA Rescue mutations with only one read supporting the variant in the DNA was 11%, with two supporting reads in the DNA 23%, with three supporting reads in the DNA 43%, and with four or more supporting reads in the DNA 94%.

3.2.2 RADIA Sensitivity on Endometrial Carcinoma Patient Data. In order to measure the sensitivity of RADIA, we considered the union of all mutations submitted by TCGA GDACs that validated as somatic as our truth set. There were 30,239 mutations that validated as somatic from TCGA. We compared our somatic mutations to this truth set and demonstrated an overall sensitivity of 84% (Figure 3B, Figure S5). Of the 4,751 calls that were missed, 1,539 (33%) were filtered by RADIA because they had a variant allele frequency less than 8% (Figure S6). In addition, 1,072 (23%) landed in blacklist regions that were not considered (Figure S6).

3.2.3 RADIA Precision on Lung Adenocarcinoma Patient Data. Finally, RADIA somatic mutations were analyzed during the course of our participation in the TCGA Lung Adenocarcinoma Analysis Working Group [14]. We ran RADIA on 230 TCGA lung adenocarcinoma triplets that we downloaded from CGHub (https://cghub.ucsc.edu, Table S2). The exomes were sequenced using the Illumina HiSeq platform, and the paired-end sequencing reads were aligned by BWA [28]. The RNA was sequenced using the Illumina HiSeq platform, and the paired-end sequencing reads were aligned by MapSplice (V2) [29]. Validation was performed by the Broad Institute on 74 genes of interest along with an additional 1,150 somatic SNVs. Validation was attempted on 2,404 RADIA somatic mutations and 2,395 (99.63%) were verified. From the DOM, 2,336 of the 2,345 mutations (99.62%)

Table 3. Validation Criteria for Endometrial Carcinoma Data.

Normal VAF	Tumor VAF			
	0%	<8%	≥8%, <20%	≥20%
= 0%	*Not Validated*	**Somatic Low**	**Somatic Med**	**Somatic High**
<3%	*Not Validated*	Ambiguous	**Somatic Med**	**Somatic High**
≥3%	*Germline/LOH*	*Germline/LOH*	*Germline/LOH*	*Germline/LOH*

Validation BAMs were used to determine the validation status for somatic mutations as shown here. A mutation is considered validated in the Somatic Low, Med, or High groups (bold), not validated in the "Not Validated" (italics) and Germline/LOH groups (italics), and Ambiguous when there was low read depth (<10 reads) or low VAFs in both the normal (<3%) and tumor (<8%) validation BAMs.

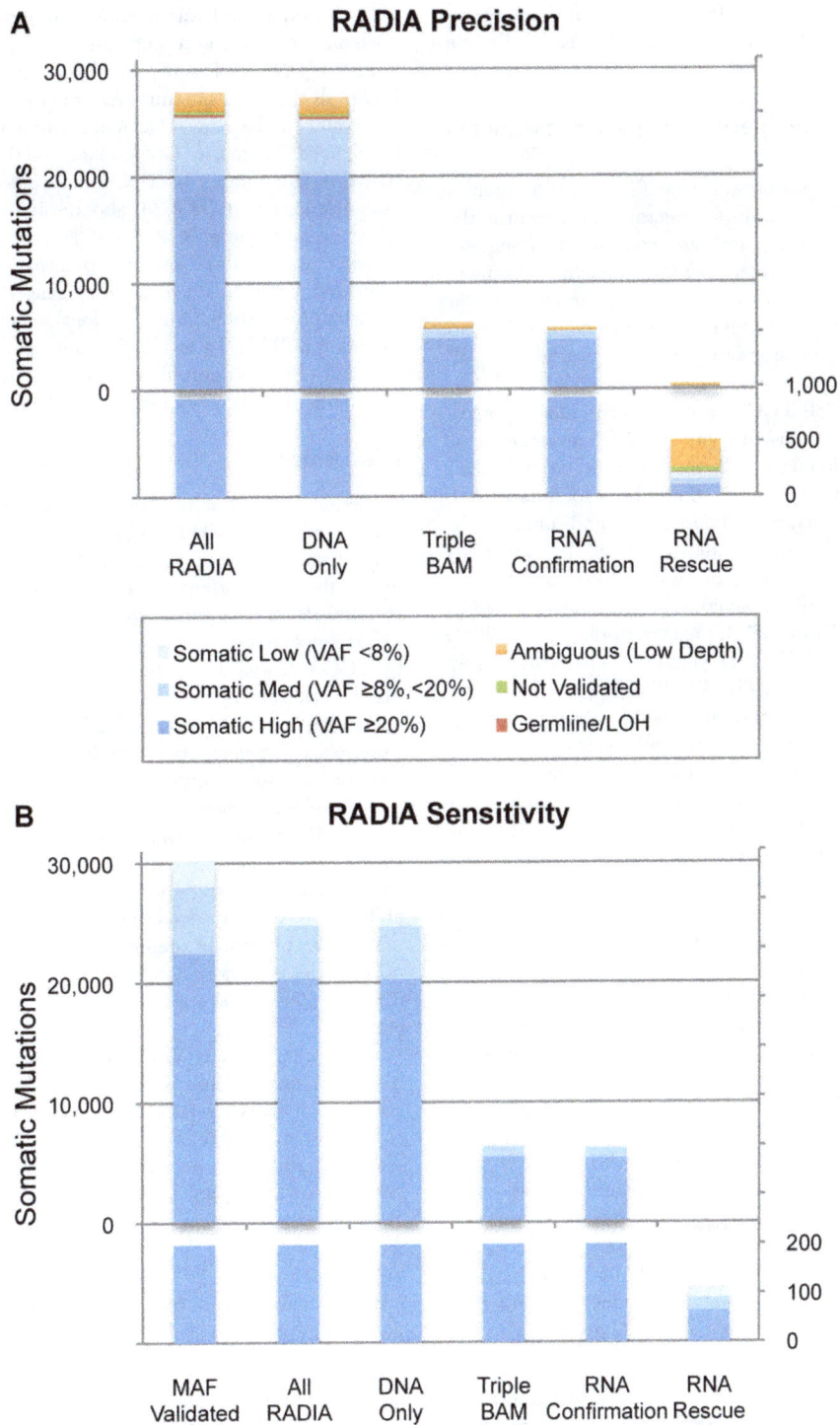

Figure 3. Precision and sensitivity of RADIA on 177 non-hypermutated endometrial carcinoma samples. Mutations are considered validated in the Somatic Low, Med, or High groups (blue), not validated in the "Not Validated" (green) and Germline/LOH (red) groups, and Ambiguous (orange) when there was low read depth (<10 reads) or ambiguity in the validation data. (A) An overall precision of 98% was demonstrated. RNA Confirmation mutations with strong DNA and RNA support validated over 99%. RNA Rescue mutations validated at 74%. (B) The union of all mutations submitted by TCGA GDACs that validated as somatic was considered as the truth set. RADIA demonstrated an overall sensitivity rate of 84%. Of the mutations that were missed, 33% occurred at low variant allele frequencies (<8%) and 23% occurred in blacklist regions that were ignored.

validated. Importantly, 469/469 (100%) of the TBM mutations consisting of 410 RNA Confirmation and 59 RNA Rescue mutations validated.

3.3 Somatic Mutations in Specific Lung Adenocarcinoma Genes

Mutations in the tumor suppressor gene *TP53* are common in the majority of human cancers. Most of the mutations occur in the DNA-Binding Domain (DBD) and are considered change-of-function mutations that alter activity of *TP53*, sometimes acting in a dominant negative manner to sequester wildtype tp53 protein *in trans* [30]. As such, many p53 mutant proteins endow cells with oncogenic characteristics by promoting cell proliferation, survival, and metastasis [31].

We ran RADIA on 230 TCGA lung adenocarcinoma triplets [14] and discovered two non-synonymous *TP53* mutations that were below the detection threshold for other mutation calling algorithms used by TCGA (Table 4). Both of the mutations were validated by the deep-sequencing validation data and confirmed as somatic in COSMIC by other studies. One of the mutations (G266E) was confirmed as somatic in another lung cancer study [32] as well as in prostate [33], pancreas [34], urinary tract [35], and hematopoietic and lymphoid [36] cancer studies. The G266E mutation occurs in the *TP53* DBD mutation hotspot frequently resulting in pathological effects [37–39]. This mutation has also been described as a gain-of-function mutation in a melanoma cell line [40]. The other *TP53* mutation (G199V) was confirmed as somatic in breast [41], ovarian [42], and medulloblastoma [43] studies. It is a known anti-apoptotic gain-of-function mutation that promotes cell survival through the signal transducer and activator of transcription-3 (STAT3) pathway [44]. Knockdown experiments of G199V p53 mutants demonstrated a level of anti-tumor activity similar to high doses of chemotherapeutic agents, suggesting that inhibition of G199V p53 mutants may be beneficial for cancer treatment [44].

Additionally, we found mutations in other well-known tumor suppressor genes such as *STK11* and *CDKN2A*. In the lung adenocarcinoma manuscript from TCGA, mutations in *STK11* and *CDKN2A* were reported in 17% and 4% of all patients, respectively [14]. *STK11* was the fourth most mutated gene and *CDKN2A* was the sixteenth [14]. The proximal-proliferative subtype in lung adenocarcinoma is characterized by an enrichment of mutations in *KRAS* along with inactivation mutations in *STK11* [14]. In the *STK11* gene, we discovered a nonsense mutation at W239* in the structurally conserved protein kinase domain that was below the detection threshold for other mutation algorithms used by TCGA. This mutation introduces an early stop codon in exon five (of ten) leading to a truncated protein. This site

is in COSMIC and was previously reported to be part of a 398 nucleotide deletion in a lung cancer study [45].

In the *CDKN2A* gene, we found one nonsense mutation at R122*, R163* and one missense mutation at R131H, R80H that were both validated by TCGA and found in COSMIC. *CDKN2A* is silenced in many CpG island methylator phenotype-high (CIMP-High) tumors by DNA methylation [14], but mutations and deletions in *CDKN2A* also result in loss of function. The nonsense mutation at R122*, R163* results in an early stop codon in exon two (of three or four, isoform dependent) leading to a truncated protein. Previous lung cancer studies [46–48] have reported frameshifts and deletions at this site. The missense mutation at R131H was also found in colon cancer [49], clear cell sarcoma [50], and chronic myeloid leukemia [51] and confirmed as somatic in biliary tract cancer [52].

Discussion

Identifying somatic mutations is a key step in characterizing the cancer genome. Until now, algorithms for mutation detection have concentrated on comparing just the normal and tumor genomes within the same individual. In the past few years, it has become common to also sequence the tumor transcriptome using RNA-Seq technologies. Large genomics studies, such as those conducted by TCGA, primarily use the RNA-Seq for gene expression, gene fusion, and splicing analyses. With the cost of sequencing steadily decreasing and the wealth of information that can by obtained from RNA-Seq, we predict that the sequencing of the tumor RNA will continue to be routine in large cancer profiling projects. We have developed a novel method called RADIA that combines the normal DNA, tumor DNA, and tumor RNA from the same individual to increase sensitivity to detect somatic mutations without compromising specificity. Here we have focused on the ability of RADIA to detect germline variants and somatic single nucleotide variants. In the future, we plan to include other classes of somatic mutations such as small insertions and deletions (INDELs), loss of heterozygosity events (LOHs) and RNA editing events.

The accurate detection of somatic mutations is complicated by biological and technical artifacts such as tumor purity and subclonality, varying allele frequencies, sequencing depths, and copy-number variation. There is a trade-off between high sensitivity and high specificity, such that it is difficult to achieve both. By including an additional dataset, we are increasing our ability to reliably detect mutations, especially at low variant allele frequencies (Figure S7) where the signal to noise ratio becomes unfavorable.

Many widely used mutation calling algorithms see a large decrease in precision as the DNA variant allele frequency declines

Table 4. RNA Rescue Mutations in Lung Adenocarcinoma not Detected by Other Methods in TCGA.

Gene	Mutation	DNA VAF	RNA VAF	Validation DNA VAF
TP53	G266E	1/7 (13%)	6/10 (60%)	47/183 (26%)
TP53	G199V	4/64 (6%)	8/57 (14%)	17/380 (4%)
CDKN2A	R131H	3/45 (7%)	22/62 (35%)	9/149 (6%)
CDKN2A	R122*/R163*	2/16 (13%)	31/34 (91%)	20/92 (22%)
STK11	W239*	1/13 (7%)	20/40 (50%)	NA

These mutations were below the detection threshold for other mutation calling algorithms used by TCGA. The ratio of reads supporting the mutations along with the variant allele frequencies are shown for both the DNA and RNA. Validation was done on four of the mutations, and the resulting validation DNA variant allele frequencies are shown.

[6,8,9,11,12]. We found that a DNA VAF of 10% gives us the best balance between sensitivity and precision. To demonstrate this point, we lowered the DNA VAF to 5% and reran RADIA on the endometrial carcinoma data from Section 3.2. We used the same validation strategy as described in Section 3.2 and compared the results to the ones with a DNA VAF of 10%. We found a slight 1% increase in overall sensitivity from 84% (at 10% VAF) to 85% (at 5% VAF) but an 8% decrease in overall precision from 97% (at 10% VAF) to 89% (at 5% VAF).

By combining the RNA with the DNA, we are able to confirm the expression of a mutation, providing insight into its likely functional effect. Confirming mutations through RNA-Seq is also advantageous for large genomic studies in providing a means for weak validation for mutations without costly resequencing for validation (Figure S8). We find that over 99% of mutations that have both strong DNA and RNA support validate upon resequencing, suggesting that if one is not using mutations in clinical practice but rather estimating overall frequencies of specific mutations in a research cohort, the extreme expense in validating every mutation may not be warranted. While the integration of RNA and DNA provides an important but limited use as a DNA variant validation technique, studying the impacts on gene expression levels may lead to a deeper understanding of the functional impact of DNA-originating variants.

Here we have outlined some of the strengths of RADIA, but approaches that use RNA-Seq for detecting variants have clear limitations [53,54]. Only expressed alleles can be evaluated, which reduces the number of genes that can be assessed. In addition, several classes of mutations, such as the introduction of premature stop codons that lead to nonsense mediated decay, cannot be verified. Expression levels can also confound the ability to detect an imbalance in the genomic VAF as influences due to feedback control to rebalance gene dosage are currently unknown.

With RADIA, we are able to detect mutations in important cancer genes such as *TP53* that were previously not identified by other algorithms because the signal was lost in the noise. Somatic mutations are commonly used to group patients into subtypes that are critical for diagnosis and treatment of the disease. Our ability to rescue back mutations for individual patients will assist in correctly identifying each patient's specific subtype and consequently their treatment options.

Supporting Information

Figure S1   Schematic of mutations detected by the DNA Only Method (DOM) and Triple BAM Method (TBM). In the first and middle columns, there is enough DNA read support for the DOM and other algorithms acting on DNA pairs to detect a mutation. In the middle and last columns, there is sufficient RNA read support for the TBM to detect a mutation. The middle column illustrates "RNA Confirmation" mutations that are detected by both the DOM and the TBM due to high read support in both the DNA and RNA. The last column represents the "RNA Rescue" mutations that have some support in the DNA and strong evidence in the RNA. The RNA Rescue mutations are typically missed by traditional mutation calling algorithms that only investigate the pairs of DNA.

Figure S2   Diagram of bamsurgeon methodology. Mutations are spiked into BAM files by selecting locations with adequate coverage, extracting the reads, and adjusting the VAF according to the desirable VAF distribution. Once the bases in the

reads are changed, the reads are remapped to the genome, replacing the reads in the original BAM file.

Figure S3   Filters applied in the Variable DNA-Constant RNA bamsurgeon simulation experiment. The DNA variant allele frequencies were distributed from 1–50% and the RNA was held constant at 25%. Most of the DOM mutations were filtered because of the low variant allele frequency and tumor strand bias. In the TBM, most of the mutations were filtered due to the minimum number of alternative alleles required to make a call (n = 4) and strand bias in the tumor DNA and RNA.

Figure S4   Filters applied in the Low Frequency DNA-Variable RNA bamsurgeon simulation experiment. The RNA variant allele frequencies were distributed from 1–50% and the DNA was held at 10% or less. Most of the DOM mutations were filtered because of the low DNA variant allele frequency and tumor strand bias. In the TBM, most of the mutations were filtered due to the minimum number of alternative alleles required to make a call (n = 4) and the low RNA variant allele frequency.

Figure S5   Distribution of overlaps between RADIA and the endometrial TCGA MAF file. The distribution of the overlaps between RADIA and the validated somatic mutations from the endometrial TCGA network MAF file.

Figure S6   Filters applied to the RADIA mutations that validated as somatic in the endometrial TCGA MAF file. Thirty-three percent of the mutations had a DNA VAF of eight percent or less while 23% landed in blacklist regions that were ignored.

Figure S7   RNA Rescue mutations are primarily at low DNA VAFs. RNA Rescue mutations are primarily found at low DNA variant allele frequencies, but they also occur at higher frequencies where they were filtered due to non-depth related artifacts (e.g. strand-bias).

Figure S8   Distribution of RNA Confirmation Calls. The total number of mutations (blue) that are covered by at least one RNA read (yellow), one RNA read supporting the alternative allele (orange), and RNA Confirmation mutations with high support in both the DNA and RNA (purple).

Table S1   Summary of TCGA samples analyzed by RADIA. RADIA has been run on over 3,300 TCGA samples across 15 different types of cancer. The RNA Rescue mutations make up two to seven percent of the total somatic mutations across the 15 types of cancer. Variant Call Format (VCF) and Mutation Annotation Format (MAF) files can be downloaded from the TCGA Data Portal (https://tcga-data.nci.nih.gov/tcga/). Open-access somatic MAFs can be visualized and downloaded via the UCSC Cancer Browser (https://genome-cancer.ucsc.edu/).

Table S2   TCGA barcodes and Universally Unique Identifiers (UUIDs) for the TCGA samples used in this study. All patients provided written informed consent in accordance with TCGA guidelines and local Institutional Review Boards (IRBs).

Acknowledgments

The results published here are in whole or part based upon data generated by the TCGA Research Network: http://cancergenome.nih.gov/. We would like to thank Sofie Salama, J. Zachary Sanborn, Christopher Wilks, and Todd Lowe for helpful discussions and feedback on this manuscript.

Availability

BAM files are available from The Cancer Genome Atlas via the UCSC Cancer Genomics Hub https://cghub.ucsc.edu/. Variant Call Format (VCF) and Mutation Annotation Format (MAF) files are available from the TCGA Data Access Portal at https://tcga-data.nci.nih.gov/tcga/. Open-access somatic MAFs can be visualized and downloaded via the UCSC Cancer Browser (https://genome-cancer.ucsc.edu/) [55]. TCGA barcodes and Universally Unique Identifiers (UUIDs) for the TCGA samples used in this study can be found in Table S2. Software available at https://github.com/aradenbaugh/radia/.

Author Contributions

Conceived and designed the experiments: AJR JZ DH. Performed the experiments: AJR SM AE. Analyzed the data: AJR SM AE JMS EAC JZ DH. Contributed reagents/materials/analysis tools: AJR SM AE. Contributed to the writing of the manuscript: AJR. Revised the manuscript: AJR SM AE JMS EAC JZ DH. Supervised the study: AJR JZ DH.

References

1. Hanahan D, Weinberg RA (2000) The hallmarks of cancer. Cell 100: 57–70.
2. Hahn WC, Counter CM, Lundberg AS, Beijersbergen RL, Brooks MW, et al. (1999) Creation of human tumour cells with defined genetic elements. Nature 400: 464–468.
3. Wang Z, Gerstein M, Snyder M (2009) RNA-Seq: a revolutionary tool for transcriptomics. Nat Rev Genet 10: 57–63.
4. Gott JM, Emeson RB (2000) Functions and mechanisms of RNA editing. Annu Rev Genet 34: 499–531.
5. Li H, Handsaker B, Wysoker A, Fennell T, Ruan J, et al. (2009) The Sequence Alignment/Map format and SAMtools. Bioinformatics 25: 2078–2079.
6. Koboldt DC, Zhang Q, Larson DE, Shen D, McLellan MD, et al. (2012) VarScan 2: somatic mutation and copy number alteration discovery in cancer by exome sequencing. Genome Res 22: 568–576.
7. Saunders CT, Wong WS, Swamy S, Becq J, Murray LJ, et al. (2012) Strelka: accurate somatic small-variant calling from sequenced tumor-normal sample pairs. Bioinformatics 28: 1811–1817.
8. Larson DE, Harris CC, Chen K, Koboldt DC, Abbott TE, et al. (2012) SomaticSniper: identification of somatic point mutations in whole genome sequencing data. Bioinformatics 28: 311–317.
9. Koboldt DC, Chen K, Wylie T, Larson DE, McLellan MD, et al. (2009) VarScan: variant detection in massively parallel sequencing of individual and pooled samples. Bioinformatics 25: 2283–2285.
10. Goya R, Sun MG, Morin RD, Leung G, Ha G, et al. (2010) SNVMix: predicting single nucleotide variants from next-generation sequencing of tumors. Bioinformatics 26: 730–736.
11. Cibulskis K, Lawrence MS, Carter SL, Sivachenko A, Jaffe D, et al. (2013) Sensitive detection of somatic point mutations in impure and heterogeneous cancer samples. Nat Biotechnol 31: 213–219.
12. Roberts ND, Kortschak RD, Parker WT, Schreiber AW, Branford S, et al. (2013) A comparative analysis of algorithms for somatic SNV detection in cancer. Bioinformatics 29: 2223–2230.
13. Kandoth C, Schultz N, Cherniack AD, Akbani R, Liu Y, et al. (2013) Integrated genomic characterization of endometrial carcinoma. Nature 497: 67–73.
14. The Cancer Genome Atlas (2014) Comprehensive molecular profiling of lung adenocarcinoma. Nature 511: 543–550.
15. Li H (2011) Improving SNP discovery by base alignment quality. Bioinformatics 27: 1157–1158.
16. 1000 Genomes Project Consortium (2010) A map of human genome variation from population-scale sequencing. Nature 467: 1061–1073.
17. 1000 Genomes Project Consortium (2012) An integrated map of genetic variation from 1,092 human genomes. Nature 491: 56–65.
18. Sherry ST, Ward MH, Kholodov M, Baker J, Phan L, et al. (2001) dbSNP: the NCBI database of genetic variation. Nucleic Acids Res 29: 308–311.
19. Kent WJ, Sugnet CW, Furey TS, Roskin KM, Pringle TH, et al. (2002) The human genome browser at UCSC. Genome Res 12: 996–1006.
20. Karolchik D, Barber GP, Casper J, Clawson H, Cline MS, et al. (2014) The UCSC Genome Browser database: 2014 update. Nucleic Acids Res 42: D764–770.
21. Musumeci L, Arthur JW, Cheung FS, Hoque A, Lippman S, et al. (2010) Single nucleotide differences (SNDs) in the dbSNP database may lead to errors in genotyping and haplotyping studies. Hum Mutat 31: 67–73.
22. Kalyana-Sundaram S, Kumar-Sinha C, Shankar S, Robinson DR, Wu YM, et al. (2012) Expressed pseudogenes in the transcriptional landscape of human cancers. Cell 149: 1622–1634.
23. Harrow J, Frankish A, Gonzalez JM, Tapanari E, Diekhans M, et al. (2012) GENCODE: the reference human genome annotation for The ENCODE Project. Genome Res 22: 1760–1774.
24. Baertsch R, Diekhans M, Kent WJ, Haussler D, Brosius J (2008) Retrocopy contributions to the evolution of the human genome. BMC Genomics 9: 466.
25. Rosenbloom KR, Sloan CA, Malladi VS, Dreszer TR, Learned K, et al. (2013) ENCODE data in the UCSC Genome Browser: year 5 update. Nucleic Acids Res 41: D56–63.
26. Cingolani P, Platts A, Wang le L, Coon M, Nguyen T, et al. (2012) A program for annotating and predicting the effects of single nucleotide polymorphisms, SnpEff: SNPs in the genome of Drosophila melanogaster strain w1118; iso-2; iso-3. Fly (Austin) 6: 80–92.

27. Kent WJ (2002) BLAT–the BLAST-like alignment tool. Genome Res 12: 656–664.
28. Li H, Durbin R (2009) Fast and accurate short read alignment with Burrows-Wheeler transform. Bioinformatics 25: 1754–1760.
29. Wang K, Singh D, Zeng Z, Coleman SJ, Huang Y, et al. (2010) MapSplice: accurate mapping of RNA-seq reads for splice junction discovery. Nucleic Acids Res 38: e178.
30. Friedman PN, Chen X, Bargonetti J, Prives C (1993) The p53 protein is an unusually shaped tetramer that binds directly to DNA. Proc Natl Acad Sci U S A 90: 3319–3323.
31. Muller PA, Vousden KH (2012) p53 mutations in cancer. Nat Cell Biol 15: 2–8.
32. Kan Z, Jaiswal BS, Stinson J, Janakiraman V, Bhatt D, et al. (2010) Diverse somatic mutation patterns and pathway alterations in human cancers. Nature 466: 869–873.
33. Lindberg J, Mills IG, Klevebring D, Liu W, Neiman M, et al. (2013) The mitochondrial and autosomal mutation landscapes of prostate cancer. Eur Urol 63: 702–708.
34. Biankin AV, Waddell N, Kassahn KS, Gingras MC, Muthuswamy LB, et al. (2012) Pancreatic cancer genomes reveal aberrations in axon guidance pathway genes. Nature 491: 399–405.
35. Gui Y, Guo G, Huang Y, Hu X, Tang A, et al. (2011) Frequent mutations of chromatin remodeling genes in transitional cell carcinoma of the bladder. Nat Genet 43: 875–878.
36. Abaan OD, Polley EC, Davis SR, Zhu YJ, Bilke S, et al. (2013) The exomes of the NCI-60 panel: a genomic resource for cancer biology and systems pharmacology. Cancer Res 73: 4372–4382.
37. Pfaff E, Remke M, Sturm D, Benner A, Witt H, et al. (2010) TP53 mutation is frequently associated with CTNNB1 mutation or MYCN amplification and is compatible with long-term survival in medulloblastoma. J Clin Oncol 28: 5188–5196.
38. Alsner J, Yilmaz M, Guldberg P, Hansen LL, Overgaard J (2000) Heterogeneity in the clinical phenotype of TP53 mutations in breast cancer patients. Clin Cancer Res 6: 3923–3931.
39. Fernandez-Cuesta L, Oakman C, Falagan-Lotsch P, Smoth KS, Quinaux E, et al. (2012) Prognostic and predictive value of TP53 mutations in node-positive breast cancer patients treated with anthracycline- or anthracycline/taxane-based adjuvant therapy: results from the BIG 02–98 phase III trial. Breast Cancer Res 14: R70.
40. Gartel AL, Feliciano C, Tyner AL (2003) A new method for determining the status of p53 in tumor cell lines of different origin. Oncol Res 13: 405–408.
41. The Cancer Genome Atlas (2012) Comprehensive molecular portraits of human breast tumours. Nature 490: 61–70.
42. Jones S, Wang TL, Ie-Shih M, Mao TL, Nakayama K, et al. (2010) Frequent mutations of chromatin remodeling gene ARID1A in ovarian clear cell carcinoma. Science 330: 228–231.
43. Robinson G, Parker M, Kranenburg TA, Lu C, Chen X, et al. (2012) Novel mutations target distinct subgroups of medulloblastoma. Nature 488: 43–48.
44. Kim TH, Lee SY, Rho JH, Jeong NY, Soung YH, et al. (2009) Mutant p53 (G199V) gains antiapoptotic function through signal transducer and activator of transcription 3 in anaplastic thyroid cancer cells. Mol Cancer Res 7: 1645–1654.
45. Davies H, Hunter C, Smith R, Stephens P, Greenman C, et al. (2005) Somatic mutations of the protein kinase gene family in human lung cancer. Cancer Res 65: 7591–7595.
46. Imielinski M, Berger AH, Hammerman PS, Hernandez B, Pugh TJ, et al. (2012) Mapping the hallmarks of lung adenocarcinoma with massively parallel sequencing. Cell 150: 1107–1120.
47. Andujar P, Wang J, Descatha A, Galateau-Salle F, Abd-Alsamad I, et al. (2010) p16INK4A inactivation mechanisms in non-small-cell lung cancer patients occupationally exposed to asbestos. Lung Cancer 67: 23–30.
48. Blons H, Pallier K, Le Corre D, Danel C, Tremblay-Gravel M, et al. (2008) Genome wide SNP comparative analysis between EGFR and KRAS mutated NSCLC and characterization of two models of oncogenic cooperation in non-small cell lung carcinoma. BMC Med Genomics 1: 25.
49. The Cancer Genome Atlas (2012) Comprehensive molecular characterization of human colon and rectal cancer. Nature 487: 330–337.

50. Takahira T, Oda Y, Tamiya S, Yamamoto H, Kawaguchi K, et al. (2004) Alterations of the p16INK4a/p14ARF pathway in clear cell sarcoma. Cancer Sci 95: 651–655.

51. Nagy E, Beck Z, Kiss A, Csoma E, Telek B, et al. (2003) Frequent methylation of p16INK4A and p14ARF genes implicated in the evolution of chronic myeloid leukaemia from its chronic to accelerated phase. Eur J Cancer 39: 2298–2305.

52. Ueki T, Hsing AW, Gao YT, Wang BS, Shen MC, et al. (2004) Alterations of p16 and prognosis in biliary tract cancers from a population-based study in China. Clin Cancer Res 10: 1717–1725.

53. Ku CS, Wu M, Cooper DN, Naidoo N, Pawitan Y, et al. (2012) Exome versus transcriptome sequencing in identifying coding region variants. Expert Rev Mol Diagn 12: 241–251.

54. Cirulli ET, Singh A, Shianna KV, Ge D, Smith JP, et al. (2010) Screening the human exome: a comparison of whole genome and whole transcriptome sequencing. Genome Biol 11: R57.

55. Zhu J, Sanborn JZ, Benz S, Szeto C, Hsu F, et al. (2009) The UCSC Cancer Genomics Browser. Nat Methods 6: 239–240.

The Putative *Leishmania* Telomerase RNA (*Leish*TER) Undergoes *Trans*-Splicing and Contains a Conserved Template Sequence

Elton J. R. Vasconcelos[1,◊], Vinícius S. Nunes[2,◊], Marcelo S. da Silva[2,3], Marcela Segatto[2], Peter J. Myler[1,4,5,*], Maria Isabel N. Cano[2*]

1 Seattle Biomedical Research Institute, Seattle, Washington, United States of America, 2 Departamento de Genética, Instituto de Biociências, Universidade Estadual Paulista (UNESP), Botucatu, São Paulo, Brazil, 3 Universidade Estadual deCampinas (UNICAMP), Campinas, São Paulo, Brazil, 4 Department of Global Health, University of Washington, Seattle, Washington, United States of America, 5 Department of Biomedical Informatics and Medical Education, University of Washington, Seattle, Washington, United States of America

Abstract

Telomerase RNAs (TERs) are highly divergent between species, varying in size and sequence composition. Here, we identify a candidate for the telomerase RNA component of *Leishmania* genus, which includes species that cause leishmaniasis, a neglected tropical disease. Merging a thorough computational screening combined with RNA-seq evidence, we mapped a non-coding RNA gene localized in a syntenic locus on chromosome 25 of five *Leishmania* species that shares partial synteny with both *Trypanosoma brucei* TER locus and a putative TER candidate-containing locus of *Crithidia fasciculata*. Using target-driven molecular biology approaches, we detected a ~2,100 nt transcript (*Leish*TER) that contains a 5′ spliced leader (SL) cap, a putative 3′ polyA tail and a predicted C/D box snoRNA domain. *Leish*TER is expressed at similar levels in the logarithmic and stationary growth phases of promastigote forms. A 5′SL capped *Leish*TER co-immunoprecipitated and co-localized with the telomerase protein component (TERT) in a cell cycle-dependent manner. Prediction of its secondary structure strongly suggests the existence of a *bona fide* single-stranded template sequence and a conserved C[U/C]GUCA motif-containing helix II, representing the template boundary element. This study paves the way for further investigations on the biogenesis of parasite TERT ribonucleoproteins (RNPs) and its role in parasite telomere biology.

Editor: Bin Tian, Rutgers New Jersey Medical School, United States of America

Funding: Funding provided by Fundação de Amparo à Pesquisa do Estado de São Paulo- FAPESP, Brazil to MINC [GRANT 2012/50263-5]. National Institute of Allergy and Infectious Diseases PHS Grant to PJM [GRANT R01 AI103858]. VSN, MS and MSS received post-doctoral and doctoral fellowships, respectively, from Fundação de Amparo à Pesquisa do Estado de São Paulo, FAPESP. EJRV is a post-doctoral fellow from Conselho Nacional de Desenvolvimento Científico e Tecnológico – CNPq, Brazil [PDE 202223/2012-4]. The funders had no role in study design, data collection and analysis, decision to publish, or preparation of the manuscript.

Competing Interests: The authors have declared that no competing interests exist.

* Email: peter.myler@seattlebiomed.org (PM); micano@ibb.unesp.br (MINC)

◊ These authors contributed equally to this work.

Introduction

Leishmania spp. are trypanosomatid protozoa, considered ancient eukaryotes whose nuclear genome is organized in linear chromosomes [1,2]. Like most eukaryotes, their chromosomes end termini are characterized by sequences known as telomeres [3]. *Leishmania* telomeres are formed by conserved 5′-TTAGGG-3′ telomeric repeats that typically end in a 3′ overhang that serves as substrate for repeat addition by telomerase [3–6], a ribonucleoprotein enzyme minimally composed of two catalytically essential subunits: the telomerase reverse transcriptase protein (TERT) and the telomerase RNA (TER) component, which contains the template that specifies the sequence of the telomeric repeats [7,8]. Enzyme activity has been detected in cell extracts of promastigotes of three *Leishmania* species, and also from different *Trypanosoma brucei* and *Trypanosoma cruzi* replicative stages [9–11]. The trypanosomatids' TERT component is one of the largest telomerase (MW ~156 kDa) described so far and the TERT enzyme from *Leishmania* species shows greater sequence similarity (86–95%) with each other than with the telomerases of other eukaryotes [12,13]. Although trypanosomatid TERT contains some important amino acid substitutions within the conserved TERT motifs, the *Leishmania* and *Trypanosoma* TERT components present all the conserved structural features shared with other TERTs, such as the N-terminus motifs that are essential for telomerase RNA (TER) binding and enzyme activity [14–16], the central domain that contains a less conserved telomerase-specific T motif, the reverse transcriptase motifs that are essential for enzyme activity, and a less conserved C-terminal domain [12,17–20]. Biochemically, *Leishmania* and *Trypanosoma* TERTs resemble other telomerases, as they are able to add TTAGGG repeats to the 3′ end of the G-rich telomeric strand and fulfill other essential criteria for telomerase activity, such as RNase A sensitivity. Enzyme processivity differs among trypanosomatids, with *T.*

brucei TERT being the most processive compared to *T. cruzi* and *Leishmania* TERTs [9–11]. The purification of *Leishmania amazonensis* TERT enzyme was only achieved using semi-purified protein extracts fractionated on a G-rich telomeric DNA affinity column (G-DNA), indicating that similar to *T. cruzi* telomerase, *Leishmania* TERT bound tightly to an antisense 2′*O*-methyl oligonucleotide complementary to the *T. brucei* TER template sequence [10,11]. This was the first hint that the *Leishmania* TER (*Leish*TER) template sequence was similar to the predicted minimum model for the RNA template region used by *T. brucei* telomerase [9].

Recently, with the characterization of the telomerase RNA molecule of *T. brucei* (*Tb*TER), it was possible to confirm that the *Tb*TER template sequence 5′-CCCTAACCCTA-3′ differs from the previously predicted template (5′-CCCTAACCC-3′) only by having two nucleotides more at its 3′ end [9,21]. In addition, *Tb*TER is processed through *trans*-splicing and was shown to interact and to copurify with TbTERT *in vivo*. Deletion or silencing of *TbTER* causes progressive shortening of telomeres and mutations in its template domain results in corresponding mutant telomere sequences, demonstrating that in *T. brucei* it is essential for telomerase activity [21,22]. Moreover, like TERs from different eukaryotes, *Tb*TER differs greatly in nucleotide sequence and size [21,22]. It shares some secondary structural elements conserved in most, but not all, eukaryotic TERs, such as a putative pseudoknot domain, that includes at the 5′ end the TER template motif, in addition to helix I and helix II, the latter of which contains a putative template boundary element (TBE) [23]. *Tb*TER also binds the core proteins of the C/D small nucleolar RNA (snoRNA) family and associates with the methyltransferase-associated protein, whose homolog also binds to mammalian TER [22].

Here, we report the identification and characterization of the putative TER components from two *Leishmania* species, *Leishmania major* (*Lm*TER) and *Leishmania amazonensis* (*La*TER), using *in silico* and experimental approaches. *Leishmania* TERs (*Leish*TER), like *Tb*TER, appear to be transcribed by RNA pol II, as they are located in the sense orientation within an mRNA directional gene cluster (DGC), and the putative mature RNA was amplified using the 39 nt spliced leader (SL) RNA sequence commonly positioned at the 5′ end of most mature trypanosomatids RNA pol II transcribed RNAs. The 3′ end of both *Leish*TERs was also amplified by 3′ RAcE using oligo-dT and total RNA, resembling the mature form of *Tb*TER [21]. *Leish*TERs also contain a template domain that is almost identical to *Tb*TER template sequence. Using immunofluorescence coupled with RNA *in situ* hybridization, we demonstrated that *Lm*TER co-localizes with LmTERT in parasite nucleus. In addition, a 5′SL capped *Lm*TER immunoprecipitates with the telomerase reverse transcriptase component, suggesting that the mature transcript is part of the ribonucleoprotein complex. We predicted a partial secondary structure of *Leishmania* TER that shows some conserved structural features shared among its *Tb*TER counterpart and other TERs.

Further experiments are required to elucidate the biological importance of the *Leish*TER molecule for parasite survival and, thus, whether it will serve as a target for anti-parasite therapy. This discovery would be of great significance because the genus *Leishmania* comprises several species that cause leishmaniasis, which are neglected tropical diseases that threaten hundreds of millions of people around the world but lack treatment options, vaccines and prophylaxis protocols (World Health Organization 2010).

Materials and Methods

in silico analyses of LeishTER candidates

We started the *in silico* screening for the *Leishmania* TER gene performing a sensitive BLASTn search [24] (-FF −W7 −m8) against *L. major*, *L. infantum*, *L. braziliensis*, *L. mexicana* and *L. tarentolae* genomes (TriTrypDB 6.0) using as query the template sequence that consisted of two tandem telomeric hexamer repeats (THR) identified previously ([6] and Cano, MIN, Personal Communication). Since the telomeric 3′ overhang may vary between *Leishmania* species [6,25], we used all six iterations of two copies of telomeric hexamer repeats (THRs) as follow: query_1: 5′ CCTAACCCTAAC 3′; query_2: 5′ CTAACCCTAACC 3′; query_3: 5′ CCCTAACCCTAA 3′; query_4: 5′ ACCCTAACCCTA 3′; query_5: 5′ AACCCTAACCCT 3′; query_6: 5′ TAACCCTAACCC 3′.

A PERL script was written to discard the great number of BLAST hits falling within the telomeres, which must have a length defined by the user. We arbitrarily chose a value of 9 kb from both chromosome ends to filter out the BLAST hits within those regions in each chromosome from all species. The template query_4 was the only one presenting a perfect match in a syntenic locus in all *Leishmania* genomes analyzed, and this region was subjected to a deeper analysis on a comparison to other trypanosomatids (See results, Table 1 and Table 2).

We ran tBLASTx [24] from the blast+ package and used its output as an input to the ACT tool [26] to compare the synteny regarding the TER locus among *L. major* and *L. mexicana* chr25, *Trypanosoma brucei* chr11 and *Crithidia fasciculata* chr28 (see results, Figure 1).

ClustalW [27] and Jalview [28] were used for the generation and visualization of global multiple alignments, respectively. The RNA secondary structures were determined by first running RNAalifold [29], using as input the.aln ClustalW file containing the expected TER sequences for all five *Leishmania* species analyzed, and then the constraints from the consensus structure were applied to the individual TER modeling by the execution of mfold with default parameters [30]. Pknots [31], KnotSeeker [32], and Turbofold [33] algorithms were applied without any success in our attempt to detect pseudoknots along the entire *Leishmania* TER sequences.

Cell lines and cell culture

Promastigotes forms of *L. major* LT252 (MHOM/IR/1983/IR) and *L. amazonensis* (MHOM/BR/73/M2269) were cultured at 26°C in M199 medium supplemented with 10% heat-inactivated fetal bovine serum as previously described [34].

L. major and L. amazonensis total RNA isolation

Total RNA from *L. major* and from *L. amazonensis* promastigotes and that obtained from nuclear protein extracts and IP isolates were isolated using TRIzol reagent (Invitrogen) according to the manufacturer's instructions. Extracted RNAs were solubilized with 50 μL of water and treated with RNase free-DNase I (Life Technologies) in 1X DNase I buffer (10 mM Tris-HCl, 2.5 mM $MgCl_2$, 0.5 mM $CaCl_2$, pH 7.6) for 15 minutes at room temperature. The reaction was inactivated by the addition of 1 μl 25 mM EDTA solution and heating for 10 min at 65°C.

Northern blot analyses

Total RNA isolated from *L. major* and from *L. amazonensis* promastigotes was fractionated in a 1.5% agarose/2.0 M formaldehyde gel electrophoresis. A 347 bp *Lm*TER fragment generated by PCR using the e+f primers (Table S1) was labeled with α-dGTP

Table 1. All extra-telomere BLASTn hits for six different combinations of the 12 nt TER template sequence in the *L. major* genome.

Query ID	Subject ID	Identity %	Alignment length	Mismatches	Gap Openings	q. start	q. end	s. start	s. end	E-value	Bit score
query_1	LmjF.28	100	12	0	0	1	12	544272	544283	3.2	24.3
query_2	LmjF.28	100	12	0	0	1	12	544273	544284	3.2	24.3
query_3	LmjF.28	100	12	0	0	1	12	544271	544282	3.2	24.3
query_4	**LmjF.25**	**100**	**12**	**0**	**0**	**1**	**12**	**333352**	**333363**	**3.2**	**24.3**
query_5	LmjF.28	100	12	0	0	1	12	544275	544286	3.2	24.3
query_6	LmjF.28	100	12	0	0	1	12	544274	544285	3.2	24.3

Table 2. Extra-telomere BLASTn hits of the TER template sequence on chromosome 25 from the other *Leishmania* species: *L. infantum* (LinJ), *L. mexicana* (LmxM), *L. tarentolae* (LtaP) and *L. braziliensis* (LbrM).

Query ID	Subject ID	Identity %	Alignment length	Mismatches	Gap Openings	q. start	q. end	s. start	s. end	E-value	Bit score
query_4	LinJ.25	100	12	0	0	1	12	322179	322190	3.2	24.3
query_4	LmxM.25	100	12	0	0	1	12	322685	322696	3.1	24.3
query_4	LtaP25	100	12	0	0	1	12	329780	329791	3.1	24.3
query_4	LbrM.25	100	12	0	0	1	12	277119	277130	3.1	24.3

All the depicted loci are syntenic to the one in *L. major* chromosome 25 shown in Table 1.

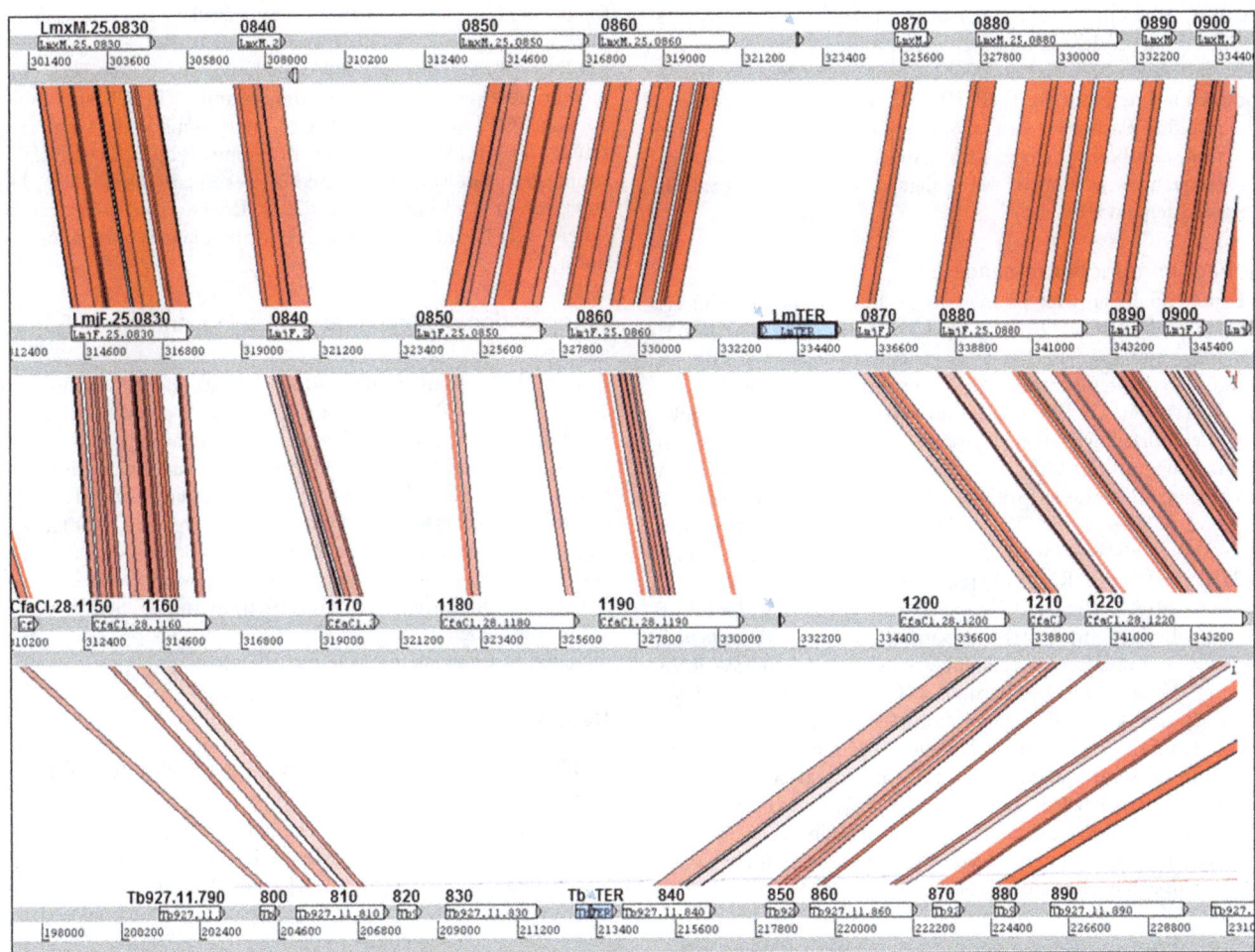

Figure 1. Protein-coding gene synteny view at the TER locus of trypanosomatid species from three different genera. tBLASTx comparisons were performed, followed by a visualization of the results using the ACT tool. Gray bars represent the forward and reverse strands of DNA, and the numbers between them correspond to the absolute coordinates within the chromosome where this syntenic locus is located: *L. mexicana* (LmxM) and *L. major* (LmjF) chromosome 25, *C. fasciculata* (CfaCl) chromosome 28 and *T. brucei* (Tb927) chromosome 11. The reddish-pink lines between sequences represent sequence similarity from tBLASTx analyses. The white box features correspond to CDSs, and the blue ones indicate the TER gene of both *L. major* and *T. brucei*. Blue arrows point to the conserved 12 nt TER-template sequence (5' ACCCTAACCCTA 3') found in all species at this particular locus (in *T. brucei* it is 11 nt, 5' CCCTAACCCTA 3'). TriTrypDB gene accession numbers are written inside each white box and the last digits placed above the boxes. The two protein-coding genes immediately upstream of *TbTER* (Tb927.11.820 and 830) appear to be *Trypanosoma*-specific, whereas *LmjF.25.0840, 0850* and *0860* are indicative of *Leishmania-/Crithidia*-specific genes. Despite the synteny disruption observed on the protein-coding content nearby the TER gene, this non-coding RNA seemed to be retained by evolutionary pressure on the same syntenic position in the three distinct genera.

[32P] and used as the specific probe. A membrane containing transferred RNAs was pre-hybridized with solution I (2X SSC, 0.5% SDS, 0.1% Ficoll, 0.1%PVP, 0.1% BSA, 0.1 mg/mL ssDNA) and hybridized with *Lm*TER probe in solution II (2X SSC, 0.5% SDS, 0.02% Ficoll, 0.02% BSA, 0.1 mg/mL ssDNA). The membrane was washed twice with wash solution I (2x SSC, 0.1% SDS) at room temperature, once with wash solution II (1x SSC, 0.1% SDS) at room temperature, once with wash solution II at 65°C and finally twice with wash solution III (0.2X SSC, 0.1% SDS) at 65°C. The membrane was exposed up to 7 days at −80°C.

5' and 3' Rapid Amplification of cDNA Ends (5' and 3' RAcE)

To map both the 5' and the 3' ends of the *Lm*TER transcript, cDNA was synthesized from total RNA (1.5 µg) using the

QIAGEN OneStep RT-PCR Kit and using respectively the combination of primers a+d [Forward - spliced leader sequence (5' RAcE) and Reverse - (5' RAcE)] and g+h [Forward (3'RAcE) and Reverse - oligo dT (3'RAcE)] (for primer sequences and names see Table S1). Thermal cycler conditions for 5' RAcE were 30 minutes at 50°C for reverse transcription, 15 minutes at 95°C for the initial PCR activation step and 40 cycles of denaturation for 45 seconds at 95°C, annealing for 1 minute at 57°C and extension for 1 minute at 72°C. Thermal cycler conditions for 3' RAcE were 30 minutes at 50°C for reverse transcription, 15 minutes at 95°C for the initial PCR activation step and 60 cycles of denaturation for 45 seconds at 95°C, annealing for 1 minute at 58°C and extension for 1 minute at 72°C.

Cloning and characterization of 5′ and 3′ RAcE products

The 5′ and 3′ RAcE products were purified from agarose gel using the Wizard SV Gel and PCR Clean-Up System (Promega) and then inserted into the TOPO TA Cloning vector (Invitrogen). The resulting plasmids were analyzed by *EcoR*I restriction enzyme digestion, and positive clones were sequenced with M13 primers. The sequences generated were aligned against the *Lm*TER predicted sequence.

Indirect immunofluorescence (IIF)

Exponentially growing promastigote cells were washed with 1X PBS (137 mM NaCl, 2.7 mM KCl, 10 mM Na_2HPO_4 and 2 mM KH_2PO_4) and fixed in 1% (v/v) formaldehyde in 1X PBS for 5 min at room temperature. Cells were then treated with 0.1% Triton-X 100 in 1X PBS for 10 min and free aldehyde molecules were neutralized with 0.1 M glycine in 1X PBS for 10 min at room temperature. Cells that were treated with RNase A (Invitrogen) were washed with 1X PBS and then incubated with 20 μg of RNase A at 37°C for 30 min. Cells not treated with RNase A were washed with 1X PBS and incubated with 1X PBS at 37°C for 30 min. RNase A-treated and -non-treated cells were washed with 1X PBS and incubated with rabbit anti-LaTERT serum, obtained from recombinant *Leishmania amazonensis* TERT N-terminal region containing a putative telomerase RNA binding domain (TRD) (Giardini & Cano, unpublished data); LaTERT and LmTERT share about 95% identity (LmTERT) [12]. α-LaTERT was diluted (1:2000) in blocking solution (4% (w/v) bovine serum albumin) for 12 h at 4°C. Goat anti-rabbit IgG (2 mg/mL) labeled with Alexa Fluor 555 (Invitrogen) was diluted 1:3000 and used as the secondary antibody. Cells were deposited on poly-L-lysine coated slides and used in fluorescence RNA *in situ* hybridization assays.

Indirect immunofluorescence coupled with RNA *in situ* hybridization

Poly-L-lysine-coated slides containing RNase A-treated and non-treated log-phase promastigote cells were dehydrated using an ethanol series (70%, 80% and 90%) and incubated with 0.3 μg.mL^{-1} of TelG-FITC PNA probe (Panagene) diluted in 1X hybridization buffer (70% formamide, 20 mM Tris-HCl, pH 7.0 and 1% BSA) at 4°C for 12 h in the dark using a 25 μl frame (Gene Frame, Pierce Biotechnology). Thereafter, the slides were washed with 1X washing buffer (50 mM Tris-HCl, pH 7.6) and dehydrated again using 70%, 80% and 90% ethanol. VECTASHIELD Mounting Medium with DAPI (Vector Labs) was used as anti-fade mounting solution and to stain nuclear and kinetoplast DNA. Finally, slides were sealed using coverslips. For these experiments, images were analyzed with a Nikon 80i fluorescence microscope and captured with a digital camera (DS-Fi1, Nikon). When necessary, images were superimposed using NIS elements software (version Ar 3.10). The parasites cultures used for the FISH-IF analysis were not synchronized since we were able to morphologically discriminate *L. major* promastigote cell cycle phases based on a previous report [35].

Immunoprecipitation (IP) and western blot analyses

Two hundred micograms of nuclear protein extract obtained according to [9] was used as input in IP assays, in conjunction with 10 μg of rabbit anti-LaTERT serum or 10 μg of the corresponding pre-immune serum as control. The IP assays were performed using Dynabeads Protein A (Novex by Life Technologies) according to the manufacturer's instructions. At the end of the assay, one-tenth of each IP eluate and 10% of the input were fractionated by 12% SDS-PAGE and transferred to nitrocellulose membranes (Bio-Rad) in transfer buffer (48 mM Tris-HCl, pH 8.3, 39 mM glycine, 20% methanol (v/v)) at 16°C. The membranes were probed with mouse anti-LaTERT and rabbit anti-LmNOP1 (control) used as primary antibodies. Goat anti-rabbit IgG (H+L) and goat anti-mouse IgG (H+L) HRP-conjugates (Bio-Rad) were used as secondary antibodies. The reactions were revealed using the ECL western blotting analysis system (GE Healthcare) according to the manufacturer's instructions.

Amplification of *Lm*TER from protein extracts and from IP eluates

Total RNA from protein extract (input) and IP eluates were obtained using TRIzol reagent (Invitrogen) as described above. A QIAGEN OneStep RT-PCR Kit was used to amplify cDNAs. For *Lm*TER amplification, we used three combinations of primers a+d, e+f, and c+d (see Table S1). The primers Forward - (RT-PCR control) and Reverse - (RT-PCR control) were used to amplify *L. major* histone H2A cDNA fragment (~150 bp), used as a control (see Table S1). Other control reactions included amplification of both cDNAs in the absence of reverse transcriptase (Superscript II, Life Technologies), in the absence of RNA, or using RNA obtained from pre-immune serum IP extracts.

Results

In silico identification of the putative TER locus within *Leishmania* genomes

We searched the five publically available *Leishmania* genomes with all six iterations of two copies of the telomeric hexamer repeat (TTAGGG) using BLASTn [24] (see Methods), based on the recently described *T. brucei* TER template, which contains an 11-nt sequence complementary to the telomeric 3′ overhang sequence [9,21,22], and the cloned telomeric terminus of *L. major* Friedlin [36] although the exact nature of the overhang varies between species [6,25]. In *L. major* (the species with the best assembled genome), after running an *ad-hoc* PERL script to eliminate BLAST hits within the telomeres, only two non-telomeric loci showed hits (Table 1). Five out of six queries matched the same region on chromosome 28, while one (query_4) showed a unique match on chromosome 25. However, only the latter (between the *LmjF.25.0860* and *LmjF.25.0870* protein-coding genes) showed matches at syntenic loci in all four other *Leishmania* species (Table 2). Thus, we concluded that this region contains the putative *TER* locus in *Leishmania*.

LmjF.25.0870 is orthologous to *Tb927.11.0850*, which is the second gene downstream of the *TER* locus in *T. brucei*, so we explored the synteny surrounding the putative *TER* locus in the genomes of three different trypanosomatid genera by using the ACT tool [26] to visualize tBLASTx comparisons of chr25 from *L. major* and *L. mexicana*, chr28 from *Crithidia fasciculata* and chr11 from *T. brucei* (Figure 1). While *L. major* and *L. mexicana* showed perfect synteny throughout the entire locus, *C. fasciculata* and *T. brucei* contained an additional protein-coding gene (*CfaC1.28.1200* and *Tb927.11.840*) immediately downstream of the *TER* locus. Conversely, *T. brucei* appeared to lack the three protein-coding genes immediately upstream of the *TER* locus in the other species, but the synteny resumed more 5′ at the *Tb927.11.0810* gene. These results support previous findings that *Crithidia* is evolutionarily closer to *Leishmania* than to *Trypanosoma* [37,38].

Multiple sequence alignment of all five *Leishmania* genomes revealed a high degree of sequence conservation throughout most of the 4.5-kb inter-CDS region containing the *TER* gene (see Figure S3), but especially surrounding the putative TER template sequence (Figure 2). Searches for non-coding RNA domains using the RNAspace webserver platform [39] revealed a~45 nt-long small nucleolar (sno) RNA domain ~300 nt downstream of the putative TER template sequence in four species (LmjF, LinJ, LmxM and LtaP) that best matches snoRNAU90 (or scaRNA7), a C/D box snoRNA found in human Cajal bodies [40].

Characterization of the *Lm*TER transcript

Northern analysis of RNA extracted from *L. major* promastigotes using a 347-nt probe specific for the putative *LmTER* locus, identified a major transcript of ~2,100 nt (Figures 3B and 3C) and a similar result was obtained for *L. amazonensis* (Figure S1). We also found that both *LmTER* and *LaTER* RNAs were expressed at similar levels in the logarithmic and stationary growth phase (Figure 3C and Figure S1). Taking into account that trypanosomatid ncRNAs transcribed by RNA polymerase II, such as snoRNAs and *Tb*TER, are *trans*-spliced and polyadenylated (Figure 3A) [21,22,41], we performed both 5′ and 3′ rapid amplification of cDNA ends (RAcE) PCR to respectively identify the SL site at the 5′ end and the polyA tail at the 3′end of the *Lm*TER transcript. First strand cDNA synthesis with internal primers b or d (Figure 3B and Table S1) followed by second strand synthesis using the SL-specific primer a (Figure 3B and Table S1) revealed products of 103 bp and 276 bp for the 5′RAcEs (Figures 3B and 3E). The 3′RAcE revealed a product of ~500 bp (Figures 3B and 3E), using oligo dT-specific primer h for the first strand cDNA synthesis and an internal primer g for the second strand (Figure 3B and Table S1). Sequencing of the 5′RAcE PCR products (Figure S2) mapped the 5′ SL site to the position 333,307 on *L. major* chromosome 25 (Figure 2), while the 3′RAcE product size is indicative of a polyA site at position 335,419 on the same chromosome. These results are consistent with SL and PolyA sites previously assigned to this genomic locus by RNA-seq experiments

(tritrypdb.org). On the other hand, RT-PCR using primers c+d, showed a 550 bp cDNA product (Figure 3D), presumably representing the longer polycistronic transcript (Figures 3A and 3B). Automated sequencing of this cDNA product confirmed they came from the targeted locus (data not shown).

*Lm*TER forms a ribonucleoprotein complex with LmTERT

Immunoprecipitation assays were conducted to identify the *Lm*TER transcript that formed a complex with the telomerase reverse transcriptase (TERT) protein component in protein parasite extracts. Western blot analysis of *L. major* S100 nuclear extracts immunoprecipitated with rabbit anti-LaTERT serum confirmed the presence of LmTERT (MW 156kDa) in both parasite nuclear extract (input 10%) and IP eluate (Figure 4A upper panel), while the LmNOP1 (MW ~70 kDa) control was present only in the parasite nuclear extract (input 10%) (Figure 4A bottom panel) and thus, was not immunoprecipitated by anti-LaTERT, confirming anti-LaTERT specificity.

Total RNA isolated from the IP eluates shown in figure 4A was used to amplify fragments of the *Lm*TER transcript using RT-PCR. The results of RT-PCR using primer pairs a+d and e+f amplified products of expected sizes, respectively ~276 bp and ~347 bp (Figures 4B and 4C). We were not able to amplify the 3′end of the IP *Lm*TER using primers g+h (data not shown), thus, we cannot certify that the *Lm*TER in complex with LmTERT is polyadenylated. In addition, the reaction with primers c+d did not amplify any product from the IP extract (data not shown), probably because primer c is located upstream of the expected SL site for the mature *Lm*TER transcript. This result gives support to the evidence that there might be only one SL site for the mature *Lm*TER (SL at position 333,307 on chromosome 25, as mentioned above) and that only the mature transcript co-immunoprecipitates with the LmTERT component. The pre-immune sera did not immunoprecipitate any of the tested proteins from the parasite extract and thus, no RNA was isolated from these reactions. As a

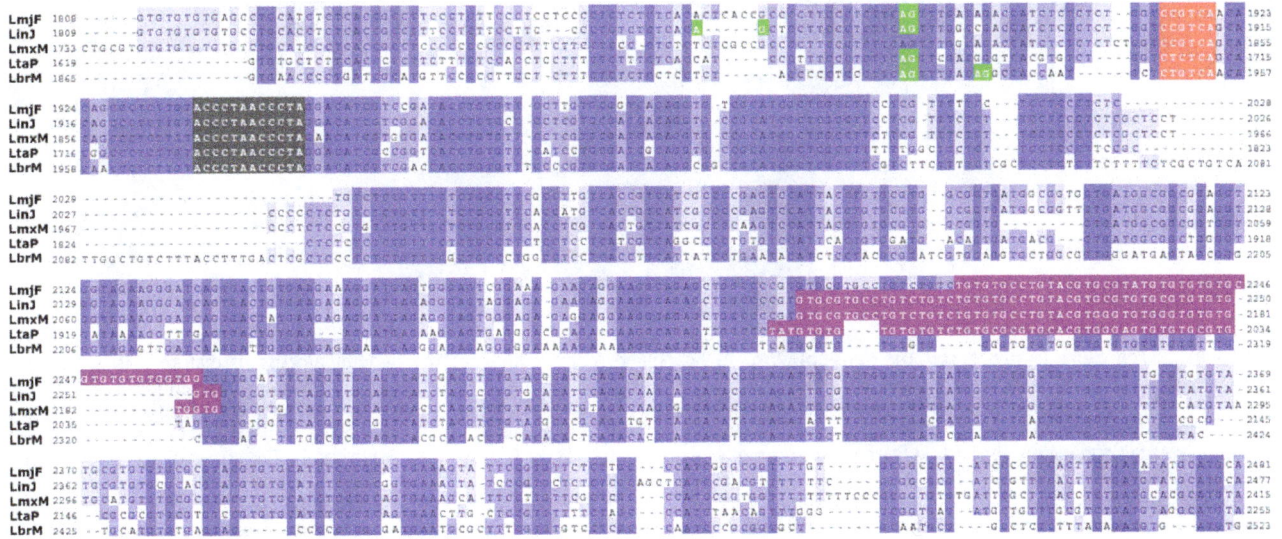

Figure 2. Global multiple alignment of the intercoding region on chromosome 25 where *LeishTER* is located. Coordinates are relative to the first base after the STOP codon of the respective CDSs: *L. major* (LmjF.25.0860), *L. infantum* (LinJ.25.0890), *L. mexicana* (LmxM.25.0860), *L. tarentolae* (LtaP25.0910) and *L. braziliensis* (LbrM.25.0740). The differently shaded colored regions represent the 12 nt template sequence (gray), 5′-C[C/T]GTCA-3′ motif that is part of the template boundary element (TBE) (pink), snoRNA domains found by the RNAspace webserver [39] (magenta, the one at position 2216–2259 in LmjF and aligned to other three species is part of snoU90 (or scaRNA7), which is a C/D box snoRNA) and splice acceptor sites (green) detected by RNA-seq (provided by Myler lab and deposited on tritrypdb.org for *L. major*). ClustalW [27] and Jalview [28] were used to align and visualize this locus, respectively. The complete alignment of the whole intercoding region is provided on Figure S3.

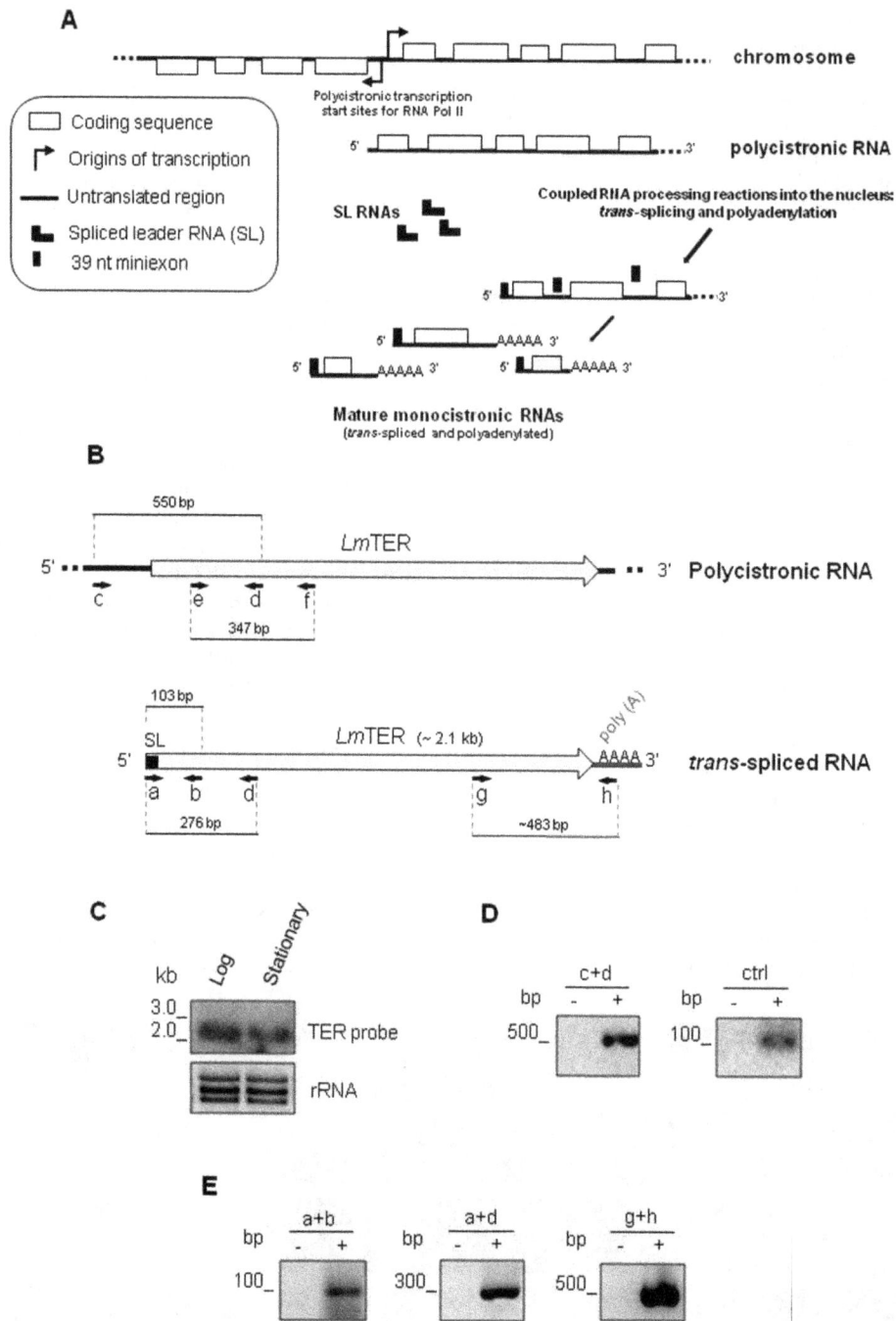

Figure 3. *LmTER* **gene is processed by** *trans-***splicing.** A) Schematic representation of the general process of mature RNA synthesis in trypanosomatids adapted from Requena, J.M. (2011) [69]. After polycistronic transcription of the DNA, RNAs are individualized into monocistronic mature units through two coupled processing reactions: *trans*-splicing and polyadenylation. The former occurs at the 5′ end of the downstream gene and consists on the addition of a capped 39 nt miniexon sequence from the SL RNA, while the latter takes place at the 3′ end of the upstream gene for the poly(A) tail generation, similar to what happens on higher eukaryotes RNAs. B) A diagram showing the *LmTER* non-coding transcript, on its immature (within the polycistronic RNA precursor) and mature (*trans*-spliced and polyadenylated) forms. a, b, c, d, g and h indicate the corresponding positions of primers used in the RT-PCR and RAcE reactions. The combination of primers e and f was used to generate the northern TER probe. The expected sizes of each amplicon are denoted. C) Total RNA (10 μg), from parasite in the logarithmic and stationary phases were separated on a 1.5% agarose/2.0 M formaldehyde gel, and the northern blot was probed with a TER specific-probe, which was generated using the primers e+f. Bottom, ethidium bromide-stained RNA gel showing rRNA, served as a loading control. D) RT-PCR using primers c+d detected a band of ~550 bp which is indicative of either a polycistronic pre-mRNA or a longer transcript, possibly the TER precursor. An amplicon of ~150 bp from Histone H2A transcript was detected as control (ctrl). E) 5′-560 Spliced form of *LmTER* confirmed by 5′RAcE-PCR using the following primer pairs: a+b, a+d. The 3′ end of *LmTER* containing the polyA tail was amplified by 3′RAcE-PCR using primers g+h. See Table S1 for a complete description of primers. Control reactions (-) were done in the presence of Taq polymerase only.

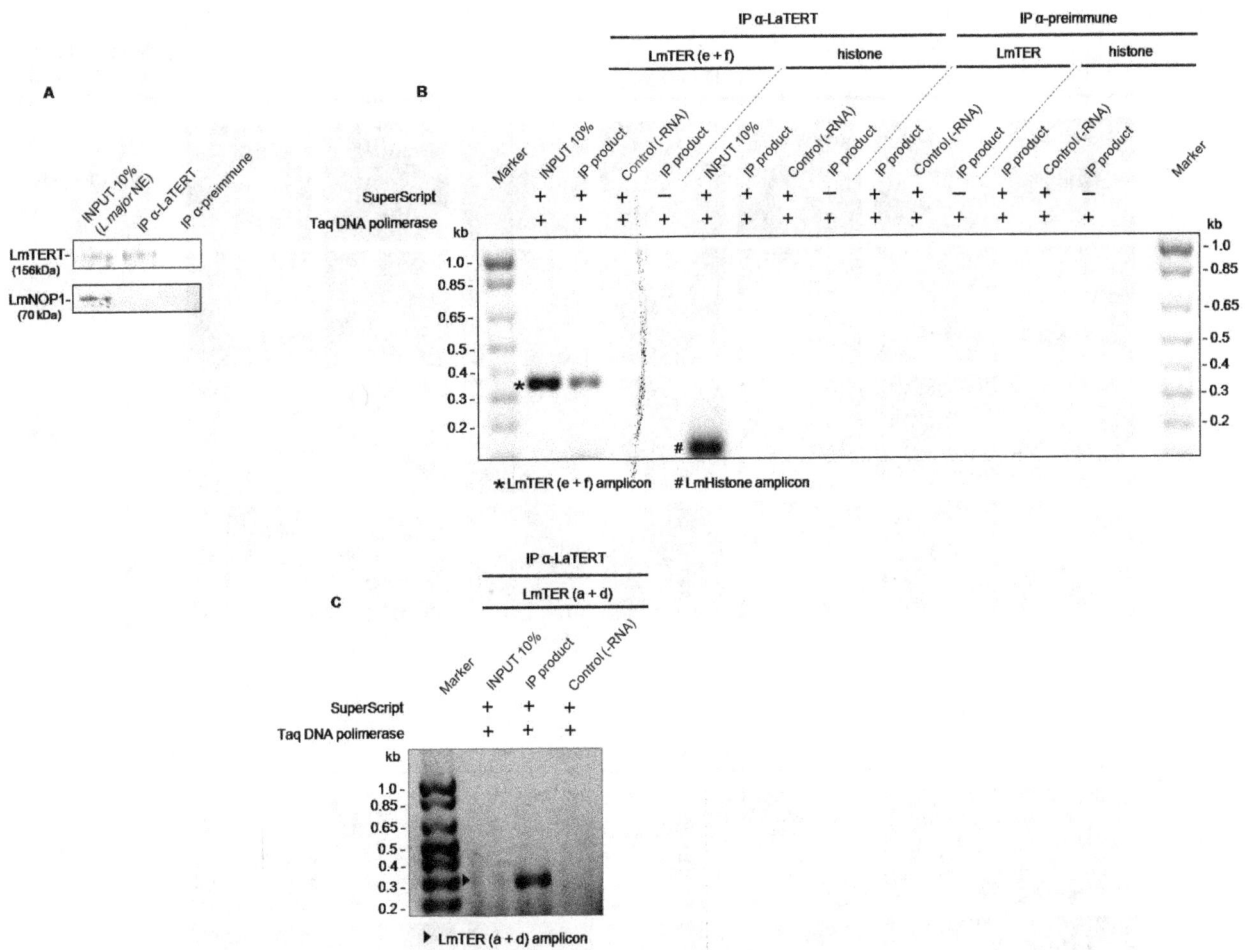

Figure 4. *Lm*TER is amplified from extracts immunoprecipitated with anti-LaTERT serum. A) Ten percent of a *L. major* nuclear extract (*L. major* NE) used as input in the IP assay and a rabbit anti-LaTERT IP eluate (10%) were fractionated in a SDS-PAGE gel and assayed in a western blot revealed with mouse anti-LaTERT (upper panel) and rabbit anti-LmNop1 (bottom panel). B) Ethidium bromide-stained agarose gel of RT-PCR of *Lm*TER and *L. major* histone cDNAs obtained from RNAs isolated from the *L. major* nuclear protein extract (input) and from *L. major* nuclear extract immunoprecipitated (IP eluate) with anti-LaTERT (IP product). *Lm*TER was amplified with primers e+f and histone H2A was amplified as a RT-PCR control. Rabbit pre-immune serum was used as an IP control. C) 5′-Spliced form of *Lm*TER was amplified by RAcE-PCR from RNAs isolated from *L. major* nuclear extract (input) and from an anti-LaTERT IP eluate using primers a+d.

control, *Leishmania* histone H2A was amplified as a~150 bp amplicon using specific primers (Table S1) only in the input sample, demonstrating the specificity of the IP assay (Figure 4B).

Together, these results confirm that the SL sequence-containing mature *Lm*TER RNA is part of the *Leishmania major* telomerase ribonucleoprotein complex.

*Lm*TER co-localizes with LmTERT in promastigotes of *L. major*

RNA Fluorescent *In Situ* Hybridization (FISH) of *L. major* promastigotes coupled with indirect immunofluorescence (IIF) was also performed and confirmed that *Lm*TER co-localizes with LmTERT. A TelG fluorescein-conjugated PNA probe containing three telomeric hexameric repeats co-localized with IIF signal obtained with rabbit anti–LaTERT serum (Figure 5A). Although there were several clusters of *Lm*TER in most cells, only a few co-localized with LmTERT. These additional clusters may be due to hybridization between the telomeric PNA probe and an immature form of *Lm*TER that is not associated with the LmTERT ribonucleoprotein complex. The FISH signal was absent in cells pretreated with RNase A, indicating that the probe was specific for

*Lm*TER RNA and did not cross-hybridize with telomeric DNA (Figure 5B). *Lm*TER was most abundant in late S/G2 and M phase of the parasite cell cycle, coincident with the timing of telomere replication in *Leishmania* promastigotes (da Silva & Cano, unpublished data) and in budding yeast [42].

*Lm*TER putative secondary structure

A putative secondary structure of *Lm*TER was obtained by using its first 5′ 139 nt as input to the mfold tool [30] (Figure 6A). Some structural features of *Lm*TER that are conserved in all TERs already described are signaled in Figure 6A, such as the single-stranded template sequence highlighted in green (5′ ACCCTAACCCTA 3′) at position 85, Helix II or TBE (Template Boundary Element), comprising the conserved short motif 5′-CCGUCA-3′ (highlighted in red), and the GC base pairing both at the proximal end of Helix II.

A comparison between our predicted initial 139 nt *Lm*TER structure with the one encompassing ~100 nt surrounding template sequence from *Tb*TER is shown in Figure 6B. Here we can observe that despite the differences, for example in the position of the template sequence between *Lm*TER and *Tb*TER,

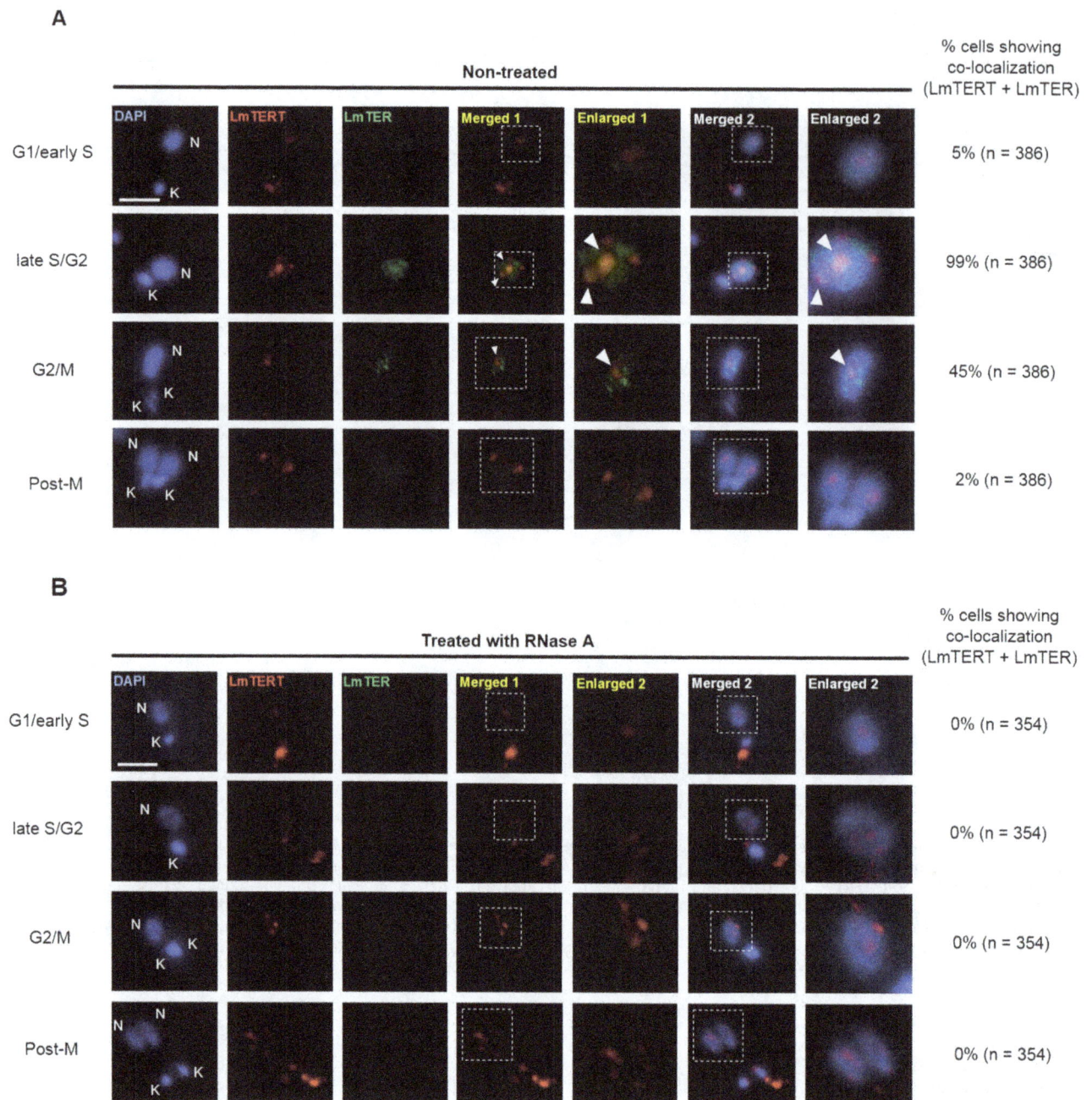

Figure 5. LmTER co-localizes with LmTERT in late S/G2 phase. RNA FISH coupled with IIF with anti-LaTERT serum. Cells were analyzed throughout the L. major cell cycle. "Merged 1" combines images from LmTERT and LmTER. "Merged 2" combines all images. Co-localization foci (white arrows) containing LmTER and LmTERT occur mainly at late S/G2 phases (A). In cells treated with RNase A (B), no RNA hybridization signal or co-localization was detected, indicating that the results shown in (A) correspond to LmTER and LmTERT co-localizing at the same foci. DAPI (blue) was used to stain DNA in kinetoplast (K) and nucleus (N). These figures contain representative cells of a series of images captured randomly to avoid bias. Scale bar represents 2 μm.

we identified similar structural features such as the TBE and two additional hairpins right downstream of the template, which maintain the general shape conservation and geometry in both molecules.

Discussion

In contrast to protein-coding DNA, non-coding RNA genes are prone to several non-lethal mutations related to no discernable

phenotype [43] principally due to compensatory nucleotide substitutions, which maintain the secondary structure of the final RNA molecule despite the modifications in the primary gene sequence. This assertion fits on what we observed for the LeishTER gene. With the exception of the TER template sequence and short stretches randomly aligned, no overall sequence similarity was found between telomerase RNA genes from the Leishmania and Trypanosoma genera (LeishTER and TbTER). More than two decades ago, in a study of ciliate TER

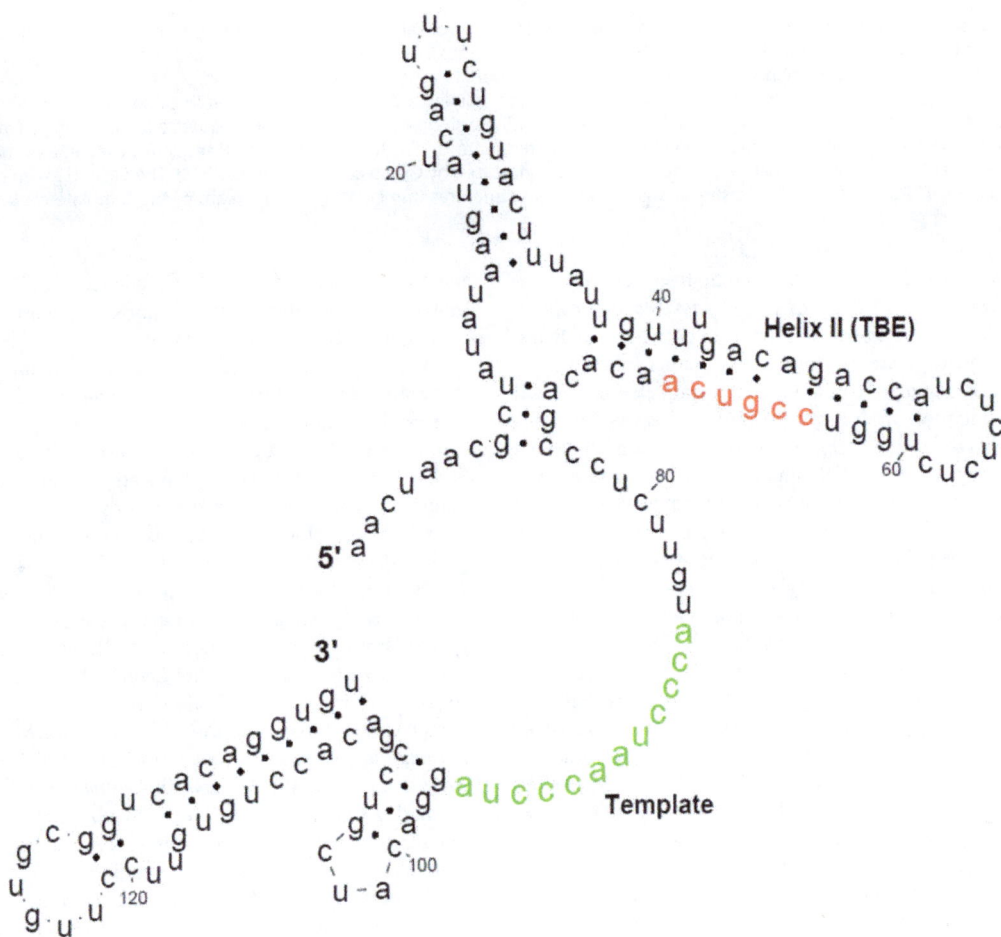

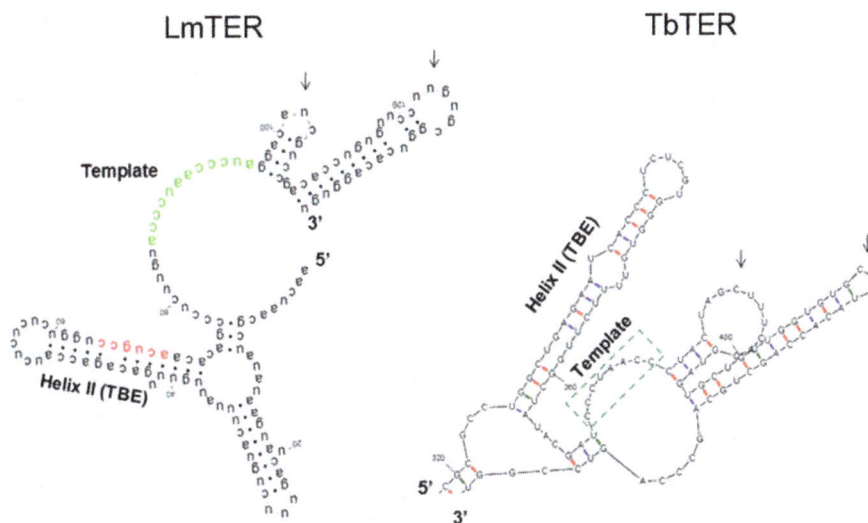

Figure 6. The predicted secondary structure model of *Lm*TER. A) Proposed secondary structure (mfold - default parameters, followed by visualization through RNAviz editor [70]) obtained from the first 139 nucleotides of *Lm*TER (39 nt SL sequence from the 5' cap processing plus 100 nt from the beginning of the gene in the genome). This folding prediction in that particular region led us to infer the existence of two crucial structured domains already detected in all other TERs reported hitherto: (*i*) Helix II, containing a CCGUCA motif (red) at its proximal 3' end, which is implicated in proper template boundary definition in *Tetrahymena thermophila*; and (*ii*) the single-stranded template sequence (green). B) The *Lm*TER structure in A was turned upside-down and compared to the ~100 nt surrounding template *Tb*TER structure (the *Tb*TER sequence used in this analysis and also the default parameters to run the mfold program were identical to those indicated by Gupta and colleagues, 2013). The dashed green box represents the template sequence in *Tb*TER. Arrows indicate similar shape of hairpin structures immediately downstream of the template on both TERs.

structure, Romero and Blackburn observed high sequence content divergence between the TER genes of *Tetrahymena thermophila* and *Euplotes crassus* (holotrichous and hypotrichous ciliates, respectively), which prevented them from either aligning the sequences or identifying common structural elements with confidence among those distinct genera [44]. This special ncRNA feature highlights the need of applying a non-trivial computational method to capture the *LeishTER* gene on focus in this study. We have successfully developed our own *in silico* approach to scan *LeishTER* candidates throughout *Leishmania* genomes based on the 12 nt TER template sequence previously assigned to *L. amazonensis* [6]. The TER template sequence is a crucial component of the TERT holoenzyme, possessing the antisense orientation of the telomeric hexamer repeat (THR) and guiding the telomerase reverse transcriptase activity on the elongation of telomeres [9,21,45]. After filtering out all THR hits that fell onto *Leishmania* telomeres, two non-telomeric loci showed hits in *L. major* genome (Table 1) and only one (on chromosome 25 between the *LmjF.25.0860* and *LmjF.25.0870* protein-coding genes) presented matches at syntenic loci in all *Leishmania* species studied herein (Table 2). This region was assigned by us as containing the putative *TER* locus in *Leishmania*.

Trypanosomatids share a remarkable degree of synteny between their genomes [46]. Therefore, we took advantage of this evolutionarily trait to perform comparative genomics analyses to reliably identify the *LeishTER* candidate locus by *in silico* methods. We verified that *LmjF.25.0870* (the gene immediately downstream of *Lm*TER) is orthologous to *Tb927.11.0850*, which is the second gene downstream of the *Tb*TER gene mapped on chr11 from *T. brucei* [21,22]. Exploring the synteny on the TER locus between three different trypanosomatid genera, it was possible to detect a lack of conservation on the TER-flanking protein-coding genes between *Trypanosoma* and *Leishmania* species, with the synteny resuming more 5' at the *Tb927.11.0810* and *LmjF.25.0830* genes, and 3' at *Tb927.11.0850* and *LmjF.25.0870* (see Results and Figure 1). We believe that this disruption of synteny for some protein-coding genes was the main adverse factor that prevented others from using the *Tb*TER location to easily map the *TER* gene within *Leishmania* genomes. This assessment shows us that trypanosomatid TER non-coding RNA gene appears to be maintained by selective pressure within a synteny-disturbed locus, which might indicate its functional relevance. It is also clear, by assessing the loci displayed in Figure 1, that *Crithidia* genus seems to be evolutionarily closer to *Leishmania* than to *Trypanosoma*, representing a middle branch between the latter genera, corroborating other studies on molecular evolution of trypanosomatids [37,38]. It is noteworthy that the *Cf*TER transcript has not yet been experimentally characterized. Therefore, similar molecular approaches performed in this work and on both *Tb*TER publications [21,22] need to be addressed to validate the *Cf*TER gene.

By thoroughly analyzing the entire *LeishTER*-containing intercoding region (>4.5 kb) within the chromosome 25 from five *Leishmania* species (LmjF, LinJ, LmxM, LtaP and LbrM), we discovered a series of interesting features: (i) The putative 12 nt

template sequence was perfectly aligned on all sequences in the global multiple alignment comparison (Figure 2 and Fig. S3), although it has been previously reported that the telomeric terminal overhang is different in length and at the 3' end nucleotides in *L. amazonensis* and *L. donovani/L. major* [6,25,36]. A possible explanation for this situation can be the occurrence of a non-conserved resection process that couples the removal of the RNA primer after DNA replication and the action of an exonuclease to generate longer 3'G-overhangs that are substrates for telomerase elongation. This is a very complex event that was poorly described in eukaryotic models and is still unknown in *Leishmania* spp. [47]. (ii) Considering that non-protein-coding regions are poorly conserved among the genomes of *Leishmania* species from different subgenera [48], we found the following overall identities among the *LeishTER*-containing intercoding regions: averages of 80.7% between species from *Leishmania (Leishmania)* subgenus (LmjF, LinJ and LmxM), 48.5% between *Leishmania (Sauroleishmania) tarentolae* and *Leishmania (Leishmania)* spp., and 49.6% between *Leishmania (Viannia) braziliensis* and *Leishmania (Leishmania)* spp. (Figure S3). It is also noteworthy that an independent and previously published large-scale mapping of conserved intercoding sequences (CICS) on the LmjF, LinJ and LbrM genomes has reported at least four short CICS within the putative *LeishTER* genes, one of which is 41 nt-long (LeishCICS-s8786) and encompasses the 12 nt TER template sequence [49], suggesting that it might be a conserved functional domain of the *Leish*TER ncRNA molecules. (iii) A novel ~3.6 kb gene (*LmjF.25.T0865*) was recently mapped within the *Lm*TER-containing intercoding region and overlaps to the *Lm*TER gene. This discovery was made together with 1,883 other new genes that were identified by a polyA-captured RNA-seq assessment in *L. major* [50]. Rastrojo and colleagues (2013) have performed no individual functional characterization on any of those novel annotated transcripts and claimed they should be considered ncRNAs until shown to be otherwise. Due to the experimental results we have gotten for *Lm*TER (Figures 3 and 4), and taking into account that there are several SL sites within the inter-CDSs region between *LmjF.25.0860* and *LmjF.25.0870* protein-coding genes (RNA-seq data info provided on tritrypdb.org), we believe that *LmjF.25.T0865* gene might reflect either a junction of two or more transcripts or a longer precursor of *Lm*TER. (iv) One ~45 nt-long small nucleolar RNA domain was detected right downstream of the TER template sequence within the transcripts of four species (LmjF, LinJ, LmxM and LtaP) (Figures 2 and Figure S3, shaded in magenta). It is part of the snoU90 (or scaRNA7), which is a C/D box snoRNA found in human Cajal bodies [40]. This finding partially corroborates the results of Gupta and colleagues (2013), which suggested that *Tb*TER is a member of the C/D box class of snoRNAs due to its affinity selection by epitope tagged TERT and SNU13 (a C/D snoRNA-binding protein), but not by NHP2 (a protein that binds H/ACA snoRNAs) [22]. Despite these findings, whether both *Tb*TER and *Leish*TER act with a function other than the TERT-associated one remains an open question.

As mentioned earlier in this section, Northern blot analyses identified a ~2,100 nt *Lm*TER transcript, but its precursor was not detected (Figure 3C) probably because the mature LmTER might be much more abundant in the total RNA extract. Thus, low levels of LmTER precursor should only be detected by a high sensitive method like RT-PCR (Figure 3D). Molecular assays supporting this hypothesis were very well conceived by the trypanosomatid research community within the past 2–3 decades [54,55]. For example, studies with the tandem array of tubulin genes revealed that polycistronic pre-mRNA precursors are visualized on Northern blots only when the transcripts accumulate after trans-splicing blockage by heat shock [54], also polycistronic HSP7O pre-mRNA appeared to be rare in the nascent RNA population, possibly because of rapid processing of the nascent RNA in the intergenic region by cleavage for trans-splicing and polyadenylation [55].

In addition, LeishTER and the Plasmodium falciparum TER (~2,200 nt, [51]) might be the longest protozoa TERs described so far, exceeding the lengths of the most commonly studied TERs in other organisms such as the ciliate Tetrahymena termophila (159 nt, AF399707.1), the budding yeast Saccharomyces cerevisiae (~1,300 nt, AM296228.1), the vertebrates Mus musculus (590 nt, MMU33831) and Homo sapiens (450 nt, AF047386.1), and the closely related species T. brucei (993 nt) [22,52]. The longest TER described so far is from Candida glabrata, a parasitic species close to S. cerevisiae that has a 2.7 kb-long TER [53]. Moreover, similar levels of LmTER mature transcript were detected during exponential promastigote growth (logarithmic and stationary phases), suggesting that the non-replicative forms of the parasite contains the same LmTER levels as the replicative ones. Although we do not have an experimental answer to this result, the most probable explanation is that LeishTER is a highly stable transcript, which is not degraded during parasite growth, as identical results were also obtained with L. amazonensis TER (Figure S1 and data not shown).

The reverse transcription polymerase chain reactions (RT-PCRs) of total RNA succeeded and the sequencing of these RT-PCR products (Figure 3E) allowed us to confirm the LmTER sequence (Fig. S2) and strongly indicated that LmTER has a 5′ SL cap added by trans-splicing at position 333,307 and a possible polyA tail added at position 335,419 from L. major chromosome 25 (Figures 2 and 3E), corroborating RNA-seq evidence for SL and polyA sites on this genomic region (tritrypdb.org).

The association of the trans-spliced LmTER form with LmTERT was first evidenced using both RT-PCR and 5′ RAcE both primed with different pairs of primers (Figures 4B and 4C) resembling its TbTER counterpart [21]. However, it was not possible to certify whether the LmTER transcript associated with LmTERT was terminated by the addition of a polyA tail because we could not amplified any product from the IP eluate by 3′ RAcE (data not shown), suggesting that the mature LmTER that associates with LmTERT is not polyadenylated. Our hypothesis to explain this result relies on the fact that different mechanisms have evolved for telomerase RNA 3′ end formation. Well-studied TERs (e.g., budding and fission yeasts and human), which are also RNA pol II-transcribed, have their 3′ end processed at sites located upstream to the mapped polyadenylation sites. Therefore, after cleavage reaction(s), they are polyA(-) transcripts [56–61]. Non-canonical 3′ end processing mechanisms, such as cleavage by RNase P, are able to process RNA pol II nascent transcripts to generate their mature 3′ ends despite the presence of nearby polyadenylation signals. It seems that a significant fraction (>25%) of long transcripts present in cells, which includes the telomerase RNAs, lack a classical polyA tail and that the selection of a proper 3′ end cleavage site represents an important step, not only for the post-transcriptional regulation of gene expression, but also to

generate the mature 3′ ends of these transcripts via multiple mechanisms, (reviewed [62]). For example, in both Schizosaccharomyces pombe and Saccharomyces cerevisiae, the functional/mature telomerase RNAs have their 3′ end processed by a specific spliceosomal cleavage mechanism and by transcription termination factors such as Nrd1 and Nab3, respectively [58,60,61,63]. We attempted without any success to find neither Nrd1/Nab3 homolog sequences in Leishmania spp. and T. brucei genomes nor short conserved cis-elements/motifs within our putative LmTER 3′ end boundary region that would be recognized by those yeast factors (data not shown). Thus, further investigations are required to determine whether the LmTER mature transcript also terminates in a non-canonical 3′ end site.

Co-localization assays using RNA-FISH coupled with IIF using a specific anti-telomerase serum, confirmed that LmTER and LmTERT partially associate principally at late S-G2 phase of the promastigote cell cycle, coinciding with the timing of parasite telomere replication (da Silva MS and Cano, unpublished data). The access of human and yeast TERs to telomerase and their substrates are regulated as a function of the cell cycle. In humans, hTER and hTERT are found in distinct nuclear foci throughout most of the cell cycle. Only during S phase, hTER, which is mainly retained in Cajal bodies, moves with TERT in the direction of telomeres. In yeast, in contrast, the mature TER (TLC1) is first exported to the cytoplasm to assemble with the telomerase holoenzyme, and then the entire RNP complex re-enters the nucleus and only in S phase TER co-localizes with a few telomeres; and like in humans, yeast TER accumulation does not also require assembly with TERT (reviewed in [61]). Despite the clear evolutionary divergence among trypanosomatids, yeast and vertebrates, it is possible that they all share common features of the telomerase RNP biogenesis pathway and regulation.

Relying on the results of the RT-PCR from the TERT-IP nuclear extracts, where we used the SL sequence as the forward primer (Figure 4C), we found that the splice site for the TERT-interacting mature LmTER is near the template sequence (as ascribed on Figure 2 and above discussed). In contrast to TbTER, which was reported to present the template sequence located far from the 5′end of TER (position 370), LmTER template is fairly close to its 5′ end (position 85) (Fig. 6), similarly to ciliates and vertebrate TERs [64]. Although we could not propose a secondary structure for the entire LmTER RNA molecule, due to its remarkable length (>2 kb) and the marked drop in computational prediction accuracy as the sequence length increases [65], we were able to detect, by using the first 5′ 139 nt from LmTER as input to the mfold tool, some bona fide structures, such as the single-stranded region encompassing the template sequence and Helix II [30]. Similar results regarding these two bona fide structures were also retrieved when we attempted to run RNAalifold [29], using a multiple alignment of all LeishTERs studied herein, and then applying the constraints from the consensus structure on the individual TER modeling by executing mfold [30] (data not shown). Figure 6A depicts both the single-stranded template sequence and Helix II, which harbors elements required for the proper template boundary definition (TBE) [66–68]. The conserved short motif 5′-CCGUCA-3′, though slightly different from the one in Tetrahymena (5′-CUGUCA-3′), is also found at the proximal 3′ end of Helix II in Leishmania TERs, as it is in ciliates [68]. Notably, we observed the exact match 5′-CUGUCA-3′ in L. braziliensis TER (Figure 2 and Figure S3, 21 nt upstream of the template sequence). More important, the GC base pairing at the proximal end of Helix II, which is essential for proper template boundary definition and required for binding of the Tetrahymena telomerase reverse transcriptase (TERT) [68], is also present in both *Plasmodium falciparum* TER (*Pf*TER) and *Leish*TERs TBEs, although both RNA secondary structures have yet to be experimentally validated.

The comparison between the generated *Lm*TER and *Tb*TER structures (Figure 6B) showed that despite the huge differences, for example in the template sequence location within both TERs, we identified similar structural features (e.g. same shape and geometry of two stem-loops right downstream of the template) that might be suggestive of common interaction pathways with their respective TERT partners and, consequently, similar functionality in the holoenzyme complex.

In this manuscript we have identified and partially characterized *Leish*TER, a long non-coding RNA that preserves some conserved features and also presents unique ones that categorize it as a strong candidate for the *Leishmania* TER component. Despite its similarities with the recently described *Tb*TER component, further structure-driven studies need to be addressed to unravel the biochemical and biophysical details of the whole *Leishmania* TERT core complex and the importance of telomerase biogenesis for parasite cell survival and homeostasis.

Supporting Information

Figure S1 Molecular validation of *L. amazonensis* TER candidate. A) 5′ Spliced form of *La*TER was confirmed by RAcE-PCR using primers a+b; arrows indicate nonspecific amplified bands. The putative 3′ end of *La*TER containing the polyA tail was also confirmed by RAcE-PCR using primers g+h. *La*TER was detected from the polycistron using primers c+d. Histone H2A was used as control (ctrl). B) Total RNA (10 μg) from parasites in the logarithmic and stationary phases of growth were separated on a 1.5% agarose/2.0 M formaldehyde gel, and the blot was probed with a *Lm*TER-specific-probe, which was generated using the combination of primers e+f. Bottom, ethidium bromide-stained RNA gel showing rRNA as the loading control. The primers used in assays shown in A) and B) are the same used in Figures 3 and 4 and are listed in Table S1.

Figure S2 *Lm*TER undergoes *trans*-splicing. cDNA prepared from wild-type *L. major* and *L. amazonensis* cells were cloned into the TOPO-TA vector (Invitrogen). The pre-*Leish*TER sequence was amplified using a sense SL RNA primer and an internal

reverse primer from the *Lm*TER sequence (as shown in Figure 3). The positions of the spliced leader (SL) (blue), Helix II structure (red), and template (green) are depicted.

Figure S3 Complete global multiple alignment and identity matrix of the entire intercoding region of *Leish*TER locus. Coordinates are relative to the first base after the stop codon of the respective CDSs: *L. major* (*LmjF.25.0860*), *L. infantum* (*LinJ.25.0890*), *L. mexicana* (*LmxM.25.0860*), *L. tarentolae* (*LtaP25.0910*) and *L. braziliensis* (*LbrM.25.0740*). The differently shaded colored regions represent the 12 nt template sequence (gray), 5′-C[C/T]GTCA-3′ motif that is part of the template boundary element (TBE) (pink), snoRNA domains reported by the RNAspace webserver (magenta), splice acceptor sites upstream of the template sequence (green) and polyA sites downstream of the template (red) identified by RNA-seq (provided by Myler lab and deposited on tritrypdb.org for *L. major*). ClustalW run locally and on the web, as well as Jalview, were used to align the sequences, build the identity matrix and visualize the alignment, respectively.

Table S1   List of primers.

Acknowledgments

We acknowledge D. R. Maldonado and J. F. da Silveira (Universidade Federal de São Paulo) for the assistance with the northern blot assays. S. Beverley (Washington University in St. Louis) and G. Ramasamy (Seattle Biomedical Research Institute) for gently providing the *Crithidia fasciculata* genome annotation. S. Subramanian (Seattle Biomedical Research Institute) for reading and commenting the manuscript.

Author Contributions

Conceived and designed the experiments: EJRV MSS VSN MINC. Performed the experiments: EJRV MSS VSN MS. Analyzed the data: EJRV MINC PJM. Contributed reagents/materials/analysis tools: MINC PJM. Wrote the paper: EJRV MINC PJM. Performed all the in silico analyses and delineated and executed the computational screening that led to the identification of the actual LeishTER locus as well as the further approaches involving genome/sequence comparisons and RNA structure predictions: EJRV.

References

1. Vickerman K, Preston TM (1970) Spindle microtubules in the dividing nuclei of trypanosomes. J Cell Sci 6: 365–383.
2. Solari AJ (1980) The 3-dimensional fine structure of the mitotic spindle in Trypanosoma cruzi. Chromosoma 78: 239–255.
3. Cano MI (2001) Telomere biology of trypanosomatids: more questions than answers. Trends Parasitol 17: 425–429.
4. Blackburn EH, Challoner PB (1984) Identification of a telomeric DNA sequence in Trypanosoma brucei. Cell 36: 447–457.
5. Fu G, Barker DC (1998) Characterisation of Leishmania telomeres reveals unusual telomeric repeats and conserved telomere-associated sequence. Nucleic Acids Res 26: 2161–2167.
6. Conte FF, Cano MI (2005) Genomic organization of telomeric and subtelomeric sequences of Leishmania (Leishmania) amazonensis. Int J Parasitol 35: 1435–1443.
7. Yu GL, Bradley JD, Attardi LD, Blackburn EH (1990) In vivo alteration of telomere sequences and senescence caused by mutated Tetrahymena telomerase RNAs. Nature 344: 126–132.
8. Feng J, Funk WD, Wang SS, Weinrich SL, Avilion AA, et al. (1995) The RNA component of human telomerase. Science 269: 1236–1241.
9. Cano MI, Dungan JM, Agabian N, Blackburn EH (1999) Telomerase in kinetoplastid parasitic protozoa. Proc Natl Acad Sci U S A 96: 3616–3621.
10. Munoz DP, Collins K (2004) Biochemical properties of Trypanosoma cruzi telomerase. Nucleic Acids Res 32: 5214–5222.
11. Giardini MA, Fernandez MF, Lira CB, Cano MI (2011) Leishmania amazonensis: partial purification and study of the biochemical properties of

the telomerase reverse transcriptase activity from promastigote-stage. Exp Parasitol 127: 243–248.
12. Giardini MA, Lira CB, Conte FF, Camillo LR, de Siqueira Neto JL, et al. (2006) The putative telomerase reverse transcriptase component of Leishmania amazonensis: gene cloning and characterization. Parasitol Res 98: 447–454.
13. Sykorova E, Fajkus J (2009) Structure-function relationships in telomerase genes. Biol Cell 101: 375–392, 371 p following 392.
14. Xia J, Peng Y, Mian IS, Lue NF (2000) Identification of functionally important domains in the N-terminal region of telomerase reverse transcriptase. Mol Cell Biol 20: 5196–5207.
15. Armbruster BN, Banik SS, Guo C, Smith AC, Counter CM (2001) N-terminal domains of the human telomerase catalytic subunit required for enzyme activity in vivo. Mol Cell Biol 21: 7775–7786.
16. Lai CK, Mitchell JR, Collins K (2001) RNA binding domain of telomerase reverse transcriptase. Mol Cell Biol 21: 990–1000.
17. Lingner J, Hughes TR, Shevchenko A, Mann M, Lundblad V, et al. (1997) Reverse transcriptase motifs in the catalytic subunit of telomerase. Science 276: 561–567.
18. Nakamura TM, Morin GB, Chapman KB, Weinrich SL, Andrews WH, et al. (1997) Telomerase catalytic subunit homologs from fission yeast and human. Science 277: 955–959.
19. Weinrich SL, Pruzan R, Ma L, Ouellette M, Tesmer VM, et al. (1997) Reconstitution of human telomerase with the template RNA component hTR and the catalytic protein subunit hTRT. Nat Genet 17: 498–502.

20. Counter CM, Meyerson M, Eaton EN, Weinberg RA (1997) The catalytic subunit of yeast telomerase. Proc Natl Acad Sci U S A 94: 9202–9207.
21. Sandhu R, Sanford S, Basu S, Park M, Pandya UM, et al. (2013) A trans-spliced telomerase RNA dictates telomere synthesis in Trypanosoma brucei. Cell Res 23: 537–551.
22. Gupta SK, Kolet L, Doniger T, Biswas VK, Unger R, et al. (2013) The Trypanosoma brucei telomerase RNA (TER) homologue binds core proteins of the C/D snoRNA family. FEBS Lett 587: 1399–1404.
23. Theimer CA, Feigon J (2006) Structure and function of telomerase RNA. Curr Opin Struct Biol 16: 307–318.
24. Altschul SF, Madden TL, Schaffer AA, Zhang J, Zhang Z, et al. (1997) Gapped BLAST and PSI-BLAST: a new generation of protein database search programs. Nucleic Acids Res 25: 3389–3402.
25. Chiurillo MA, Beck AE, Devos T, Myler PJ, Stuart K, et al. (2000) Cloning and characterization of Leishmania donovani telomeres. Exp Parasitol 94: 248–258.
26. Carver TJ, Rutherford KM, Berriman M, Rajandream MA, Barrell BG, et al. (2005) ACT: the Artemis Comparison Tool. Bioinformatics 21: 3422–3423.
27. Thompson JD, Higgins DG, Gibson TJ (1994) CLUSTAL W: improving the sensitivity of progressive multiple sequence alignment through sequence weighting, position-specific gap penalties and weight matrix choice. Nucleic Acids Res 22: 4673–4680.
28. Waterhouse AM, Procter JB, Martin DM, Clamp M, Barton GJ (2009) Jalview Version 2—a multiple sequence alignment editor and analysis workbench. Bioinformatics 25: 1189–1191.
29. Bernhart SH, Hofacker IL, Will S, Gruber AR, Stadler PF (2008) RNAalifold: improved consensus structure prediction for RNA alignments. BMC Bioinformatics 9: 474.
30. Zuker M (2003) Mfold web server for nucleic acid folding and hybridization prediction. Nucleic Acids Res 31: 3406–3415.
31. Rivas E, Eddy SR (1999) A dynamic programming algorithm for RNA structure prediction including pseudoknots. J Mol Biol 285: 2053–2068.
32. Sperschneider J, Datta A (2008) KnotSeeker: heuristic pseudoknot detection in long RNA sequences. Rna 14: 630–640.
33. Harmanci AO, Sharma G, Mathews DH (2011) TurboFold: iterative probabilistic estimation of secondary structures for multiple RNA sequences. BMC Bioinformatics 12: 108.
34. Kapler GM, Coburn CM, Beverley SM (1990) Stable transfection of the human parasite Leishmania major delineates a 30-kilobase region sufficient for extrachromosomal replication and expression. Mol Cell Biol 10: 1084–1094.
35. Ambit A, Woods KL, Cull B, Coombs GH, Mottram JC (2011) Morphological events during the cell cycle of Leishmania major. Eukaryot Cell 10: 1429–1438.
36. Chiurillo MA, Ramirez JL (2002) Charaterization of Leishmania major Friedlin telomeric terminus. Mem Inst Oswaldo Cruz 97: 343–346.
37. Fernandes AP, Nelson K, Beverley SM (1993) Evolution of nuclear ribosomal RNAs in kinetoplastid protozoa: perspectives on the age and origins of parasitism. Proc Natl Acad Sci U S A 90: 11608–11612.
38. Jackson AP (2010) The evolution of amastin surface glycoproteins in trypanosomatid parasites. Mol Biol Evol 27: 33–45.
39. Cros MJ, de Monte A, Mariette J, Bardou P, Grenier-Boley B, et al. (2011) RNAspace.org: An integrated environment for the prediction, annotation, and analysis of ncRNA. Rna 17: 1947–1956.
40. Darzacq X, Jady BE, Verheggen C, Kiss AM, Bertrand E, et al. (2002) Cajal body-specific small nuclear RNAs: a novel class of 2'-O-methylation and pseudouridylation guide RNAs. Embo J 21: 2746–2756.
41. Kolev NG, Franklin JB, Carmi S, Shi H, Michaeli S, et al. (2010) The transcriptome of the human pathogen Trypanosoma brucei at single-nucleotide resolution. PLoS Pathog 6: e1001090.
42. Bianchi A, Shore D (2008) How telomerase reaches its end: mechanism of telomerase regulation by the telomeric complex. Mol Cell 31: 153–165.
43. Mattick JS (2009) The genetic signatures of noncoding RNAs. PLoS Genet 5: e1000459.
44. Romero DP, Blackburn EH (1991) A conserved secondary structure for telomerase RNA. Cell 67: 343–353.
45. Greider CW, Blackburn EH (1989) A telomeric sequence in the RNA of Tetrahymena telomerase required for telomere repeat synthesis. Nature 337: 331–337.
46. El-Sayed NM, Myler PJ, Blandin G, Berriman M, Crabtree J, et al. (2005) Comparative genomics of trypanosomatid parasitic protozoa. Science 309: 404–409.
47. Longhese MP, Bonetti D, Manfrini N, Clerici M (2010) Mechanisms and regulation of DNA end resection. EMBO J 29: 2864–2874.
48. Laurentino EC, Ruiz JC, Fazelinia G, Myler PJ, Degrave W, et al. (2004) A survey of Leishmania braziliensis genome by shotgun sequencing Molecular and Biochemical Parasitology 137: 81–86.
49. Vasconcelos EJ, Terrao MC, Ruiz JC, Vencio RZ, Cruz AK (2012) In silico identification of conserved intercoding sequences in Leishmania genomes: unraveling putative cis-regulatory elements. Mol Biochem Parasitol 183: 140–150.
50. Rastrojo A, Carrasco-Ramiro F, Martin D, Crespillo A, Reguera RM, et al. (2013) The transcriptome of Leishmania major in the axenic promastigote stage: transcript annotation and relative expression levels by RNA-seq. BMC Genomics 14: 223.
51. Chakrabarti K, Pearson M, Grate L, Sterne-Weiler T, Deans J, et al. (2007) Structural RNAs of known and unknown function identified in malaria parasites by comparative genomics and RNA analysis. Rna 13: 1923–1939.
52. Hukezalie KR, Wong JM (2013) Structure-function relationship and biogenesis regulation of the human telomerase holoenzyme. Febs J 280: 3194–3204.
53. Kachouri-Lafond R, Dujon B, Gilson E, Westhof E, Fairhead C, et al. (2009) Large telomerase RNA, telomere length heterogeneity and escape from senescence in Candida glabrata. FEBS Lett 583: 3605–3610.
54. Muhich ML, Boothroyd JC (1988) Polycistronic transcripts in trypanosomes and their accumulation during heat shock: evidence for a precursor role in mRNA synthesis. Mol Cell Biol 8: 3837–3846.
55. Huang J, van der Ploeg LH (1991) Maturation of polycistronic pre-mRNA in Trypanosoma brucei: analysis of trans splicing and poly(A) addition at nascent RNA transcripts from the hsp70 locus. Mol Cell Biol 11: 3180–3190.
56. Mitchell JR, Cheng J, Collins K (1999) A box H/ACA small nucleolar RNA-like domain at the human telomerase RNA 3' end. Mol Cell Biol 19: 567–576.
57. Fu D, Collins K (2003) Distinct biogenesis pathways for human telomerase RNA and H/ACA small nucleolar RNAs. Mol Cell 11: 1361–1372.
58. Leonardi J, Box JA, Bunch JT, Baumann P (2008) TER1, the RNA subunit of fission yeast telomerase. Nat Struct Mol Biol 15: 26–33.
59. Egan ED, Collins K (2010) Specificity and stoichiometry of subunit interactions in the human telomerase holoenzyme assembled in vivo. Mol Cell Biol 30: 2775–2786.
60. Noel JF, Larose S, Abou Elela S, Wellinger RJ (2012) Budding yeast telomerase RNA transcription termination is dictated by the Nrd1/Nab3 non-coding RNA termination pathway. Nucleic Acids Res 40: 5625–5636.
61. Egan ED, Collins K (2012) Biogenesis of telomerase ribonucleoproteins. Rna 18: 1747–1759.
62. Wilusz JE, Spector DL (2010) An unexpected ending: noncanonical 3' end processing mechanisms. Rna 16: 259–266.
63. Chapon C, Cech TR, Zaug AJ (1997) Polyadenylation of telomerase RNA in budding yeast. Rna 3: 1337–1351.
64. Chen JL, Greider CW (2004) An emerging consensus for telomerase RNA structure. Proc Natl Acad Sci U S A 101: 14683–14684.
65. Proctor JR, Meyer IM (2013) COFOLD: an RNA secondary structure prediction method that takes co-transcriptional folding into account. Nucleic Acids Res 41: e102.
66. Autexier C, Greider CW (1995) Boundary elements of the Tetrahymena telomerase RNA template and alignment domains. Genes Dev 9: 2227–2239.
67. Lai CK, Miller MC, Collins K (2002) Template boundary definition in Tetrahymena telomerase. Genes Dev 16: 415–420.
68. Richards RJ, Theimer CA, Finger LD, Feigon J (2006) Structure of the Tetrahymena thermophila telomerase RNA helix II template boundary element. Nucleic Acids Res 34: 816–825.
69. Requena JM (2011) Lights and shadows on gene organization and regulation of gene expression in Leishmania. Front Biosci (Landmark Ed) 16: 2069–2085.
70. De Rijk P, Wuyts J, De Wachter R (2003) RnaViz 2: an improved representation of RNA secondary structure. Bioinformatics 19: 299–300.

4

Exome-Wide Somatic Microsatellite Variation Is Altered in Cells with DNA Repair Deficiencies

Zalman Vaksman[1], Natalie C. Fonville[1], Hongseok Tae[1¤], Harold R. Garner[1,2*]

1 Virginia Bioinformatics Institute, Virginia Tech, Blacksburg, Virginia, 24061, United States of America, 2 Genomeon LLC, Floyd, Virginia, 24091, United States of America

Abstract

Microsatellites (MST), tandem repeats of 1–6 nucleotide motifs, are mutational hot-spots with a bias for insertions and deletions (INDELs) rather than single nucleotide polymorphisms (SNPs). The majority of MST instability studies are limited to a small number of loci, the Bethesda markers, which are only informative for a subset of colorectal cancers. In this paper we evaluate non-haplotype alleles present within next-gen sequencing data to evaluate somatic MST variation (SMV) within DNA repair proficient and DNA repair defective cell lines. We confirm that alleles present within next-gen data that do not contribute to the haplotype can be reliably quantified and utilized to evaluate the SMV without requiring comparisons of matched samples. We observed that SMV patterns found in DNA repair proficient cell lines without DNA repair defects, MCF10A, HEK293 and PD20 RV:D2, had consistent patterns among samples. Further, we were able to confirm that changes in SMV patterns in cell lines lacking functional BRCA2, FANCD2 and mismatch repair were consistent with the different pathways perturbed. Using this new exome sequencing analysis approach we show that DNA instability can be identified in a sample and that patterns of instability vary depending on the impaired DNA repair mechanism, and that genes harboring minor alleles are strongly associated with cancer pathways. The MST Minor Allele Caller used for this study is available at https://github.com/zalmanv/MST_minor_allele_caller.

Editor: Michael Shing-Yan Huen, The University of Hong Kong, Hong Kong

Funding: This work was funded by the Virginia Bioinformatics Institute Medical Informatics Systems Division director's funds, Virginia Bioinformatics Institute Genomics Research Lab Small Grant (CLF-1172), high performance computing was supported by a grant from the National Science Foundation (OCI-1124123) and NSF S-STEM grant (DUE-0850198). This work was supported by these 4 funds. The first two funds were internal university funds (Virginia Tech), and the latter two were from the National Science Foundation (NSF). None had a role in the study design, data collection and analysis, decision to publish, or preparation of the manuscript. These funds supported portions of author salaries and benefits, computer costs, laboratory supplies, sequencing services, publication costs, and overhead (indirect expenses to the university).

Competing Interests: HT currently works at Caris Life Sciences; the work for this paper was done when he was employed at Virginia Tech and is in no way connected to his current employment. Harold Garner is owner and founder of Genomeon, however Genomeon was not involved in funding or directing this work.

* Email: garner@vbi.vt.edu

¤ Current address: Bioinformatics group, Caris Life Sciences, Phoenix, Arizona, 85040, United States of America.

Introduction

Microsatellites (MSTs) are regions of repetitive DNA at which 1–6 nucleotides are tandemly repeated; and are present ubiquitously throughout the genome, both in gene and intergenic regions. Observations of somatic variation in MSTs have demonstrated that MST mutation rates are between 10 and 1000 time higher than that of surrounding DNA [1,2], rendering microsatellites mutational "hot-spots" [3,4]. The increased mutational rate of MSTs is thought to be primarily due DNA polymerase slippage and mis-alignment of the slipped structure due to local homology [5–7]. This difference in primary mutational mechanism suggests that, unlike non-repetitive DNA whose mutational spectrum is primarily SNPs, microsatellites are more prone to INDELs [4,7,8]. Specifically MSTs are prone to INDELs that are 'in-phase' or result in expansion or contraction by complete repeat units. For example, a dimer microsatellite will typically expand or contract by 2N nucleotides while a trimer will expand or contract by 3N [1].

MSTs are found in and around a significant number of coding and promoter regions and specific microsatellite variations have been linked to over 40 disorders, such as the CAG microsatellite whose expansion is associated with Huntington's disease and the CGG repeat whose expansion is associated with Fragile X [1,9]. In addition, a more general increase in MST instability has been associated with colon cancer, which, if detected, results in better prognosis and can influence treatment [10,11]. Currently, MST instability is clinically defined based on the results of a kit that tests somatic variation of 18–21 "susceptible" loci (PowerPlex 21, Promega). Although the test has been shown to be effective for identifying MST unstable colon cancer [12], it is significantly less effective for most other disorders including other cancers [13–15]. The ability to capture and discern variation patterns exome-wide would provide a more accurate and useful clinical data for a broader range of disorders. In recent reports next-gen sequencing has been used to uncover MST instability in intestinal and endometrial cancers by observing genotype changes in MSTs between tumor and healthy tissue [14,15].

The goal of this research was to identify patterns of somatic variation in MSTs as a possible marker for genomic instability. We hypothesize that the variable nature of MSTs and the quantification of minor allele content makes them ideal candidates for in-depth next-gen analysis and that somatic variation of microsatellite loci can be quantified using high-depth sequencing. A broadening of the definition of MST instability to include changes in somatic variability and using an exome/genome-wide approach may enable a more accurate diagnosis of patients then what is currently provided by PowerPlex 21.

Somatic variability, novel genomic polymorphisms that arise within a cell population not found in the progenitors, plays a critical role in cellular reprogramming leading to the development and progression of cancer [16]. Suppression of mutations is essential for genomic stability, therefore cells have evolved multiple mechanisms to repair damaged or unpaired nucleotides [17,18]. Currently the only established DNA repair defect that that has been directly linked to MST instability is mismatch repair (MMR). MMR impairments have been shown to increase somatic variation at MSTs in both cell lines and tumors [19–21]. Although the role other DNA repair mechanisms such as inter-strand crosslink repair (as seen in Fanconi anemia genes) and homologous recombination (HR) play in MST instability is less clear, both are important for genomic and chromosomal stability (reviewed by [22,23]).

In this study we first show that we can robustly detect signatures of MST mutation bias and somatic variation occurring in cell lines in next-gen data including a high frequency of in-phase INDELs. We are then able to construct a pattern of somatic MST variation (SMV) by using DNA repair proficient cell lines. Our results indicate that ~5% of microsatellite loci show somatic variation, i.e. have at least one additional non-haplotype allele present. Finally, we are able to differentiate between cell lines with known defects in various DNA repair mechanisms (mismatch repair, DNA crosslink repair, homologous recombination), which correlate with an altered distribution of loci with non-haplotype alleles. These findings suggest that signatures that distinctly define specific defective DNA repair mechanisms can be gleaned from next-gen sequencing data and that this information has the potential to be utilized for detection of individuals with altered levels of somatic variation that are at increased risk of disease or the evaluation of patient's tumor that may yield clinically actionable information.

Methods

Cells, DNA prep and sequencing

HEK (human embryonic kidney) and MCF10A (immortalized breast epithelial) and HEK293 (human embryonic kidney) cells were obtained from ATCC. PD20 and PD20 RV:D2 (FANCD2 and FANCD2 retrovirally corrected) cell lines were obtained from the Fanconi Anemia Foundation (Eugene OR). Sequencing data for Capan-1 cells was previously published by Barber and coworkers [12].

PD20, PD20 RV:D2 and HEK293 cells were grown at 37°C with 5% CO_2, in DMEM supplemented with 10% FBS (Invitrogen) and 1X pen/strep (Invitrogen) to 80% confluence. MCF10A cells were grown to confluence in DMEM/F12 medium (Invitrogen, Carlsbad, CA), supplemented with 5% horse serum (Invitrogen), antibiotics- 1X Pen/Strep (Invitrogen), 20 ng/mL EGF (Peprotech, Rocky Hill, NJ), 0.5 mg/mL hydrocortisone (Sigma), 100 ng/mL cholera toxin (Sigma), and 10 µg/mL insulin (Sigma) at 37°C with 5% CO_2. All cell lines were collected by

trypsinazation and prepared for DNA extraction. DNA was extracted using the Qiagen DNAeasy kit (Qiagen) as per manufacturers instructions.

Since PD20 RV:D2 were derived from PD20 cells by retroviral insertion of the corrected FANCD2 gene we confirmed the maintenance of the corrected version using the sequencing data. Further, a comparison of growth-curves showed an order of magnitude more cells 48 hours after exposure to the DNA interstrand cross-linker Cisplatin, confirming a partial rescue phenotype.

Sequencing and analysis pipeline

Exome paired-end libraries were prepared using the Agilent (Chicago, IL) SureSelectXT Human All Exon V4 capture library. 2×100 bp reads were obtained using an Illumina (San Diego, CA) HiSeq 2500 instrument in Rapid Run mode on a HiSeq Rapid v1 flowcell. Indexed reads were de-multiplexed with CASAVA v1.8.2.

Paired-end sequencing reads were trimmed using fastX_Toolkit and aligned to HG19/GRCh37 human reference genome (http://www.genome.ucsc.edu) using BWA-mem. The output was then sorted, indexed and PCR duplicates were removed using SAMTOOLS [24]. Bam files were then locally realigned and target loci marked using GATK IndelRealigner and TargetIntervals. MST alleles were retrieved and analyzed using software described in the next section.

Microsatellite minor-allele software

A catalogue of MST loci was generated from the HG19/GRCh37 reference genome using Tandem Repeats Finder [25] (with the following parameters: 2.7.7.80.10.18.6). The list was filtered to remove any loci that were shorter than 8 nucleotides, had less than 3 copies of a given motif unit or were below 85% sequence purity. Duplicated loci were identified based on sequence purity and sequence length and were removed.

MSTs were analyzed using a custom MST minor-allele caller based on GenoTan and ReviSTER software [26,27], which were developed by this group to improve MST haplotype predictions (https://github.com/zalmanv/MST_minor_allele_caller). The minor-allele caller extracts marked MSTs from bam files using SAMTOOLs. MST loci are called based on predicted alignments and an adjustable length flanking sequence (this study used either 5 or 7 nucleotide sequence). Reads with low base call scores (below a base score of 28) for nucleotides within the repeats and those with mapping quality score below 10% were eliminated. Alleles are initially called only when two or more reads, verified in both directions of a paired-end run, have the same sequence. All alleles for a given locus are binned with the number of supporting paired-end reads. The final number of alleles is computed based on a user specified minimal requirement of substantiating reads (for this study the minimum number of substantiating reads is either 2 or 3 reads per allele). If more than one allele per locus was found, zygosity and the sequence length difference from the most common allele were recorded. Heterozygotic loci were called using the following criteria as described and confirmed in the GenoTan and ReviSTER manuscripts [26,27]: 1) it is the second most common allele, 2) The number of confirming reads is greater than 25% of the total reads for the locus or greater than 50% of the depth for the most common allele, if the total is below 25% of the total depth.

In addition to MST loci, we also generated a somatic variability profile for non-MST loci. To make the data comparable we randomly selected 3 million loci, each consisting of 15 nucleotides segments, from the HG19 genome. We then filtered out any loci that intersected with our MST and were left with over 2 million loci. The same pipeline as for MSTs was used to generate the data for non-MST loci. This data yielded information on the number of loci with minor alleles and type of mutation (SNPs and INDELs).

Sequence validation and allele calls validated by independent Sanger sequencing method

The MST minor-allele caller we use in this paper is a modified version of a published and experimentally verified code, however to further validate the multi-allele capability of the modified code 30 loci, including 17 showing multiple alleles, were verified using Sanger sequencing. Figure S1A in file S1 shows the data from the minor-allele caller output at one of these loci, chr10:72639137-72639161, at which we would predict at least 3 alleles to be present in this sample (MCF10A) with lengths of 21, 23, and 25 nucleotides. Sanger sequencing confirmed that multiple alleles were present, with the alleles being greater than 21 nucleotides long (figure S1B in file S1). Of the 30 loci 28 loci verified the genotype and 14 of 17 loci with minor alleles also had visible minor alleles by Sanger sequencing.

Modeling error rates to establish rules that differentiate errors from high confidence minor alleles

Two methods were used to generate models of NGS runs for chromosomes 17 and 21; 1) Wgsim (https://github.com/lh3/wgsim) a commonly used paired-end read generator and 2) in-house designed generator. Both methods were set to have a per nucleotide error rate between 0.5% and 5%. The major difference between the two methods was that wgsim was used to obtain modeling data with fairly similar coverage (read depth) across the reference chromosome while the lab-designed algorithm allowed for a more variable coverage as is observed in a typical next-gen sequencing run. The generated fastq files were run through the same pipeline as actual real sequencing data. The accuracy of the pipeline was analyzed by the verification of the predicted alignment. Predicted error rates ranged between 1.3% and 1.9%, with the majority of errors due to misalignments.

Results

We modified a previously published and verified MST genotyper [26] to enumerate all possible alleles present within next-gen data, as opposed to only capturing the most common (haplotype) alleles. We first characterized the error which may cause false positive allele calls via a parametric sensitivity study conducted on in-silico generated data, and showed that our measure can then be used to accurately quantify minor alleles and thus be used to distinguish between mutational mechanisms that are exhibited in different cell lines. To accomplish this, we establish a baseline SMV profile from DNA repair proficient cell lines, and compared this to what is seen in cell lines with various DNA repair defects.

Characterizing the effect of sequencing error on minority allele calling

This analysis evaluates each MST locus to establish the one or two alleles that define the genotype, then it robustly calls additional non-haplotype or 'minor' alleles that are present at lower frequency within next-gen data. However, the accuracy of such minority allele calls can be significantly affected by sequencing errors found within the raw reads that map to each locus. To minimize the number of false positive 'alleles', we first established the minimal number of reads necessary for confirming

an allele in the presence of typical next-gen errors. It has been established by a number of studies that 3 reads mapped to a loci is sufficient to properly call major alleles [28–30]. To corroborate this, we created an in-silico sequencing data set for chromosomes 21 and 17, with randomly generated errors ranging from 0.5% to 5% which mimicked next-gen sequencing data in both the error types that were created and read coverage per locus (results depicted in figure 1).

We first determined the parameters required to optimize the measurement of the fraction of loci without minor alleles in sequencing data with the above-mentioned error rates. Alignment and zygosity calling accuracy is displayed in table S1 in file S1. The sequencing data generator produced between 8 and 10.5 million reads that contained over 58,000 targeted MSTs. Over 98.5% of the reads mapped correctly with an accuracy of over 99.8% in coding regions (regions captured by exome sequencing). The accuracy of zygosity calls was over 99.98% for all error rates. Next we varied the minimum number of reads covering a locus required to call an allele. Changing the threshold from 2 confirming reads (figure 1A) to 3 confirming reads (figure 1B) statistically and significantly decreased the fraction of loci with more alleles than the haplotype number (1 if homozygotic or 2 if heterozygotic). Using a threshold of 2 confirming reads per allele, the fraction of loci without minor alleles identified (due to sequencing errors being interpreted as alleles) was 19–62% for simulated data sets with error rates ranging between 5%–0.5% respectively (figure 1A), indicating that requiring only 2 reads to identify an allele leads to a high level of false alleles. By increasing the threshold to 3 confirming reads the percent of loci without minor alleles increases to 73–99% for the same data set (figure 1B). By increasing to 4 confirming reads per allele we further increase the number of loci without minor alleles 87%–99% (figure S2A in file S1). However, at error rates close to the actual HiSeq rates (of ~1%), we only saw a modest increase in the number of loci without minor alleles, a change from 97% (3 reads per allele) to 99% (4 reads per allele). This is in contrast to an increase from 61% with 2 reads per allele to 97% with 3 confirming reads per allele.

We next examined how sequencing error might affect the number of alleles present in our data. To do this we used modeling data with error rates similar to the actual HiSeq error rate (1%) and 2.5% error (figure 2), and determined the average read depth per locus with increasing alleles. For the in-silico generated data, we found a linear increase in the total read depth as the number of alleles increased (using 2–4 confirming reads per allele) up to 8 alleles (figures 2 and Figure S2 in File S1). A comparison of these results to actual sequencing data from our cell lines (discussed in more detail later) shows that when 3 or more reads are required to confirm an allele, the number of alleles called for a given read depth is greater than what would be expected from error, even at a rate of 2.5% which is substantially more than the observed next-gen error rate of 1% (figure 2B and Figure S2B in File S1), i.e. more alleles are called at a lower read depth in the actual data than would be present due to error. Based on these results, requiring a minimum of 3 reads covering a locus to confirm an allele minimizes the number of 'false' alleles being identified due to sequencing error.

Polymerase slippage vs. nucleotide misincorporation

Another potential source of error in calling alleles from sequencing data is amplification errors induced during the library preparation process [31]. These errors would likely be present at higher frequency than errors generated during sequencing [31,32]; therefore cannot be minimized by solely increasing the minimum

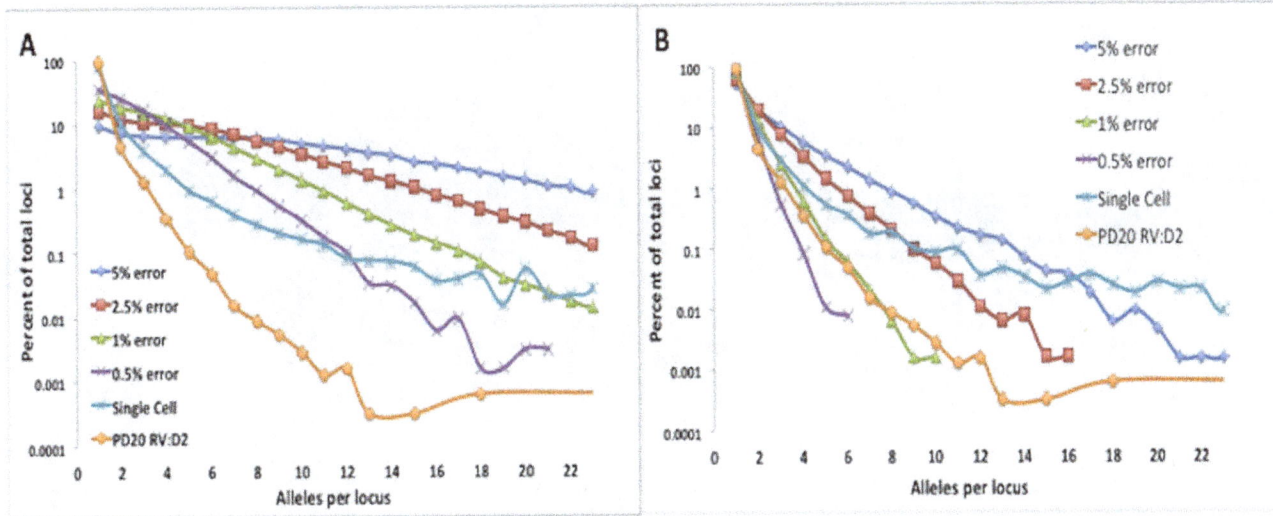

Figure 1. Effects of sequencing error and the minimum number of reads required to call an allele on the number of alleles called in sequencing data. Modeling data with different error frequencies (0.5%–5%) showed an increase in loci with multiple alleles as error increased when both 2 (A) and 3 (B) reads were minimally required to call an allele. In contrast, standard exome sequencing data from DNA repair proficient cells (PD20 RV:D2 cells) and exome sequencing after whole genome amplification from a single cell were insensitive to the cut-off used.

read coverage (as above). Somatic mutation of MSTs is primarily associated with polymerase slippage [33,34], which is thought to cause the characteristic INDEL bias [31,35,36]. In contrast, nucleotide mis-incorporation errors during *in-vitro* amplification would be predicted to lead primarily to SNPs in sequencing data [37]. Both of the mentioned DNA synthesis methods would lead to an increase in the number of loci with non-haplotype alleles, however with a predicted variation pattern that is distinctly different. To differentiate between the two predicted SMV patterns including minority alleles, and to assess the influence of nucleotide mis-incorporation/amplification error on our results, we compared a standard exome sequence from cells which are proficient for DNA repair (described later) that did not undergo whole genome amplification (WGA) with data from the sequencing of a single cell [38] which would be expected to have no somatic variation within the sample, but has necessarily undergone WGA to generate the quantity of DNA necessary for sequencing. Therefore, for the WGA sample, presumably all non-haplotype alleles present are due to amplification error. As expected, genome amplification increases the number of loci with non-haplotype alleles (figure 1) to 11.3% and 7% of the total with a threshold of 2 and 3 reads, respectively. The DNA repair proficient cells, which did not undergo extensive amplification, were only decreased by 1.7%, from 7% to 5.3%, by altering the minimum read cutoff. From this it can be concluded that neither errors during library prep nor during the sequencing run account for more than 4 percent of the total non-haplotype alleles detected.

Approximately 85% of mutations found within microsatellite loci in the WGA single-cell data were SNPs, which is expected as a consequence of polymerase errors during amplification. These results were comparable to those predicted by our model, which showed that ~88% of the total minor alleles were composed of alleles carrying SNPs rather than INDELs (Figure 3). In contrast, SNPs account for only 36% (±3.4%) of the total minor alleles in DNA repair proficient cell lines. In addition, although for all the DNA repair proficient cell lines the most common MST motifs with minor alleles observed were mono-nucleotide repeats found within 56%–66% of loci, loci containing tri-nucleotide motifs

accounted for over 55% of the total loci with minor alleles in the WGA data (table S2 in file S1). These results further support the hypothesis that this approach can differentiate between distinct MST mutational profiles: INDELs, particularly at mono-nucleotide runs predominantly reflect DNA repair proficient biological SMV whereas SNPs in MSTs, particularly at tri-nucleotide motif containing loci are predominantly amplification-induced errors or potentially due to altered DNA maintenance capacity. This is further supported by a similar study that has found that the majority of MSTs that are variable within the normal population (individuals sequenced as part of the 1,000 Genomes Project) are predominantly INDELs at mono-nucleotide runs [30].

MST vs non-MST regions

MSTs are considered to be more susceptible to mutations than the surrounding non-repetitive DNA regions [3,14,39]. Because of this, one could expect that non-MST regions would have less somatic variability (non-MST equivalent of SMV) than MST regions. In order to perform a fair comparison with the MST data, 2 million segments consisting of 15 nucleotides each were randomly selected throughout the genome. The same analysis as was performed on loci containing MSTs was also applied to these non-MST regions. It was found that for these non-MST loci the average fraction of loci that were homozygotic was 98.9% with a standard deviation of 0.2, while only 96.7% of the MST containing loci was homozygotic. Even more significant, only 2% (standard deviation of 0.2) of the non-MST loci (homozygotic and heterozygotic) had minor alleles, while 5.1% of the MST loci harbored minor alleles (table 1). Further, a comparison of SNP and INDEL distributions indicated that, unlike MST regions where INDEL variations prevail (64%), SNPs account for the majority (96.9%) of the differences in minor alleles at non-MST loci (table 2). Taken together, these results confirm that, consistent with the literature, MSTs are more susceptible to mutation [2–4,34].

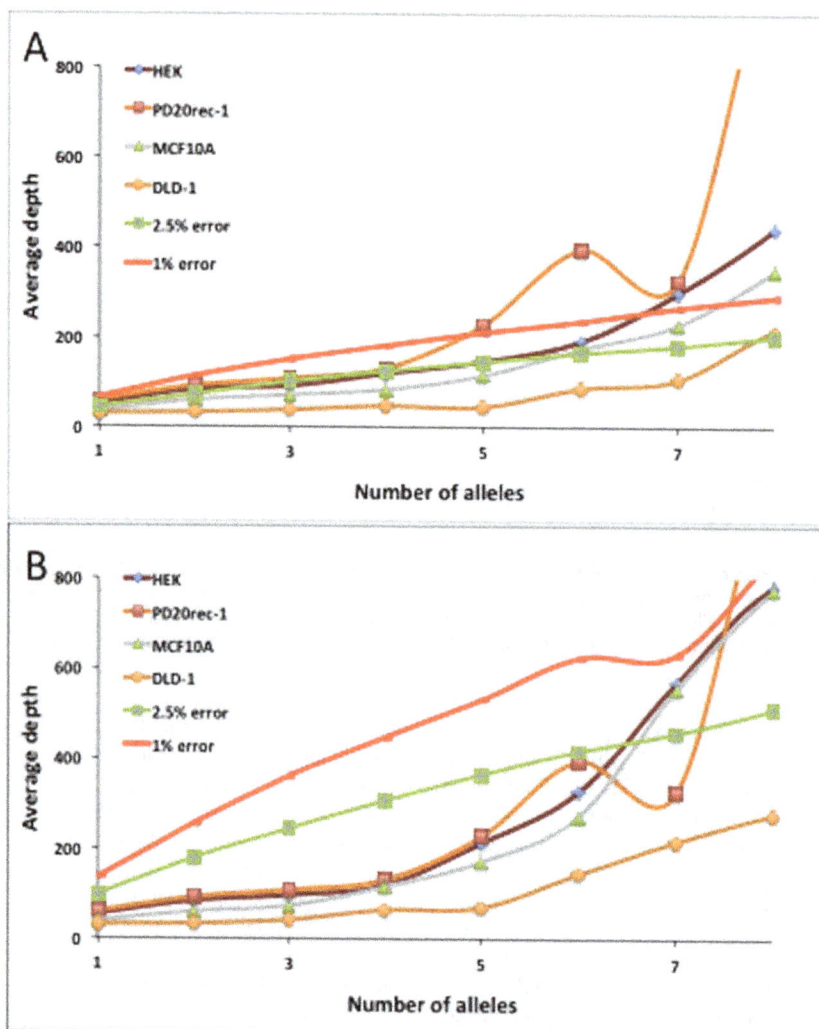

Figure 2. Variation in average depth per locus cannot explain the number of loci with minor alleles. The average read depth at loci with increasing numbers of alleles using A) 2 and B) 3 confirming reads per allele for in-silico generated data using 1% and 2.5% induced error rate for 4 different cell lines.

Reproducibility within a cell line

The objective of this study is to characterize the pattern of SMV from DNA repair proficient cells and then compare to cell populations in which DNA repair is compromised. SMV changes associated with disease will likely be subtle and require highly reproducible control data. To test the reproducibility of SMV measurements within a cell line, two biological replicate cultures of PD20 RV:D2 (PD20 RV:D2-1 and PD20 RV:D2-2) cells were grown separately and sequenced. PD20 RV:D2 are fibroblasts derived from an individual with Fanconi Anemia subgroup D2, retrovirally complimented with a functional copy of FANCD2 [40]. Using a minimum read depth cutoff of 15 to genotype a given loci, we successfully called over 280 K and 250 K loci (at an average depth of 52 and 45 reads per locus) for PD20 RV:D2-1 and 2 respectively. Both samples showed a similar SNP to INDEL ratio, with INDELs making up over ~67% of the minor alleles (table 2). A genotype analysis showed that approximately 96.8% of called loci were homozygous while heterozygosity was observed in ~3.2% of the loci called (table 1). Comparison of those loci that were called in both samples shows that haplotype discordance (i.e. homo- or heterozygotic using standard genotyping) was 1.1%

(table 3), of which 92% were due the fraction of reads supporting a second allele being below the haplotype threshold (see method) and was therefore counted as a minor allele instead of a second haplotype allele, as is the convention in established genotype callers. Only 173 discordant loci were due to sequence differences between the two samples.

For the purpose of this study SMV is defined by the presence of variant MST alleles that are supported by a minimum of 3 confirming reads but do not contribute to haplotype. An analysis of variant MST alleles found a total of 5.4% and 5.3% of MST loci in the PD20 RV:D2-1 and 2 samples, respectively, had 1 or more minor alleles (table 1). The concordance of loci without minor alleles in either sample is 93.9% while 3.4% of loci have at least one minor allele in both samples. By concordance we mean a locus has minor alleles or the same haplotype in multiple samples. Conversely, discordance, where a locus in only one of the compared samples had minor alleles, was 2.7% (table 3). To confirm the significance of these values, we calculated the probabilities of concordance and discordance based on a cohort of randomly selected loci (5.4% and 5.3% of a total samples), which was <0.25% concordant, and compared with our results.

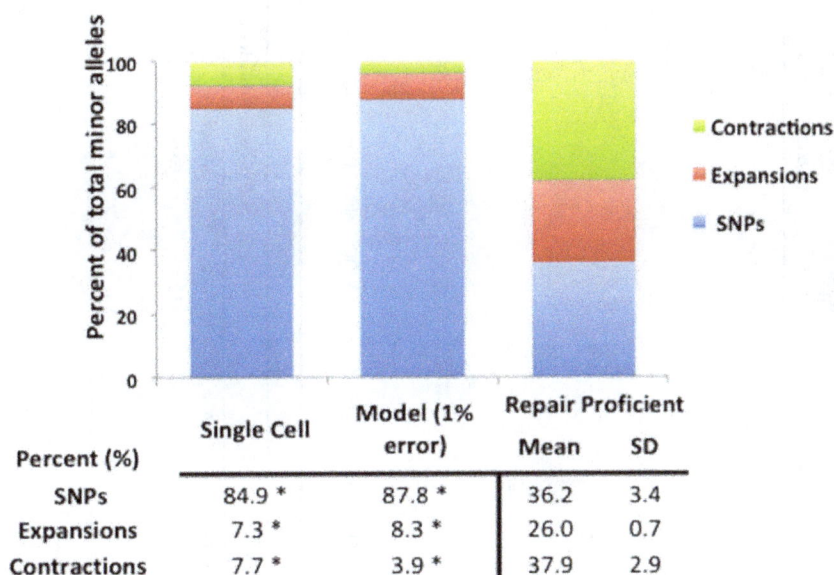

Percent (%)	Single Cell	Model (1% error)	Repair Proficient	
			Mean	SD
SNPs	84.9 *	87.8 *	36.2	3.4
Expansions	7.3 *	8.3 *	26.0	0.7
Contractions	7.7 *	3.9 *	37.9	2.9

Figure 3. DNA repair proficient cells vary significantly from the *in-silico* modeling and single cell sequencing analysis with respect to SNPs and INDELs. The percent of SNPs, expansion and contractions for single cell sequencing and the *in-silico* model as well as the mean and standard deviation for the control cell lines. * significant difference p<0.01.

Using a Pearson's goodness of fit X^2, we verified that the concordant loci are not randomly distributed (p<0.0001). To determine within cell line reproducibility we compared the percent of loci having minor alleles by chromosome as a whole and binned into a million base regions. A linear regression model comparing the percent of loci with minor alleles for each chromosome (as depicted in figure S3 in file S1) shows a significant correlation ($R^2 = 0.85$ and p<0.001) between two independently cultured samples (Figure 4A). Similarly, a comparison of the binned chromosome also shows a significant correlation ($R^2 = 0.60$ and p<0.001, figure 4B). Visualization of the distribution of fraction of MST loci showing somatic variation in a representative chromosome (chr1), depicted in figure 5, indicates specific chromosomal regions that may harbor SMV "hot-spots". An evaluation of MST loci in translated (exon) regions found over 820 genes containing MSTs with a minimum of 2 minor alleles in both PD20 RV:D2 samples, with some of genes found within segments of chromosome 1 with increased SMV depicted in figure 5 (a complete list of exonal MSTs with the minor alleles called, for all cell lines discussed in this paper are available in File S2).

Taken together these results support our hypothesis that this method truly reflects SMV rather than error generated during sequencing and that the results are highly reproducible. The data further suggests that within an individual or cell line, specific genomic regions may contain MSTs that are more susceptible to somatic variability.

Reproducibility between cell lines

To begin to establish a SMV baseline for DNA repair proficient cells, we compared the haplotype, minor allele and SNP/INDEL distributions for two DNA repair proficient cell lines and the PD20 RV:D2 cells discussed above. MCF10A cells are immortalized breast epithelial cells derived from a healthy human female and HEK293 cells are a human embryonic kidney cell line derived from a healthy male fetus. Sequencing produced over 45 million reads with over 170 K microsatellite loci called at an average depth of 42 reads per locus for HEK293 cells and over 190 K

microsatellite loci called at an average depth of 39 reads per locus for MCF10A cells. Considering major alleles only, 96.4% and 97.0% of all MST loci, respectively, are homozygotic (table 1). The average fraction of loci with minor alleles for all three cell lines was 5.1% with a standard deviation of 0.4%. Although MCF10A cells had fewer loci with minor alleles than the PD20 RV:D2 and HEK293 cells (4.5% compared with 5.3% and 5.4% respectively, table 1), and showed a difference in the fraction of secondary alleles with SNPs compare to INDELS (table 2), MCF10A was not considered an outlier (using Grubb's test for outliers). When we compared the haplotype and minor allele concordance between two non-related cell lines, MCF10A and PD20 RV:D2, we found that 3.8% of loci have different genotypes with only 60% due to haplotype differences. For those loci with minor alleles, discordance is 4.0% and concordance is only 2.0%, the result is significantly above what would be anticipated by chance with Pearson's X^2 (i.e. <0.3%,). Interestingly, a full factorial comparison of the fraction of loci with minor alleles for each chromosome (as depicted in figure S4 in file S1), using a linear regression model, found a non-significant correlation ($R^2 = 0.061$ and p<0.23, figure 4C). However, a correlation using the 1 million base bins is significant with an R^2 value of 0.33 and a p<0.0001 (figure 4D), supporting the concept that certain regions contain minor allele susceptibility hot spots. These results demonstrate substantial reproducibility between unrelated independently grown DNA repair proficient cell lines even when the samples are derived from different tissues of origin. These results also suggest that a baseline profile of SMV can be established for DNA repair proficient cells to compare to cell lines with DNA repair defects.

SMV in cells with compromised DNA repair capacity

Thus far we have established that (1) three DNA repair proficient cell lines show similar SMV with low variability both within and between cell lines and that (2) we can differentiate between different SMV trends based on the ratio of INDELs to SNPs. However, the larger goal of this study is to compare SMV

Table 1. Exome sequencing data indicates that MST and non-MST haplotype and somatic polymorphism are reproducible in DNA repair proficient cell lines.

Percent (%)	Microsatellite loci				Repair Proficient		Non-Microsatellite loci				Repair Proficient	
	PD20 RV:D2-1	PD20 RV:D2-2	MCF10A	HEK293	Mean	SD	PD20 RV:D2-1	PD20 RV:D2-2	MCF10A	HEK293	Mean	SD
Homo-zyg	96.8	96.8	96.4	97.0	96.7	0.3	99.0	99.0	98.6	99.1	98.9	0.2
Hetero-zyg	3.2	3.2	3.6	3.0	3.3	0.3	1.0	1.0	1.4	0.9	1.1	0.2
Multi-alleles	5.4	5.3	4.5	5.3	5.1	0.4	1.7	2.0	2.1	2.1	2.0	0.2

Table 2. MST and non-MST containing loci from exome sequencing of DNA repair proficient cells, but not from sequencing of a single cell after whole genome amplification, show the expected high ratio of INDELs (expansions and contractions) to SNPs.

Percent (%)	Microsatellite loci				Repair Proficient		Non-microsatellite loci				Repair Proficient	
	PD20 RV:D2-1	PD20 RV:D2-2	MCF10A	HEK293	Mean	SD	PD20 RV:D2-1	PD20 RV:D2-2	MCF10A	HEK293	Mean	SD
SNPs	33.6	32.7	41.4	36.9	36.2	3.4	96.9	96.6	96.8	97.2	96.9	0.2
Expansions	26.2	27.0	25.3	25.5	26.0	0.7	1.3	1.6	1.6	1.4	1.5	0.1
Contractions	40.3	40.3	33.3	37.5	37.9	2.9	1.8	1.8	1.6	1.3	1.6	0.2

Table 3. Percent concordance/discordance of haplotype and loci with minor alleles for cell lines.

| | Genotype | More then haplotype alleles | | Haplotype Allele number |
	Discordance	Concordance	Discordance	Concordance
PD20 RV:D2-1 & -2	1.06	3.43	2.69	93.88
PD20rec-1 & PD20	1.15	2.50	3.07	94.43
PD20rec-1 & MCF10A	3.79	1.99	3.95	94.10
PD20rec-1 & Capan-1	2.68	1.92	12.68	85.40
MCF10A & Capan-1	2.19	1.24	13.62	85.10

patterns between cell lines representative of healthy individuals and those that may have altered DNA repair capacity. To test this, we evaluated 3 cell lines commonly used to study DNA repair and stability. DLD-1 cells are MST instability (MSI) high colon cancer cell line, impaired in Mismatch repair (MMR), selected as positive controls for this study [41]. Capan-1 cells were sequenced previously [12] and are a BRCA2- cell line that can propagate in culture. PD20 cells are from a FANCD2(-) cell line from which the PD20 RV:D2 cells were derived [40]. Both the Capan-1 cells and the PD20 cells have mutations in genes that are involved in

normal DNA repair (homologous recombination and interstrand crosslink repair, respectively).

For DLD-1 and PD20 cells, the number of loci that passed filters ranged between 185 K and 260 K with an average depth of between of 56 and 62 reads per locus respectively. Only 124 K loci were called for Capan-1 cells, with an average depth of 71 reads per locus. To capture MST differences between the DNA repair proficient and DNA repair defective cell lines we first evaluated haplotypes and the presence of minor alleles for each cell line. Both DLD-1 and Capan-1 cells significantly differ with respect to

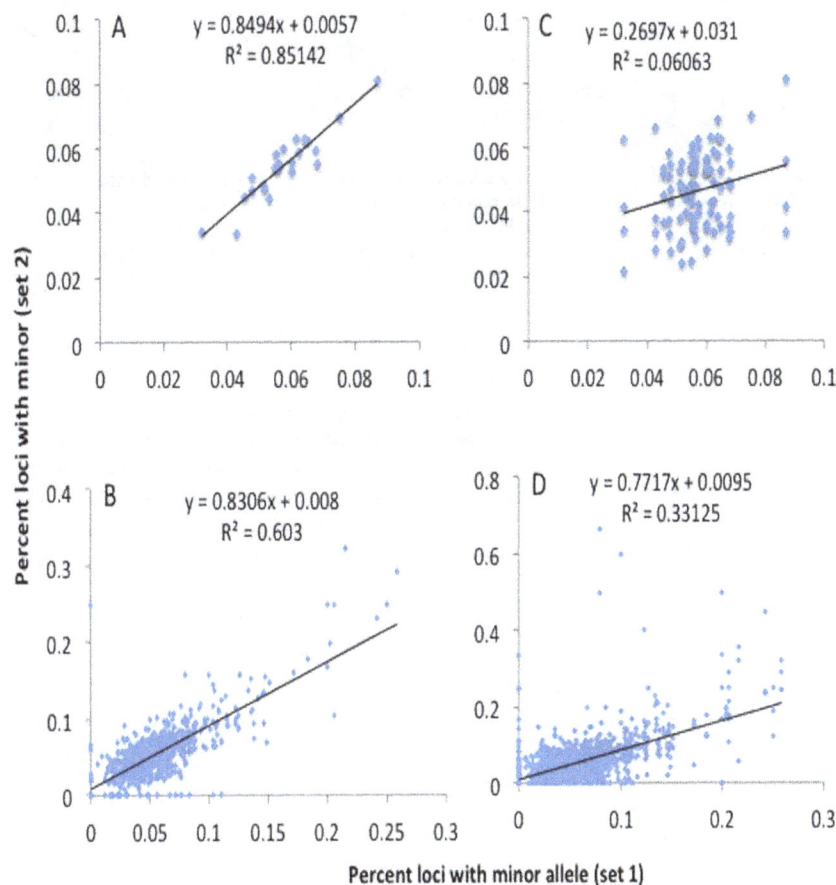

Figure 4. A regression analysis indicates a significant within and between cell line correlation in the fraction of loci with one or more minor alleles. Full factorial plots of the fraction of loci with minor alleles by chromosome, regression line and correlation coefficient for A) PD20 RV:D2-1 and 2 C) PD20 RV:D2-1, 2, MCF10A and HEK293. Also full factorial plots of the fraction of loci with minor alleles for the corresponding 1 million base segments of all the chromosomes, a regression line and the correlation coefficient for B) PD20 RV:D2-1 and 2 D) PD20 RV:D2-1, 2, MCF10A and HEK293.

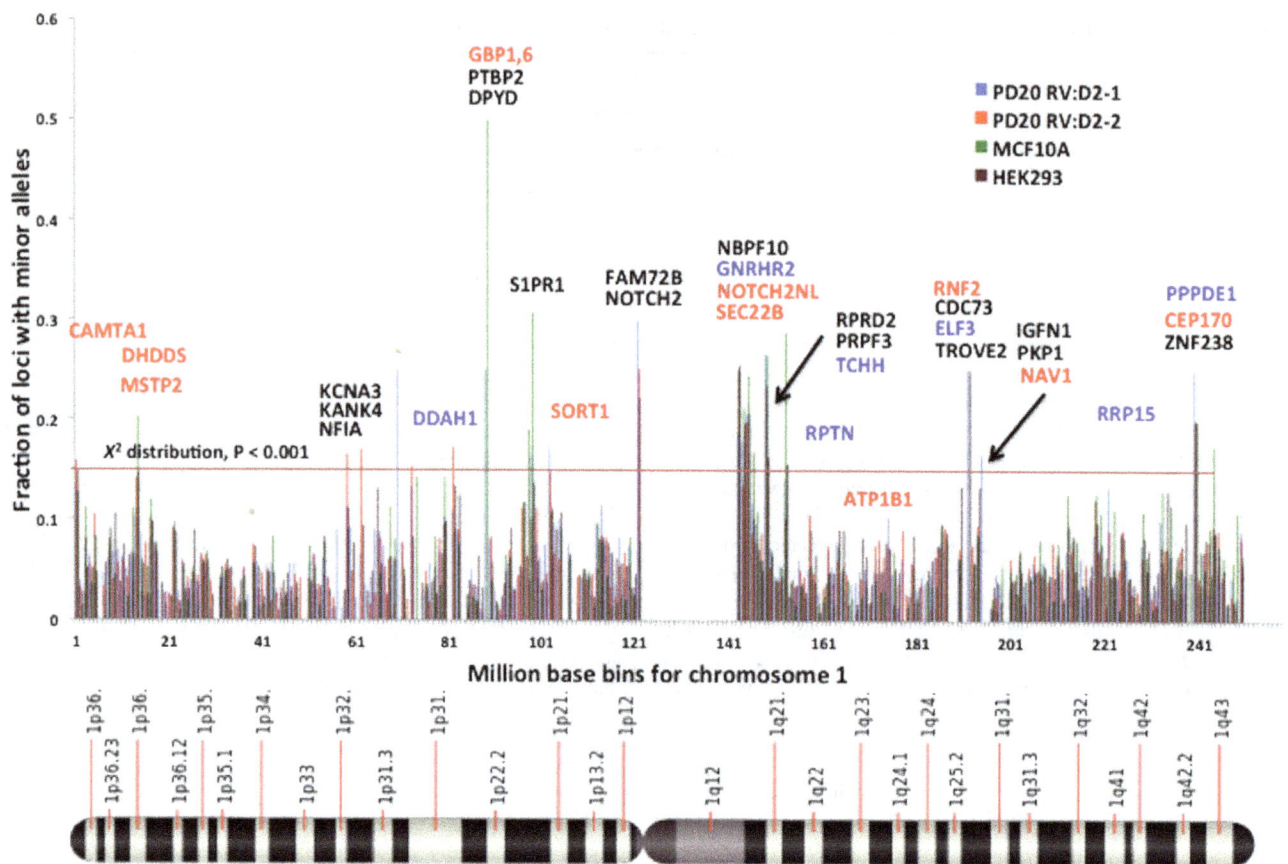

Figure 5. The distribution of MST loci showing somatic variability for chromosome 1 binned into 1 million base regions in PD20 and the derived PD20 RV:D2 cell line. The horizontal line demarcates outlier segments, based on a X^2 distribution. All genes shown were found to contain exonal MSTs that with at least 2 minor alleles in both PD20 RV:D2 samples and were found in regions that exceeded the demarcated level. Genes shown in red were found to contain exonal MSTs with at least 2 minor alleles in all 4 DNA repair proficient cell line samples and those shown in blue were found in 3 of the 4 samples. The chromosome image shown at the bottom was obtained from http://en.wikipedia.org/wiki/Chromosome_1 _(human).

haplotype distribution from DNA repair proficient cells (table 4). Capan-1 cells showed a significant decrease in heterozygotic loci, 2.1% compare to 3.3% for DNA repair proficient, which was anticipated due to the known trend for loss of heterozygosity in these cells as reported in the literature due to gene conversion in the absence of BRCA2 [42,43]. In contrast, there was an increase (5.5%) in hetereozygotic loci in DLD-1 cells, which can potentially be attributed to increased mutation due to the MMR defects responsible for the MSI in DLD-1 cells. Surprisingly, haplotype distribution analysis at non-MST loci shows that DLD-1 cells, but not Capan-1 differ significantly from DNA repair proficient (1.8% compared to 1.2% for DLD-1 and Capan-1 respectively). This was unexpected because neither mutation mechanism (homologous recombination nor MMR) would necessarily be restricted to MST vs non-MST regions. A comparison of SNPs and INDELs in the DNA repair impaired cell lines showed Capan-1 cells significantly differed from the DNA repair proficient mean in the fraction of SNPs, with 47% and 91% for MST and non-MST loci respectively (table 5). Conversely, DLD-1 and PD20 cells were not found to be different from DNA repair proficient cell lines. For the DNA repair proficient cells the mean fraction of loci with minor alleles was 5.1% with a SD of 0.4%. Capan-1 cells showed again, a greater susceptibility to mutation with a significant increase (6.2%) in the number of loci with minor alleles (table 4). In contrast, PD20 and DLD-1 cells both show a significant

decrease in loci with minor alleles, 3.1% and 3.2% respectively. This was surprising, particularly because the PD20 cells showed a decrease with respect to their corrected cell line PD20 RV:D2. Concordance of loci with minor alleles between the two related cell lines, PD20 and PD20 RV:D2, was 2.5% while discordance was 3.1%, which was significantly above chance (Pearson's X^2). However, it was greater than the concordance between PD20 RV:D2 and MCF10A, which is to be expected since PD20 and PD20 RV:D2 are related strains (Table 3).

Because Capan-1 cells displayed the highest disparity in mutation rate from DNA repair proficient cell lines, including changes in SNP:INDEL ratios, we decided to check the concordance of genotype and minor allele containing loci between them and PD20 RV:D2s (table 3). Genotype concordance for the loci that were found in both samples, was over 97.3%, even higher than when we compared PD20 RV:D2 with MCF10As. When comparing the loci with minor alleles ~2% of the total had minor alleles in both samples (were concordant) however 12% were found to have minor alleles in only one samples, meaning discordance (table 3). Although this is strikingly different, for the PD20 RV:D2 cells to MCF10A comparison, the concordance rate is still significantly greater than expected by chance. Very similar results were obtained when Capan-1 cells were compared to MCF10A cells. These results offer additional support the

Table 4. Haplotype distribution and somatic polymorphism rate differ in DNA repair defective cell lines compared to DNA repair proficient cell lines.

Percent (%)	Repair Proficient		Microsatellite loci Repair impaired cell lines			Repair Proficient		Non-microsatellite loci repair impaired cell lines		
	Mean	SD	PD20	DLD-1	Capan-1	Mean	SD	PD20	DLD-1	Capan-1
Homo-zyg	96.7	0.3	97.2 #	94.5 #	97.9 #	98.9	0.2	98.8	98.2 #	99.2
Hetero-zyg	3.3	0.3	2.8	5.5 #	2.1 #	1.1	0.2	1.2	1.8 #	0.8
Multi-alleles	5.1	0.4	3.1 #	3.2 #	6.2 #	2.0	0.2	1.2 #	1.2 #	3.7 #

significantly different p<0.01 - z-test.

Table 5. SNP and INDEL fractions differ in DNA repair defective cell lines compared to DNA repair proficient cells.

Percent (%)	Repair Proficient		Microsatellite loci Repair impaired cell lines			Repair Proficient		Non-microsatellite loci repair impaired cell lines		
	Mean	SD	DP20	DLD-1	Capan-1	Mean	SD	PD20	DLD-1	Capan-1
SNPs	36.2	3.4	35.7	36.9	47.6 #	96.9	0.2	95.4 #	94.9 #	90.8 #
Expansions	26.0	0.7	26.3	29.7	21.2 #	1.5	0.1	2.1 #	2.2 #	2.8 #
Contractions	37.9	2.9	38.0	33.3	31.2	1.6	0.2	2.5 #	2.9 #	6.4 #

significantly different p<0.01 - z-test.

hypothesis that some MST loci are more susceptible to mutations than others.

For DLD-1 cells, the increase in heterozygotic loci coupled with the significant reduction in the number of minor alleles is counterintuitive. This suggests the possibility of a proliferation of a small number of subpopulations. If our hypothesis is correct we would anticipate two things to occur: 1) an increase the average depth of reads that define the second allele and 2) an increase in the read depth supporting minor alleles without an increase in the number. To test our hypothesis we first compared the fraction of total reads covering the second allele regardless of haplotype and reads covering only minor alleles. As depicted in figure 6, DLD-1 cells show greater than a 4% increase with respect to the DNA repair proficient average in the fractional coverage of the second allele and more than 8% increase (figure 6A and B) for the percent coverage supporting minor alleles. Both were statistically significant. Neither Capan-1 nor PD20 were found to be different from the DNA repair proficient group for either of these parameters. These results suggest a population bottleneck where only a small number of distinct subpopulations are the predominant contributors of the reads captured by the sequencer.

SMV in exons

MSTs are present ubiquitously throughout the genome and are found in over 16% of exons [1]. Although MST expansions or contractions in promoter and interexonal regions can affect transcription, mutations in exons are the most frequently implicated in downstream effects, consistent with exons being under significant selective pressure. An analysis of heterozygotic loci found that exons had significantly less heterozygotic loci, a reduction of over 1.2% compared to untranslated regions (2.4% and 3.8% respectively, figure 7A). However the difference in the fraction of loci with minor alleles in exons and untranslated regions was not significant (5.1% and 5.6%, figure 7B). In the previous sections we showed that DLD-1 cells, a strain defective in MMR, was found, unexpectedly, to have a significant reduction in the number of MST loci with minor alleles and an increase in heterozygotic loci. Based on this comparison it appears that the results are due to the increased difference between translated and untranslated regions. As shown in figure 8A, the fraction of MST loci with minor alleles in exons is 1.1% (compared to 4.7% in untranslated regions) while the fraction of loci that are heterozygotic is 1.7%, compared to 7.9% in untranslated regions

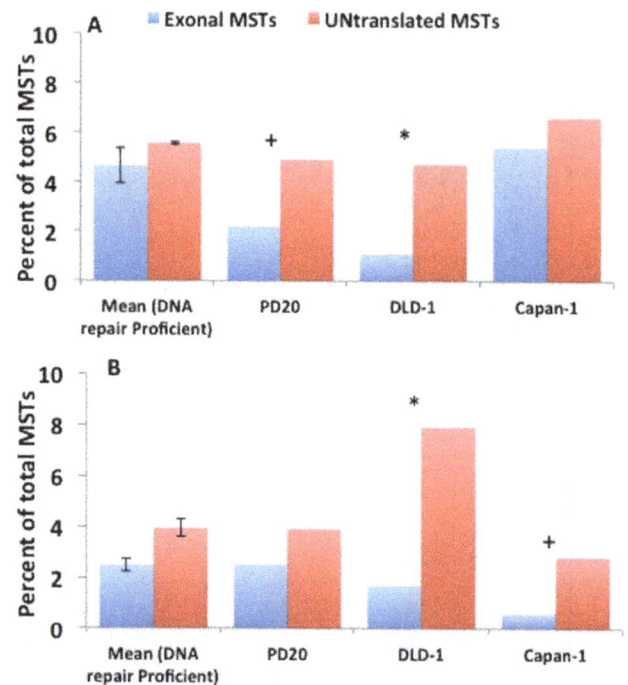

Figure 7. A comparison of the percent of heterozygotic loci and loci exhibiting SMV in exons and untranslated genomic regions in DNA repair proficient and impaired cell lines. A) The percent of MST loci that for which minor alleles were found and B) percent of heterozygotic MST loci, in exons and untranslated regions. Depicted in both figures are the means for the DNA repair proficient cell lines and the individual percentage for PD20, DLD-1 and capan-1 cell lines. (+) $p < 0.05$ as compared to DNA proficient cells and (*) $p < 0.001$ as compared to DNA proficient cells in measurement of the difference between exons and untranslated regions.

(figure 8B). These results further support hypothesis that DLD-1 cells have undergone a population bottleneck.

To determine the potential genetic implications of minor allele hot spots, we focused on the analysis of genes affected, specifically we inspected genes containing MST loci found in exons that with 2 or more alleles that did not contribute to the haplotype (minor alleles). This data is provided in a spreadsheet (file S2). The

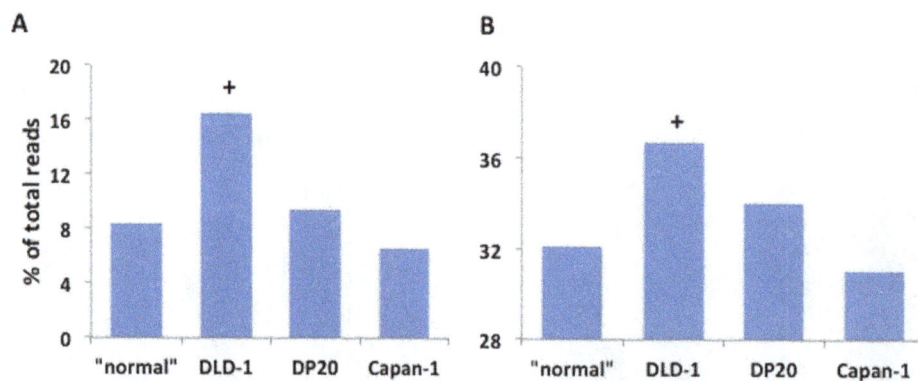

Figure 6. An increase in the fraction of reads substantiating the second alleles if present, and all minor alleles. The average fraction of reads representing A) all minor alleles (only for loci with minor alleles) and B) the second allele in both heterozygotic and homozygotic loci that have at least one minor allele, for DLD-1, PD20 and Capan-1 cells were compared to the average of the DNA repair proficient cell lines. The ($^+$) denotes a significant difference from DNA repair proficient ($p < 0.01$) with z-test.

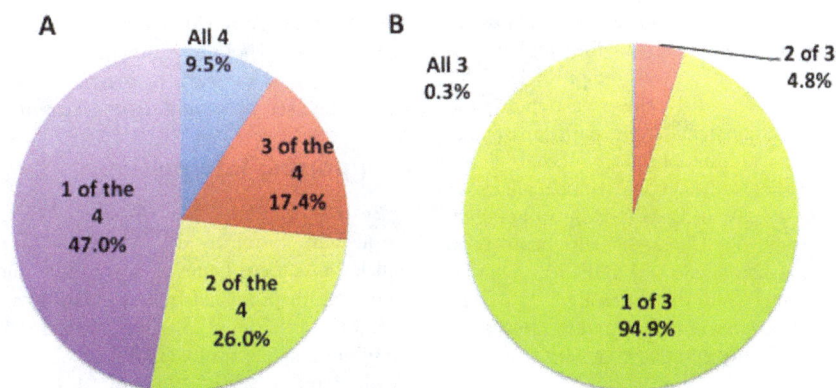

Figure 8. The distribution of genes that show SMV in DNA repair deficient cell lines appears random while those in the DNA repair proficient cell lines show significant similarity. The percent of genes with MSTs that with MSTs that have a minimum of 2 minor alleles in A) DNA repair proficient cell lines and B) DNA repair deficient cell lines that are found in all the or some of the sequenced samples. In figure B) the genes that are present in all three DNA repair deficient cell lines is 0.3% and the slice of the pie chart is not visible due to the small percentage.

spreadsheet lists the MST loci (based on the HG19 genome), gene name, cell genotype, total number of alleles, variants called and other pertinent information. Of the 2603 genes whose exons harbor minor allele containing loci found in at least one of the 4 DNA repair proficient samples sequenced 47% were found to have 2 or more minor alleles in more then one sample and 9.5% were found in all 4 samples (figure 7A). A Genome Ontology (GO) analysis of the 247 genes harboring MSTs with multiple minor alleles in all 4 samples found only a borderline (p<0.01, we use a lower p then 0.05 to compensate for the number of comparisons) significant enrichment of GOTERM categories that included transcription factors, regulators, repressors and DNA binding genes. In addition, there was no significant enrichment for any KEGG pathway categories or cataloged disorders. Conversely, of the ~1100 minor allele harboring genes found in the DNA repair impaired cell lines, only 3 (0.27%) were found in all three cell lines while 95% are in only 1 of the three cell lines (figure 7B), which suggests this concordance pattern was primarily random. Further, no genes with multiallelic MSTs were found in all of the sequenced samples and only 18 were found in 6 of the 7 cell line samples. A KEGG pathway enrichment analysis of the minor allele harboring genes found in the DNA repair impaired cell lines suggests a pattern associated with various cancer pathways. Significant KEGG terms enriched were general cancer, colorectal cancer, myeloma, cervical cancer and cell adhesion (with p<0.001). Together, these results support the hypothesis that specific MST loci in repair proficient cells are more susceptible to somatic mutations but the genes associated with them are not associated with any specific categorized pathway. In contrast, for cells that have impairments in DNA repair pathways, somatic mutations in MSTs appear in higher frequency in loci that are specific to the DNA repair deficiency, and these mutations are implicated in disease, specifically cancer.

Discussion

Somatic mutation can lead to subpopulations of cells carrying mutated alleles. These are examined in cancers, as tumors can be considered to contain subpopulations of cells, i.e. the tissues are not gnomically homogenous [44,45]. Tumors usually carry an allele or set of alleles that confirm their abnormal growth. These alleles, when detected in the tumor but not parent cells, can be the basis for important clinical treatment decisions [11,38,45]. In cell

populations with increased somatic mutation rates, like those with altered DNA repair capacity, there may be a concordant increase in subpopulation diversity. As a subpopulation propagates the mutations become more abundant, which becomes detectable in next-gen sequencing data [31,32]. A major assumption of our analysis is that an increase in the number of alleles detected in next-gen sequencing data is reflective of an increase in cell subpopulations or somatic mutation present in the sequenced sample. In this paper we evaluate allele frequencies at MSTs in various cell populations as a quantifiable indicator of variation.

The data presented here evaluate both the standard genotype and minor alleles that are present in next-gen data to establish a baseline for SMV in DNA repair proficient cells and compare this to cells with altered DNA repair capacity. The focus on cell lines with known etiologies is to establish the viability and robustness of our approach. The results show the utility in identifying the consequences of DNA repair impairments on genomic stability. There are several major objectives/findings from this analysis including (1) complimenting genomic analysis away of matched DNA samples with in-sample quantification of variation, (2) demonstrating that DNA repair proficient cells and those with different defects in DNA repair can have different SMV profiles that may be potential markers for these defects and (3) a quantitative measure of the fraction of loci that exhibit minor alleles may be reflective of subpopulations of cells with different genomic content, potentially those cells that may contribute to tumor formation. MST instability is important in the prognosis and selection of treatment for various cancers, and better, more accurate identification methods are always being sought [10,11].

These data demonstrate that the SNP:INDEL ratio at MSTs can be used to distinguish between different *in-vivo* mutational mechanisms and PCR amplified genomes. Both the WGA single cell sample and the Capan-1 cell line showed an increase in SNPs compared to INDELs at MST loci, however the fractions differed greatly. This is consistent with what was expected from both nucleotide mis-incorporation errors by polymerases (WGA single cell sample) and defects in DNA repair (Capan-1). Neither DLD-1 nor PD20 cells, which are defective in MMR and interstrand cross-link repair, respectively, had a significant alteration of the ratio of SNPs:INDELs at MST loci.

Capan-1 cells displayed a reduction of heterozygotic loci as compared to DNA repair proficient cell lines. This was expected since Capan-1 cells are a BRCA2- cells (impaired in homologous recombination) and have been shown to exhibit a loss of heterozygocity [43]. However, our analysis also indicates a significant increase in the fraction of loci with minor alleles. This could be due to two reasons: 1) Capan-1 cells are a hypotriploid with over 35 structural rearrangements (www.path.cam.ac.uk/%7epawefish/index.html) and with multiple chromosomal regions having more than three copies [46,47]. The minor alleles in Capan-1 cells can therefore be part of the genotype rather than somatic variation. Conversely, 2) Capan-1 cells have been reported to have an extremely high rate of INDELs and SNPs, significantly higher than expected from the hyperploidy [12]. The results shown here could be due to increased mutation rate shown with this cell line [12] and further support general genomic instability in Capan-1 cells.

Unexpectedly, although DLD-1 cells are a MST unstable cell line, they did not display either of our predicted markers for increase in MST mutation rate: 1) an increase in the number of minor alleles, as was seen with Capan-1 cells, or 2) a decrease in the number heterozygotic loci and the number of minor alleles, as we found in Capan-1 and PD20 cells (table 5). Conversely, DLD-1 cells showed both a significant increase in the number of heterozygotic loci and a reduction in the fraction of loci with more than two alleles. Further, they displayed a great reduction in both the fraction of loci with minor alleles and heterozygotic loci in exons (conserved chromosomal regions). We hypothesize that this is the result of defective MMR leading to an increase in mutations that have become fixed in the population. Alternatively, this may have resulted from a bottleneck in the growth of the cell population. If this was the case, the increase in heterozygotic loci allele may be a product of a limited set of surviving cell subpopulations. If a subpopulation with an un-repaired mutation, reached a sufficient proportion of the population due to the bottleneck it would generate sufficient reads for the locus to be mistakenly called heterozygotic. This point is reinforced by the significant increase in the portion of the total number of reads covering the second allele while the fraction of loci with minor alleles and the number of minor alleles per locus are decreased. This is important to note because it suggests that we can not only distinguish between different mutational mechanisms using the minor alleles in next-gen sequencing, but may also be able to identify cells that have experienced a growth-limiting condition as we expand this work in the future.

The work presented here is a proof-of concept of an approach to assess somatic variation in MSTs using next-gen sequencing. Using this analysis we were able to establish a SMV profile in DNA repair proficient cell lines which we can use to compare to cells with potential or known alterations in DNA repair capacity to begin to evaluate exome or whole genome sequenced samples without requiring a matched genomic sample as baseline. Based on the results presented here this approach can be used to ascertain both scientifically and clinically relevant information.

Scientifically, even with known mutations the consequences on the genome as a whole is still relatively unknown. Clinically, somatic variation is a measure of genomic stability and this approach might be used as an addition to current MST instability criteria.

Supporting Information

File S1 Contains the following files: **Figure S1.** Sanger sequencing confirms the prediction of the at least 3 different alleles, in a locus found to have minor alleles in nextGen data. A) The output produced by our caller (locus is shown in the first 5 columns in line 1) predict 3 different length alleles using a minimum of 2 reads to confirm an allele. The major allele is 23 nts with 2 minor alleles, 25 and 21 nts long. B) The sequencing chromatogram. The black arrows are showing the start point of different alleles. **Figure S2.** Effects of sequencing error and the minimum number of reads required to call an allele on of the number of alleles called in sequencing data. (A) Modeling data with different error frequencies (0.5%–5%) showed an increase in loci with multiple alleles as error increased when 4 reads were minimally required to call an allele. (B) The average read depth at loci with increasing numbers of alleles using 4 confirming reads per allele for in-silico generated data using 1% and 2.5% error rate and 4 different cell lines. **Figure S3.** The distribution of MST loci showing somatic variability by chromosome for both PD20 RV:D2 samples. **Figure S4.** The distribution of MST loci showing somatic variability by chromosome, for both PD20 RV:D2, MCF10A and HEK293 cell lines. **Table S1.** In-silico model mapping and genotyping accuracy. **Table S2.** The total minor alleles sorted by MST motif length indicate that single cell exome amplification alters the distributions observed in DNA repair proficient cell lines.

File S2 Genomic data file.

Acknowledgments

We thank the system administrators in the VBI computational core (Michael Snow, Dominik Borkowski, David Bynum, Douglas McMaster, and Vedavyas Duggirala) for technical support. We also acknowledge members of the VBI Genomics Research Lab (Saikumar Karyala, Jennifer Jenrette, Megan Friar, and Kris Lee) for the library prep, and sequencing of genomic and Sanger validation samples.

Author Contributions

Conceived and designed the experiments: ZV HRG. Performed the experiments: ZV NCF. Analyzed the data: ZV NCF HRG. Contributed reagents/materials/analysis tools: ZV HT. Wrote the paper: ZV HRG. Wrote the software: ZV HT.

References

1. Gemayel R, Vinces MD, Legendre M, Verstrepen KJ (2010) Variable tandem repeats accelerate evolution of coding and regulatory sequences. Annu Rev Genet 44: 445–477.

2. Fonville NC, Ward RM, Mittelman D (2011) Stress-induced modulators of repeat instability and genome evolution. J Mol Microbiol Biotechnol 21: 36–44.

3. Bagshaw AT, Pitt JP, Gemmell NJ (2008) High frequency of microsatellites in S. cerevisiae meiotic recombination hotspots. BMC Genomics 9: 49.

4. Payseur BA, Jing P, Haasl RJ (2011) A genomic portrait of human microsatellite variation. Mol Biol Evol 28: 303–312.

5. Delagoutte E, Goellner GM, Guo J, Baldacci G, McMurray CT (2008) Single-stranded DNA-binding protein in vitro eliminates the orientation-dependent impediment to polymerase passage on CAG/CTG repeats. J Biol Chem 283: 13341–13356.

6. Hile SE, Eckert KA (2008) DNA polymerase kappa produces interrupted mutations and displays polar pausing within mononucleotide microsatellite sequences. Nucleic Acids Res 36: 688–696.

7. Ananda G, Walsh E, Jacob KD, Krasilnikova M, Eckert KA, et al. (2013) Distinct mutational behaviors differentiate short tandem repeats from microsatellites in the human genome. Genome Biol Evol 5: 606–620.

8. Leclercq S, Rivals E, Jarne P (2010) DNA slippage occurs at microsatellite loci without minimal threshold length in humans: a comparative genomic approach. Genome Biol Evol 2: 325–335.

9. Budworth H, McMurray CT (2013) Bidirectional transcription of trinucleotide repeats: roles for excision repair. DNA Repair (Amst) 12: 672–684.

10. Xiao H, Yoon YS, Hong SM, Roh SA, Cho DH, et al. (2013) Poorly differentiated colorectal cancers: correlation of microsatellite instability with clinicopathologic features and survival. Am J Clin Pathol 140: 341–347.

11. Hong SP, Min BS, Kim TI, Cheon JH, Kim NK, et al. (2012) The differential impact of microsatellite instability as a marker of prognosis and tumour response between colon cancer and rectal cancer. Eur J Cancer 48: 1235–1243.

12. Barber LJ, Rosa Rosa JM, Kozarewa I, Fenwick K, Assiotis I, et al. (2011) Comprehensive genomic analysis of a BRCA2 deficient human pancreatic cancer. PLoS One 6: e21639.

13. Lacroix-Triki M, Lambros MB, Geyer FC, Suarez PH, Reis-Filho JS, et al. (2010) Absence of microsatellite instability in mucinous carcinomas of the breast. Int J Clin Exp Pathol 4: 22–31.

14. Yoon K, Lee S, Han TS, Moon SY, Yun SM, et al. (2013) Comprehensive genome- and transcriptome-wide analyses of mutations associated with microsatellite instability in Korean gastric cancers. Genome Res 23: 1109–1117.

15. Kim TM, Laird PW, Park PJ (2013) The landscape of microsatellite instability in colorectal and endometrial cancer genomes. Cell 155: 858–868.

16. Poduri A, Evrony GD, Cai X, Walsh CA (2013) Somatic mutation, genomic variation, and neurological disease. Science 341: 1237758.

17. Harris RS, Kong Q, Maizels N (1999) Somatic hypermutation and the three R's: repair, replication and recombination. Mutat Res 436: 157–178.

18. Kunz C, Saito Y, Schar P (2009) DNA Repair in mammalian cells: Mismatched repair: variations on a theme. Cell Mol Life Sci 66: 1021–1038.

19. Baptiste BA, Ananda G, Strubczewski N, Lutzkanin A, Khoo SJ, et al. (2013) Mature microsatellites: mechanisms underlying dinucleotide microsatellite mutational biases in human cells. G3 (Bethesda) 3: 451–463.

20. Shah SN, Hile SE, Eckert KA (2010) Defective mismatch repair, microsatellite mutation bias, and variability in clinical cancer phenotypes. Cancer Res 70: 431–435.

21. Eckert KA, Mowery A, Hile SE (2002) Misalignment-mediated DNA polymerase beta mutations: comparison of microsatellite and frame-shift error rates using a forward mutation assay. Biochemistry 41: 10490–10498.

22. Roy R, Chun J, Powell SN (2012) BRCA1 and BRCA2: different roles in a common pathway of genome protection. Nat Rev Cancer 12: 68–78.

23. Kottemann MC, Smogorzewska A (2013) Fanconi anaemia and the repair of Watson and Crick DNA crosslinks. Nature 493: 356–363.

24. Li H, Handsaker B, Wysoker A, Fennell T, Ruan J, et al. (2009) The Sequence Alignment/Map format and SAMtools. Bioinformatics 25: 2078–2079.

25. Benson G (1999) Tandem repeats finder: a program to analyze DNA sequences. Nucleic Acids Res 27: 573–580.

26. Tae H, Kim DY, McCormick J, Settlage RE, Garner HR (2013) Discretized Gaussian mixture for genotyping of microsatellite loci containing homopolymer runs. Bioinformatics.

27. Tae H, McMahon KW, Settlage RE, Bavarva JH, Garner HR (2013) ReviSTER: an automated pipeline to revise misaligned reads to simple tandem repeats. Bioinformatics 29: 1734–1741.

28. McIver LJ, McCormick JF, Martin A, Fondon JW 3rd, Garner HR (2013) Population-scale analysis of human microsatellites reveals novel sources of exonic variation. Gene 516: 328–334.

29. McIver LJ NCF, Karunasena E, Garner HR (Submitted) Microsatellite genotyping reveals a signature in breast cancer exomes. Breast Cancer Research and Treatment.

30. Fonville NC LJM, Vaksman Z, Garner HR (Submitted) Microsatellites in the exome are predominantly single-allelic and invariant. Genome Biology.

31. Schmitt MW, Kennedy SR, Salk JJ, Fox EJ, Hiatt JB, et al. (2012) Detection of ultra-rare mutations by next-generation sequencing. Proc Natl Acad Sci U S A 109: 14508–14513.

32. Gundry M, Vijg J (2012) Direct mutation analysis by high-throughput sequencing: from germline to low-abundant, somatic variants. Mutat Res 729: 1–15.

33. Kruglyak S, Durrett RT, Schug MD, Aquadro CF (1998) Equilibrium distributions of microsatellite repeat length resulting from a balance between slippage events and point mutations. Proc Natl Acad Sci U S A 95: 10774–10778.

34. Jarne P, Lagoda PJ (1996) Microsatellites, from molecules to populations and back. Trends Ecol Evol 11: 424–429.

35. Kanagawa T (2003) Bias and artifacts in multitemplate polymerase chain reactions (PCR). J Biosci Bioeng 96: 317–323.

36. Meyerhans A, Vartanian JP, Wain-Hobson S (1990) DNA recombination during PCR. Nucleic Acids Res 18: 1687–1691.

37. Brodin J, Mild M, Hedskog C, Sherwood E, Leitner T, et al. (2013) PCR-induced transitions are the major source of error in cleaned ultra-deep pyrosequencing data. PLoS One 8: e70388.

38. Hou Y, Song L, Zhu P, Zhang B, Tao Y, et al. (2012) Single-cell exome sequencing and monoclonal evolution of a JAK2-negative myeloproliferative neoplasm. Cell 148: 873–885.

39. Mestrovic N, Castagnone-Sereno P, Plohl M (2006) Interplay of selective pressure and stochastic events directs evolution of the MEL172 satellite DNA library in root-knot nematodes. Mol Biol Evol 23: 2316–2325.

40. Ohashi A, Zdzienicka MZ, Chen J, Couch FJ (2005) Fanconi anemia complementation group D2 (FANCD2) functions independently of BRCA2- and RAD51-associated homologous recombination in response to DNA damage. J Biol Chem 280: 14877–14883.

41. Chen TR, Hay RJ, Macy ML (1983) Intercellular karyotypic similarity in near-diploid cell lines of human tumor origins. Cancer Genet Cytogenet 10: 351–362.

42. Holt JT, Toole WP, Patel VR, Hwang H, Brown ET (2008) Restoration of CAPAN-1 cells with functional BRCA2 provides insight into the DNA repair activity of individuals who are heterozygous for BRCA2 mutations. Cancer Genet Cytogenet 186: 85–94.

43. Butz J, Wickstrom E, Edwards J (2003) Characterization of mutations and loss of heterozygosity of p53 and K-ras2 in pancreatic cancer cell lines by immobilized polymerase chain reaction. BMC Biotechnol 3: 11.

44. Tang DG (2012) Understanding cancer stem cell heterogeneity and plasticity. Cell Res 22: 457–472.

45. Schor SL (1995) Fibroblast subpopulations as accelerators of tumor progression: the role of migration stimulating factor. EXS 74: 273–296.

46. Sirivatanauksorn V, Sirivatanauksorn Y, Gorman PA, Davidson JM, Sheer D, et al. (2001) Non-random chromosomal rearrangements in pancreatic cancer cell lines identified by spectral karyotyping. Int J Cancer 91: 350–358.

47. Grigorova M, Staines JM, Ozdag H, Caldas C, Edwards PA (2004) Possible causes of chromosome instability: comparison of chromosomal abnormalities in cancer cell lines with mutations in BRCA1, BRCA2, CHK2 and BUB1. Cytogenet Genome Res 104: 333–340.

5

Sequencing, Annotation and Analysis of the Syrian Hamster (*Mesocricetus auratus*) Transcriptome

Nicolas Tchitchek[1], David Safronetz[2], Angela L. Rasmussen[1], Craig Martens[3], Kimmo Virtaneva[3], Stephen F. Porcella[3], Heinz Feldmann[2], Hideki Ebihara[2]*[9], Michael G. Katze[1,4]*[9]

1 Department of Microbiology, University of Washington, Seattle, Washington, United States of America, 2 Laboratory of Virology, Division of Intramural Research, National Institute of Allergy and Infectious Diseases, National Institutes of Health, Rocky Mountain Laboratories, Hamilton, Montana, United States of America, 3 Genomics Unit, Research Technologies Section, National Institute of Allergy and Infectious Diseases, National Institutes of Health, Rocky Mountain Laboratories, Hamilton, Montana, United States of America, 4 Washington National Primate Research Center, University of Washington, Seattle, Washington, United States of America

Abstract

Background: The Syrian hamster (golden hamster, *Mesocricetus auratus*) is gaining importance as a new experimental animal model for multiple pathogens, including emerging zoonotic diseases such as Ebola. Nevertheless there are currently no publicly available transcriptome reference sequences or genome for this species.

Results: A cDNA library derived from mRNA and snRNA isolated and pooled from the brains, lungs, spleens, kidneys, livers, and hearts of three adult female Syrian hamsters was sequenced. Sequence reads were assembled into 62,482 contigs and 111,796 reads remained unassembled (singletons). This combined contig/singleton dataset, designated as the Syrian hamster transcriptome, represents a total of 60,117,204 nucleotides. Our *Mesocricetus auratus* Syrian hamster transcriptome mapped to 11,648 mouse transcripts representing 9,562 distinct genes, and mapped to a similar number of transcripts and genes in the rat. We identified 214 quasi-complete transcripts based on mouse annotations. Canonical pathways involved in a broad spectrum of fundamental biological processes were significantly represented in the library. The Syrian hamster transcriptome was aligned to the current release of the Chinese hamster ovary (CHO) cell transcriptome and genome to improve the genomic annotation of this species. Finally, our Syrian hamster transcriptome was aligned against 14 other rodents, primate and laurasiatheria species to gain insights about the genetic relatedness and placement of this species.

Conclusions: This Syrian hamster transcriptome dataset significantly improves our knowledge of the Syrian hamster's transcriptome, especially towards its future use in infectious disease research. Moreover, this library is an important resource for the wider scientific community to help improve genome annotation of the Syrian hamster and other closely related species. Furthermore, these data provide the basis for development of expression microarrays that can be used in functional genomics studies.

Editor: Vincent Laudet, Ecole Normale Supérieure de Lyon, France

Funding: This work was supported by the Intramural Research Program of the National Institute of Allergy and Infectious Diseases, National Institutes of Health and the NIAID Regional Centers of Excellence (U54 AI081680) to MGK. The funders were not involved in the study design, data collection and analysis, or preparation of the manuscript. NIAID was solely involved in the decision to publish.

Competing Interests: The authors have declared that no competing interests exist.

* Email: ebiharah@niaid.nih.gov (HE); honey@uw.edu (MK)

⑨ These authors contributed equally to this work.

Introduction

The Syrian hamster (golden hamster, *Mesocricetus auratus*) has recently been used as an experimental rodent model for important infectious diseases including Ebola and other viral hemorrhagic fevers [1–8]. For instance, Syrian hamsters infected with mouse-adapted Ebola virus (EBOV) manifest many of the clinical and pathological findings observed in EBOV-infected non-human primates (NHPs) and humans, including systemic viral replication, suppression of the innate immune response, an uncontrolled inflammatory response, and disseminated intravascular coagulation syndrome [9]. The Syrian hamster is emerging as a promising

model for leishmaniasis [10] and dyslipidaemia research [11,12]. The Syrian hamster is also an important animal model in neurosciences research [13,14]. For instance, this species has been widely used in the studies of circadian rhythms [15], cardiomyopathy [16], aggression [17], reproduction [18], and sensory systems [19].

Genotyping of *Mesocricetus auratus* is currently under way at the Broad Institute (NCBI-BioProject accession: PRJNA77669) but not yet published. So far, only 860 cDNA sequences from the Syrian hamster are available in the NCBI-dbEST database [20], where 728 sequences have been collected in the context of testis organs [21] and 125 sequences have been collected in the context of embryonic cells [22]. More recently, while Schmucki et al. analyzed the liver transcriptome of the Syrian hamster with a focus on lipid metabolism [23] the data is not publicly available as of this writing.

Drafts of the genome and transcriptome of Chinese hamster ovary (CHO) cells have recently been published [24,25], although it should be noted that CHO cells represent cells in an immortalized condition and therefore will likely contain genetic mutations not present in natural conditions. The current release of the CHO cell draft genome is composed of 109,152 scaffolds and 265,786 contigs representing a total length of 2,318,115,958 nucleotides. Preliminary gene annotation of the CHO cell genome was performed using vertebrate experimental data and cross-species comparisons. The current release of the CHO cell transcriptome comprises 121,636 transcript fragments representing a total length of 179,731,611 nucleotides. More recently, Lewis et al. compared the genome of CHO cells and the genome of the Chinese hamster obtained from tissues, and they showed a significant proximity between these different conditions [26]. Further efforts will be continued regarding the update of the CHO and Chinese hamster genomes and transcriptomes.

The aims of our study were: (i) to provide to the scientific community a large panel of annotated mRNA sequences from the *Mesocricetus auratus* transcriptome; (ii) to provide new biological insights and knowledge about the *Mesocricetus auratus* species; and (iii) to use this data to allow the design of a future gene expression microarray. Here we sequenced a normalized 3′ mRNA fragment primed cDNA library produced from pooled RNA isolated from the major organs of adult female Syrian hamsters following strategies in common-use described elsewhere [27,28]. We reasoned that pooling a large variety of different organs of animals will provide a large pool of mRNA fragments to sequence and annotate. Sequencing reads were de novo assembled into contigs. The combined contig and unassembled read (singleton) dataset, designated as the Syrian hamster transcriptome, was annotated based on the mouse and rat transcriptomes. We identified the most highly covered and the most highly expressed transcripts in our Syrian hamster transcriptome and performed a functional enrichment analysis to identify which canonical pathways and biological functions were most significantly represented. In order to contribute to the annotation efforts of the Chinese hamster species, we aligned our Syrian hamster transcriptome to the current version of the CHO cell genome and transcriptome. Finally, we aligned our Syrian hamster transcriptome to 14 other primate species and analyzed the genomic divergence of our transcripts in order to gain insights into the genomic evolution of the Syrian hamster.

Results

Sample collection and sequencing of a cDNA library produced from female Syrian hamster organs

The brains, lungs, spleens, kidneys, livers, and hearts were collected from three adult female Syrian hamsters. Total RNAs were isolated, pooled, and contaminating genomic DNA removed. Following adaptor ligation, cDNAs were 3′ fragment-sequenced on a Roche 454 GS FLX Titanium instrument. The sequencing generated 1,283,840 reads with an average length of 344 bases. Reads were trimmed for quality and reads shorter than 40 bases were discarded, resulting in 1,212,395 sequence reads available for further assembly and analysis. **Figure 1A** shows the length distribution of reads before assembly. Consistent with most of the publicly available transcriptome libraries [29], we observed that our reads ranged between 200 and 600 nucleotides in length.

Library assembly

Quality-filtered reads were assembled into contigs. Resulting contigs and unassembled reads (singletons) were quality filtered and contigs or singletons shorter than 50 bases were discarded. Among the 1,212,395 reads, 62,482 contigs and 111,796 singletons were generated. **Figure 1B** shows the length distributions of the 174,278 combined contig/singleton dataset. The lengths of the singletons ranged from 50 to 614, with a median length of 187.50 bases. The lengths of the contigs ranged from 50 to 4,054, with a median length of 473.50 bases. We observed that most of the reads ranging between 75 and 400 nucleotides were assembled. Short reads are subject to noise and have low quality scores, making them more difficult to assemble. On the other hand, larger reads are difficult to assemble in this context because our library was targeted against 3′ mRNA priming. The final dataset (contigs plus singletons) represents a total of 60,117,204 nucleotides and is designated as the Syrian hamster transcriptome.

Library annotation

The Syrian hamster transcriptome was aligned to the mouse and rat transcriptome references (**Table 1**). Amongst the 174,278 contigs and singletons, 41,651 (23.90%) were significantly aligned (expected value cutoff of 10) to the mouse transcriptome and 26,258 (15.07%) were significantly aligned to the rat transcriptome. Of these, 11,648 transcripts (representing 9,562 genes) contained functional annotation in the mouse transcriptome, and 7,223 transcripts (representing 7,137 genes) were functionally annotated in the rat transcriptome (**Table 1**). Therefore, 11,648 Syrian sequence fragments or transcripts are now annotated by way of homology with the mouse genome.

We also investigated the positioning of the mRNA encoded contigs and singletons of our Syrian hamster transcriptome against other species' different transcript regions such as, 5′ untranslated regions (5′ UTR), coding regions, or 3′ untranslated regions (3′ UTR). With respect to the mouse transcriptome reference, 4,314 fragments of our Syrian hamster transcriptome (10.36%) aligned to 5′ UTRs, while 6,493 fragments of our dataset (15.59%) aligned to coding regions. In addition, 26,764 fragments of the Syrian hamster transcriptome (64.26%) aligned to 3′ UTRs (**Figure 2A**). A further 4,080 fragments of the Syrian hamster transcriptome (9.80%) aligned between 5′ UTRs, coding regions, and 3′ UTRs of the mouse transcriptome. Based on the rat transcriptome reference, 521 fragments of the Syrian hamster transcriptome (1.98%) aligned to 5′ UTRs while, 5,568 fragments of the Syrian hamster transcriptome (21.20%) aligned to coding regions. In addition, 13,371 fragments of the Syrian hamster transcriptome (50.92%) aligned to 3′ UTRs (**Figure 2B**). Finally, a total of 6,798 fragments of the Syrian hamster transcriptome (25.89%) aligned between 5′ UTRs, coding regions, and 3′ UTRs of the rat transcriptome. As expected from the experimental design of our library, the majority of our Syrian hamster transcriptome sequences aligned to 3′ UTRs of mouse and rat annotated transcripts.

Regarding the publicly available mouse and rat genome datasets, 45,804 of our Syrian hamster transcriptome fragments aligned to either the mouse or rat transcriptomes, and 22,105 of these same sequences aligned to both transcriptomes simultaneously, suggesting commonly occurring transcripts. Our Syrian hamster transcriptome dataset was 65.10% similar to the mouse transcriptome and 64.46% similar to the rat transcriptome. These similarities increased to 74.48% and 74.26% for the mouse and rat respectively, when comparisons were restricted to coding regions-only within those two reference genomes.

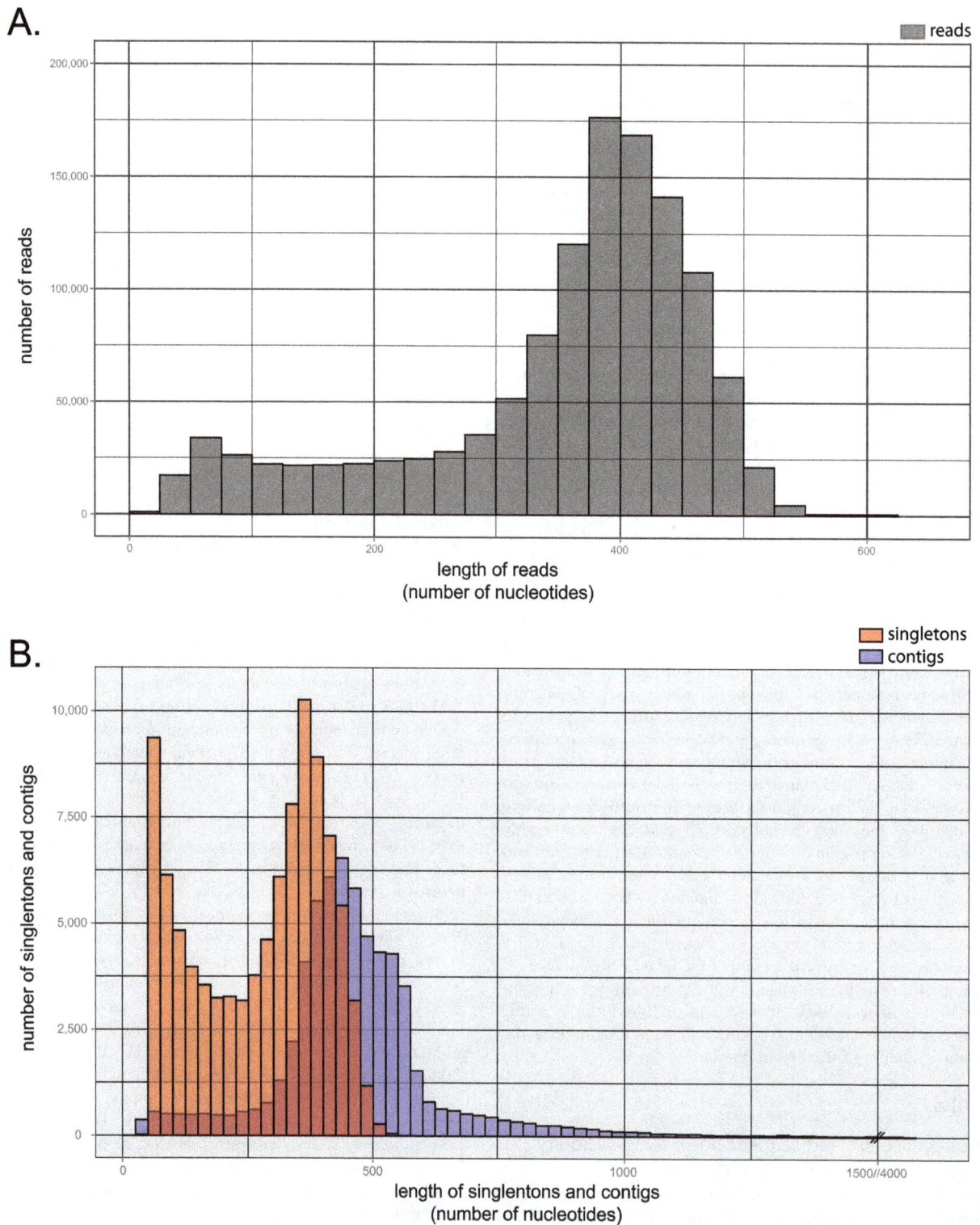

Figure 1. Histograms showing the length distribution of the reads and the length distribution of the singletons and contigs. (A) The length distribution of the reads is shown in a gray histogram. Bins of the histogram have been set to 50 nucleotides. The lengths of the reads range from 40 to 631, with a median length of 387 and a mean length of 352. The reads represents a total of 426,683,712 nucleotides bases. (B) The length distribution of the 111,796 singletons is shown in a red histogram while the length distribution of the 62,482 contigs is shown in a blue histogram. Bins of the histograms have been set to 25 nucleotides. The lengths of the singleton sequences range from 50 to 614, with a median length of 187 and a mean length of 265. The lengths of the contig sequences range from 50 to 4,054, with a median length of 473 and a mean length of 487. Our Syrian hamster transcriptome represents a total of 60,117,204 nucleotides bases.

Table 1. Transcriptome references and alignment statistics.

Species name	# of genes	# of transcripts	# and % of alignments	# and % of mapped genes	# and % of mapped transcripts
Mouse (Mus musculus)	38,293	92,484	41,651 (23.90%)	9,562 (22.96%)	11,648 (12.59%)
Rat (Rattus norvegicus)	26,405	29,189	26,258 (15.07%)	7,137 (27.18%)	7,223 (24.75%)
Chinese Hamster Ovary cells (Cricetulus griseus)	NA	121,636*	7,845 (4.50%)	NA	4,390 (3.61%)
Chimpanzee (Pan troglodytes)	28,012	29,160	2,884 (1.65%)	1,631 (56.55%)	1,643 (5.63%)
Ferret (Mustela putorius furo)	23,811	23,963	16,169 (9.28%)	4,169 (25.78%)	4,187 (17.47%)
Gorilla (Gorilla gorilla gorilla)	29,216	35,727	8,319 (4.77%)	2,733 (32.85%)	2,735 (7.66%)
Guinea pig (Cavia porcellus)	25,028	26,129	15,014 (8.61%)	4,050 (26.97%)	4,155 (15.90%)
Human (Homo sapiens)	62,316	213,551	23,020 (13.21%)	5,409 (23.50%)	7,254 (3.40%)
Kangaroo rat (Dipodomys ordii)	26,405	29,189	2,103 (1.21%)	1,252 (59.53%)	1,252 (4.29%)
Macaque (Macaca mulatta)	30,246	44,725	13,792 (7.91%)	3,804 (27.58%)	4,163 (9.31%)
Orangutan (Pongo abelii)	28,443	29,447	15,331 (8.80%)	3,929 (25.63%)	3,952 (13.42%)
Pig (Sus scrofa)	25,322	30,586	10,910 (6.26%)	2,978 (27.30%)	3,051 (9.98%)
Pika (Ochotona princeps)	23,028	23,028	1,575 (0.90%)	989 (62.79%)	989 (4.29%)
Rabbit (Oryctolagus cuniculus)	23,394	28,188	4,344 (2.49%)	1,946 (44.80%)	2,007 (7.12%)
Shrew (Sorex araneus)	19,134	19,139	1,330 (0.76%)	759 (57.07%)	759 (3.97%)
Squirrel (Ictidomys tridecemlineatus)	22,398	23,572	7,730 (4.44%)	2,723 (35.23%)	2,733 (11.59%)
Tree Shrew (Tupaia belangeri)	20,820	20,824	1,786 (1.02%)	1,091 (61.09%)	1,091 (5.24%)

For each transcriptome reference used in this study, the name of the species, the number of genes available, and the number of transcripts available are indicated. *The number of available transcripts indicated for the Chinese hamster ovary cells represents the number of available transcript fragments available and not the number of distinct transcripts. Moreover, for each transcriptome reference used in this study, the number of aligned contigs and singletons, the number of mapped transcripts and the number of mapped genes are indicated. The percentages of mapped transcripts and mapped genes relative to the total number of transcripts and genes available on the transcriptome references are provided. Moreover the percentage of alignments relative to the total number of contigs and singletons in our library (174,278) is also provided.

In the mouse genome, we found that 214 of those transcripts mapped at 90% of their lengths to either contigs or singletons in the Syrian hamster transcriptome (**Table S1**). Among these highly covered mouse transcripts were genes associated with a range of cellular activities involving, but not limited to inflammation, cell death, metabolism, and initiation of translation. These results

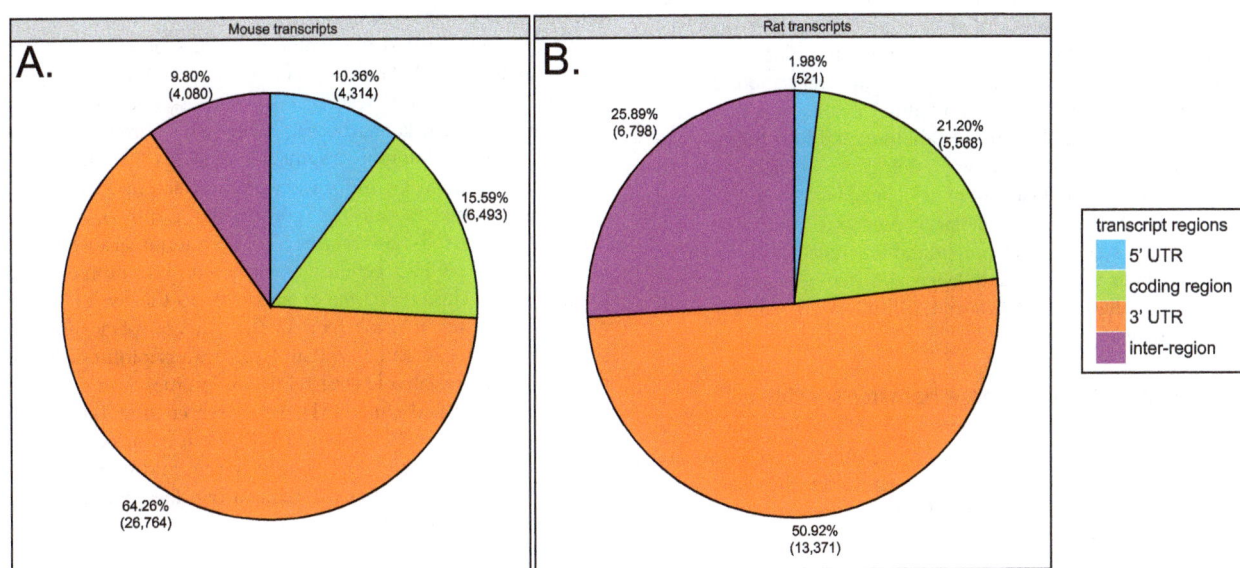

Figure 2. Pie diagrams showing the alignment positions of the contigs and singletons on the mouse and rat transcript regions. (A) Pie diagram showing the distribution of alignment positions of the 41,651 contigs and singletons on the mouse transcripts regions (5′ UTR, coding region, 3′ UTR, or inter-region). (B) Pie diagram showing the distribution of alignment positions of the 26,258 contigs and singletons on the rat transcripts regions. For each species and transcript region the number and percentage of aligned sequences are indicated.

suggest that highly covered transcripts, representative of a wide variety of cellular processes, were obtained through our methodology.

Over-expressed sequence reads and over-represented canonical pathways

In order to obtain further biological insight into our Syrian hamster transcriptome, we next identified over-expressed genes based on the number of individual reads that mapped to mouse-annotated genes (**Table 2**). We found that 20 mouse genes contained at least 600 x read depth, and 49 mouse genes contained at least 500 fold read depth.

Most of the mouse genes showing high read depth were annotated as being involved in fundamental cellular processes such as cell morphology and organization, cell cycle progression, cell function and maintenance, transcription, protein synthesis and turnover, cell death, and molecular transport. Genes associated with cell type or tissue-specific functions were not significantly over-represented, consistent with our method of generating cDNA reads from pooled, multiple organ tissues. Our aim in this study was to sequence and annotate a large number of hamster mRNA 3′ fragments as a preliminary effort towards generation of an expression array, our observation that the distribution of reads were across common cellular functions, suggests our assembly is not overly biased against a specific cell or tissue type.

We also performed a functional enrichment analysis of our Syrian hamster transcriptome. Based on the list of 9,562 mouse genes that were mapped to our contigs and singletons, we identified the over-represented canonical pathways in our library (**Table 3**). "Protein ubiquitination" (**Figure 3A**, p-value = 1.99E-18) and "molecular mechanisms of cancer" (**Figure 3B**, p-value = 5.01E-14) were the two most over-represented canonical pathways. However, there was also significant enrichment of many other canonical pathways related to biochemical, cellular, and disease-associated cellular processes. These included a multitude of signaling pathways, including RhoGTPase, protein kinase A, integrin, Rac, ERK/MAPK, mTOR, PI3K/Akt, PTEN, insulin, WNT/b-catenin, growth factor (VEGF, NGF, HGF, FGF, GM-CSF), and cellular junction signaling pathways. All of these pathways are biologically essential for intra- and intercellular communication and have known pleiotropic effects on transcription and translation, cellular proliferation, development, differentiation, cytoskeletal dynamics, cellular morphology, cell death, metabolism, and host responses to stress or infection. Consistent with this data, we also observed enrichment of functional categories associated with these biological activities (**Table 3**). The biological functions associated with "cardiovascular system development and function" (p-values range from 1.05E-03 to 4.15E-17) and "nervous system development and function" (p-values range from 1.29E-03 to 1.46E-19) were statistically over-represented.

Comparison with the Chinese hamster species

In order to contribute to the annotation efforts for the Chinese hamster (*Cricetulus griseus*) species, we aligned our Syrian hamster transcriptome to the current draft versions of the CHO cell genome and its transcriptome (**Table 1**).

We found that 7,845 fragments in our Syrian hamster transcriptome aligned to the CHO cell transcriptome (**Table 1**) and 85,652 aligned to the CHO cell genome (**Table S2**). On the other hand, 4,390 transcript fragments from the CHO cell dataset mapped to the Syrian hamster transcriptome (**Table 1**). Our aligned Syrian hamster transcriptome showed 85.14% similarity with the CHO cell transcriptome, an expectedly higher value than

what we saw for the same comparison with the mouse and rat transcriptomes.

Cross-species comparison

In order to obtain further insights about the genomic evolution of the Syrian hamster we aligned our Syrian hamster transcriptome to 14 other transcriptomes, all of which are publicly available on the Ensembl database [30] (**Table 1**). This compendium of transcriptome references included the human (*Homo sapiens*), chimpanzee (*Pan Troglodytes*), gorilla (*Gorilla gorilla gorilla*), macaque (*Macaca mulatta*), and orangutan (*Pongo abelii*) sequences, as well as the ferret (*Mustela putorius furo*), guinea pig sequences (*Cavia porcellus*), and pig (*Sus scrofa*). As expected, the greatest number of aligned sequences occurred with the mouse and rat species transcriptomes (**Table 1**). The human and the non-human primate species also showed high numbers of aligned sequences, possibly due to the current high quality assembly and annotation of those genomes. The CHO, ferret, pig, rabbit (*Oryctolagues cuniculus*), and squirrel (*Ictidomys tridecemlineatus*) species showed intermediary numbers of aligned sequences, while the guinea pig, kangaroo rat (*Dipodomys ordii*), pika (*Ochotona princeps*), shrew (*Sorex araneus*) and tree shrew (*Tupaia belangeri*) had the lowest numbers of aligned sequences. Of 174,278 Syrian hamster transcriptome fragments 50,433 aligned to at least one transcript reference while 61 fragments from our dataset aligned in common across all of these transcriptome references. Importantly, 76,175 of our Syrian hamster transcriptome fragments did not align to any of the 17 transcriptomes tested, nor to the CHO cell genome. It is important to note that some of the variability seen in our transcriptome comparisons may be due to differences in genome quality, assembly and annotation for the reference genomes tested.

Figure 4A is a distogram showing the results of our analysis of transcript sequences shared in common. The kangaroo rat, pika, shrew, and tree shrew had the lowest amount of commonly aligned sequences, amongst themselves and with the other species. The mouse and rat species showed the highest number of aligned sequences, presumably because of both their relatedness and genome quality/completeness.

We then investigated the evolutionary divergence between the Syrian hamster and the 13 species with the largest numbers of mapped sequences and the largest degrees of shared sequences (i.e. excluding the pika, kangaroo rat, shrew, tree shrew). We found that 611 transcriptomic fragments (**Table S3**) have been significantly aligned on the transcriptome references of these 13 most related species and we constructed a phylogenetic tree (**Figure 4B**).The Syrian hamster transcriptome branched most closely with the CHO genome as expected. The mouse and rat transcriptome clustered together and close to the Syrian hamster and CHO cluster, as expected. All the primate species formed a super group, while the ferret and pig transcriptomes clustered together as the rabbit and squirrel transcriptomes. Consistent with a recently published study [31], we observed that the genomic divergence between the Syrian and Chinese hamsters is comparable to the divergence seen between the rat and mouse. Also, as expected, we observed that the Guinea pig does not cluster with the rodent species [32,33].

Discussion

Here we present the assembly and analysis of a Syrian hamster transcriptome derived from the pooled RNAs from brains, lungs, spleens, kidneys, livers, and hearts of three adult females. The 3′ poly-T primed cDNAs that were sequenced on a long read-format

Table 2. List of the top 50 expressed genes in the library.

Ensembl Gene ID	Associated Gene Name	Description	Count
ENSMUSG00000028647	Mycbp	c-myc binding protein	1120
ENSMUSG00000020594	Pum2	pumilio 2 (Drosophila)	1017
ENSMUSG00000008575	Nfib	nuclear factor I/B	945
ENSMUSG00000022010	Tsc22d1	TSC22 domain family, member 1	895
ENSMUSG00000062078	Qk	quaking	861
ENSMUSG00000078578	Ube2d3	ubiquitin-conjugating enzyme E2D 3	795
ENSMUSG00000026621	Mosc1	MOCO sulphurase C-terminal domain containing 1	710
ENSMUSG00000028161	Ppp3ca	protein phosphatase 3, catalytic subunit, alpha isoform	707
ENSMUSG00000028790	Khdrbs1	KH domain containing, RNA binding, signal transduction associated 1	695
ENSMUSG00000006740	Kif5b	kinesin family member 5B	684
ENSMUSG00000031627	Irf2	interferon regulatory factor 2	682
ENSMUSG00000036781	Rps27l	ribosomal protein S27-like	660
ENSMUSG00000026655	Fam107b	family with sequence similarity 107, member B	658
ENSMUSG00000006373	Pgrmc1	progesterone receptor membrane component 1	652
ENSMUSG00000060961	Slc4a4	solute carrier family 4 (anion exchanger), member 4	641
ENSMUSG00000024750	Zfand5	zinc finger, AN1-type domain 5	639
ENSMUSG00000028788	Ptp4a2	protein tyrosine phosphatase 4a2	634
ENSMUSG00000019943	Atp2b1	ATPase, Ca++ transporting, plasma membrane 1	605
ENSMUSG00000097347	AC121292.1		603
ENSMUSG00000004980	Hnrnpa2b1	heterogeneous nuclear ribonucleoprotein A2/B1	600
ENSMUSG00000093904	Tomm20	translocase of outer mitochondrial membrane 20 homolog (yeast)	593
ENSMUSG00000068823	Csde1	cold shock domain containing E1, RNA binding	586
ENSMUSG00000020315	Spnb2	spectrin beta 2	579
ENSMUSG00000068798	Rap1a	RAS-related protein-1a	579
ENSMUSG00000020390	Ube2b	ubiquitin-conjugating enzyme E2B	570
ENSMUSG00000026064	Ptp4a1	protein tyrosine phosphatase 4a1	570
ENSMUSG00000020053	Igf1	insulin-like growth factor 1	569
ENSMUSG00000027706	Sec62	SEC62 homolog (S. cerevisiae)	553
ENSMUSG00000064373	Sepp1	selenoprotein P, plasma, 1	549
ENSMUSG00000014956	Ppp1cb	protein phosphatase 1, catalytic subunit, beta isoform	538
ENSMUSG00000007850	Hnrnph1	heterogeneous nuclear ribonucleoprotein H1	536
ENSMUSG00000031207	Msn	moesin	518
ENSMUSG00000020152	Actr2	ARP2 actin-related protein 2	515
ENSMUSG00000022261	Sdc2	syndecan 2	514
ENSMUSG00000047187	Rab2a	RAB2A, member RAS oncogene family	512
ENSMUSG00000004936	Map2k1	mitogen-activated protein kinase kinase 1	510
ENSMUSG00000026576	Atp1b1	ATPase, Na+/K+ transporting, beta 1 polypeptide	506
ENSMUSG00000022234	Cct5	chaperonin containing Tcp1, subunit 5 (epsilon)	504
ENSMUSG00000001175	Calm1	calmodulin 1	502
ENSMUSG00000069662	Marcks	myristoylated alanine rich protein kinase C substrate	490
ENSMUSG00000017776	Crk	v-crk sarcoma virus CT10 oncogene homolog (avian)	484
ENSMUSG00000038014	Fam120a	family with sequence similarity 120A	484
ENSMUSG00000036478	Btg1	B cell translocation gene 1, anti-proliferative	483
ENSMUSG00000027177	Hipk3	homeodomain interacting protein kinase 3	478
ENSMUSG00000043991	Pura	purine rich element binding protein A	474
ENSMUSG00000022283	Pabpc1	poly(A) binding protein, cytoplasmic 1	471
ENSMUSG00000031342	Gpm6b	glycoprotein m6b	471
ENSMUSG00000050608	Minos1	mitochondrial inner membrane organizing system 1	471
ENSMUSG00000018446	C1qbp	complement component 1, q subcomponent binding protein	469

Table 2. Cont.

Ensembl Gene ID	Associated Gene Name	Description	Count
ENSMUSG00000026568	Mpc2	mitochondrial pyruvate carrier 2	461

For each of the top 50 expressed genes in the library, based on the mouse annotations, the Ensembl mouse gene identified, the associated gene name, description, and the number of count (number of time that the genes have been mapped by the reads) are indicated.

Roche 454 were assembled into contigs, such that 22,105 of these contigs or singletons were annotated based on homology with both the mouse and rat transcriptomes, while 45,804 contigs or singletons were annotated based upon homology to one or the other mouse or rat transcriptomes. We identified 214 quasi-complete transcript sequences based on homology with mouse mRNAs and their annotations. In addition, we aligned our Syrian hamster transcriptome to the CHO cell transcriptome in order to further annotate our hamster species, and we observed a transcriptome similarity of 85.14% between the two.

When compared to a large compendium of transcriptome references, comprised of rodent, primate, and laurasiatheria species, using 661 Syrian hamster transcriptome fragments that aligned in common, the Syrian hamster transcriptome was found to be evolutionarily closest to the CHO genome and in close proximity to the mouse and rat species. The branch pattern and branch length between the Syrian and Chinese hamster transcriptomes was found to be similar to that observed between the mouse and the rat species. This observation was also described by Ryu et al. [31], but those previous efforts focused on mitochondrial gene sequences for their phylogeny analysis.

In the Syrian hamster transcriptome, we were able to identify a number of genes involved in a broad spectrum of fundamental biological processes. In addition to the 214 quasi-complete transcripts, identified based on mouse annotations and the most highly expressed transcripts, functional analysis of the entire set of sequence fragments in the Syrian hamster transcriptome that mapped to mouse genes revealed that a number of critical biological pathways are well-represented, including many related to key processes that are potentially perturbed or induced during infection. Among the most significantly enriched canonical pathways were several involved with protein synthesis, turnover, and antigen processing (protein ubiquitination, EIF2 signaling), metabolism and stress responses (mitochondrial dysfunction, NRF2-mediated oxidative stress response, PI3K/Akt, and mTOR signaling), and inflammatory and immune responses (production of NO and reactive oxygen species by macrophages, CXCR4 signaling, IL-1 signaling, and IL-3 signaling). The aim of this study

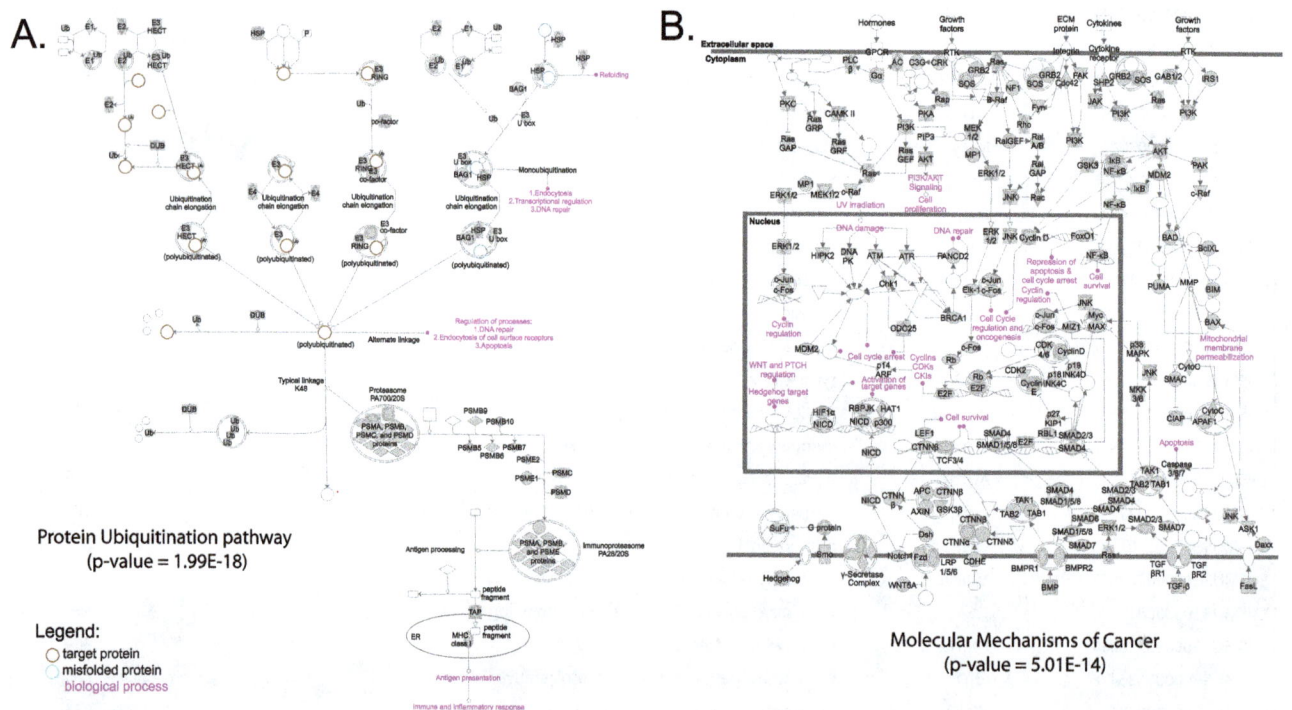

Figure 3. Schematic representation of the top two over-represented canonical pathways in our transcriptome assembly. (A) Representation of the "Protein Ubiquitination" canonical pathway. (B) Representation of the "Molecular Mechanisms of Cancer" canonical pathway. Both pathways have been generated based on mouse annotations. Transcripts involved in these pathways are indicated by different node shapes and associations are indicated by different edge shapes. Legends for the different nodes and edges are given in **Figure S1**. For both pathways, transcripts present in our library are indicated in gray. Associated p-values showing the statistical over-representation significance of the canonical pathways are also indicated.

Table 3. Functional enrichment of the mouse genes mapped by our transcriptome assembly.

Rank	Biological Function [p-value range]	Canonical pathway (p-value)
1	Organismal Surviva [1.11E-03 – 4.03E-26]	Protein Ubiquitination Pathway (1.99E-18)
2	Nervous System Development and Function [1.29E-03 – 1.46E-19]	Molecular Mechanisms of Cancer (5.01E-14)
3	Organ Morpholog [1.32E-03 – 4.20E-19]	Integrin Signaling (3.16E-13)
4	Tissue Morphology [1.08E-03 – 1.07E-18]	EIF2 Signaling (3.98E-12)
5	Cardiovascular System Development and Function [1.05E-03 – 4.15E-17]	Epithelial Adherens Junction Signaling (2.51E-11)

List of the top 5 biological functions and the top 5 canonical pathways found as statistically over-represented based on the list of 9,546 mouse genes mapped by our transcriptome assembly. The range of p-values is indicated for the biological functions and the p-value is indicated for each canonical pathways.

was to collect and annotate a large panel of transcripts regardless of tissue origin. These observations suggest that we have generated a representative transcriptome of the Syrian hamster. Therefore this transcriptome data could be used to generate a biologically meaningful first-generation expression DNA microarray for analysis of Syrian hamster response to disease, including those infectious agents known to alter immune and pro-inflammatory responses. Mechanisms of transcriptome regulation in the Syrian hamster, by way of these important pathways can now be monitored and analyzed further.

Only ~20% of the fragments in the Syrian hamster transcriptome aligned to the mouse and rat transcriptomes and even less aligned to the CHO cell transcriptome. This low percentage is due in part to species specificity, alignment stringency, but also to the fact that transcriptome references are far from being completely known and annotated. For instance, some classes of non-coding transcripts are now increasingly recognized as major components of regulation, and are widely expressed, but are poorly characterized and annotated. The transcriptome references that we used mainly contain known and annotated transcripts and our assembly may contain many expression contigs and singletons currently unknown and un-annotated in these other genomes.

The CHO cell genome is a useful tool for further improving the quality of our Syrian hamster transcriptome annotation for functional genomics work [24,25]. CHO cells have been used in a variety of genetic, cell biology, and pharmacology studies. They also are the mammalian cell line of choice for producing large quantities of recombinant proteins in large amounts or in or

industrial laboratory settings. Although Chinese and Syrian hamsters are phylogenetically distinct within the rodent subfamily *Cricetinae* [34,35], our data confirm that they are more closely related to one another as compared to other muroid rodents.

Through our work, we have increased the number of contig sequences available in the public domain for the Syrian hamster from 860 to 174,278, where 50,433 (28.93% of the Syrian hamster transcriptome) aligns to at least one transcriptome reference. Moreover, 85,652 (49.14% of the Syrian hamster transcriptome) fragments have aligned to the draft CHO genome, leading to an overall total of 98,103 (56.29%) annotated Syrian hamster transcripts. As a note, the work performed by Schmucki et al in [23] focused on transcriptome analysis of lipid metabolism in the golden hamster liver, and no contigs or other sequences have been released to the public domain to date.

With additional funding, future plans are in place for Illumina-based RNA sequencing using paired-end technology to add and improve on our current contig assembly. These efforts will improve our coverage of the Syrian hamster transcriptome, as well as permit more comprehensive and robust phylogenetic comparisons with other species. These combined efforts will lead to a better understand of the Syrian hamster transcriptome under a variety of infectious agent models related to human disease and pathogenesis.

Conclusions

The Syrian hamster is becoming an increasingly popular model for a variety of diseases, in particular, diseases known to infect non-

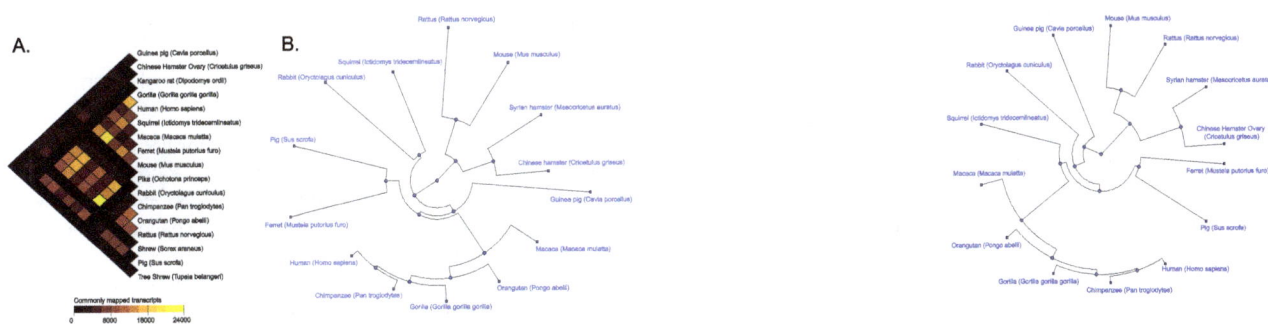

Figure 4. Distogram showing the commonly mapped transcripts and phylogenetic tree showing the divergences amongst the different species. (A) Distogram showing the number of transcripts commonly mapped by the Syrian hamster transcriptome between the different species used in this study. Each cell of the distogram represents the number of transcripts commonly mapped by two different species using a gradient color. (B) Phylogenetic tree showing the genomic divergence between a subset of the different species used in this study. Each leaf of the tree represents a different species and the distances of the edges are proportional to the genomic distances between the species. Genomic distances have been calculated based on the list of 611 Syrian hamster contigs and singletons that have been commonly aligned on the transcriptome references of the 13 species having the highest number of commonly aligned sequences.

human primates and humans. This Syrian hamster transcriptome discussed here represents a critical step forward in providing the tools necessary for advancing functional genomics in this important animal model.

Material and Methods

Animal housing

All hamsters were housed in individually ventilated cages (IVCs). All hamsters are co-housed, unless scientifically justified and approved by the Institutional Animal Care and Use Committee (IACUC) or deemed necessary for veterinary reasons. Housing density is determined by the guidelines outlined in the Guide for the Care and Use of Laboratory Animals and the Association for the Assessment and Accreditation of the Laboratory Animal Care, International (AAALAC). Food and sterile or acidified water were provided ad libitum. Hamster diets were consist of pellets containing a variety of foods such as grains and dried vegetables along with some seeds. Water was provided by either water bottles or water pouches. The light/dark cycle was 14 hours light, 10 hours dark.

RNA extraction

Three adult female Syrian hamsters were euthanized (exsanguinated while under isoflurane sedation) and six tissues – liver, lung, heart, brain, kidney, and spleen – were harvested from each hamster. All animal studies conformed to the guidelines set forth by the National Institutes of Health (NIH) and were reviewed and approved by the Institutional Animal Care and Use Committee (IACUC) at Rocky Mountain Laboratories, Division of Intramural Research, National Institute of Allergy and Infectious Diseases, NIH. One hundred mg of hamster tissue was homogenized with a Qiagen TissueLyzer II (Qiagen, Valencia, CA) in 1 mL Trizol (Invitrogen, Carlsbad, CA) following manufacturer's recommendations. To each aliquot 200 μL of 1-bromo-3-chloropropane (Sigma-Alrich) was added, the mixture was vortexed for 15 seconds and centrifuged at 4°C at 16,000x for 15 minutes. The aqueous phase was removed and passed through a Qiagen QiaShredder column to fragment remaining gDNA in the sample. The Qiagen AllPrep DNA/RNA 96 method was then performed including on-column Dnase 1 treatment to obtain high quality RNA with no genomic DNA contamination (Qiagen, Valencia, CA). RNA yield was determined by spectrophotometry (A260/A280) and RNA quality was determined using an Agilent 2100 Bioanalyzer (Agilent Technologies, Santa Clara, CA). The average RNA integrity number (RIN) for all 18 RNAs (3 animals times 6 tissues) was 6.4. An RNA aliquot from each organ of each animal was pooled and a total of 170 μg of RNA was prepared for sequencing.

Library construction and 454 sequencing.

The eighteen total RNA samples (6 tissues times 3 animals) were pooled equally into one pool. The total RNA pool underwent additional cleaning using the mirVana isolation kit following manufacturer's recommendations (Ambion). Poly A RNA cDNA was synthesized according to a standard protocol using an oligo(dT)-linker primer for first strand synthesis. The N0 cDNA was PCR amplified during 18 cycles using a high fidelity DNA polymerase. Normalization was carried out by one cycle of denaturation and renaturation of the cDNA, resulting in N1-cDNA. Reassociated ds-cDNA was separated from the remaining ss-can (normalized cDNA) by passing the mixture over a hydroxylapatite column. After hydroxylapatite chromatography, the ss-cDNA was amplified with 15 PCR cycles. For 454

sequencing, cDNA in the size range of 500–700 bps was eluted from a preparative agarose gel. An aliquot of the size fractionated cDNA was analyzed on a 1.5% agarose gel. 454 adaptors were ligated to the size fractionated N1 cDNA and 3' fragment sequenced on a Roche 454 using GS FLX technology with Titanium series chemistry following manufacturer's recommendations.

GS FLX sequencing generated 1,283,840 reads with an average length of 344 bases. Raw reads were trimmed for quality and reads shorter than 40 bases were discarded. The sequencing resulted in 1,212,395 reads of a total length 426,683,712 bases.

Library assembly

The trimmed and filtered reads was assembled using MIRA [36] (version 3) with the following parameters: mira –job = denovo,est,accurate,454 454_SETTINGS -CL:qc = no:cpat = yes -AL:mo =40:mrs =90. MIRA assembly produced 62,567 contigs and 125,228 singletons. There were 85 contigs and 13,432 singletons discarded due to poor quality (repetitive or poly-T sequence) or short read length (<50 bases), resulting in 62,482 contigs and 111,796 singletons for a total of 174,278 Syrian hamster transcriptome sequences totaling 60,117,204 bases.

Transcriptome and genome references

The transcriptome references used in this study were retrieved from the Ensembl Database [30] via the Biomart interface. Transcriptome references used in this study were obtained from the release 71 of the Ensembl database. The draft version of the CHO genome and transcriptome were retrieved from the Pre Ensembl website.

Alignments

Syrian hamster transcriptome sequences were aligned to transcriptome references using BLAST [37]. An Expect value cutoff parameter of 10 was used and alignments results were filtered in order to only keep sequences aligned at least at 80%.

Similarities of the assembled library with the transcriptome references

The similarities to the mouse, rat and other transcriptome references were calculated based on BLAST results. For all Syrian hamster transcriptome sequences that aligned to the transcriptomes, we calculated the ratio between the total number of correct nucleotide matches and the total combined length of our Syrian hamster transcriptome, which is 60,117,204 bases.

Identification of over-represented canonical pathways and biological functions

Functional enrichment of canonical pathways and biological functions was performed using Ingenuity Pathways Analysis (Ingenuity Systems, Inc.). Canonical pathways refer to pathways curated by Ingenuity as part of its knowledgebase, based on extensive characterization in the peer-reviewed literature published using human, mouse, and rat experimental models. These typically represent common properties of a particular signaling module, mechanism, or pathway. IPA examines differentially expressed transcripts in the context of known biological functions, mapping each gene identifier to its corresponding molecule in the Ingenuity Pathways Knowledge Base (IPKB). For all analyses, the p-values – representing the statistical over-representation significance – were generated using the right-tailed Fisher's Exact Test [38] and were adjusted using the Benjamini-Hochberg Multiple Testing correction [39].

Distrogram construction

The distogram represented in figure 4A was constructed using the "squash" package [40] of the R suite [41]. This representation is a color-coded, rotated triangular matrix indicating the distance between every pair of species in term of number of aligned sequences shared.

Phylogenetic tree construction

To construct the phylogenetic tree, we used the 611 transcriptomic sequences that have been significantly aligned on the transcriptome reference of the 13 species having the largest numbers of mapped sequences and the largest degrees of shared sequences. Sequences and matched transcripts were aligned using the Needleman-Wunsch multiple alignment algorithm [42], using the multialign function in MATLAB (open and extend gap penalties have been taken into consideration). Genomic divergences between sequences were calculated using the Jukes-Cantor method [43], based on the 'NUC44' scoring matrix. Indel mismatches have not been taken into consideration for the computation of genomic divergences. The phylogenetic tree was constructed by using the neighbor-joining method (NJ) [44].

Supporting Information

Figure S1. Legend for the IPA canonical pathways representations. Figure showing the annotations of the different node and edge shapes in the representations of the canonical pathways obtained from Ingenuity Pathway Analysis (IPA).

Table S1 List of highly-covered transcripts by the Syrian hamster transcriptome. Table showing the 214 highly-covered transcripts (transcripts mapped at least at 90% by the contigs and singletons) based on the mouse annotations. For each highly-covered transcript, the Ensembl Mouse gene identified, the associated gene name as well as the gene description is indicated.

Table S2 Alignment positions of the Syrian hamster transcriptome over the Chinese Hamster Ovary cell genome. Table showing the alignment positions of the Syrian hamster transcriptome sequences to the draft of the Chinese hamster genome. We found that 85,652 contigs and singletons of our library have been aligned on the Chinese Hamster Ovary cell genome draft. For each aligned Syrian hamster transcriptome sequence, the CHO genome segment, the start alignment position, the end alignment position, and the strand are indicated.

Table S3 List of sequences used to infer the phylogenic tree. Table providing the list of contigs and singletons used to construct the phylogenic tree shown in figure 4. All the 611 contigs and singletons in this table have been significantly aligned on the transcriptomes of species having the highest number of commonly aligned sequences.

Acknowledgments

The authors thank Marcus J. Korth for valuable feedback on the manuscript.

Author Contributions

Conceived and designed the experiments: DS ALR HF HE MGK. Performed the experiments: KV DS. Analyzed the data: NT CM. Contributed reagents/materials/analysis tools: DS KV SFP. Wrote the paper: NT DS ALR CM SFP HE.

References

1. Wahl-Jensen V, Bollinger L, Safronetz D, de Kok-Mercado F, Scott DP, et al. (2012) Use of the Syrian hamster as a new model of ebola virus disease and other viral hemorrhagic fevers. Viruses 4: 3754–3784. doi:10.3390/v4123754.
2. Xiao SY, Guzman H, Zhang H, Travassos da Rosa AP, Tesh RB (n.d.) West Nile virus infection in the golden hamster (Mesocricetus auratus): a model for West Nile encephalitis. Emerg Infect Dis 7: 714–721. doi:10.3201/eid0704.010420.
3. Tesh RB, Guzman H, da Rosa AP, Vasconcelos PF, Dias LB, et al. (2001) Experimental yellow fever virus infection in the Golden Hamster (Mesocricetus auratus). I. Virologic, biochemical, and immunologic studies. J Infect Dis 183: 1431–1436. doi:10.1086/320199.
4. Aguilar P V, Barrett AD, Saeed MF, Watts DM, Russell K, et al. (2011) Iquitos virus: a novel reassortant Orthobunyavirus associated with human illness in Peru. PLoS Negl Trop Dis 5: e1315. doi:10.1371/journal.pntd.0001315.
5. Shinya K, Makino A, Tanaka H, Hatta M, Watanabe T, et al. (2011) Systemic dissemination of H5N1 influenza A viruses in ferrets and hamsters after direct intragastric inoculation. J Virol 85: 4673–4678. doi:10.1128/JVI.00148-11.
6. Roberts A, Lamirande EW, Vogel L, Jackson JP, Paddock CD, et al. (2008) Animal models and vaccines for SARS-CoV infection. Virus Res 133: 20–32. doi:10.1016/j.virusres.2007.03.025.
7. Safronetz D, Zivcec M, Lacasse R, Feldmann F, Rosenke R, et al. (2011) Pathogenesis and host response in Syrian hamsters following intranasal infection with Andes virus. PLoS Pathog 7: e1002426. doi:10.1371/journal.ppat.1002426.
8. Zivcec M, Safronetz D, Haddock E, Feldmann H, Ebihara H (2011) Validation of assays to monitor immune responses in the Syrian golden hamster (Mesocricetus auratus). J Immunol Methods 368: 24–35. doi:10.1016/j.jim.2011.02.004.
9. Ebihara H, Zivcec M, Gardner D, Falzarano D, LaCasse R, et al. (2013) A Syrian golden hamster model recapitulating ebola hemorrhagic fever. J Infect Dis 207: 306–318. doi:10.1093/infdis/jis626.
10. Gomes-Silva A, Valverde JG, Ribeiro-Romão RP, Plácido-Pereira RM, DA-Cruz AM (2013) Golden hamster (Mesocricetus auratus) as an experimental model for Leishmania (Viannia) braziliensis infection. Parasitology: 1–9. doi:10.1017/S0031182012002156.
11. Briand F (2010) The use of dyslipidemic hamsters to evaluate drug-induced alterations in reverse cholesterol transport. Curr Opin Investig Drugs 11: 289–297.
12. Castro-Perez J, Briand F, Gagen K, Wang S-P, Chen Y, et al. (2011) Anacetrapib promotes reverse cholesterol transport and bulk cholesterol excretion in Syrian golden hamsters. J Lipid Res 52: 1965–1973. doi:10.1194/jlr.M016410.
13. Morin LP, Wood RI (2001) A stereotaxic atlas of the golden hamster brain. Academic Press San Diego:
14. Van Hoosier GL, McPherson CW (1987) The Laboratory Hamsters. Access Online via Elsevier.
15. Monecke S, Brewer JM, Krug S, Bittman EL (2011) Duper: a mutation that shortens hamster circadian period. J Biol Rhythms 26: 283–292. doi:10.1177/0748730411411569.
16. Nigro V, Okazaki Y, Belsito A, Piluso G, Matsuda Y, et al. (1997) Identification of the Syrian hamster cardiomyopathy gene. Hum Mol Genet 6: 601–607.
17. Ricci LA, Schwartzer JJ, Melloni RH (2009) Alterations in the anterior hypothalamic dopamine system in aggressive adolescent AAS-treated hamsters. Horm Behav 55: 348–355. doi:10.1016/j.yhbeh.2008.10.011.
18. Chelini MOM, Palme R, Otta E (2011) Social stress and reproductive success in the female Syrian hamster: endocrine and behavioral correlates. Physiol Behav 104: 948–954. doi:10.1016/j.physbeh.2011.06.006.
19. Rawji KS, Zhang SX, Tsai Y-Y, Smithson LJ, Kawaja MD (2013) Olfactory ensheathing cells of hamsters, rabbits, monkeys, and mice express α-smooth muscle actin. Brain Res 1521: 31–50. doi:10.1016/j.brainres.2013.05.003.
20. Boguski MS, Lowe TM, Tolstoshev CM (1993) dbEST–database for "expressed sequence tags". Nat Genet 4: 332–333. doi:10.1038/ng0893-332.
21. Oduru S, Campbell JL, Karri S, Hendry WJ, Khan SA, et al. (2003) Gene discovery in the hamster: a comparative genomics approach for gene annotation by sequencing of hamster testis cDNAs. BMC Genomics 4: 22. doi:10.1186/1471-2164-4-22.
22. Landkocz Y, Poupin P, Atienzar F, Vasseur P (2011) Transcriptomic effects of di-(2-ethylhexyl)-phthalate in Syrian hamster embryo cells: an important role of early cytoskeleton disturbances in carcinogenesis? BMC Genomics 12: 524. doi:10.1186/1471-2164-12-524.

23. Schmucki R, Berrera M, Küng E, Lee S, Thasler WE, et al. (2013) High throughput transcriptome analysis of lipid metabolism in Syrian hamster liver in absence of an annotated genome. BMC Genomics 14: 237. doi:10.1186/1471-2164-14-237.

24. Xu X, Nagarajan H, Lewis NE, Pan S, Cai Z, et al. (2011) The genomic sequence of the Chinese hamster ovary (CHO)-K1 cell line. Nat Biotechnol 29: 735–741. doi:10.1038/nbt.1932.

25. Hammond S, Swanberg JC, Kaplarevic M, Lee KH (2011) Genomic sequencing and analysis of a Chinese hamster ovary cell line using Illumina sequencing technology. BMC Genomics 12: 67. doi:10.1186/1471-2164-12-67.

26. Lewis NE, Liu X, Li Y, Nagarajan H, Yerganian G, et al. (2013) Genomic landscapes of Chinese hamster ovary cell lines as revealed by the Cricetulus griseus draft genome. Nat Biotechnol 31: 759–765. doi:10.1038/nbt.2624.

27. Liu H, Wang T, Wang J, Quan F, Zhang Y (2013) Characterization of liaoning cashmere goat transcriptome: sequencing, de novo assembly, functional annotation and comparative analysis. PLoS One 8: e77062. doi:10.1371/journal.pone.0077062.

28. Ji P, Liu G, Xu J, Wang X, Li J, et al. (2012) Characterization of common carp transcriptome: sequencing, de novo assembly, annotation and comparative genomics. PLoS One 7: e35152. doi:10.1371/journal.pone.0035152.

29. Nagaraj SH, Gasser RB, Ranganathan S (2007) A hitchhiker's guide to expressed sequence tag (EST) analysis. Brief Bioinform 8: 6–21. doi:10.1093/bib/bbl015.

30. Flicek P, Ahmed I, Amode MR, Barrell D, Beal K, et al. (2013) Ensembl 2013. Nucleic Acids Res 41: D48–55. doi:10.1093/nar/gks1236.

31. Ryu SH, Kwak MJ, Hwang UW (2013) Complete mitochondrial genome of the Eurasian flying squirrel Pteromys volans (Sciuromorpha, Sciuridae) and revision of rodent phylogeny. Mol Biol Rep 40: 1917–1926. doi:10.1007/s11033-012-2248-x.

32. D'Erchia AM, Gissi C, Pesole G, Saccone C, Arnason U (1996) The guinea-pig is not a rodent. Nature 381: 597–600. doi:10.1038/381597a0.

33. Graur D, Hide WA, Li WH (1991) Is the guinea-pig a rodent? Nature 351: 649–652. doi:10.1038/351649a0.

34. Romanenko SA, Volobouev VT, Perelman PL, Lebedev VS, Serdukova NA, et al. (2007) Karyotype evolution and phylogenetic relationships of hamsters (Cricetidae, Muroidea, Rodentia) inferred from chromosomal painting and banding comparison. Chromosome Res 15: 283–297. doi:10.1007/s10577-007-1124-3.

35. Trifonov VA, Kosyakova N, Romanenko SA, Stanyon R, Graphodatsky AS, et al. (2010) New insights into the karyotypic evolution in muroid rodents revealed by multicolor banding applying murine probes. Chromosome Res 18: 265–275. doi:10.1007/s10577-010-9110-6.

36. Chevreux B, Pfisterer T, Drescher B, Driesel AJ, Müller WEG, et al. (2004) Using the miraEST assembler for reliable and automated mRNA transcript assembly and SNP detection in sequenced ESTs. Genome Res 14: 1147–1159. doi:10.1101/gr.1917404.

37. Altschul SF, Gish W, Miller W, Myers EW, Lipman DJ (1990) Basic local alignment search tool. J Mol Biol 215: 403–410. doi:10.1016/S0022-2836(05)80360-2.

38. Fisher RA (1922) On the Interpretation of χ2 from Contingency Tables, and the Calculation of P. J R Stat Soc 85: 87–94. doi:10.2307/2340521.

39. Benjamini Y, Hochberg Y (1995) Controlling the False Discovery Rate: A Practical and Powerful Approach to Multiple Testing. J R Stat Soc Ser B 57: 289–300. doi:10.2307/2346101.

40. Eklund A (2012) squash: Color-based plots for multivariate visualization. Available: http://cran.r-project.org/package=squash.

41. R Development Core Team R (2011) R: A Language and Environment for Statistical Computing. R Found Stat Comput 1: 409. doi:10.1007/978-3-540-74686-7.

42. Needleman SB, Wunsch CD (1970) A general method applicable to the search for similarities in the amino acid sequence of two proteins. J Mol Biol 48: 443–453.

43. Jukes TH, Cantor CR (1969) Evolution of Protein Molecules. Munro HN, editor Academy Press.

44. Saitou N, Nei M (1987) The neighbor-joining method: a new method for reconstructing phylogenetic trees. Mol Biol Evol 4: 406–425.

45. Kodama Y, Shumway M, Leinonen R (2012) The Sequence Read Archive: explosive growth of sequencing data. Nucleic Acids Res 40: D54–6. doi:10.1093/nar/gkr854.

Novel Biogenic Aggregation of Moss Gemmae on a Disappearing African Glacier

Jun Uetake[1,2]*, Sota Tanaka[3], Kosuke Hara[4], Yukiko Tanabe[5], Denis Samyn[6], Hideaki Motoyama[3], Satoshi Imura[3], Shiro Kohshima[7]

1 Transdisciplinary Research Integration Center, Minato-ku, Tokyo, Japan, 2 National Institute of Polar Research, Tachikawa, Tokyo, Japan, 3 Faculty of Science, Chiba University, Chiba, Chiba, Japan, 4 Graduate School of Science, Kyoto University, Kyoto, Japan, 5 Institute for Advanced Study, Waseda University, Shinjuku-ku, Tokyo, Japan, 6 Department of Mechanical Engineering, Nagaoka University of Technology, Nagaoka, Nigata, Japan, 7 Wildlife Research Center, Kyoto University, Kyoto, Kyoto, Japan

Abstract

Tropical regions are not well represented in glacier biology, yet many tropical glaciers are under threat of disappearance due to climate change. Here we report a novel biogenic aggregation at the terminus of a glacier in the Rwenzori Mountains, Uganda. The material was formed by uniseriate protonemal moss gemmae and protonema. Molecular analysis of five genetic markers determined the taxon as *Ceratodon purpureus*, a cosmopolitan species that is widespread in tropical to polar region. Given optimal growing temperatures of isolate is 20–30°C, the cold glacier surface might seem unsuitable for this species. However, the cluster of protonema growth reached approximately 10°C in daytime, suggesting that diurnal increase in temperature may contribute to the moss's ability to inhabit the glacier surface. The aggregation is also a habitat for microorganisms, and the disappearance of this glacier will lead to the loss of this unique ecosystem.

Editor: Lucas C.R. Silva, University of California Davis, United States of America

Funding: This study was partially supported by Ministry of Education, Culture, Sports, Science and Technology Grant-in-Aid for Scientific Research (A) No. 22241005 and by Proposal for Seeds of Transdisciplinary Research from the Transdisciplinary Research Integration Center, and by an NIPR publication subsidy. The funders had no role in study design, data collection and analysis, decision to publish, or preparation of the manuscript.

Competing Interests: The authors have declared that no competing interests exist.

* Email: juetake@nipr.ac.jp

Introduction

Many psychrophilic and psychrotolerant microorganisms inhabit supraglacial environments, which have recently been recognized as an important biome [1]. Cryonite granules, dark spherical aggregates typically 1 mm in diameter have been frequently observed on ablation zones of glaciers in many parts of the world [2]. These cryoconites consist of mineral particles, organic matter, and microorganisms, which are mainly formed by the aggregation of filamentous cyanobacteria [3]. These cryoconite harbor a diverse range of microorganisms, and studies of their molecular diversity have revealed microbial communities of bacteria [4] and archaea [5,6], as well as algae, fungi, amoebas, and invertebrates such as tardigrades [5]. These microbial communities play an important role in the carbon and nitrogen cycles on the glacier [1,7].

Other types of biological aggregations have been reported from supraglacier ecosystems in Iceland and Alaska, namely, globular moss aggregations known as 'glacier mice' [8,9] or 'moss polster' [10]. These are lenticular moss cushions (0.02 to 0.1 m in diameter) and are composed of a moss envelope covering an internal clast formed from glacial sediment and airborne particles [11]. These moss cushions are expected to impact the ecology and nutrient cycle of the supraglacial ecosystem [11], and also provide a favorable habitat for a variety of invertebrates, including Collembola, Tardigrades, and Nematoda [12].

Previous biological studies have frequently examined mid-latitude and polar glaciers, however, the tropical glaciers are have been studied rarely, except for New Guinea [13]. In equatorial Africa, glaciers persist in three major mountain regions (Mt. Kilimanjaro in Tanzania, Mt. Kenya in Kenya, and the Rwenzori Mountains in Uganda), which have not been previously been targeted in surveys of glacier biology. The Rwenzori glaciers are shrinking rapidly and are expected to disappear by 2020 due to climatic warming [14] and/or lowered humidity and lowered cloudiness [15], as measured by aerial photography and satellite imagery [14].

During a biological field survey on a glacier near the summit of Mt. Stanley, the highest peak in Uganda and in the Rwenzoris, we found a large, black bioaggregation (average long and short axes: 18.1 mm and 12.7 mm) in the supraglacial environment, greater than cryoconites. Examination revealed that these granular structures were formed by filamentous moss gemmae and protonema, not cyanobacteria. This is the first report to describe this habitat for such a structure, which we classified as a "glacial moss gemmae aggregation" (GMGA). In order to identify the material we measured the structure (size and mass) and isolated the dominant moss species using both culture and molecular techniques. Furthermore, we examined the photosynthetic activity of isolates under various temperature and radiation conditions.

Materials and Methods

Glacier characteristics and sampling

The Rwenzori Mountains (5109 m above sea level) contain the third-highest mountain in Africa, and are straddling the equator along the border of Uganda and the Democratic Republic of the Congo (Fig. 1). Since the LIA (Lac Gris Stage: 19th centry or just before), glaciers in the Rwenzoris have been shrinking; in 1906 the Elena glacier was estimated to cover 6.5 km^2 [16] and by 2003 it had decreased to approximately 1 km^2 [14]. During this period, glaciers on Mts. Emin, Gessi, and Luigi disappeared completely, leaving only glaciers remaining on three major peaks: Mts. Speke, Baker, and Stanley. In this study, we surveyed Stanley Plateau, the largest glacier on Mt. Rwenzori. Stanley Plateau is a flat sloped glacier that flows from Mt. Stanley's Alexandra Peak, and is around 1 km long and 0.1–0.3 km wide (Fig. 2a).

In February 2012 and 2013, we collected surface ice samples, including biological debris, at three sites: ST1 (N00°22′31.3″, E29°52′40.26″), ST2 (N00°22′34.74″, E29°52′37.2″) and ST3 (N00°22′52.32″, E29°52′24.6″). At each site, 5 samples were collected from different 0.1×0.1 m areas and stored in 50 ml plastic bottles for cell counts, and 5 samples from different areas (samples size not measured) and placed in 8 ml plastic bottles for DNA analysis and isolation. In the glacier foreland, located about 10 m from ST1, we collected shoots of bryophyte on dried GMGAs and placed them in 8 ml plastic bottles with RNAlater (Life Technologies, Carlsbad, USA). All samples were collected using pre-cleaned stainless steel scoops and spoons.

Samples for cell counts were fixed with 3% formaldehyde and stored at room temperature. All other samples were kept cold around 0°C in large stainless steel vacuum flasks with glacial ice samples until transport to Kasese, Uganda, the closest city to Rwenzori Mountains National Park. There, samples for molecular analysis were kept frozen around −20°C and samples for isolation were kept cold around 0°C, until they could be transported to the lab for analysis at the National Institute of Polar Research (Tokyo, Japan).

In the field, the internal temperature of the GMGAs at ST1 was measured using a waterproof temperature logger (R-52i; T&D, Matsumoto, Japan between 9–13 Feb. 2013) with 0.6 m sensor probe (TR-5106; T&D, Matsumoto, Japan). Probes were inserted into the center of two GMGAs and were monitored by camera (Optio WG-2: Ricoh, Tokyo, Japan) at intervals to ensure that the measuring apparatus was not disturbed.

Microscopic observation of biological materials on ice surface

The samples were cold-preserved prior to isolation and identification of species. After formaldehyde fixation, 0.1–0.4 ml of 12–60 fold diluted samples were filtered through a hydrophilic polytetrafluoroethylene membrane (Omnipore JGWP01300; Merck Millipore, Billerica, USA) with diameter 13 mm and pore size 0.2 μm. We observed and counted cell concentrations from one-quarter of the membrane, using a fluorescent microscope (IX71 and 81; Olympus, Tokyo, Japan).

18S r RNA gene molecular cloning

DNA of approximately 0.3 g was extracted from samples ST1, ST2, and ST3 using the Fast DNA SPIN Kit for soil (MP Biomedicals, Santa Ana, USA) according to the manufacturer's instructions. Extracted DNA was diluted to 1.62 ng/μl with water (Ambion Nuclease-Free Water; Life Technologies, Carlsbad, USA). Five aliquots from each site were combined for DNA amplification. Thermal cycling was performed with an initial denaturation step at 98°C for 3 min, followed by 25 cycles of denaturation at 98°C for 10 s, annealing at 55°C for 30 s, and elongation at 72°C for 1.5 min, using Ex Taq HS DNA polymerase (Takara, Shiga, Japan) and the primer pair of Euk A (5′-ACCTGGTTGATCCTGCCAGT-3′) and EukB (5′-GATCCTTCTGCAGGTTCACCTAC′). Cycling was completed by a final elongation step at 72°C for 3 min. The PCR-amplified DNA fragments were cloned into the pCR4 vector of the TOPO TA cloning kit (Invitrogen, Carlsbad, USA). Clones obtained from the libraries were sequenced using the 3130×l Genetic Analyzer (Life technologies, Carlsbad, USA) at the National Institute of Polar Research. All sequences were assembled using CodonCode Aligner (CodonCode Corporation, Centerville, USA) and assembled full-length sequences of 18S rRNA were aligned with the eukaryotic Silva database [17] using mothur ver. 1.27.0 [18]. Tentative chimeric sequences were removed using both the reference and *de novo* modes of Uchime [19] implemented in mothur software package. All good-quality sequences with more than 97% similarity were clustered into operational taxonomic units (OTU).

Isolation of moss and molecular identification

Fragments of cold-preserved GMGA samples were inoculated in liquid Bold's basal medium (BBM) [20] in a laminar flow bench and incubated at 4°C for 1 month. Protonemata that grew directly from observed gemmae (Fig. 2 c,d) were transplanted to fresh BBM liquid medium and a 1-month-incubation was repeated. Isolated protonemata were kept in BBM and 1.5% agar medium before extraction and analysis. DNA of a single cluster of protonema in liquid medium was extracted with the Fast DNA SPIN Kit for soil, and 4 different regions (18S rRNA; chloroplast genes, *trn*L, *rps*4 and *atpB-rbc*L intergenic spacer; and mitochondria gene, *nad*5) were amplified by using Ex Taq HS DNA polymerase (Takara, Shiga, Japan). Thermal cycling for 18S rRNA was carried out following Remias *et al.* [21], with 35 cycles of denaturation at 98°C for 10 s, annealing at 54°C for 30 s, and elongation at 72°C for 1 min 45 s, using the primer pair NS1 (5′-GTAGTCATATGCTTGTCTC-3′) and 18L(5′-CACCTACG-GAAACCTTGTTACGACTT-3′). For chloroplast *trn*L, thermal cycling was performed with 35 cycles of denaturation at 98°C for 10 s, annealing at 60°C for 30 s, and elongation at 72°C for 1 min 45 s and primer pair trnC (5′- CGAAATCGGTAGACGC-TACG-3′) and trnF (5′-ATTTGAACTGGTGACACGAG-3′), following Taberlet *et al.* [22]. For chloroplast *rps*4, thermal cycling was performed according to Nadot *et al.* [23] and Souza-Chies *et al.* [24], with 35 cycles of denaturation at 98°C for 10 s, annealing at 55°C for 30 s, and elongation at 72°C for 1 min 45 s using primer pair rps5 (5′-ATGTCCCGTTATCGAGGACCT-3′) and trnS (5′-TACCGAGGGTTCGAATC-3′). For chloroplast *atpB-rbc*L intergenic spacer, thermal cycling was performed according to Chiang *et al.* [25], with 35 cycles of denaturation at 98°C for 10 s, annealing at 49°C for 30 s, and elongation at 72°C for 30 s using primer pair *atpB*-1 (5′-ACATCKAR-TACKGGACC-3′) and *rbc*L-1 (5′-AACACCAGCTTTRAATC-CAA-3′). Lastly, for mitochondria nad5, thermal cycling was performed according to Shaw *et al.* [26], with 35 cycles of denaturation at 98°C for 10 s, annealing at 52°C for 30 s, and elongation at 72°C for 1 min 45 s using primer set nad5F4 (5′-GAAGGAGTAGGTCTCGCTTCA-3′) and nad5R3 (5′-AAAACGCCTGCTGTTACCAT-3′). Some of shoots of Bryophyta on dried GMGAs were picked up by tweezers and DNA was analyzed by same method as isolated protonema.

Figure 1. Location map of Rwenzori Mountains in Republic of Uganda.

Photosynthetic rate of GMGA and isolate

Photosynthetic rate at 7 different incubation temperatures (5, 10, 15, 20, 25, 30, and 40°C) was measured using a pulse amplitude modulation (PAM) fluorometer (Water-PAM, Waltz, Effeltrich, Germany) with Win-control software for control and analysis following Tanabe *et al.* [27]. PAM fluorometer is useful to measure the electron transport rate (ETR) of isolate underdifferent incubation factor [27]. For incubation temperatures of 5, 10, and 15°C, photosynthetically active radiation (PAR) intensities were 3, 64, 94, 144, 215, 305, 422, 687, and 1000 mmol photons/m^2/s. After these measurments, we had changed to another PAM device, because this device is obviously unstable only under 40°C due to mechanical trouble. Then, we used another device and measured again from 20°C. Results of 20, 25, 30°C are almost same as previous analysis, and measurement was stable at 40°C in next time. For incubation temperatures of 20, 25, 30, and 40°C, PAR intensities were 8, 62, 92, 140, 209, 297, 412, 674, and 986 mmol photons/m^2/s. After a 30 s exposure, a saturating pulse of > 2000 mmol photons/m^2/s was applied for 0.4 sat 5°C in a temperature-controlled incubator. The gain value of the photo-electric multiplier (PM-Gain) was set to 3 for all measurements. After incubation of each sample at 5°C for 60 min in dark conditions, a tissue sample of GMGA and isolated protonemata were transferred to the quartz cuvette of the fluorometer. After measurement, incubation temperature was raised by 5°C to 10°C and incubated for 1 h, after which the temperature was raised by 5°C again and incubated for 1 h repeatedly until incubation temperature was 40°C. Light curves were obtained by running a rapid light curve protocol in Win-control software. The photosynthetic rate expressed as relative electron transport rate *rETR* [28] was as follows:

Figure 2. Research site and glacier moss gemmae aggregation (GMGA). a) Stanley Plateau glacier and Margarita Peak from Mt. Baker, b) Glacier ice surface covered by GMGAs, c) GMGAs (grid cells beneath GMGA are 1×1 mm), d) Cross section of GMGA (scale bar: 2 mm).

$$rETR = (Fm\text{-}Fm')/Fm' \times PAR. \qquad (1)$$

Here, F and Fm' are the transient and maximum fluorescence levels at certain actinic light intensities at a given time and (Fm'–F)/Fm' indicates Photosystem II (PSII) yield. Non-photochemical quenching (NPQ) was calculating by the following equation:

$$NPQ = (Fm\text{-}Fm')/Fm' \times PAR, \qquad (2)$$

where Fm is the maximum fluorescence level of non-illuminated samples.

Ethics Statement

Uganda Wildlife Authority and Uganda National Council for Science and Technology authorized all field researchs in Rwenzori Mountains National Park.

Results

Morphological features of GMGA

We found ellipsoidal blackish bioaggregations covering the glacier surface at ST1 (Fig. 2b). We collected and sampled 96 of these bioaggregations, as well as measured their long and short axes, thickness, and mass (Fig. 2c,d). The average long axis, short axis, thickness, and mass were 18.7 mm, 12.7 mm, 8.3 mm, and 1.6 g, respectively (Fig. 3), and these aggregations were clearly larger than cryoconites (average diameter: 1.1 mm; 3). Short axis length was well correlated with long axis length ($R^2 = 0.705$), but not as well correlated with thickness and mass ($R^2 = 0.543$ and 0.616, respectively). This means that the structure of this

bioaggregation is not spherical but is instead flattened. The bioaggregation was composed of many gemmae and protonema. The gemmae were germinating and developing filamentous protonema, and gemmae were formed repeatedly on protonema. Many moss gemmae were observed especially on the surface of the bioaggregation (Fig. 4a,b), from the top 1–2 mm of the cross section (Fig. 2d). The main framework of these structures was formed by moss gemmae, so we named this structure as "glacial moss gemmae aggregation (GMGA)".

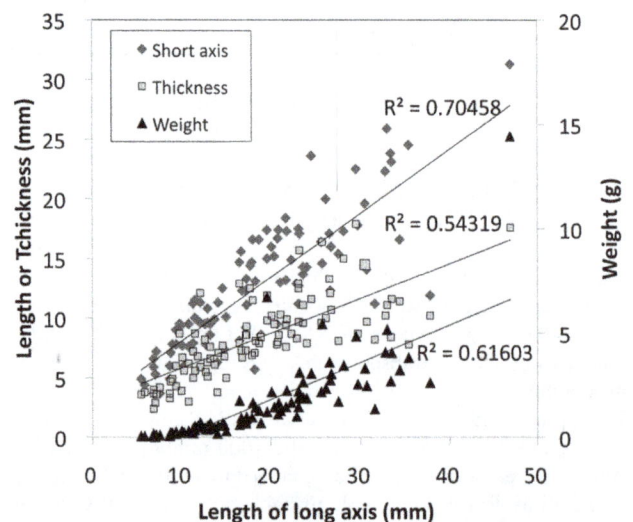

Figure 3. Size (long and short axes, thickness) and mass distribution of GMGAs.

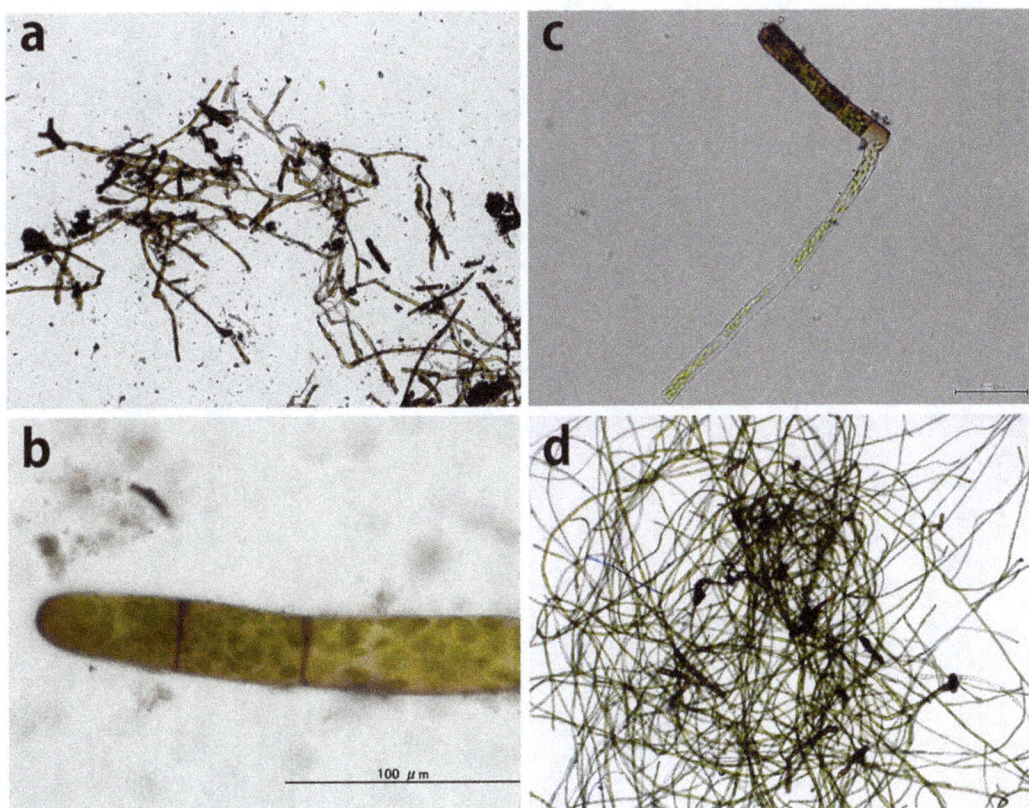

Figure 4. Cells of moss gemmae and protonemata. a,b) Moss protonemal cells formed the main frame of the GMGA (scale bar: 100 μm), c) Moss protonema grew from gemmae below 4°C (scale bar: 100 μm), d) protonemal cells for molecular identification after incubation below 4°C for 1 month in liquid Bold's basal medium.

The gemmae are filamentous, composed of 1–2 rows of 2–20 cells with slightly thickened brownish cell walls, 100–200 um long in maximum. These morpholocial characteristics were well agree with rhizoidal gemmae of cosmopolitan moss, *Ceratodon purpureus* (Hedw.) Brid. described by Imura and Kanda (1986) based on Antarctic specimens [29].

Molecular identification of moss species

We obtained a total of 81 clones from the GMGAs by 18S rRNA gene PCR-cloning. Sixty-three clones (77.8%) were clustered into the same OTU (AB858433: Table 1). The remaining 18 clones were of cercozoa, green algae, and fungi.

The 18S rRNA, *rps*4, *trn*L, *atp*B-*rbc*L intergenic spacer and *nad*5 gene sequence of the isolated protonemata (Fig. 4c,d) that grew from the observed gemmae and shoots on dried GMGA were summarized in Table 1. These high-percentage matches (more than 99.9% similarity) from five different regions of the protonemata show that the isolated moss is indeed *C. purpureus*. Also results from four different regions of shoots on dried GMGA show that this specie belonging to genus: *Bryum*, however we could not identify species level from these regions.

Optimum temperature and PAR of GMGA and isolated protonemata of *C. purpureus*

Internal (center) temperature changes was measured during the 2013 field season of two *in situ* GMGAs in ST1 and one dried GMGA found on a rock in glacier foreland (Fig. 5). Temperature change data show clear diurnal cycles with daily exposure to below-freezing temperatures daily. Maximum temperatures

reached 8–10°C for *in situ* GMGAs and above 20°C for dried GMGA. The daytime increase in temperature was due to absorption of thermal radiation, but was variable due to decreases in radiation from frequent cloud cover and cooling by glacier ice.

A photosynthetic light curve was measured using a PAM fluorometer under different temperatures of GMGA and the two isolates (Fig. 6). The ETR of GMGA and isolates is high between 20–30°C, and highest at 25°C for GMGA. Electron transport was detected in all three samples even at temperatures as low as 5°C, but ETR was zero or extremely low at 40°C. These results indicate that the optimum temperatures of GMGA and isolated *C. purpureus* are around 25°C. The ETR of GMGA and the isolates was high at low PAR levels (305 and 422 μmol/m^2/s) at 5–15°C; however, ETR was high at a medium PAR level (687 μmol/m^2/s) at 20–30°C (Fig. 5). Therefore, the optimal PAR value for *in situ* temperatures (0–10°C) is likely between 305–422 μmol/m^2/s. The highest ETR value in GMGAs was 674 μmol/m^2/s, which was approximately twice that of the other two isolates (Fig. 6).

Distribution of GMGAs on the glacier

GMGAs of *Ceratodon purpureus* were observed only at site ST1, the glacier terminus (Fig. 7). The organic carbon mass (62.72±19.39 g/m^2) was highest at the terminus. This record is higher than current highest glacial organic carbon mass (38.5±12.4 g/m^2) from Qiyi Glacier, China [30]. Moreover, organic carbon mass at our sites without observed GMGAs (ST2:20.73±8.91 g/m^2, ST3:23.55±6.89 g/m^2) were roughly equal to the average high organic carbon mass at Qiyi Glacier (mean: 25.4±16.5 g/m^2) [26].

Table 1. Sequences list of five different genetic regions from three different sample types (1: GMGA_cloning, 2: isolate protonema and 3: dried GMGA_cloning) and their closest relatives.

Gene type	Genetic region	Sample type	Accession number	Length (bp)	Closest relative specie	Accession number of relative	Identity (%)	sequence match (bp)
Ribosamal RNA gene	18S rRNA	1: GMGA_cloning	AB858433	1819	Ceratodon sp. AM2008N12	KC291530	99.8	1721/1724
					Ceratodon purpureus	Y08989	99.7	1751/1757
		2: isolate protonema			Ceratodon purpureus	KC291530	99.9	1677/1678
					Ceratodon purpureus	Y08989	99.7	1673/1679
					GMGA_cloning (this study)	AB858433	100	1679/1679
		3: dried GMGA_cloning	AB872997	1697	Bryum caespiticium	AF023703	100	1697/1697
Chloroplast gene	rps4	2: isolate protonema	AB848717	674	Ceratodon purpureus	FJ572605	100	623/623
					Ceratodon purpureus	FJ572589	100	625/625
					Ceratodon purpureus	AF435271	100	561/561
					Ceratodon purpureus	AY908122	100	652/652
					Trichodon cylindricus*	AY908125	94.1	622/661
		3: dried GMGA_cloning	AB872999	684	Bryum cyathiphyllum	AF521683	99.9	667/668
	trnL	2: isolate protonema	AB848718	482	Ceratodon purpureus	FJ572485	100	482/482
					Ceratodon purpureus	AF435310	100	482/482
					Glyphomitrium humillimum*	EU246911	94.2	438/465
		3: dried GMGA_cloning	AB873000	520	Bryum cyathiphyllum	AY150351	100	492/492
	atpB-rbcL intergenic spacer	2: isolate protonema	AB980065	637	Ceratodon purpureus	AY881031	100	621/621
					Ceratodon purpureus	AY881034	100	621/621
					Ceratodon purpureus	AY881052	100	598/598
					Cheilothela chloropus*	AY881063	89.9	571/635
Mitocondorial gene	nad5	2: isolate protonema	AB848719	1112	Ceratodon purpureus	AY908859	99.9	1093/1094
					Ceratodon purpureus	AY908862	99.9	1090/1091
					Trichodon cylindricus*	AY908863	99.5	1089/1095
		3: dried GMGA_cloning	AB872998	1107	Bryum argenteum	AY908945	100	1082/1082

*show second highly related species.

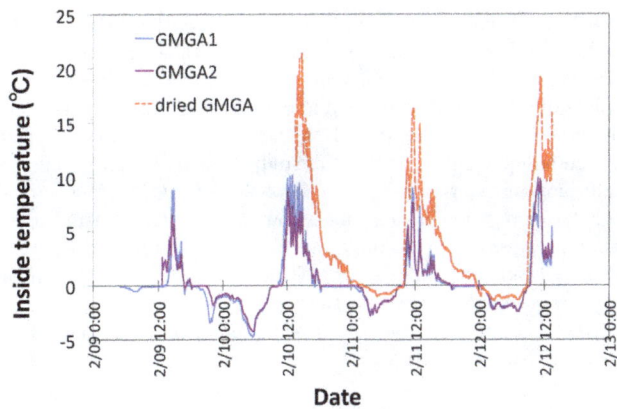

Figure 5. Internal changes in temperature of 2 GMGA and 1 dried GMGA left on a rock during the 2013 research period (February 9–12, 2013).

Discussion

Glacial Moss Gemmae Aggregation (GMGA) is a novel moss aggregation

Mosses, in the form of "glacier mice", have been previously recorded from supraglacial habitats [11,12]; however, the structure of GMGA is completely different from that of "glacier mice". Whereas "glacier mice" are formed by the moss shoots, that level of cellular differentiation was not detected in GMGAs. These findings report first description of developing moss gemmae and protonema in the supraglacial environments. *Ceratodon purpureus*, which formed the GMGA observed in this study, is a cosmopolitan moss species widely distributed throughout entire continents [31] and is known to grow in extreme environments (i.e. polluted sites including highway shoulders and on coal and heavy metal mine tailings) [32]. *Ceratodon purpureus* also occurs in the cryosphere in high alpine areas, Antarctica [33–35]. In the Rwenzori Mountains, unfortunately inhabitation of *C. purpureus* around glacier had not directly observed by authors, however, *C. purpureus* has been detected at elevations from 2800 m to 3700 m [36] and *C. purpureus* specimen (PC0106302) taken at just below the Speke Glacier (4480 m a.s.l.) are stored in Muséum National d'Histoire Naturelle, Paris, France. These evidences would show that possibility of dispersal of spore or gemma from near glacier and deposition on the glacier surface by local wind circulation.

Adaptation of GMGA isolate to warmer temperature

The optimum temperature of polar mosses are widly distributed from 2°C to 35°C according to species [37]. The optimum temperature for the *C. purpureus* isolates (25°C; Fig. 6) is normal value even in polar region, but this was higher than that for *C. purpureus* in Antarctica as previously reported, which was 15°C in the liquid and agar cultures [33]. Moreover, another study showed that the optimum temperature for photosynthesis in *C. purpureus* is around 15°C, but significant carbon fixation occurs at 5°C [35]. Although the measurement of optimum temperature by measuring fluorescence of chlorophyll used in the present study is an indirect measurement of growth (e.g., [33]), these values reflect photosynthetic ability at each temperature (e.g., [35]). Therefore, the populations of *C. purpureus* from this Ugandan glacier have likely adapted to a higher optimum temperature than Antarctic populations.

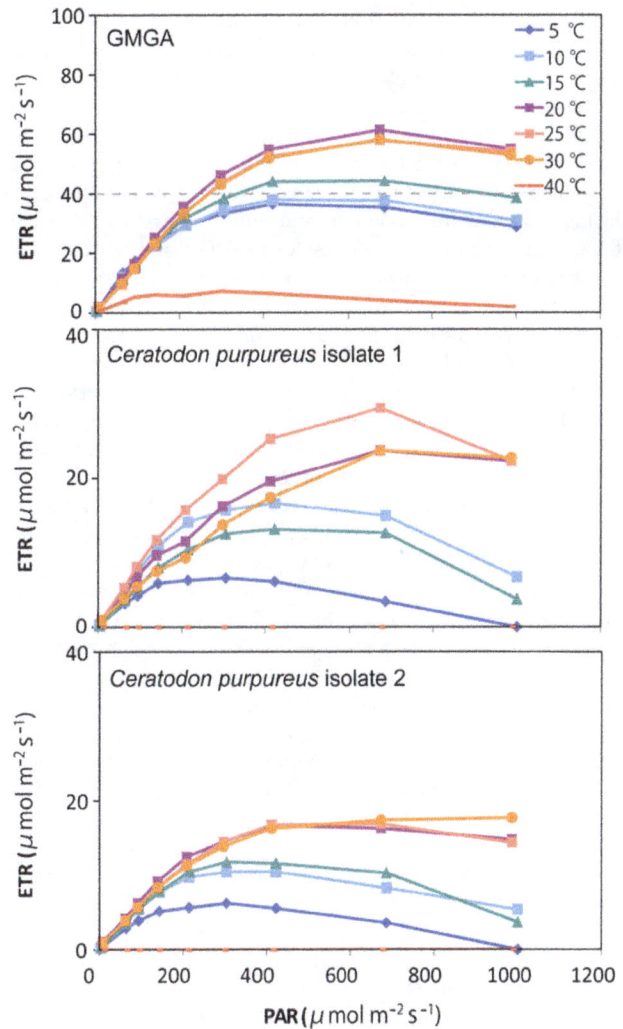

Figure 6. Photosynthetic light curves of GMGA and the two isolates under different incubation temperatures (from 5°C to 40°C) using a pulse amplitude modulation fluorometer.

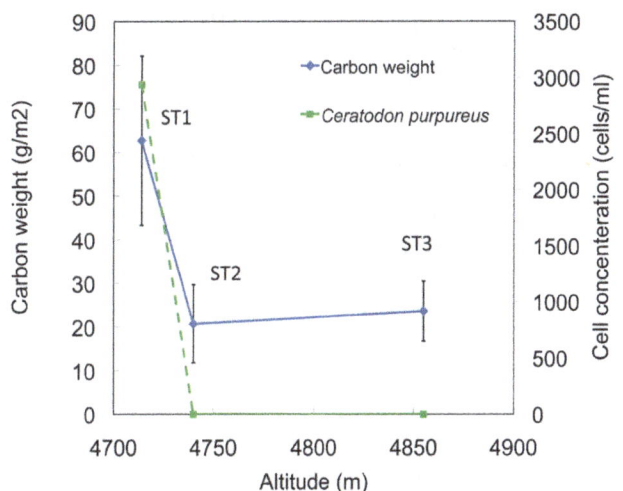

Figure 7. Distribution of carbon mass and *Ceratodon purpureus* cell concentrations on Stanley Plateau Glacier.

Cold and light stress in isolated *C. purpureus*

Polar mosses tend to adapt broad range of favorable temperature. For example, relationship between net assimilation rate (NAR) and temperature of *Drepanocladus uncinatus* in Signy Island show that optimum temperature is 20°C and NAR are more than 40% from 0°C to 30°C (5°C interval) by using cold incubation sample (5°C in light/−5°C in dark) [37]. Relatively higher temperature optimum and broad range of favorable temperature is similar to our result of GMGA and *C. purpureus* isolates. Optimum temperature shift with environmental temperature change were reported from some of experiments [38,39], but *Drepanocladus uncinatus* in Signy Island [37] did not correspond to these studies. Also in our study, optimum temperature did not shift to lower temperature inspite of preincubation temperature at 4°C. The internal temperature of the GMGAs (Fig. 5) was below the optimum (Fig. 6), therfore, the daytime internal temperature of GMGA (10°C) is not optimum but favorable for growth of Ugandan glacial *C. purpureus*. In this temperature range of internal GMGAs (−5°C to 10°C), the optimum PAR is below the warmer temperature (Fig. 6). This phenomenon may be able to explain photoinhibition from cold stress. In low-temperature conditions, PSII is inhibited due to decreased rate of repair of damaged D1 protein and increased excitation pressure [40].

Polar mosses are able to survive short-term freezing and thawing cycles in summer and prolonged freezing in winter. *Bryum argenteum* taken from tropical to polar origin showed no damage after 10 days with a temperature regime of 5°C in light/−5°C in dark, and grew slowly under these conditions [37]. In Rwenzori, *C. purpureus* live under similar dirnal cycle (from −5°C to 10°C) through year and GMGAs structure would be formed slowly. After being deposited and frozen at below −20°C for 2.5 years, regeneration of the *C. purpureus* from east Antarctica is very active [33]. *Ceratodon purpureus* protonemata from GMGA can be isolated and grown at 4°C after half a year of cryopreservation at −80°C. Therefore, isolate of *C. purpureus* in this study have potential to survive both short-term and prolonged freezing stress.

PAR in this natural environment is higher than optimum PAR at 5–15°C. Although we did not measure PAR directly, an automatic weather station with a radiation meter was installed beside the Stanley Plateau (N0°22′34.55″, E 29°52′43.24″; 4750 m above sea level) by the Stations at High Altitude for Research on the Environment project [41]. According to their data, and assuming no significant divergence with today's conditions, the maximum diurnal seasonal shortwave radiation (c.a. 400 W/m^2) occurs around 2:00 PM. PAR (400–700 nm) generally comprises 50% of total solar radiation reaching the Earth's surface. Assuming a conversion rate of 1 W/m^2 to 4.57 μmol/m^2/s, our estimated maximum PAR is 914 μmol/m^2/s when maximum shortwave radiation is 400 W/m^2. This value is higher than the optimum PAR in any temperature and this may cause photoinhibition in low temperature.

These results indicate that both low temperature and high radiation on the glacier are stress factors for *C. purpureus*. In acidic rivers in Japan, *Dicranella heteromalla* (Hedw.) Schimp. remains in a prolonged protonema stage for several growing seasons without producing shoots or sporophytes [42], which researchers concluded was due to the water's extraordinarily low pH (1.9–2.1). A similar prolonged protonema phase of the moss *Scopelophila cataractae* (Mitt.) Broth was reported in copper-rich sites as well in Japan [43]. Therefore, the low temperature and high radiation stress on the glacier, may keep *C. purpureus* in the gemmae and protonemal stage instead of developing into shoots.

Higher photosynthesis activity of GMGA than of isolates

The value of ETR from GMGAs was twofold that of isolates (Fig. 6), possibly due to differences in nutrient condition and the effects of other photosynthetic microorganisms. Growth conditions of moss may be more suitable in GMGA than in the artificial medium (liquid BBM) used in this study, because GMGA contains sufficient nutrients for effective growth. Yet, GMGA is not a simple aggregation of only moss, but also contains many other microorganisms. For example, we observed *Cylindrocystis brebissonii* cells and red snow algae [44], which is related to the green algae commonly found in supraglacial environments, in the GMGAs. Both moss and green algae affect the total photosynthetic activity of GMGAs.

Possible process of GMGA formation

If biological material in sites without GMGAs form a thick deposition layer (more than a few millimeters), the temperature below the surface would increase to above 0°C, the same as in GMGAs. If this is so, then the invasion of gemmae of *C. purpureus* adapted to warmer temperatures on the cold glacier surface can be attributed to this increased subsurface temperature. Similar temperature increases occur in other glaciers (e.g., Qiyi Glacier); however, GMGA-like structures and growth of moss gemmae have not been found on any other glacier, despite studies of glacier biology being conducted around the world [45]. This may relate to possible inhabitation of *C. purpureus* near the glacier and unique features of the Rwenzori; namely, the lack of a clear seasonal temperature cycle. On Stanley Plateau, diurnal temperature change in all seasons is in a range of approximately 0°C to 5°C [41]. Consequently, long periods of freezing do not exist, permitting microorganisms to grow throughout the year. Therefore, we suppose that at least these two factors (the internal temperature rise and the long growth season) may contribute to the formation of GMGAs.

However, GMGAs are disproportionally dominant near the glacier terminus. The high number of GMGAs observed at ST1 must be supported by factors specific to that site. Although our data does not let us reach firm conclusions about the distribution of GMGAs, downward transpotation by surface melting water and availability of sunlight is a likely candidate factor.

Glacier surface melt of Stanley Plateau homogenously spread over the ice area and remarkable water channels on surface are few. Takeuchi [46] speculated cryoconite granules (around 1 mm diameter) are more stable from meltwater than unicellular microorganisms due to larger size character. GMGAs are much larger diameter than typical cryoconite (Fig. 3) seems to be more stable on the ice. Also GMGAs penetrate few mm into ice due to radiation warming (Fig. 2b). These also prevent to wash this material to downward. Therefore, downward transportation of well-developed GMGAs seems unlikely happened.

During the biological growth season, supraglacial light conditions generally change based on depth of snow cover. In the early melt season seasonal snow is removed by melting to expose the glacial ice surface to sunlight at lower elevations only. The snow line then retreats until reaching equilibrium line altitudes at the end of melt season. This leaves the entire surface of the ablation zone snow-free, making it an available habitat for photosynthetic microorganisms. As a result, the biodiversity of glacier microbial communities changes with elevation [44,46–48].

In early February 2013, the entire glacier surface was covered by snow except for a steep slope at the glacier terminus, where ST1 is located. The snow cover near ST2 was 0.85 m deep, but near ST1, zero or a few centimeters of snow cover were observed. Because snow blocks radiation, these differences in snow depth cause variable light conditions and GMGA internal temperatures. Although the precise factors causing this difference in snow depth are unknown, slope angle and wind erosion likely to cause gradients in snow cover. We observed similar types of steep slopes on all edges of this glacier and other glaciers located beyond the ridge (Margarita Glacier: Fig. 2a); however, further observations and measurements must be necessary.

Direct ecological linkage between glacier and glacier foreland

In the glacier foreland immediately adjacent to the glacier terminus, we found an abundance of dried GMGAs on rock surfaces, which were likely left on the freshly bared subglacial rocks after glacier retreat. The temperature of these dried GMGAs on the rocks reaches approximately 20°C and conditions appear much drier than on the glacier, where water is supplied by melting (Fig. 5). Dried GMGAs create a soil-like structure on the abiotic rock surface with gametophyte of different dominant Bryophyta (*Bryum* sp.). The succession of dominant bryophyte species from *C. purpureus* gemmae to *Bryum* sp. shows that GMGAs had changed after leaving the glacier. Previous studies conceptually proposed the linkage between glacier and glacier foreland by nutrient connection [49] and outwash of cryoconite granules [2]. Otherwise, our findings indicate that GMGAs accumulate as a soil-like structure on the abiotic rock surface, directly linking the glacier and glacier foreland ecosystems.

Furthermore, recently another linkage between glacier and glacier foreland was found from subglacial environment [50,51]. In Canadian Arctic, varaeties of mosses regenerate from old populations, which had been entombed in subglacial environment from Little Ice Age (LIA) and recently released onto ice –free glacier foreland due to glacial retreat [50]. These evidences show

releases of developed biological material from both supraglacier and subglacier supply the stable ecological substructure to glacier foreland ecology.

If the glacier disappears due to climate change and/or albedo reduction, this unique glacial ecosystem and its contribution to the glacier foreland will also disappear. Many other tropical glaciers that are expected to disappear in the near feature [52,53], which may also contain unique biota that are under threat. In this respect, the tropical glacial ecosystem is an urgent subject of study to understand the biodiversity.

Acknowledgments

The authors thank A. Wada of the Greenleaf Tourist Club for management of local transportation in Uganda. We also thank the guides and porters of the Rwenzori Mountaineering Service for guidance and transport of research equipment on the mountain, and K. Watanabe and M. Mori for assistance in laboratory experiments. I am also indebted to three reviewers (Dr. Peter Convey, Dr. Catherine La Farge and Dr. Nicoletta Cannone) for valuable suggestions, which greatly improved this article. This study was partially supported by Ministry of Education, Culture, Sports, Science and Technology Grant-in-Aid for Scientific Research (A) No. 22241005, and by Proposal for Seeds of Transdisciplinary Research from the Transdisciplinary Research Integration Center, and by an NIPR publication subsidy. Field research was supported by the Uganda Wildlife Authority, the Uganda National Council for Science and Technology, and Dr. S. Anguma of Mbarara University of Science and Technology.

Author Contributions

Conceived and designed the experiments: JU YT. Performed the experiments: JU ST KH YT DS. Analyzed the data: JU YT. Contributed reagents/materials/analysis tools: JU YT HM SI SK. Wrote the paper: JU YT DS SK.

References

1. Anesio AM, Hodson AJ, Fritz A, Psenner R, Sattler B (2009) High microbial activity on glaciers: importance to the global carbon cycle. Glob Chang Biol 15: 955–960.
2. Wharton RA, McKay CP, Simmons GM, Parker BC (1985) Cryoconite holes on glaciers. Bioscience 35: 499–503.
3. Takeuchi N, Nishiyama H, Li Z (2010) Structure and formation process of cryoconite granules on Ürümqi glacier No. 1, Tien Shan, China. Ann. Glaciol 51: 9–14.
4. Edwards A, Anesio AM, Rassner SM, Sattler B, Hubbard B, et al. (2011) Possible interactions between bacterial diversity, microbial activity and supraglacial hydrology of cryoconite holes in Svalbard. ISME J 5: 150–60.
5. Cameron KA, Hodson AJ, Osborn AM (2011) Structure and diversity of bacterial, eukaryotic and archaeal communities in glacial cryoconite holes from the Arctic and the Antarctic. FEMS Microbiol Ecol 82: 254–267.
6. Hamilton TL, Peters JW, Skidmore ML, Boyd ES (2013) Molecular evidence for an active endogenous microbiome beneath glacial ice. ISME J 7: 1402–1412.
7. Telling J, Anesio AM, Tranter M, Irvine-Fynn T, Hodson A, et al. (2011) Nitrogen fixation on Arctic glaciers, Svalbard. J Geophys Res 116: 2–9.
8. Eythórsson J (1951) Correspondence. Jökla-mýs. J Glaciol 1 (9): 503.
9. Benninghoff WS (1955) Correspondence. Jokla mys. J Glaciol 2: 514–515.
10. Heusser CJ (1972) Polsters of the moss Drepanocladus berggrenii on Gilkey Glacier, Alaska. Bull. Torrey Bot. Club 99: 34–36.
11. Porter PR, Evans AJ, Hodson AJ, Lowe AT, Crabtree MD (2008) Sediment-moss interactions on a temperate glacier: Falljokull, Iceland. Ann Glaciol 48 (1): 25–31.
12. Coulson SJ, Midgley NG (2012) The role of glacier mice in the invertebrate colonisation of glacial surfaces: the moss balls of the Falljökull, Iceland. Polar Biol 35: 1651–1658.
13. Hope GS, Peterson JA, Radok U, Allison I (1976) The Equatorial Glaciers of New Guinea: Results of the 1971–1973 Australian Universities' Expeditions to Irian Jaya: Survey, Glaciology, Meteorology, Biology and Palaeoenvironments, A.A. Balkema, Rotterdam, 81–92.
14. Taylor RG, Mileham L, Tindimugaya C, Majugu A, Muwanga A, et al. (2006). Recent glacial recession in the Rwenzori Mountains of East Africa due to rising air temperature. Geophys Res Lett 33: 2–5.
15. Mölg T, Rott H, Kaser G, Fischer A, Cullen NJ (2006) Comment on "Recent glacial recession in the Rwenzori Mountains of East Africa due to rising air temperature" by Richard G. Taylor, Lucinda Mileham, Callist Tindimugaya,

Abushen Majugu, Andrew Muwanga, and Bob Nakileza. Geophys Res Lett 33: 33–36.
16. Kaser G, Osmaston H (2002) Tropical glaciers. Cambridge University Press, Cambridge, UK. 63–116.
17. Quast C, Pruesse E, Yilmaz P, Gerken J, Schweer T, et al. (2013) The SILVA ribosomal RNA gene database project: improved data processing and web-based tools. Nucl Acids Res 41 (D1): D590–D596.
18. Schloss PD, Westcott SL, Ryabin T, Hall JR, Hartmann M, et al. (2009) Introducing mothur: open-source, platform-independent, community-supported software for describing and comparing microbial communities. Appl Environ Microbiol 75: 7537–7541.
19. Edgar RC, Haas BJ, Clemente JC, Quince C, Knight R (2011) UCHIME improves sensitivity and speed of chimera detection, Bioinformatics 27 (16): 2194–2200. Available at: doi:0.1093/bioinformatics/btr381.
20. Andersen R (2005) Algal Culturing Techniques. Elsevier, Amsterdam, Netherlands. 437.
21. Remias D, Schwaiger S, Aigner S, Leya T, Stuppner H, et al. (2012) Characterization of an UV- and VIS-absorbing, purpurogallin-derived secondary pigment new to algae and highly abundant in *Mesotaenium berggrenii* (Zygnematophyceae, Chlorophyta), an extremophyte living on glaciers. FEMS Microbiol Ecol 79: 638–648.
22. Taberlet P, Gielly L, Pautou G, Bouvet J (1991) Universal primers for amplification of three non-coding regions of chloroplast DNA. Plant Mol Bio 17: 1105–1109.
23. Nadot S, Bittar G, Carter L (1995) A Phylogenetic Analysis of Monocotyledons Based on the Chloroplast Gene rps 4, Using Parsimony and a New Numerical Phenetics Method. Mol Phylogenet Evol 4: 257–282.
24. Souza-Chies T, Bittar G, Nadot S, Carter L, Besin E, et al. (1997) Phylogenetic analysis ofIridaceae with parsimony and distance methods using the plastid generps4. Plant Syst Evol 204: 109–123.
25. Chiang T, Schaal BA, Peng C (1998) Universal primers for amplification and sequencing a noncoding spacer between the atpB and rbcL genes of chloroplast DNA. Bot. Bull. Acad. Sin. 39: 245–250.
26. Shaw AJ, Cox CJ, Boles SB (2003) Polarity of peatmoss (Sphagnum) evolution: who says bryophytes have no roots? Am J Bot 90: 1777–1787.
27. Tanabe Y, Shitara T, Kashino Y, Hara Y, Kudoh S (2011) Utilizing the effective xanthophyll cycle for blooming of Ochromonas smithii and O. itoi (Chrysophy-

ceae) on the snow surface. PLoS One 6 (2), e14690. doi:10.1371/journal.-pone.0014690.

28. McMinn A, Hegseth EN (2004) Quantum yield and photosynthetic parameters of marine microalgae from the southern Arctic Ocean, Svalbard. J Mar Biol Ass UK 84: 865–871.

29. Imura S, Kanda H (1986) The gemmae of the mosses collected from the Syowa Station area, Antarctica. Mem. Natl. Inst. Polar Res. 44: 241–246.

30. Takeuchi N, Matsuda Y, Sakai A, Fujita K (2005) A large amount of biogenic surface dust (cryoconite) on a glacier in the Qilian Mountains, China. Bull Glaciol Res 22: 1–8.

31. McDaniel SF, Shaw J (2005) Selective sweeps and intercontinental migration in the cosmopolitan moss *Ceratodon purpureus* (Hedw.) Brid. Mol. Ecol. 14: 1121–32.

32. Shaw J, Jules ES, Beer SC (1991) Effects of metals on growth, morphology, and reproduction of *Ceratodon purpureus*. Bryologist 94(3): 270–277.

33. Kanda H (1979) Regenerative development in culture of Antarctic plants of *Ceratodon purpureus* (HEDW.) BRID. Mem Natl Inst Polar Res Special issue 11: 58–69.

34. Imura S, Kanda H (1986) The gemmae of the mosses collected from the Syowa Station area, Antarctica. Mem Natl Inst Polar Res 24: 241–246.

35. Lewis Smith RI (1999) Biological and environmental characteristics of three cosmopolitan mosses dominant in continental Antarctica. J Veg Sci 10: 231–242.

36. Hauman L (1942) Les Bryophytes des hautes altitudes au Ruwenzori. Bulletin du Jardin botanique de l'État a Bruxelles 16: 311–353. (in French).

37. Oechel WC, Sveinbjörnsson B (1978) Primary Production Processes in Arctic Bryophytes at Barrow, Alaska. Vegetation and Production Ecology of an Alaskan Arctic Tundra, Ecological Studies Volume 29, Springer-Verlag, New Tork, USA. 269–298.

38. Hicklenton PR, Oechel WC (1976) Physiological aspects of the ecology of Dicranum fuscescens in the subarctic. I: Acclimation and acclimation potential of CO_2 exchange in relation to habitat, light, and temperature, Canadian Journal of Botany 54(10): 1104–1119.

39. Longton RE (1988) Biology of polar bryophytes and lichens. Cambridge University Press, Cambridge, UK. 141–210.

40. Sonoike K (1998) Various Aspects of Inhibition of Photosynthesis under Light/Chilling Stress: "Photoinhibition at Chilling Temperatures" versus "Chilling Damage in the Light". J Plant Res 111: 121–129.

41. Lentini G, Cristofanelli P, Duchi R, Marinoni A, Verza G, et al. (2011) Mount Rwenzori (4750 m a.s.l., Uganda): meteorological characterization and Air-Mass transport analysis. Geografia Fisica e Dinamica Quaternaria 34 (3): 183–193.

42. Higuchi S, Kawamura M, Miyajima I, Akiyama H, Kosuge K, et al. (2003) Morphology and phylogenetic position of a mat-forming green plant from acidic rivers in Japan. J Plant Res 116: 461–467.

43. Satake K, Nishikawa M, Shibata K (1990) A copper-rich protonemaI colony of the moss *Scopelophila cataractae*. J Bryol 1928: 109–116.

44. Takeuchi N, Uetake J, Fujita K, Aizen VB, Nikitin SD (2006) A snow algal community on Akkem glacier in the Russian Altai mountains. Ann Glaciol 43: 378–384.

45. Hodson A, Anesio AM, Tranter M, Fountain A, Osborn M, et al. (2008) Glacial Ecosystems Ecol Monogr 78: 41–67.

46. Takeuchi N (2001) The altitudinal distribution of snow algae on an Alaska glacier (Gulkana Glacier in the Alaska Range). Hydrol Process 15: 3447–3459.

47. Segawa T, Takeuchi N, Ushida K, Kanda H, Kohshima S (2010) Altitudinal Changes in a Bacterial Community on Gulkana Glacier in Alaska. Microbes Environ 25: 171–182.

48. Uetake J, Yoshimura Y, Nagatsuka N, Kanda H (2012) Isolation of oligotrophic yeasts from supraglacial environments of different altitude on the Gulkana Glacier (Alaska). FEMS Microbiol Ecol 82: 279–286.

49. Stibal M, Tranter M, Telling J, Benning LG (2008) Speciation, phase association and potential bioavailability of phosphorus on a Svalbard glacier. Biogeochemistry 90: 1–13.

50. La Farge C, Williams KH, England JH (2013) Regeneration of Little Ice Age bryophytes emerging from a polar glacier with implications of totipotency in extreme environments. Proc Natl Acad Sci USA 110: 9839–9844.

51. Thompson LG, Mosley-Thompson E, Davis ME, Zagorodnov VS, Howat IM, et al. (2013) Annually resolved ice core records of tropical climate variability over the past ~1800 years. Science 340: 945–50.

52. Rabatel A, Francou B, Soruco A, Gomez J, Cáceres B, et al. (2013) Current state of glaciers in the tropical Andes: a multi-century perspective on glacier evolution and climate change. Cryosph 7: 81–102.

53. Thompson LG, Brecher HH, Mosley-Thompson E, Hardy DR, Mark BG (2009) Glacier loss on Kilimanjaro continues unabated. Proc Natl Acad Sci USA 106: 19743–19744.

Whole-Genome Sequencing of the World's Oldest People

Hinco J. Gierman[1], Kristen Fortney[1], Jared C. Roach[2], Natalie S. Coles[3,4], Hong Li[2], Gustavo Glusman[2], Glenn J. Markov[1], Justin D. Smith[1], Leroy Hood[2], L. Stephen Coles[3,4], Stuart K. Kim[1]*

1 Depts. of Developmental Biology and Genetics, Stanford University, Stanford, CA, United States of America, 2 Institute for Systems Biology, Seattle, WA, United States of America, 3 Gerontology Research Group, Los Angeles, CA, United States of America, 4 David Geffen School of Medicine, University of California Los Angeles, Los Angeles, CA, United States of America

Abstract

Supercentenarians (110 years or older) are the world's oldest people. Seventy four are alive worldwide, with twenty two in the United States. We performed whole-genome sequencing on 17 supercentenarians to explore the genetic basis underlying extreme human longevity. We found no significant evidence of enrichment for a single rare protein-altering variant or for a gene harboring different rare protein altering variants in supercentenarian compared to control genomes. We followed up on the gene most enriched for rare protein-altering variants in our cohort of supercentenarians, TSHZ3, by sequencing it in a second cohort of 99 long-lived individuals but did not find a significant enrichment. The genome of one supercentenarian had a pathogenic mutation in DSC2, known to predispose to arrhythmogenic right ventricular cardiomyopathy, which is recommended to be reported to this individual as an incidental finding according to a recent position statement by the American College of Medical Genetics and Genomics. Even with this pathogenic mutation, the proband lived to over 110 years. The entire list of rare protein-altering variants and DNA sequence of all 17 supercentenarian genomes is available as a resource to assist the discovery of the genetic basis of extreme longevity in future studies.

Editor: Patrick Lewis, UCL Institute of Neurology, United Kingdom

Funding: This work was supported by the Ellison Medical Foundation/American Federation for Aging Research Fellowship, Stanford Dean's Fellowship, The Paul Glenn Foundation Biology of Aging Seed Grant, National Institute of General Medical Sciences Center for Systems Biology (P50 GM076547) and the University of Luxembourg – Institute for Systems Biology Program. The funders had no role in study design, data collection and analysis, decision to publish, or preparation of the manuscript.

Competing Interests: The authors have declared that no competing interests exist.

* Email: stuartkm@stanford.edu

Introduction

Supercentenarians are the world's oldest people, living beyond 110 years of age [1]. As would be expected for people that reach this age, supercentenarians have escaped many age-related diseases [2–5]. For example, there is a 19% lifetime incidence of cancer in centenarians compared to 49% in the normal population [6]. Similarly, supercentenarians have a lower incidence of cardiovascular disease and stroke than controls [5].

The genetic component of human lifespan based on twin studies has been estimated to be around 20–30 percent in the normal population [7], but higher in long-lived families [8–10]. Furthermore, siblings, parents, and offspring of centenarians also live well beyond average [11,12]. Lifestyle choices in terms of smoking, alcohol consumption, exercise, or diet does not appear to differ between centenarians and controls [13]. Taken together, these findings provide ample evidence that extreme longevity has a genetic component .

Several gene association studies have compared cohorts of long-lived subjects to controls. Analysis of candidate genes has shown that polymorphisms in the Insulin-like Growth Factor 1 Receptor gene (IGF1R) and the FOXO3 transcription factor gene are associated with extreme longevity [14,15]. Genome-wide association studies have shown that the ApoE4 haplotype is depleted in centenarians [16–18]. Sebastiani et al. compiled a list of 281

independent single-nucleotide polymorphisms (SNPs) that showed strong associations with extreme longevity (though none were genome-wide significant except for an ApoE SNP) [17]. They then showed that a genetic signature that combines information from these 281 SNPs is predictive for extreme longevity, indicating that at least some of these SNPs are truly associated with longevity. However, specific variants associated with longevity have not yet been identified [18,19].

More recently, studies have begun to use whole-exome sequencing and whole-genome sequencing (WGS) of centenarians to find variants associated with extreme longevity [19–21]. Ye et al. compared the genome sequence of a pair of 100-year-old twins to a pair of 40-year-old twins and found no evidence of accumulation of somatic mutations during aging [20]. By sequencing blood cells of a supercentenarian, Holstege et al. first identified somatic mutations and then used this information to infer clonal lineages in hematopoietic stem cells. They found that white blood cells in this individual were derived from only two clones of hematopoietic stem cells [21].

Here, we have sequenced the genomes of 17 supercentenarians. We limited the majority of our analyses to the thirteen genomes from Caucasian females. From this small sample size, we were unable to find rare protein-altering variants significantly associated with extreme longevity. However, we did find that one supercentenarian carries a pathogenic variant associated with arrhythmogenic right ventricular cardiomyopathy (ARVC), which had little or no effect on his/her health as this person lived over 110 years.

Materials and Methods

Ethics Statement, Supercentenarian Recruitment and Age Validation

Supercentenarian subjects, their family members, or their caretakers provided written informed consent. The study was approved by the Stanford University Institutional Review Board (IRB-19119) and by the Western Institutional Review Board (WIRB protocol #20101350). Supercentenarians were considered validated (i.e., 110 years or older) if they possessed each of the following documents: (1) A birth certificate, a baptismal certificate, or Census Record dating back to the original time of birth; (2) A marriage certificate in the case of married women not using their maiden names; (3) a current government-issued photo ID, such as a driver's license or passport. Supercentenarian health status and medical history for major age-related diseases were based on interviews conducted with subjects and/or their caretakers.

DNA Isolation, PCR and Sanger Sequencing

Whole-blood samples were drawn into PAXgene (Qiagen) blood tubes from which high molecular weight DNA was isolated. DNA samples were quantified using a dsDNA Broad-Range Assay on a Qubit Fluorometer (Life Technologies) and checked for size and degradation on an agarose gel. For Sanger sequencing, samples were amplified by nested PCR and variants were validated by forward and reverse reads. Primers were designed with Primer3 [22] and 10 ng was amplified with Phusion High-Fidelity Polymerase (Thermo Scientific). PCR bands were either column-purified or cut out from an agarose gel and purified with a Qiaquick Gel Extraction kit (Qiagen). PCR product was Sanger sequenced at Sequetech, Inc. Reads were trimmed by 10 bp at the 5′ end and at a 0.01 error probability limit and then aligned to the human genome reference sequence build GRCh37 (hg19) using Geneious software. For sequencing of TSHZ3 in the Georgia Centenarian Study samples, all coding regions were sequenced except the first 13 amino acids (i.e., exon 1). None of the rare protein altering variants found in the 13 supercentenarians or the 4,300 NHBLI controls were located in exon 1. All experiments were performed according to manufacturer's protocol unless otherwise indicated.

Ancestry and Relatedness

Principal component analysis (PCA) of ancestry was done by analyzing the intersection of all genotyped SNPs from 1184 individuals from 12 different populations from HapMap Phase 3 [23] and the 17 supercentenarians. Only bi-allelic SNPs that had at least one non-reference allele in the 17 supercentenarians were used, resulting in a subset of 1.2 million SNPs. Genome-wide Complex Trait Analysis (GCTA) software was used to perform the PCA [24]. All pairs of 17 supercentenarians were tested for relatedness using Estimation of Recent Shared Ancestry (ERSA) [25,26].

Whole-Genome Sequencing and Analysis Pipelines

All DNA samples were submitted for WGS to 40x coverage by Complete Genomics, Inc. (CGI). Standard protocols were used to map reads and call variants using CGI pipeline 2.0.2 [27]. To analyze variants, we first produced a cross-reference matrix out of CGI variant files using custom Perl scripts [28] and the CGI command line tool CGAtools (listvar, testvar). To reduce platform errors and biases, we removed any variant with >50% double no-call rate in a control set of public genomes sequenced on the same CGI platform. We used 54 of the unrelated HapMap genomes (for variant analysis) or the 34 PGP genomes (for the RVT1 burden test). The 54 HapMap Genomes were obtained as part of the public CGI Diversity panel of 69 and the 34 PGP genomes were obtained from the Personal Genome Project [29]. The baseline characteristics of the 34 PGP genomes are listed in Table S1. Next, we used ANNOVAR and its build GRCh37 (hg19) database files [30] and custom scripts to annotate protein-altering variants: missense, frameshift, non-frameshift indels, stop-gain, stop-loss, and splice-site disruption. Splice-site variants were those disrupting the canonical splice-donor (GU) or splice-acceptor (AG) site of the RefSeq sequence. To test for enrichment of a rare protein-altering variant, we used the 379 European individuals from the 1000Genome (1000G EUR) Project Phase 1 (April 2012) build database as controls [31]. We included all protein-altering variants and did not require missense SNPs to be predicted as damaging by, e.g., SIFT or PolyPhen-2.

To filter out common variants, we used dbSNP version 131 [32]. This version was released on February 2010, and lacks most low-frequency variants deposited by large consortia like NHLBI and 1000G in later versions. Rare variants were tested for enrichment in cases (13 Caucasian female supercentenarians) vs. controls (379 European individuals from the 1000G Project) using Fisher's Exact Test. We repeated our analysis with reduced stringency by lowering the quality score threshold, but we did not see any significantly enriched variant or gene. Consistent with previous reports, Sanger sequencing of candidate rare protein-altering variants from WGS showed that 30 percent were likely sequencing errors [33,34].

Next, we applied a collapsing test to determine if any gene showed an enrichment of rare protein-altering variants in supercentenarian vs. control genomes. We started with the set of protein-altering variants in autosomal RefSeq genes observed in supercentenarians and controls (34 Caucasians from the PGP), and filtered to retain only rare variants with a minor allele frequency (MAF) <1.5% in 1000G EUR, and with an empirical MAF<10% in our samples. For each gene, we computed the RVT1 statistic [35] to determine whether the burden of mutations differed in supercentenarians and controls using R scripts [36]. RVT1 performs a logistic regression to model phenotype (case/control status) as a function of the proportion of rare variants seen in each genome. We repeated our burden test using a 5% instead of 1.5% as the 1000G EUR MAF cutoff, and again saw no significantly enriched gene. For the recessive model test, we compared all subjects having two or more variants per gene and scored significance using Fisher's Exact test.

Cohorts used to follow-up TSHZ3 variants

Samples from the Georgia Centenarian Study [37] were obtained from Coriell as DNA samples (Coriell ID: AGPLONG3). All Caucasian samples (n = 100) were analyzed and used for PCR and Sanger sequencing as described above. Two of our supercentenarians had previously participated in the Georgia Centenarian Study; their samples were identified by genotyping and removed from the cohort (NG18205, NG20051). In addition, we checked that none of the other supercentenarians with a protein-altering variant in TSHZ3 was present in the Georgia cohort by Sanger sequencing several loci in the Georgia cohort. We added a female Caucasian centenarian sample from our own study (age 100), bringing the total to 99. For controls, we used exome data for 4,300 Caucasians obtained from the NHLBI Exome Variant Server [38].

Table 1. Characteristics of supercentenarians.

Age	Age at Draw	Sex	Race	Major Age-related Diseases	Hearing	Vision	Dental	Communi-cation	Mobility
116	114	F	CAU	None	••	••	•••	•••	•
114	110	F	HIS	None	••	•	•	••	••
114	112	F	CAU	None	•	•	•	•	•
114	112	F	CAU	None	•		•	•	••
114	114	F	CAU	None	•		•••	•	•
114	110	F	HIS	None	••	••	•••	••	•••
113	111	F	CAU	None	•••	•••	•••	••	••
113	112	F	CAU	None	•	•		••	••
113	113	F	AA	None	•••	•••	•••	••	••
112	110	F	CAU	None	••	••	•••	•	••
111	110	F	CAU	Alzheimer's	••	•	•	•	•
111	110	F	CAU	None	•	•••	•	••	••
111	110	F	CAU	None		•	•	••	••
111	110	F	CAU	None	••	•	•••	•••	•••
111	110	M	CAU	Cancer	•••	•••	•••	••	•••
111	111	F	CAU	None	•	•	•	•	••
110	110	F	CAU	None	•	•	•	•	••

Age is age at death or last reported age alive. Age at (blood) draw was validated as described in methods. Sex is female (F) or male (M). Race (or ethnicity) is Caucasian (CAU), Hispanic (HIS) or African-American (AA). Major age-related diseases were known events of cancer, cardiovascular disease, stroke, Alzheimer's or type 2 diabetes at blood draw (i.e. enrollment). Functional status is indicated as: ••• (good), •• (moderate) or • (poor). Hearing: ••• good in both ears; •• good in one, impaired in other ear; • impaired in both ears. Vision: ••• could read newspaper; •• could watch television; • could do neither. Teeth: ••• had teeth of their own; •• no teeth of their own. Communication: ••• talked independently and coherently; •• slow speech, needed interpreter; • incoherent or no communication. Mobility: ••• could walk; •• uses wheelchair; • bed confined.

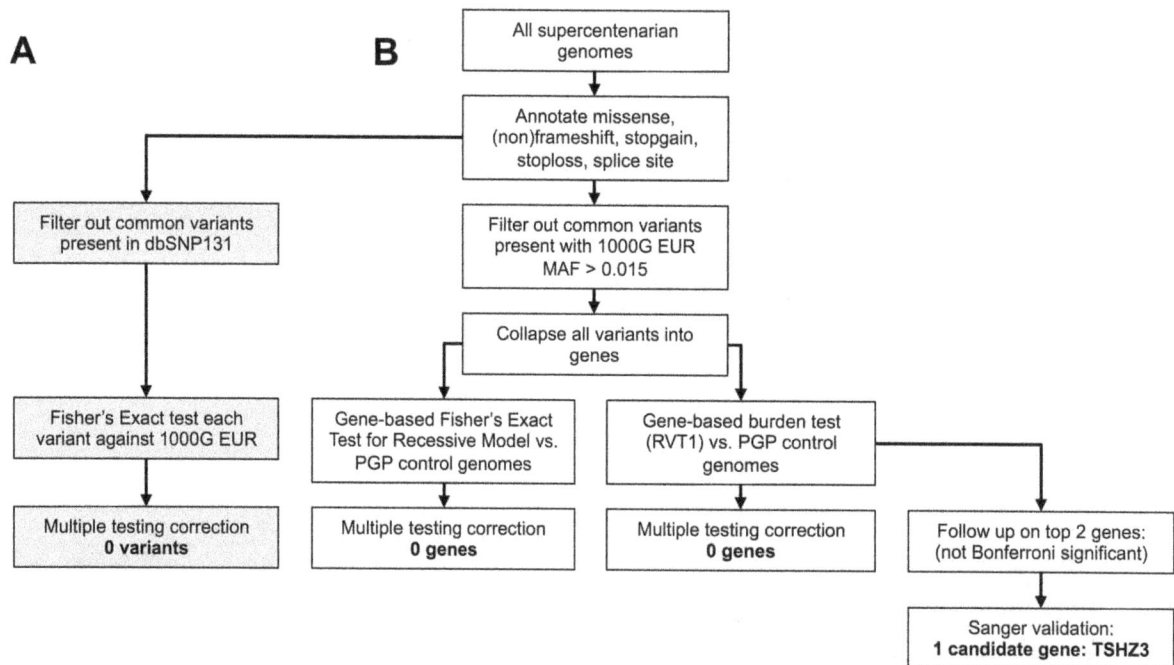

Figure 1. Pipeline to test supercentenarians for enrichment of rare protein-altering variants or genes harboring them. All female Caucasian supercentenarian genomes were annotated for protein-altering variants. (A) To test for enrichment of a single variant, we filtered against dbSNP131 and compared each remaining rare protein-altering variant against 1000G EUR. No single variant was significantly enriched. (B) To test for enrichment of a gene with rare protein-altering variants, we collapsed all variants in to their respective genes and filtered against 1000G EUR (MAF< 0.015). We tested for enrichment against 34 control genomes from PGP using the RVT1 burden test or a gene-based Fisher's Exact (for recessive model). No gene was significantly enriched for rare protein-altering variants in supercentenarians. We then Sanger validated TSHZ3 as the best candidate from our burden-test for follow-up.

Analysis of Pathogenic Variants

We used the recently published list from the American College of Medical Genetics and Genomics (ACMG) of potentially lethal pathogenic variants in 56 genes recommended for reporting to subjects [39]. All 17 supercentenarian genomes were annotated as described above, except without filtering for common variants. ClinVar and Human Gene Mutation Database (HGMD) were used to identify known pathogenic variants in the supercentenarian genomes in all 56 genes identified by the ACMG [40,41]. Besides the known pathogenic variants, new variants can be expected to be pathogenic in 45 of the 56 genes if the new variant clearly strongly reduces or eliminates protein function, such as frameshift, stop-gain, stop-loss, or splice-site mutations [42]. Any variant suggested to be benign based on annotation in ClinVar or HGMD was removed. The scoring of variants as either pathogenic or benign was also checked using Locus Specific Databases (LSDB). Pathogenic annotation of the c.631-2A>G mutation in DSC2 was confirmed in the Arrhythmogenic Right Ventricular Dysplasia/Cardiomyopathy (ARVD/C) database [43], which is part of the Leiden Open Variation Database [44].

Data Access

Upon acceptance for publication, the complete genome sequence for the 17 supercentenarians will be deposited in dbGAP and Google Genomics.

Results

The Supercentenarian Cohort

We recruited 17 supercentenarians and validated their age of 110 years or greater (see Methods). Their mean age at time of

death was 112 years and the subject that lived the longest died at the age of 116 years. At the time of her death, she was the world's oldest person and remains in the top ten of oldest people in recorded history [45]. We determined the medical history and health status of supercentenarians at the time of enrollment by interviewing them, their family, and caretakers. Many of the supercentenarians were cognitively and physically functional to a high degree well into old age. For example, one of our subjects worked as a pediatrician until the age of 103. Another subject drove a car until the age of 107. Table 1 gives an indication of some of the aspects of the supercentenarian health at the time of blood draw.

Among the 17 supercentenarians, at least one subject had a previous case of cancer and one was diagnosed with Alzheimer's disease. To the best of our knowledge, none of the supercentenarians were known to have cardiovascular disease, stroke or diabetes at the time of enrollment. In contrast, people in the US at age 85 often have had at least one major age-related disease. For example, 45 percent of 85-year olds have been diagnosed with cancer and 35 percent have had an incidence of cardiovascular disease [5]. The low rate of disease in our cohort of supercentenarians is consistent with previous reports showing that supercentenarians delay or escape most age-related diseases [5].

We isolated DNA from whole blood and sent the samples to Complete Genomics for WGS. Samples were sequenced to a read depth of 40x, and 94.1% of the genomes and 94.8% of the exomes had a read depth of at least 20x (Figure S1). To confirm the self-reported ancestry of all subjects, we performed a Principal Component Analysis (PCA) on the genomes of our 17 supercentenarians and that of 1184 HapMap individuals with known ancestry to serve as controls (Figure S2). This analysis confirmed

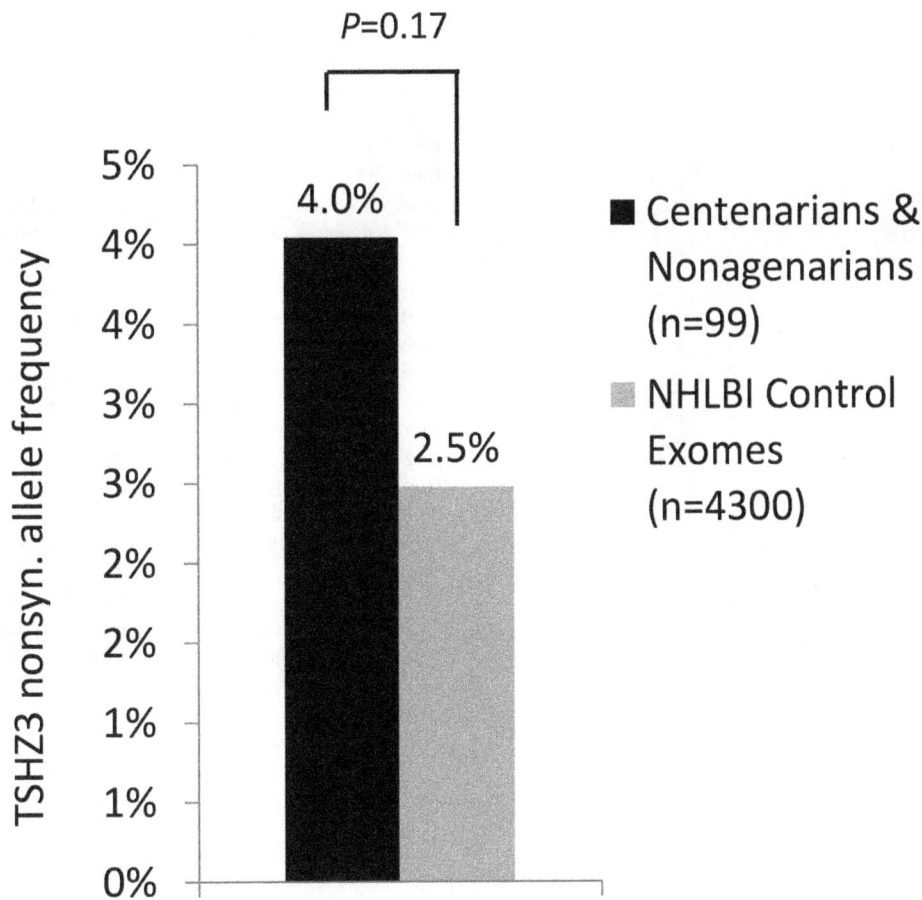

Figure 2. Rare protein-altering variants in TSHZ3 in the Georgia Centenarian cohort versus NHLBI cohort. To see if TSHZ3 is enriched for rare protein-altering variants in long-lived individuals, Sanger sequencing was performed on TSHZ3 in 99 Caucasians with extreme longevity (age 98–105). There was not a significant enrichment comparing the allele frequency of all rare protein-altering variants in the centenarians (4.0%; black bar) to 4300 Caucasian controls from the NHLBI exome project (2.5%; white bar). Both cohorts were annotated for protein-altering variants and filtered against 1000G EUR (MAF<0.015).

that 14 supercentenarians were of European ancestry, one was African American, and two were Hispanic. To prevent confounding our analyses due to differences in race or sex, we used only supercentenarian genomes that were both Caucasian and female for our main analyses. This left us 13 genomes for the main analysis with one male, two Hispanic, and one African-American genome reserved for follow-up analyses.

Next, we checked the genomes of our supercentenarians for unknown relatedness to each other, as any close relationship would confound analyses for enrichment of shared rare variants. We checked for shared regions of identity-by-descent using Maximum-

likelihood Estimation of Recent Shared Ancestry [25,26]. The results indicated that none of the 17 supercentenarians were within five degrees of relationship of any other supercentenarian, which means that at least 97 percent of any of the supercentenarian genomes was not identical-by-descent to any of the other supercentenarian genomes.

Table 2. Baseline statistics of follow-up cohorts.

	Georgia Centenarian Study	NHLBI Controls
Sample size, n	99	4300
Age, mean (range)	101 (98–105)	(≥18)
Females, n (%)	82 (83%)	2428 (56%)

Ages for Georgia Centenarian Study subjects were obtained from Corriell website. Number of females from NHLBI cohort was derived for X chromosome genotypes. Age information for NHLBI controls was obtained from www.nhlbi.nih.gov/recovery/media/NHLBI_DNA_cohort.htm.

Table 3. Protein-altering variants in TSHZ3 in Georgia Centenarian cohort.

Position on Chr19	Ref/Var	AA Pos	AA1/AA2	Supercent	Cent	Nona	1000G EUR MAF
31769738	G/A	321	R/W	0	1	0	novel
31769366	C/T	445	V/M	1	0	0	0.0013
31769293	T/C	469	E/G	1	2	1	0.01
31769021	T/C	560	M/V	0	1	0	novel
31768639	G/A	687	P/L	1	0	0	novel
31768594	A/C	702	L/W	0	0	1	novel
31768267	G/A	811	T/M	0	1	0	novel
31768178	C/T	841	E/K	1	0	0	novel
31767599	C/T	1034	E/K	0	1	0	novel

Position (bp) on chromosome 19 (Chr19) of variant, reference (Ref) and Variant (Var) allele, Amino Acid (AA) position, AA1 (ref), AA2 (var), Supercentenarian carriers (shown for reference), Centenarians carriers, Nonagenarians carriers, Minor allele frequency (MAF) in 1000G EUR.

Are Supercentenarians Enriched for a Rare Protein-Altering Variant?

For people born around 1900, the odds of living to 110 are estimated to be less than 10^{-5} per birth [46], hence we assume that any genetic variant that contributes strongly to extreme longevity would also be rare. One possibility is that a specific mutation could alter the protein-coding region in a gene and confer a significant increase in longevity. Such a mutation could act in a dominant or recessive fashion, and might be shared by a significant fraction of the supercentenarian genomes but not by control genomes. We created a computational pipeline to determine whether our supercentenarian genomes are enriched for such a variant compared to controls (Figure 1). We annotated the variants in all of the female Caucasian genomes and retained those predicted to alter a protein. The polymorphism could be a single nucleotide polymorphism (SNP) or an insertion/deletion (Indel). The polymorphism could change the protein-coding sequence by causing a missense, frameshift, non-frameshift indel, nonsense (i.e., stop-gain), stop-loss, or splice-site disruption (Table S2). To identify rare variants, we filtered out common variants by removing any variant present in the public database dbSNP build 131. We then compared the frequency of the rare protein-altering variants in the supercentenarian genomes with that in the 379 European individuals in the 1000Genomes Project (1000G EUR) using a Fisher's Exact Test. In total, there were 13,892 rare protein-altering variants screened in the supercentenarian genomes. To adjust for multiple hypothesis testing, we applied a Bonferroni correction using a threshold of $P<0.05/13,892 = 3.6 \times 10^{-6}$. A variant that was present in four supercentenarian genomes but absent in all genomes in 1000G EUR would have a P-value of 7.4×10^{-07} and would have been detected by our method. Using high quality sequence calls, preliminary analysis suggested that one novel variant was shared by three supercentenarian subjects but not by the control genomes; however, Sanger sequencing subsequently showed that this was a sequencing error in the supercentenarian data. To increase our sensitivity for finding a longevity variant, we repeated the analysis including low quality calls. This yielded three additional novel variants in the supercentenarian genomes. However, Sanger sequencing showed that each of the three variants was a sequencing error. Even though the overall error rates for SNPs in WGS data (>40x coverage) are under 1% [27], the process of screening for apparent rare protein-altering variants also enriches for sequencing errors [33]. Therefore, we conclude that we found no evidence for a statistically significant enrichment of a specific protein-altering variant in the female Caucasian supercentenarian genomes compared with controls. Table S2 contains a list of the rare coding variants found in our 17 supercentenarian and 34 PGP control genomes.

Are Supercentenarians Enriched for a Gene with Rare Protein-Altering Variants?

Another possibility is that there may be a gene that confers extreme longevity when it is altered by any one of a number of protein alterations. Many of the supercentenarians may carry variants in the same gene, but the variant in each supercentenarian may be different. The variants could act in a dominant fashion and affect only one of the two alleles. Or else they could act in a recessive fashion such that both alleles would be affected, either with the same variant (homozygous) or with different mutations in each allele (compound heterozygous). Therefore, we asked whether any of the genes in the female Caucasian supercentenarian genomes was enriched for harboring

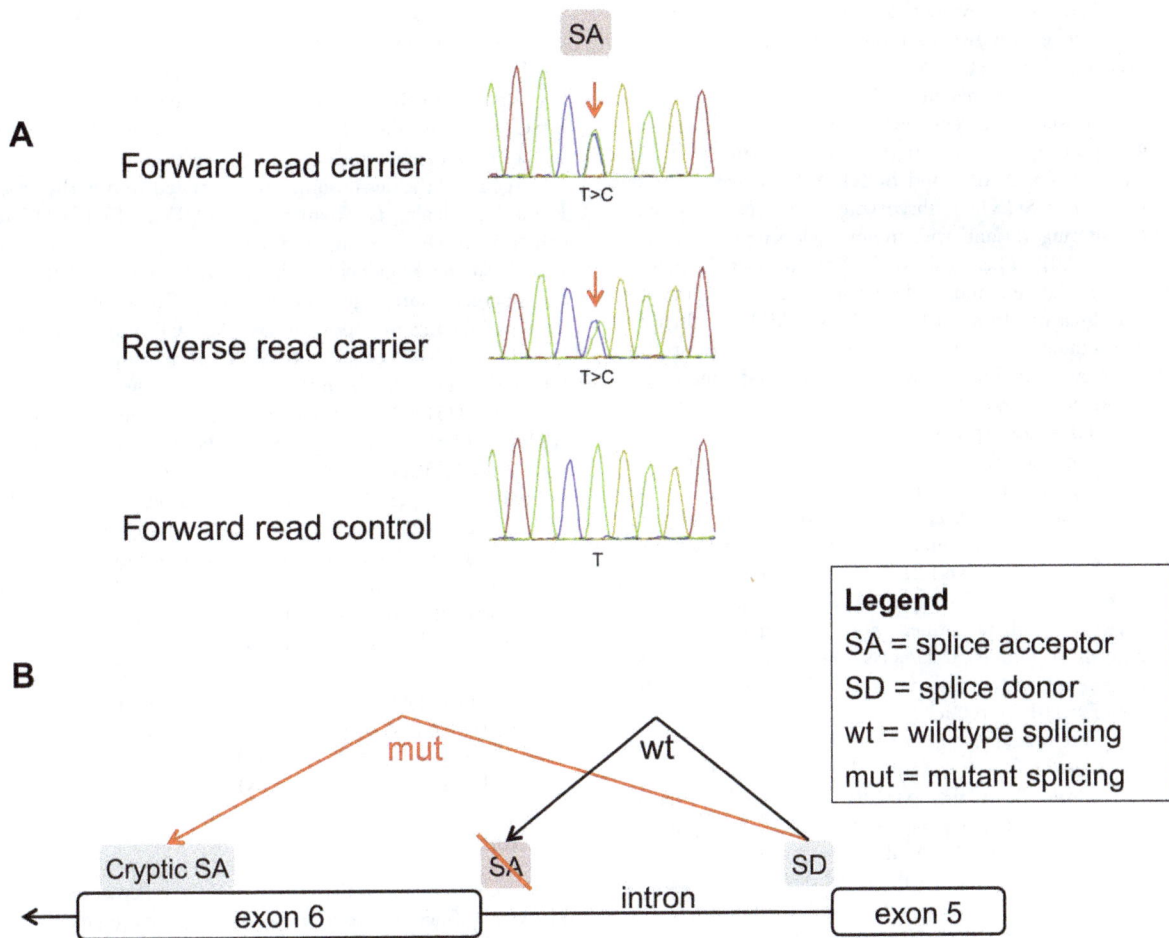

Figure 3. A supercentenarian with a known pathogenic mutation implicated in cardiomyopathy. (A) Sanger validation confirmed that one supercentenarian possessed a known pathogenic mutation in a splice acceptor site of Desmocollin-2 (DSC2), a component of the myocardial desmosome. (B) This rare mutation has been reported in 2 independent cases of Arrhythmogenic Right Ventricular Cardiomyopathy and has been shown to cause cryptic splicing and mRNA degradation [54,55].

rare protein-altering variants (either one or two copies) when compared to control genomes. Although the 1000G are a large group of controls, they cannot be used for a gene-based test as only the frequency of each variant is known, and not the individual genotypes. Therefore, as controls we used WGS of 34 Caucasian individuals (ages 21–79) from the Personal Genome Project (PGP) that were sequenced on the same platform as the supercentenarians [29].

We created a pipeline that used the annotated supercentenarian and PGP genomes from the previous analyses as input (see also Figure 1). Next, we filtered out common variants, which we defined as having a minor allele frequency of 1.5% or higher in the 1000G EUR (i.e., Caucasian populations in the 1000G). For each gene and each genome, we counted the number of rare protein-altering variants. We then computed the RVT1 statistic [35] to determine whether any gene showed a different burden of variants in supercentenarians vs. controls.

There were 10,508 genes with at least one rare, protein-altering variant in controls or supercentenarians. We used a Bonferroni threshold of $P < 0.05/10,508 = 4.7 \times 10^{-06}$ to correct for multiple hypothesis testing. We were thus powered to detect genes altered in seven supercentenarians, if the gene harbored no alleles in any of the 34 controls. None of the genes showed a genome-wide significant enrichment using the Bonferroni threshold (Table S3).

Furthermore, we performed pathway analysis but failed to find a genetic pathway that showed a significant difference between supercentenarians and controls; specifically, we performed Gene Set Enrichment Analysis [47] using the results of the gene burden test, but no KEGG [48] pathway or Gene Ontology [49] category was significant at a false discovery rate <25%. To increase our sensitivity, we repeated our analyses including low-quality calls. This time, two genes initially appeared to be enriched for rare protein-altering alleles in the supercentenarian genomes, but Sanger sequencing showed that many of the variants were WGS errors.

We also specifically tested a recessive model for a gene conferring exceptional longevity, in which both alleles of a gene might harbor mutations. Supercentenarians would be enriched for carrying two or more different variants in such a gene (consistent with compound heterozygosity, if the mutations are out of phase), but controls would only carry zero or one mutation, but not two. The RVT1 test performs a logistic regression on the proportion of rare variants and hence might detect a bias in supercentenarians (two alleles in the gene) vs. the controls (one or zero alleles). But the RVT1 test was not specifically designed to compare the number of compound heterozygous cases and controls. We performed a gene-based test to compare the number of cases and controls carrying at least two variants in the same gene applying Fisher's

Exact Test to compute P values. We found that no gene was significantly enriched for two or more mutations after multiple testing correction (Table S4).

Although none of the genes showed a significant enrichment in the female Caucasian supercentenarian genomes, we nevertheless decided to follow up on the top three genes from the RVT1 burden test: TSHZ3, NAB2, and SCN11A (each with nominal $P = 4.3 \times 10^{-4}$). For SCN11A, three control genomes contained rare protein-altering variants with minor allele frequencies below 0.05 (but above 0.01). This result weakens the distinction between the supercentenarian genomes and the control genomes, and thus this gene was discarded from further analysis. NAB2 was discarded when Sanger sequencing showed that two out of four variants were sequencing errors. For TSHZ3, Sanger sequencing validated all four protein-altering variants, and this gene was chosen as a candidate for follow-up experiments.

To validate the result from the analysis of the supercentenarian genomes, we examined whether TSHZ3 is enriched for rare protein-altering variants in a cohort of 99 people aged 98–105 years from the Georgia Centenarian Study compared to 4,300 control exomes from the NHLBI Exome Variant Server [38] (Table 2). We obtained DNA samples of Caucasian nonagenarians and centenarians and performed Sanger sequencing of the TSHZ3 gene in all long-lived subjects. We used the same filter as for the genome-wide burden test of the supercentenarian genomes (MAF>0.01 in 1000G EUR.). We discovered a higher frequency of protein-altering alleles in the TSHZ3 sequence from 99 long-lived genomes (8 variants; 4%) than in the 4,300 Caucasian controls from the NHLBI cohort (213 variants; 2.5%), but this difference was not statistically significant (P = 0.17; Figure 2; Table 3). Analysis of a larger cohort of supercentenarians may show that the small difference in variants in TSHZ3 compared to controls is statistically significant.

In summary, the results from all three analyses do not show a statistical enrichment for a gene harboring rare protein-altering variants in female Caucasian supercentenarians compared to controls.

Do Supercentenarians Carry Pathogenic Alleles?

WGS has revealed that seemingly healthy individuals can carry pathogenic mutations that are potentially fatal [50]. Based on their extreme longevity, supercentenarians can be viewed as extremely healthy individuals. We asked whether these extremely healthy individuals might also carry pathogenic mutations. To do this, we analyzed all 17 supercentenarian genomes for the presence of pathogenic alleles as defined by the recent publication of the American College of Medical Genetics and Genomics (ACMG) [39]. The ACMG recommends that these mutations be reported to the patient, even if they are incidental findings. Their paper was a concerted and systematic effort resulting in a list of 56 genes, which are known to harbor strongly pathogenic mutations known to be fatal.

Two supercentenarians possessed a variant that was annotated as being pathogenic by the Human Gene Mutation Database (HGMD) or ClinVar. The first supercentenarian carried a missense SNP (L1564P) in the Breast Cancer Associated 1 (BRCA1) gene. Although null mutations in BRCA1 are pathogenic, the pathogenicity of L1564P is unclear. The L1564P variant appeared in the breast cancer of a 33-year old female along with another missense SNP (Q1785H) [51]. Using an *in vitro* assay, it was found that both missense SNPs in this breast cancer were mild alleles that partially reduced, but did not eliminate, BRCA1 protein function [52]. The L1564P mutation, the Q1785H mutation or both together may have caused breast cancer in this

one individual. Hence, the pathogenicity of the L1564P mutation in our supercentenarian remains unclear.

The second supercentenarian possessed a known pathogenic SNP (rs397514042) that disrupts a splice-site in Desmocollin-2 (DSC2). Desmocollin-2 is part of the myocardial desmosome structure in the heart. Loss-of-function mutations in DSC2 and other genes of the desmosome are associated with Arrhythmogenic Right Ventricular Cardiomyopathy (ARVC) [53]. rs397514042 causes an A -> G change in the splice acceptor site of exon 6 of DSC2. Sanger sequencing validated the presence of this SNP in the supercentenarian genome (Figure 3). The variant is annotated as a pathogenic mutation in HGMD, ClinVar, and the Locus Specific Database (LSDB) ARVD/C, which is part of the Leiden Open Variation Database (LOVD).

The rs397514042 SNP has been observed in two patients with ARVC [54,55]. Heuser et al. further showed that the mutant allele (rs397514042) leads to a decrease in DSC2 mRNA and protein in the patient compared to the reference allele. In zebrafish lacking DSC2, expression of the wild-type human allele rescued the mutant phenotype and led to normal desmosomes, but expression of the mutant human allele corresponding to rs397514042 did not fully rescue the mutant phenotype and resulted in malformed desmosomes. Although the evidence suggests that this SNP can be highly pathogenic, its penetrance is unknown. The supercentenarian subject carrying rs397514042 was asymptomatic to the best of our knowledge and died from a cause unrelated to cardiomyopathy. We conclude that at least 1 out of 17 supercentenarians possessed a known pathogenic SNP.

Discussion

We have sequenced the genomes of 17 supercentenarians (over 110 years of age) to see if we could uncover the genetic basis for their extreme longevity. We analyzed rare protein-altering variants, but found no strong evidence for enrichment of either a single variant or a single gene harboring different variants in female Caucasian supercentenarians compared to controls. From our gene-based analysis, the gene showing the most enrichment for protein-altering variants in supercentenarians compared to controls was the TSHZ3 transcription-factor gene. Because it was the top hit, we pursued this gene further in a study consisting of 99 genomes from subjects aged 98–105 years old. We found that TSHZ3 carried protein-altering variants in more of the long-lived subjects than the controls, although this difference was not statistically significant (P = 0.17).

A larger sample size would be required to establish whether the difference in frequency of protein-altering variants in TSHZ3 between subjects with extreme longevity compared to controls is statistically significant. We did not analyze single nucleotide variants in non-coding DNA in the supercentenarians because of the large number of non-coding variants compared to coding variants. Our analysis of putative rare protein-altering variants in the whole genome sequencing data led us to test a number of candidates, of which 30% were subsequently determined to be false positive variant calls in WGS data. This high false discovery rate is consistent with previous reports [34] and is largely due to a selection bias as sequencing errors often appear as rare protein-altering variants [33].

Our analyses show that it is extremely unlikely that there is a single gene harboring rare protein-altering variants shared by all supercentenarians but no controls. It is not surprising that a highly complex trait such as longevity is not explained by a single Mendelian gene.

To our surprise, we discovered that one of our supercentenarians carried a known pathogenic allele in the DSC2 gene associated with arrhythmogenic right ventricular cardiomyopathy (ARVC). This is a potentially fatal condition, causing affected individuals to die of sudden cardiac death. This example points out an important aspect about policy regarding the reporting of pathogenic mutations found in genomic sequences. The American College of Medical Genetics and Genomics identified a set of genes that can cause pathology when disrupted. But what is often not known is how frequently people with the variant have pathology (i.e., the penetrance). Our example shows that the DSC2 pathogenic mutation rs397514042 did not cause a fatal cardiomyopathy during the proband's over 110 years of life. Thus, the presence of this mutation in the DNA sequence of a young person today should be reported to him/her and their families with caution, as it may or may not result in arrhythmogenic right ventricular cardiomyopathy. Generally, variants that are annotated as pathogenic are of unknown penetrance [56].

The full set of protein-coding variants are given in Table S2 and the full-genome sequence from this paper are publicly available via dbGAP and Google Genomics. By making our data available as a public resource, we hope it can be included in future meta-analyses of supercentenarian genomes. Supercentenarians are extremely rare and their genomes could hold secrets for the genetic basis of extreme longevity.

Supporting Information

Figure S1 Genome coverage for supercentenarians. Average genome coverage is shown for the whole genome (dark grey) and exome (light grey) of all 17 supercentenarians. Coverage is shown for $\geq 1x$ and $\geq 20x$ coverage.

Figure S2 Principal Component Analysis of supercentenarian ancestry. PCA was performed on all 17 supercentenarians (black dots) and HapMap genotypes. All Caucasian supercentenarians (CAU) clustered with Caucasian HapMap individuals, while the two supercentenarians of Hispanic ethnicity clustered with Mexican HapMap individuals and the African-American supercentenarian (AA) clustered with African HapMap individuals. HapMap populations are ASW (African ancestry in Southwest USA), CEU (Utah residents with Northern and Western European ancestry from the CEPH collection), CHB (Han Chinese in Beijing, China), CHD (Chinese in Metropolitan Denver, Colorado), GIH (Gujarati Indians in Houston, Texas), JPT (Japanese in Tokyo, Japan), LWK (Luhya in Webuye, Kenya), MXL (Mexican ancestry in Los Angeles, California), MKK (Maasai in Kinyawa, Kenya), TSI (Toscani in Italy) and YRI (Yoruba in Ibadan, Nigeria). See insert for color codes.

Table S1 Baseline statistics for 34 Caucasian PGP genomes.

Table S2 All variants in protein coding regions with genotypes for all 17 supercentenarian and 34 PGP control genomes.

Table S3 Burden of rare protein-altering variants per gene in supercentenarians and controls.

Table S4 Gene-based Fisher's Exact test for recessive model of rare protein-altering variants in supercentenarians and controls.

Author Contributions

Conceived and designed the experiments: SKK LSC LH HJG KF. Performed the experiments: HJG KF JCR HL GG GJM JDS. Analyzed the data: HJG KF JCR HL GG GJM JDS. Contributed reagents/materials/analysis tools: NSC LSC. Contributed to the writing of the manuscript: HJG KF JCR NSC HL GG GJM JDS LH LSC SKK.

References

1. Coles LS, Muir ME, Young RD (2014) Validated worldwide supercentenarians, living and recently deceased. Rejuvenation Res 17: 80–83. doi:10.1089/rej.2014.1553
2. Evert J, Lawler E, Bogan H, Perls T (2003) Morbidity profiles of centenarians: survivors, delayers, and escapers. J Gerontol A Biol Sci Med Sci 58: 232–237.
3. Terry DF, Wilcox MA, McCormick MA, Perls TT (2004) Cardiovascular disease delay in centenarian offspring. J Gerontol A Biol Sci Med Sci 59: 385–389.
4. Willcox DC, Willcox BJ, Wang N-C, He Q, Rosenbaum M, et al. (2008) Life at the extreme limit: phenotypic characteristics of supercentenarians in Okinawa. J Gerontol A Biol Sci Med Sci 63: 1201–1208.
5. Andersen SL, Sebastiani P, Dworkis DA, Feldman L, Perls TT (2012) Health span approximates life span among many supercentenarians: compression of morbidity at the approximate limit of life span. J Gerontol A Biol Sci Med Sci 67: 395–405. doi:10.1093/gerona/glr223.
6. Pavlidis N, Stanta G, Audisio RA (2012) Cancer prevalence and mortality in centenarians: a systematic review. Crit Rev Oncol Hematol 83: 145–152. doi:10.1016/j.critrevonc.2011.09.007.
7. Herskind AM, McGue M, Holm NV, Sørensen TI, Harvald B, et al. (1996) The heritability of human longevity: a population-based study of 2872 Danish twin pairs born 1870–1900. Hum Genet 97: 319–323.
8. Schoenmaker M, de Craen AJM, de Meijer PHEM, Beekman M, Blauw GJ, et al. (2006) Evidence of genetic enrichment for exceptional survival using a family approach: the Leiden Longevity Study. Eur J Hum Genet EJHG 14: 79–84. doi:10.1038/sj.ejhg.5201508.
9. vB Hjelmborg J, Iachine I, Skythe A, Vaupel JW, McGue M, et al. (2006) Genetic influence on human lifespan and longevity. Hum Genet 119: 312–321. doi:10.1007/s00439-006-0144-y.
10. Sebastiani P, Perls TT (2012) The genetics of extreme longevity: lessons from the new England centenarian study. Front Genet 3: 277. doi:10.3389/fgene.2012.00277.
11. Perls TT, Wilmoth J, Levenson R, Drinkwater M, Cohen M, et al. (2002) Life-long sustained mortality advantage of siblings of centenarians. Proc Natl Acad Sci U S A 99: 8442–8447. doi:10.1073/pnas.122587599.
12. Perls T, Kohler IV, Andersen S, Schoenhofen E, Pennington J, et al. (2007) Survival of parents and siblings of supercentenarians. J Gerontol A Biol Sci Med Sci 62: 1028–1034.
13. Rajpathak SN, Liu Y, Ben-David O, Reddy S, Atzmon G, et al. (2011) Lifestyle factors of people with exceptional longevity. J Am Geriatr Soc 59: 1509–1512. doi:10.1111/j.1532-5415.2011.03498.x.
14. Suh Y, Atzmon G, Cho M-O, Hwang D, Liu B, et al. (2008) Functionally significant insulin-like growth factor I receptor mutations in centenarians. Proc Natl Acad Sci U S A 105: 3438–3442. doi:10.1073/pnas.0705467105.
15. Willcox BJ, Donlon TA, He Q, Chen R, Grove JS, et al. (2008) FOXO3A genotype is strongly associated with human longevity. Proc Natl Acad Sci U S A 105: 13987–13992. doi:10.1073/pnas.0801030105.
16. Nebel A, Kleindorp R, Caliebe A, Nothnagel M, Blanché H, et al. (2011) A genome-wide association study confirms APOE as the major gene influencing survival in long-lived individuals. Mech Ageing Dev 132: 324–330. doi:10.1016/j.mad.2011.06.008.
17. Deelen J, Beekman M, Uh H-W, Helmer Q, Kuningas M, et al. (2011) Genome-wide association study identifies a single major locus contributing to survival into old age; the APOE locus revisited. Aging Cell 10: 686–698. doi:10.1111/j.1474-9726.2011.00705.x.
18. Sebastiani P, Solovieff N, Dewan AT, Walsh KM, Puca A, et al. (2012) Genetic signatures of exceptional longevity in humans. PloS One 7: e29848. doi:10.1371/journal.pone.0029848.
19. Sebastiani P, Riva A, Montano M, Pham P, Torkamani A, et al. (2012) Whole genome sequences of a male and female supercentenarian, ages greater than 114 years. Front Genet 2: 90. doi:10.3389/fgene.2011.00090.
20. Ye K, Beekman M, Lameijer E-W, Zhang Y, Moed MH, et al. (2013) Aging as accelerated accumulation of somatic variants: whole-genome sequencing of

centenarian and middle-aged monozygotic twin pairs. Twin Res Hum Genet Off J Int Soc Twin Stud 16: 1026–1032. doi:10.1017/thg.2013.73.

21. Holstege H, Pfeiffer W, Sie D, Hulsman M, Nicholas TJ, et al. (2014) Somatic mutations found in the healthy blood compartment of a 115-yr-old woman demonstrate oligoclonal hematopoiesis. Genome Res 24: 733–742. doi:10.1101/gr.162131.113.

22. Untergasser A, Cutcutache I, Koressaar T, Ye J, Faircloth BC, et al. (2012) Primer3–new capabilities and interfaces. Nucleic Acids Res 40: e115. doi:10.1093/nar/gks596.

23. International HapMap 3 Consortium, Altshuler DM, Gibbs RA, Peltonen L, Altshuler DM, et al. (2010) Integrating common and rare genetic variation in diverse human populations. Nature 467: 52–58. doi:10.1038/nature09298.

24. Yang J, Lee SH, Goddard ME, Visscher PM (2011) GCTA: a tool for genome-wide complex trait analysis. Am J Hum Genet 88: 76–82. doi:10.1016/j.ajhg.2010.11.011.

25. Huff CD, Witherspoon DJ, Simonson TS, Xing J, Watkins WS, et al. (2011) Maximum-likelihood estimation of recent shared ancestry (ERSA). Genome Res 21: 768–774. doi:10.1101/gr.115972.110.

26. Li H, Glusman G, Hu H, Shankaracharya, Caballero J, et al. (2014) Relationship estimation from whole-genome sequence data. PLoS Genet 10: e1004144. doi:10.1371/journal.pgen.1004144.

27. Drmanac R, Sparks AB, Callow MJ, Halpern AL, Burns NL, et al. (2010) Human genome sequencing using unchained base reads on self-assembling DNA nanoarrays. Science 327: 78–81. doi:10.1126/science.1181498.

28. Perl Development Team (2013) The Perl Programming Language. Available: http://www.perl.org/. Accessed 27 February 2014.

29. Ball MP, Thakuria JV, Zaranek AW, Clegg T, Rosenbaum AM, et al. (2012) A public resource facilitating clinical use of genomes. Proc Natl Acad Sci 109: 11920–11927. doi:10.1073/pnas.1201904109.

30. Wang K, Li M, Hakonarson H (2010) ANNOVAR: functional annotation of genetic variants from high-throughput sequencing data. Nucleic Acids Res 38: e164–e164. doi:10.1093/nar/gkq603.

31. Consortium T 1000 GP (2012) An integrated map of genetic variation from 1,092 human genomes. Nature 491: 56–65. doi:10.1038/nature11632.

32. Sherry ST, Ward M-H, Kholodov M, Baker J, Phan L, et al. (2001) dbSNP: the NCBI database of genetic variation. Nucleic Acids Res 29: 308–311. doi:10.1093/nar/29.1.308.

33. MacArthur DG, Balasubramanian S, Frankish A, Huang N, Morris J, et al. (2012) A systematic survey of loss-of-function variants in human protein-coding genes. Science 335: 823–828. doi:10.1126/science.1215040.

34. Han J, Ryu S, Moskowitz DM, Rothenberg D, Leahy DJ, et al. (2013) Discovery of novel non-synonymous SNP variants in 988 candidate genes from 6 centenarians by target capture and next-generation sequencing. Mech Ageing Dev. doi:10.1016/j.mad.2013.01.005.

35. Morris AP, Zeggini E (2010) An evaluation of statistical approaches to rare variant analysis in genetic association studies. Genet Epidemiol 34: 188–193. b013e3280d942c4.

36. R Core Team (2013) R: A Language and Environment for Statistical Computing. Available: http://www.r-project.org/. Accessed 27 February 2014.

37. Poon LW, Clayton GM, Martin P, Johnson MA, Courtenay BC, et al. (1992) The Georgia Centenarian Study. Int J Aging Hum Dev 34: 1–17. doi:10.2190/8M7H-CJL7-6K5T-UMFV.

38. Tennessen JA, Bigham AW, O'Connor TD, Fu W, Kenny EE, et al. (2012) Evolution and functional impact of rare coding variation from deep sequencing of human exomes. Science 337: 64–69. doi:10.1126/science.1219240.

39. Green RC, Berg JS, Grody WW, Kalia SS, Korf BR, et al. (2013) ACMG recommendations for reporting of incidental findings in clinical exome and genome sequencing. Genet Med Off J Am Coll Med Genet 15: 565–574. doi:10.1038/gim.2013.73.

40. Landrum MJ, Lee JM, Riley GR, Jang W, Rubinstein WS, et al. (2014) ClinVar: public archive of relationships among sequence variation and human phenotype. Nucleic Acids Res 42: D980–985. doi:10.1093/nar/gkt1113.

41. Cooper DN, Krawczak M (1996) Human Gene Mutation Database. Hum Genet 98: 629.

42. Richards CS, Bale S, Bellissimo DB, Das S, Grody WW, et al. (2008) ACMG recommendations for standards for interpretation and reporting of sequence variations: Revisions 2007. Genet Med Off J Am Coll Med Genet 10: 294–300. doi:10.1097/GIM.0b013e31816b5cae.

43. Van der Zwaag PA, Jongbloed JDH, van den Berg MP, van der Smagt JJ, Jongbloed R, et al. (2009) A genetic variants database for arrhythmogenic right ventricular dysplasia/cardiomyopathy. Hum Mutat 30: 1278–1283. doi:10.1002/humu.21064.

44. Fokkema IFAC, Taschner PEM, Schaafsma GCP, Celli J, Laros JFJ, et al. (2011) LOVD v.2.0: the next generation in gene variant databases. Hum Mutat 32: 557–563. doi:10.1002/humu.21438.

45. Gerontology Research Group (2014) Table A - Verified Supercentenarians. Available: http://grg.org/Adams/A.HTM. Accessed 21 July 2014.

46. Schoenhofen EA, Wyszynski DF, Andersen S, Pennington J, Young R, et al. (2006) Characteristics of 32 supercentenarians. J Am Geriatr Soc 54: 1237–1240. doi:10.1111/j.1532-5415.2006.00826.x.

47. Subramanian A, Tamayo P, Mootha VK, Mukherjee S, Ebert BL, et al. (2005) Gene set enrichment analysis: A knowledge-based approach for interpreting genome-wide expression profiles. Proc Natl Acad Sci U S A 102: 15545–15550. doi:10.1073/pnas.0506580102.

48. Kanehisa M, Goto S, Sato Y, Furumichi M, Tanabe M (2012) KEGG for integration and interpretation of large-scale molecular data sets. Nucleic Acids Res 40: D109–D114. doi:10.1093/nar/gkr988.

49. Gene Ontology Consortium (2013) Gene Ontology annotations and resources. Nucleic Acids Res 41: D530–535. doi:10.1093/nar/gks1050.

50. Dewey FE, Grove ME, Pan C, Goldstein BA, Bernstein JA, et al. (2014) Clinical interpretation and implications of whole-genome sequencing. JAMA J Am Med Assoc 311: 1035–1045. doi:10.1001/jama.2014.1717.

51. Panguluri RC, Brody LC, Modali R, Utley K, Adams-Campbell L, et al. (1999) BRCA1 mutations in African Americans. Hum Genet 105: 28–31.

52. Carvalho MA, Marsillac SM, Karchin R, Manoukian S, Grist S, et al. (2007) Determination of cancer risk associated with germ line BRCA1 missense variants by functional analysis. Cancer Res 67: 1494–1501. doi:10.1158/0008-5472.CAN-06-3297.

53. Van Tintelen JP, Hofstra RM, Wiesfeld AC, van den Berg MP, Hauer RN, et al. (2007) Molecular genetics of arrhythmogenic right ventricular cardiomyopathy: emerging horizon? Curr Opin Cardiol 22: 185–192. doi:10.1097/HCO.0-b013e3280d942c4.

54. Heuser A, Plovie ER, Ellinor PT, Grossmann KS, Shin JT, et al. (2006) Mutant Desmocollin-2 Causes Arrhythmogenic Right Ventricular Cardiomyopathy. Am J Hum Genet 79: 1081–1088. doi:10.1086/509044.

55. Baskin B, Skinner JR, Sanatani S, Terespolsky D, Krahn AD, et al. (2013) TMEM43 mutations associated with arrhythmogenic right ventricular cardio-myopathy in non-Newfoundland populations. Hum Genet 132: 1245–1252. doi:10.1007/s00439-013-1323-2.

56. Bick AG, Flannick J, Ito K, Cheng S, Vasan RS, et al. (2012) Burden of Rare Sarcomere Gene Variants in the Framingham and Jackson Heart Study Cohorts. Am J Hum Genet 91: 513–519. doi:10.1016/j.ajhg.2012.07.017.

8

Early Chordate Origin of the Vertebrate Integrin αI Domains

Bhanupratap Singh Chouhan[1], Jarmo Käpylä[2], Konstantin Denessiouk[1], Alexander Denesyuk[1], Jyrki Heino[2], Mark S. Johnson[1]*

1 Structural Bioinformatics Laboratory, Biochemistry, Department of Biosciences, Åbo Akademi University, Turku, Finland, **2** Department of Biochemistry, University of Turku, Turku, Finland

Abstract

Half of the 18 human integrins α subunits have an inserted αI domain yet none have been observed in species that have diverged prior to the appearance of the urochordates (ascidians). The urochordate integrin αI domains are not human orthologues but paralogues, but orthologues of human αI domains extend throughout later-diverging vertebrates and are observed in the bony fish with duplicate isoforms. Here, we report evidence for orthologues of human integrins with αI domains in the agnathostomes (jawless vertebrates) and later diverging species. Sequence comparisons, phylogenetic analyses and molecular modeling show that one nearly full-length sequence from lamprey and two additional fragments include the entire integrin αI domain region, have the hallmarks of collagen-binding integrin αI domains, and we show that the corresponding recombinant proteins recognize the collagen GFOGER motifs in a metal dependent manner, unlike the α1I domain of the ascidian *C. intestinalis*. The presence of a functional collagen receptor integrin αI domain supports the origin of orthologues of the human integrins with αI domains prior to the earliest diverging extant vertebrates, a domain that has been conserved and diversified throughout the vertebrate lineage.

Editor: Edward F. Plow, Lerner Research Institute, United States of America

Funding: These studies received financial support from the National Doctoral Network in Informational and Structural Biology (BSC), the Academy of Finland, Sigrid Juselius Foundation, Joe, Pentti and Tor Borg Memorial Fund, and the Abo Akademi Center of Excellence in Cell Stress and Aging. The funders had no role in study design, data collection and analysis, decision to publish, or preparation of the manuscript.

Competing Interests: The authors have declared that no competing interests exist.

* Email: johnson4@abo.fi

Introduction

Integrins are multi-domain cell-surface receptors that fulfill numerous function roles at the level of cell-cell communication and interactions between cells and proteins of the extracellular matrix (for a review, see [1]). Integrins have an early origin, preceding the first metazoans [2], with most component domains identifiable in bacterial sequences (see e.g. [3–5]; reviewed in [6]) and, despite multicellular species that do not have integrins (e.g. fungi and plants), integrins were likely necessary and greatly facilitated the development and diversification of multicellular animals. The bidirectional signaling mediated by integrins enables changes relative to the external environment when instigated by cytoplasmic events in individual cells or promotes cellular changes as a result of ligand binding to the external ectodomain. Consider, for example, the dynamic processes involved in tissue remodeling and wound repair, where e.g. cells accumulate on collagen fibers of the ECM and cells of the immune system bind at sites of inflammation, but where these cells also will need to detach and relocate.

In humans there are 24 integrin heterodimers that have been observed to form from 18 α subunits and 8 β subunits [7]. Half of the α subunits have an extra "inserted" I domain [8] or "A" domain [9] (see Fig. 1). Of the nine integrins with αI domains, five have immune system functions: αLβ2, αMβ2, αDβ2, αXβ2 and

αEβ7; and four are collagen receptors: α1β1, α2β1, α10β1 and α11β1. The first X-ray structures of integrins deposited within the Protein Data Bank (PDB; [10]) have focused on the αI domain in human integrin α subunits: e.g. αM (PDB code: 1IDO and 1JLM; [11,12]) and αL (1LFA; [13]) of the immune system type; and α2 without (PDB code: 1A0X; [14]) and with (1DZI; [15]) collagen-like triple-helical GFOGER peptide bound. In 2010, the αXI domain was solved within the ectodomain context of the αβ subunit complex (3K6S; [16]).

The αI domains are Rossmann folds, but more specifically they belong to the von Willebrand factor type A-like fold (vWA-like, the SCOP database, [17]) and the sequences are categorized to von Willebrand factor type A protein ECM (vWA_ECM) in the NCBI Conserved Domain Database (CDD, [18]). The integrin αI domains (cd01469 sequence cluster; vWA_integrins_α_subunit) are only one of nine domain subfamilies (CDD ID: cd01450, vWFA_subfamily_ECM) that includes at least 110 different eukaryotic domains [6].

All integrin β subunits contain a βI-like domain (Fig. 1) and, for example, in the αVβ3 integrin that does not have an inserted αI domain, protein ligands bind via the RGD sequence motif (and variants; see e.g. [19]) located on external loops where the aspartic acid binds to the metal ion dependent adhesion site (MIDAS) of the βI-like domain and arginine binds to the β-propeller domain of

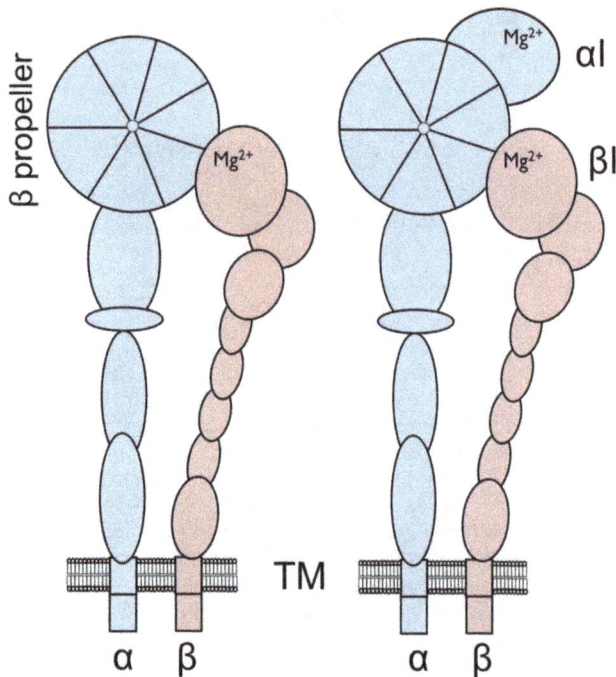

Figure 1. Schematic representation of integrin heterodimers. Integrins are large heterodimeric, bi-directionally signaling, cell surface receptors that consist of a large extracellular ectodomain, a transmembrane region and relatively short intracellular "tails" (right). (A) The constituent α and β subunits are non-covalently associated and the α subunit (ca. 1100 residues) is generally larger than the β subunit (ca. 800 residues). (B) Half of the human integrin α subunits – α1, α2, α10 and α11 of the collagen receptors and αD, αX, αL, αM and αE of the leukocyte clade – contain an additional domain known as the "inserted" αI domain, which buds out between the second and third repeat of the β-propeller domain located at the α subunit N-terminus. The αI domain is a member of the von Willebrand factor A domain family present in many other proteins, including all integrin β subunits and many proteins related to the extracellular matrix, and it is known to adapt the Rossmann fold. The αI domain contains the highly solvent-exposed MIDAS site (Mg^{2+}) where natural ligands bind via a negatively-charged amino acid glutamate. The βI-like domain is located towards the N-terminus in β subunits and acts as the recognition site for external ligands in those integrin heterodimers that do not have the αI domain (A), but binds a glutamate residue – an intrinsic ligand – from the αI domain in the collagen receptor and leukocyte clade α subunits (B).

the α subunit (1L5G; [20]). MIDAS in the αI domain is also key to ligand recognition and function of integrins with αI domains as seen in the three-dimensional structures of α2I-GFOGER [15] and α1I-GLOGEN (PDB code: 2M32, [21]) where the glutamate of the triple-helical collagen-like peptides bind at a coordinating position to a divalent metal cation. Similarly, glutamate e.g. from ICAM1 (1MQ8, [22]; 3TCX, [23]), ICAM3 (1T0P, [24]) and ICAM5 (3BN3, [25]) bind to MIDAS of the αLI domain. The collagen binding integrins and those that recognize leukocytes also have recognizable differences, having the αC helix containing a key tyrosine residue (Y285 in the α2I domain; 1AOX) and present only in the collagen receptor αI domains [14] – an easy-to-scan sequence feature observable in alignments (Fig. 2; [26,27]), observed in the ligand-free structures of the α1I and α2I domains but unraveled (Y285 moves by over 17 Å forming a hydrogen bond with S316; 1DZI) after the conformational changes accompanying ligand binding.

Integrin sequences with αI domains have not been observed in echinoderms [4] nor in the genome [28] of the earliest-diverging chordate – *Branchiostoma floridae*, the lancelet [6,29], but integrins do make their initial appearance in another early chordate species, with one αI domain sequence identified in the tunicate *Halocynthia roretzi* [30] and eight α subunits with αI domains identified [29,31] among the genomic sequencing data [32] of *Ciona intestinalis*. Tunicate integrins with αI domains are not orthologues of the nine human integrin α subunits with I domains [4,6,29,31], and none of the tunicate sequences contain the αC helix that characterizes the human collagen receptor integrins (Fig. 2).

The I domain leads to a dramatic alteration to the integrin ligand-recognition structure in that it shifts the ligand recognition site (see Fig. 1) from a narrow space where an exposed loop on the protein ligand that can cross-link MIDAS of the βI-like domain with the β-propeller domain to a more exposed site that recognizes larger, tubular-shaped and bulkier domain ligands, e.g. collagen fibers bundled into large macroscopic structures and immuno-globulin-fold ICAM domains. With the α2I domain, other, opportunistic ligands such as a snake venom metalloproteinase and echovirus 1 [33–36] very likely bind to the αI domain, covering the MIDAS site, but not directly via a ligand-metal interaction at MIDAS. In integrins with an αI domain, the βI-like domain of the β subunit assumes a new role, by binding a negatively-charged residue (e.g. E336 in α2I) from the α subunit as an "intrinsic ligand", helping to stabilize one of several conformations in the dynamic, mechanical responses to bidirectional signaling [16,36–38].

Here, we have sought to clarify the origins of the integrin α subunits having I domains with features characteristic of the human receptors. In searching for integrin sequences throughout the chordates we identified three sequences from lamprey and possibly one from hagfish that have the hallmarks of αI domains. Furthermore, three fragments from a shark genome study [39], seen earlier [26], two of which have the αC helix, are clearly derived from integrins orthologous to human integrins and now, with the genome published, at least four complete α subunits of integrins with I domains are identifiable. Here, we characterize the features of those sequences and their likely structures and place them within the contextual framework for integrin evolution that has unfolded over the past 25 years.

Results

Searches Identify Likely αI Domain Sequences in Cartilaginous Fish and Tunicates

Orthologues of the human integrin α subunits with I domains are found in species extending from the bony fish (Osteichthyes) through to the mammals [6,29,40]. Thus, we can bracket the appearance of the integrin α subunits with I domains, having features found in the human receptors, to ancestors of species that appeared since the divergence of the tunicate ascidians and before the appearance of the bony fish. Only a few extant representative groups have diverged after the tunicates and before the bony fish, and some genomic data are available for two Agnathostomes (jawless vertebrates) – *Eptatretus burgeri* (inshore hagfish) and *Petromyzon marinus* (sea lamprey) and from cartilaginous fish (Chondrichthes; sharks/rays/skates/chimaera).

We have been regularly searching genomic sequencing data for integrins sequences in order to clarify the origins of different features, especially integrin α subunits, individual domains and αI domains in particular. We conducted searches [26] of all the available genomic assemblies and ESTs from species that diverged

A

		MIDAS		αC helix	Intrinsic Ligand

```
              MIDAS                              αC helix         Intrinsic
                                                                   Ligand
Hsa α1     DIVIVLDGSNSIYP......QTMTA......VTDGESH......IQRFSIAILGSYNRGNLSTEKFVEE......FALEAT
Gga α1     DIVIVLDGSNSIYP......QTMTA......VTDGESH......IQRFAIAILGSYSRGNLSTEKFVEE......FALEAT
Xtr α1     DIVIVLDGSNSVYP......QTMTA......VTDGESH......IQRFSIAILGSYNRGNLSTETLVEE......FALEAT
Dre α1     DIVIVLDGSNSIYP......KTMTA......VTDGESH......IERFAVAVLGDYNRQNKSID[6]EE......FALEAT
Cmi α1     DIMIVLDGSNSIYP......QTKTA......VTDGESH......ITMFAIAVLGSYNRGNQSTVKFLKE......FALEAT
Hsa α2     DVVVVCDESNSIYP......LTNTF......VTDGESH......ILRFGIAVLGYLNRNALDTKNLIKE......FSIEGT
Gga α2     DIVVVCDESNSIYP......LTNTF......VTDGESH......ITRFGIAVLGYLIRNELDTKNLIKE......FSIEGT
Dre α2     DIAIVLDGSNSIYP......ETNTF......VIDGESH......ITRFGIAVLGYYIRNDIDTSKLIAE......FNIEGV
Cmi α2     DIVIVLDGSNSIWP......ETNTA......VTDGESH......IIRFGIAVLGYYNRVGIDTSNLIKE......FSIEGT
Hsa α10    DVVIVLDGSETKTA......ETKTA......VTDGESH......VTRYGIAVLGHYLRRQRDPSSFLRE......FGLEGS
Dre α10    DIVIVLDGSNSIYP......ETRTA......VTDGESH......ITRYAIAVLGHYIRRQQDPETFINE......FSLEGT
Hsa α11    DIVIVLDGSNSIYP......ETRTA......ITDGESH......VTRYAVAVLGYYNRRGINPETFLNE......FSLEGT
Gga α11    DIIIVLDGSNSIYP......ETRTA......ITDGESH......VTRYAVAVLGYYNRRGINPEAFLNE......FSLEGT
Xtr α11    DIVIVLDGSNSIYP......ETRTA......ITDGESH......ITRYAVAVLGYYNRRGINPEAFLNE......FSLEGG
Dre α11B   DIVIVLDGSNSIYP......ETRTA......ITDGESH......ITLYGIAVLGYYNRRGINPEAFLRE......FSLEGT
Dre α11A   DIVIVLDGSNSIYP......ETRTA......ITDGESH......ITRYAIAVLGYYNRRGINPEAFLNE......FSLEGT
Cmi α11    DIVIVLDGSNSIYP......ETNTA......ITDGESH......ITRYSIAVLGYYNRRGINPTHFLKE......FSLEGT
Pma f1     DIVFVLDGSNSIYP......MERGN......VTDGESH......ITRYAIAVLRSYSSNADDVARLINE......FSLEGT
Pma f2     DIVIVLDGSNSIYP......RTASA......VTDGESH......ITRYAIAVLGYYKRKNIDPSNFISE......FSLEGT
Pma f3     DIVLVLDGSNSIWP......VTNTA......VTDGESS......ITRFGIAVLDYYISSNMNVEKLQAE......YSLEGT
Hsa αD     DIVFLIDGSGSIDQ......LTFTA......ITDGQKY......IIRYAIGVG-----HAFQGPTARQE......YAVEGT
Hsa αX     DIVFLIDGSGSISS......FTYTA......ITDGESK......IIRYAIGVG-----LAFQNRNSWKE......FAIEGT
Dre αX-like DIAFLLDGSGSVDP......WTFTA......ITDGESN......IIRYAIGVG-----NAFNKYSARNE......IAIEGT
Hsa αM     DIAFLIDGSGSIIP......RTHTA......ITDGEKF......VIRYVIGVG-----DAFRSEKSRQE......FAIEGT
Hsa αL     DLVFLFDGSMSLQP......LTNTF......ITDGEAT......IIRYIIGIG-----KHFQTKES-QE......YVIEAT
Dre αL     NLVFLFDGSRSMKP......LTNTH......ITDGDPT......ILRYIIGVG-----GLAN--LARLT......YNIEGS
Hsa αE     EIAIILDGSGSIDP......VTKTA......LTDGGIF......VERFAIGVG-----EEFKSARTARE......ISMEGT
Xtr αE     EIAIVLDGSGSISE......VTKTA......LTDGDIF......IERFVIGVG-----EAFQKEKALKT......IGIEGT
Dre αE     EIAFVLDGSGSIQD......LTKTA......LSDGKIL......VTRYSIGVG-----DGIKNKDAIKE......IGTEGT
Cmi αE     EIAIILDGSGSIDA......VTKTA......VTDGEIY......VERFAIGVG-----DATKKPKPVEE......VGIEGT
Ebu f      DIVVLFDGSRSVTD......GTNAY......ISDGESD......DALN----------------------......------
Hro α1     DVLFVLDGSGSVGK......TTYTG......LTDGQAK......IATFAVGVG----------EYDISE......FVLEGG
Ci α1      DLIFLIDESTSVLE......GTATG......LTDGKSQ......IVMFAIGVG----------KVVMGE......ASLESQ
Ci α2      DMLFVLDGSGSVGK......TTYTA......LTDGLST......ITTFAVGVG----------EANEKE......FVLEGA
Ci α3      DLVYVVDGSNSISD......NTFTS......ITDGKAN......ITVYAIGVA----------LKSDAE......SSGEGQ
Ci α4      DIIILLDGSTSVFP......QTFIH......ITDGEAT......IILTAVGIG----------SSVNE......------
Ci α5      DIIFVVDESCTVNR......GTYIG......LTDGRAD......IVTVSVGVG----------DKINE......VKLEGA
Ci α6      DIIFVVDGSGSVDV......LTYIG......LTDGAAT......IVTVSVGVG----------SRVDE......VKLEGD
Ci α7      DIMFVLDDSSSVDD......GTYIS......LTDGGAS......IVLVSVGVG----------TSVNN......LTARTN
Ci α8      DIIFVVDESGSVDT......LTYIG......LTDGRAT......IVTVSVGVG----------SGIIE......VKLEGQ
```

B

C

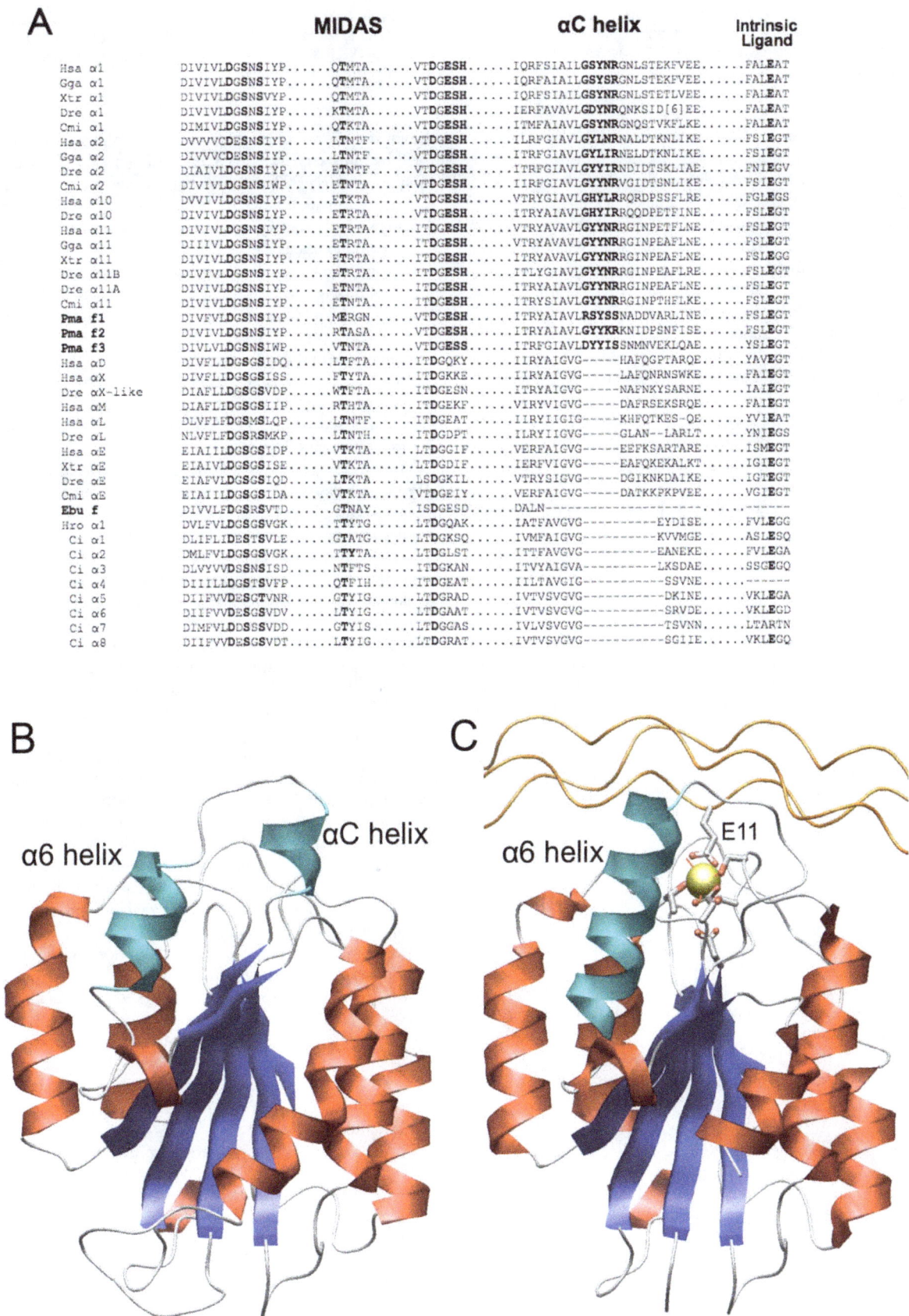

Figure 2. Key features of the integrin αI domain. (A) Alignment of representative sequences, including the three sea lamprey fragments, one short EST fragment derived from the inshore hagfish genome, and four sequences from the elephant shark genome (highlighted in bold). The residues DxSxS...D...T of MIDAS (in bold) function to bind directly or via water molecules to the metal ion where natural ligands bind via a glutamate

residue. The sequence ESH (bold) is characteristic of collagen-binding αI domains; the αC helix (bold) is a distinctive hallmark of the collagen receptor α subunits. The intrinsic glutamate ligand (bold) of the αI domain binds to MIDAS of the βI-like domain in integrins that have the inserted αI domain. Structure of the α2I domain without (B) (PDB code: 1AOX; [14]) and with (C) bound GFOGER tripeptide (PDB code: 1DZI; [15]). The peptide binds to the metal (yellow sphere) at MIDAS via glutamate E11 of the peptide. Consequently, the αC helix unravels and the α6 helix lengthens.

after the urochordates and before the bony fish: including *P. marinus*, *E. burgeri*, *Callorhinchus milii*, (chimaera; elephant shark; Australian ghost shark), *Raja erinacea* (little skate) and *Squalus acanthias* (dogfish shark). Although our intuition is that orthologues of human αI domains should be found in cartilaginous fish, our searches of the chimaera, skate and shark assemblies only yielded three short fragments. Two sequence fragments from *C. milii* were very similar to portions of the human integrin I domains α1 (AAVX01128089.1; 55 residues; 76% identical) and α2 (AAVX01352230.1; 55 residues; 71% identical), beginning by matching the αI domain αC helix; a third fragment from *C. milii* matched repeat 5 of the β-propeller domain of human α2 (AAVX01625876.1; 52 residues; 63% identical). Now, with the publication of the genome sequence of *C. milii* [39], there are at least four orthologues of the corresponding human integrin subunits: collagen-binding α1, α2 and α11, and αE from the leukocyte clade (fig. 3).

Searches [26] also identified three sequence fragments from the sea lamprey genome [41]. With more recent updates these fragments include Pma_f1 having two splice variants (ENSPMAP00000003339, 617 amino acids; ENSPMAP00000003342, 582 amino acids), Pma_f2 (ENSPMAP00000008300, 478 amino acids) and Pma_f3 (ENSPMAP00000003839, 1099 amino acids), which is nearly full-length and missing about 120 residues (compared to the α10I and α11I domains) corresponding to the first two repeats from the N-terminus of the β-propeller domain [26]. In this study we have considered the sequence of the larger 617 amino acid splice variant of Pma_f1.

Additionally, one short 133-residue fragment (Ebu_f) of a possible αI domain from the hagfish *E. burgeri* genome [42] was identified by Blast searches (NCBI service) using human αI domain sequences as the query. When compared with the nine human integrin αI domains, the sequences derived from the sea lamprey genome were found to contain the signature αC helix located towards the C-terminal region of the αI domain – the hallmark of the collagen-receptor integrin subunits (Fig. 2). The αC helix region is not found in either the immune system I domains nor in the sequences of the nine tunicate integrin αI domains. The short fragment that is derived from the hagfish terminates just prior to the αC helix (Fig. 1) but sequence searches suggested it may be most similar to a leukocyte clade member, the integrin αLI domain.

Agnathostome αI Domain Sequences Cluster with Human αI Domains

Here, we have constructed three separate sets of phylogenetic trees from sequence alignments and based on three different tree reconstruction methods. The sequences include representatives from 15 chordate species containing the αI domain (Table 1 and Table S1 in File S1.doc). In addition to sequences from nine human α subunits, sequences are included from other mammals, chicken, a frog (*Xenopus laevis*), four bony fish species (*Tetraodon nigroviridis*, green spotted pufferfish; *Oreochromis niloticus*, Nile tilapia; *Danio rerio*, zebrafish; *Cyprinus carpio*, common carp), four sequences from the elephant shark *C. milii* and sequences from the tunicates *H. roretzi* and *C. intestinalis*. The sets of trees differ in having (a) 69 sequences including the near full-length lamprey Pma_f3 sequence; (b) 72 partial sequences that include

the maximum common regions – 406 to 409 residues shared by the 3 lamprey sequences Pma_f1-3; and (c) 73 αI domain regions of approximately 200 residues that include the hagfish sequence fragment Ebu_f and Pma_f1-3. The 701-residue sequence fragment of α1 from *C. intestinalis* is included in the αI domain and common regions trees.

Phylogenetic trees were inferred from sequence pairwise distances (using either the JTT distance matrix [43]; or the Whelan and Goldman (WAG) matrix [44]) obtained from the aligned sequences and based on implementations of the Maximum Likelihood (ML; Fig. 3), Bayesian (Fig. S1 in File S1.doc) and Neighbor Joining (NJ; Fig. S2 in File S1.doc) methods as described in the Materials and Methods. Multivariate 3D plots were prepared based on the JTT distance data and lend support to the tree topologies (Fig. 4).

The clusterings represented by trees constructed using the ML (Fig. 3) and Bayesian (Fig. S1 in File S1.doc) methods reflect the identical segregation of major groups and most details within the groups also match, and are in agreement with published trees ([4,6,29,31,40,45–51] among others). In general, the tunicate sequences cluster as a single clade and as an apparent outlier to the remaining integrin I domains. The vertebrate integrin sequences segregate into two major clusters corresponding to the immune system or leukocyte clade integrins and those of the collagen receptors, and both clusters subdivide according to the generally accepted subgroups (Fig. 3A). Fish sequences exhibit subtype pairs (e.g. zebrafish α11A and α11B) and the fish cluster branching after the αE and αL branches appear to have diverged prior to the αM-αD-αX diversification found in mammals. Some discrepancies do appear, e.g. α1/α2 subunit clustering in the NJ tree (Fig. S2A in File S1.doc; also indicated by poor bootstrap replication) and when trees are based on the shorter, less-informative sequence fragments i.e. the αI domain region. The fragments from the elephant shark *C. milii* that were observed by us earlier clearly corresponded to orthologues of the human α11I and α2I domains. Three full-length sequences now available from the published genome sequence cluster appropriately as outliers to the α1, α2 and α10 collagen receptor integrins, prior to the bony fish representatives, consistent with them being true orthologues of these vertebrate integrins. Similarly, the αE sequence of *C. milii* appears to be a true orthologue since it also branches just prior to the zebrafish sequence in the αE cluster. Thus, it appears that true orthologues of at least four integrins with αI domains, from both collagen receptors and from the immune system integrins, found in species from bony fish to human are also present in the cartilaginous fish.

The ML tree based on the largest common fragment from the three lamprey sequences (Fig. 3A) places the lamprey Pma_f3 sequence after the α1/α2 divergence and as an outlier of the α10/α11 clade, in agreement with the Bayesian (Fig. S1A in File S1.doc) and NJ (Fig. S2A in File S1.doc) trees. The bootstrap reproducibility of the ML and NJ trees are near 100% (1000 replicates) for nodes where Pma_f3 branches. The posterior probabilities assigned to the branches in the Bayesian tree are 100% for most branches and for the node leading to the Pma_f3 branch.

Similarly, the ML, Bayesian and NJ trees (Fig. 3B, Figs. S1B and S2B in File S1.doc) based on the largest common region

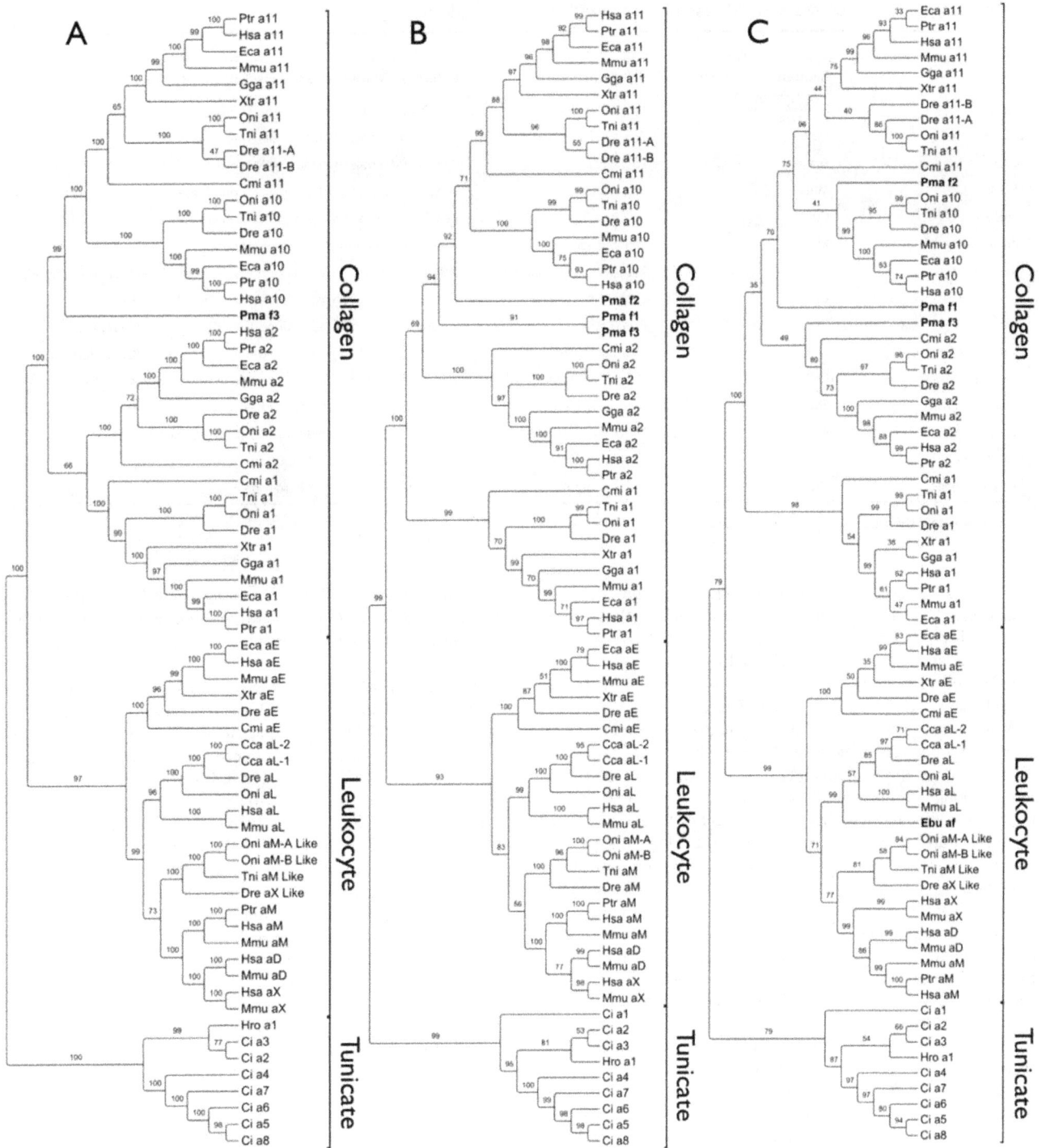

Figure 3. Phylogenetic analysis of integrin sequences with the Maximum Likelihood method. (A) Tree based on the full-length sequence alignment of integrin α subunits derived from the species listed in Table 1. This dataset contains the nearly full-length integrin α subunit from the sea lamprey Pma_f3 (highlighted in bold). (B) Tree based on the aligned common sequence region in all three lamprey sequence fragments Pma_f1, Pma_f2 and Pma_f3 (highlighted in bold). The common region of the α subunit includes three of seven beta propeller repeats (a small portion of repeat number 2, repeat 3 and repeat 4) and the integrin αI domain; the alignment spans about 550 positions. (C) Tree based on the alignment of the integrin αI domain sequences; this dataset includes the three lamprey αI domain sequences Pma_f1, Pma_f2 and Pma_f3 (highlighted in bold) and the hagfish fragment Ebu_f (highlighted in bold). The sequence alignment or the αI domains spans about 250 positions including gaps. Trees were constructed using MEGA by implementing the Whelan and Goldman substitution matrix with frequency model and gamma distribution with invariant sites (WAG+I+G+F). Statistical support for each phylogenetic tree was obtained with 1000 bootstrap replicates and the percentage bootstrap support value is indicated at each node.

Table 1. Chordate genomes and EST assemblies utilized for the integrin phylogenetic analysis.

Organism	Sequence code used	Scientific name	Subphylum/Superclass/Class/Subclass/Order
Human	Hsa	*Homo sapiens*	Vertebrata/Tetrapoda/Mammalia/Theria/Primates
Chimpanzee	Ptr	*Pan troglodytes*	Vertebrata/Tetrapoda/Mammalia/Theria/Primates
Horse	Eca	*Equus caballus*	Vertebrata/Tetrapoda/Mammalia/Theria/Perissodactyla
Mouse	Mmu	*Mus musculus*	Vertebrata/Tetrapoda/Mammalia/Theria/Rodentia
Chicken	Gga	*Gallus gallus*	Vertebrata/Tetrapoda/Aves/-/Galliformes
African clawed frog	Xtr	*Xenopus laevis*	Vertebrata/Tetrapoda/Amphibia/-/Anura
Green spotted pufferfish	Tni	*Tetraodon nigroviridis*	Vertebrata/Osteichthyes/Actinopterygii/Neopterygii/Tetraodontiformes
Nile tilapia	Oni	*Oreochromis niloticus*	Vertebrata/Osteichthyes/Actinopterygii/Neopterygii/Perciformes
Zebrafish	Dre	*Danio rerio*	Vertebrata/Osteichthyes/Actinopterygii/Neopterygii/Cypriniformes
Common carp	Cca	*Cyprinus carpio*	Vertebrata/Osteichthyes/Actinopterygii/Neopterygii/Cypriniformes
Elephant shark	Cmi	*Callorhinchus milii*	Vertebrata/Chondrichthyes/Chondrichthyes/Holocephali/Chimaeriformes
Inshore hagfish	Ebu	*Eptatretus burgeri*	Vertebrata/-/Myxini/-/Myxiniformes
Sea lamprey	Pma	*Petromyzon marinus*	Vertebrata/-/Cephalaspidomorphi/-/Petromyzontiformes
Vase tunicate	Ci	*Ciona intestinalis*	Tunicata/-/Ascidiacea/-/Enterogona
Sea pineapple	Hro	*Halocynthia roretzi*	Tunicata/-/Ascidiacea/-/Pleurogona

"-" indicates that the classification is not available.

shared by all three lamprey sequences places the three lamprey sequences as an outlier of the α10/α11 clade, where Pma_f1 and Pma_f3 cluster together and adjacent to Pma_f2. There is clearly more noise in the trees overall, reflected in differences within the branch orders among the trees and with the full-length trees, and less reliable bootstrap and probability indicators at nearby nodes.

Although the alignments of the sequences corresponding to the shorter αI domain regions are very reliable, the similarity differences over the αI domain are less discriminating than those from the longer sequences. The trees based only on the αI domain regions (Fig. 3C, and Figs. S1C and S2C in File S1.doc) reflect the general features of the other trees based on the longer sequences, but the level of noise is even higher and there are more discrepancies, e.g. in the collagen integrin subdivisions. Nonetheless, the lamprey sequences cluster with the collagen receptor αI domains, although their locations are more variable compared to the full-length and common-segment trees, but then the support for the trees in the vicinity of the lamprey sequences is also poor. The features of all three sets of trees are also reflected in the multivariate plots (Fig. 4).

The hagfish fragment (Ebu_f) ends prior to the αC helical region (Fig. 2). A search of the fragment using the Blast server ([52]; blast.ncbi.nlm.nih.gov) identifies as the closest matches multiple αL integrins, and in all three trees (Fig. 3C, and Figs. S1C and S2C in File S1.doc) the sequence branches off with the immune cell receptor αLI domains, and this is consistent with the multivariate analysis of the distance data (Fig. 4C). Thus, the short fragment from the hagfish (Ebu_f) may be a homologue of the leukocyte specific integrin α subunit, but one must be cautious given the short fragment and lack of other clear distinguishing features in the sequence.

Functional Residues are Shared between Human and Lamprey αI Domains

Key residues involved in αI domain recognition of the collagen-like GFOGER and GLOGEN tripeptides were identified from known representative three-dimensional structures of complexes using Surf2 (MS Johnson, unpublished), and then we examined the similarities and differences among equivalent residues in the human set of integrin αI domains and the residues present in the agnathostome sequence fragments (Table 2, and Tables S2 and S3 in File S1.doc).

The integrin αI domain provides a highly-exposed surface for ligand recognition. The central metal is presumably Mg^{2+} at the MIDAS site and binds glutamic acid of ligands, although Co^{2+} used in the crystallization is present in the α2I domain and binds E11 from one chain of the GFOGER tripeptide ligand in the complex structure (Fig. 5; [15]). Similarly, a glutamate of the GLOGEN tripeptide binds to the metal ion at MIDAS in the α1I domain structure of the complex [21] but the peptide is rotated about the glutamate with respect to the α2I-GFOGER tripeptide complex structure, which may suggest that different collagen recognition sequences bind at different rotational positions on the surface of a particular collagen-binding αI domain. In the leukocyte clade αLI domain structures with bound ICAM-1 D1 (3TCX; [23]), ICAM-3 (1T0P; [24]) and ICAM-5 (3BN3; [25]), immunoglobulin-like fold domains bind to αLI respectively via E34, E37 and E37 to the metal at MIDAS.

The residues from the human α2I domain within 4.2 Å of the tripeptide are shown in Table 2 along with the equivalent sequences in the other eight human αI domains, the three lamprey sequences and the hagfish fragment. Similarly, the nearby residues in the α1I-GLOGEN [21] and αLI-ICAM3 [24] complexes are compared with the other sequences (Tables S2 and S3 in File S1.doc). Residues of MIDAS are absolutely conserved with the exception of Pma_f1, where there is no nearby equivalent residue to T221 in the α2I domain. Glutamate in the sequence "MER" in Pma_f1 may be able to fulfill that role in binding metal, but this is solely based on modeling of the structure and has not yet been tested experimentally. There are clear differences with the leukocyte αI domains as well as similarities. D219 and equivalent residues in collagen-binding αI domains are important for collagen selectivity [53], where residue swaps at this position, e.g. D219R in α2I and R218D in α1I, exchange the collagen preferences of α2I (the wild type prefers collagen I-III) and α1I/α10I (prefer collagen types IV and VI). This position is

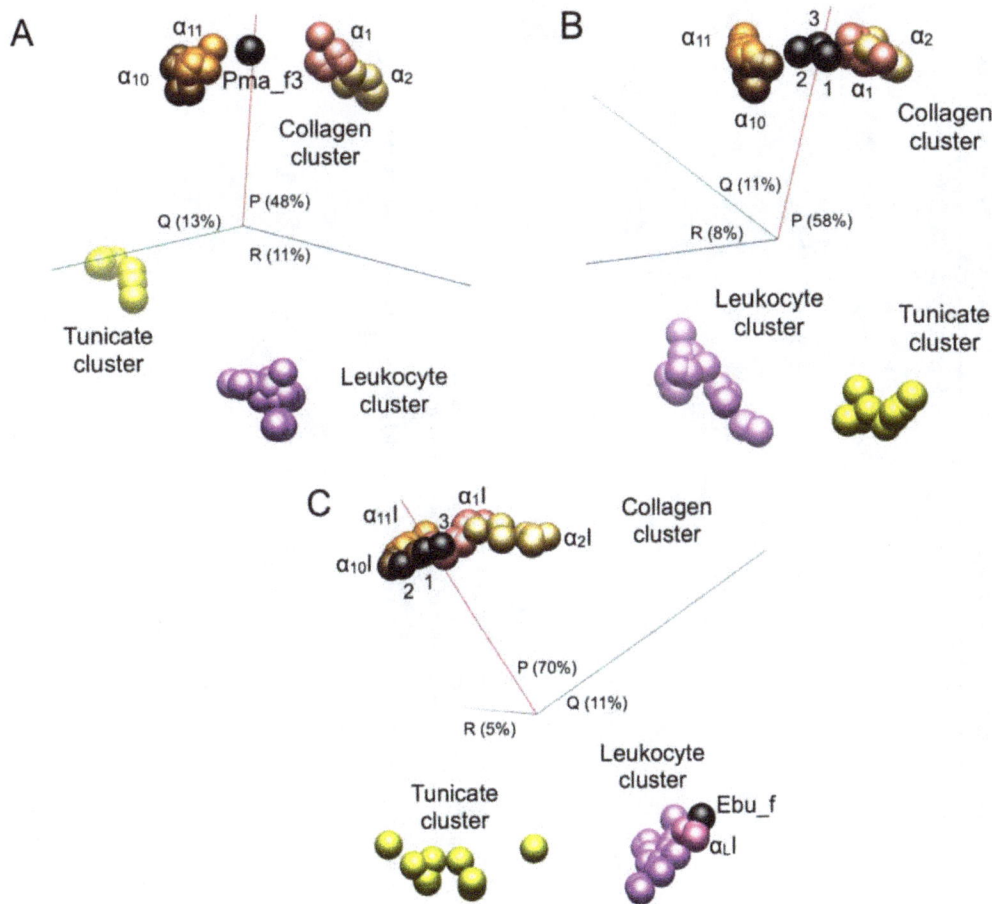

Figure 4. Multivariate plots reflect the details of the phylogenetic analyses. (A) Full-length sequences of the integrin α subunit, (B) sequence regions shared in common with Pma_f1-3, and (C) the αI domain region. The plots were based on distances (JTT scoring) obtained from the sequence alignments. The plots show the relationships among the sequences for the three most informative dimensions and the percentage variance accounted for along the axis is indicated.

absent – a gap – in the leukocyte sequences and in the sequence of Ebu_f. Two residues from the αC helix, Y285 and L286, have equivalent residues in the collagen receptor αI domains and Pma_f1-3, but they are absent in the leukocyte domains; the Ebu_f sequence fragment ends prior to this region.

Residues from the lamprey sequences clearly look most similar to the collagen receptor αI domain residues involved in binding than to the corresponding residues of the leukocyte clade (Table 2). The similarity is reiterated in the corresponding analysis made for α1I-GLOGEN interactions (Table S2 in File S1.doc) and αLI-ICAM3 interactions (Table S3 in File S1.doc), suggesting that the lamprey sequences should recognize multiple collagen subtypes just as the human collagen receptor αI domains do. The sequence ESH (also see Fig. 2) in α2I domain surrounds R12$_B$ in the GFOGER peptide complex and H118 from α1I domain forms a key interaction with N213$_C$ of the GLOGEN tripeptide in the complex; this sequence is conserved in Pma_f1 and Pma_f2, and ESS in Pma_f3, and ESD in Ebu_f, but less conserved in the leukocyte αI domains (Table 2).

In order to evaluate the potential of the lamprey αI domain sequences for binding collagen, structures were modeled for the three lamprey αI domains with GFOGER triple-helical peptide based on the α2I complex structure (1DZI; [15]) and a wider set of known X-ray structures of αI domains was used to optimize the alignments for structure modeling.

Structural models were built for the lamprey sequences and a comparison of the key features of the X-ray structure of the α2I-GFOGER complex (Fig. 5A and C) and the structural model built for Pma_f3 (Fig. 5B and D) show extensive similarities. Pma_f3 is overall 44% identical with the α2I domain sequence and only one two-position deletion is present in Pma_f3, mapping to the opposite end of the αI domain from MIDAS. Of 18 residues from α2I domain, 16 within 4.2 Å of GFOGER and two other residues that are part of the MIDAS motif, 12 of 18 residues are identical in Pma_f3 (Fig. 2 and Table 2) and, correspondingly, 14 of 18 residues are identical between Pma_f3 and the α11I domain. This includes all five metal-binding residues at MIDAS (i.e. D151, S153, S155, T221 and D254) – all are fully conserved in αI domains, even in the tunicates (Fig. 2) and in some other non-integrin proteins with vWFA domains. Two of three residues important for binding R12$_B$ of the GFOGER tripeptide to α2I are also conserved and the replacement of serine for histidine in Pma_f3 would also support interactions with arginine R12$_B$ of the peptide. In the model constructed for Pma_f3 (fig. 5B and D), the sequence features at the ligand binding site in the vicinity of where R12$_C$ binds to α2I are unique, as it is for the other αI domains, but many features are seen in common with one or more of the human collagen-binding αI domains. In the human collagen receptors, the residue at the position equivalent to D219 in the α2I domain (R218 in α1I) largely determines collagen subtype preferences

Table 2. Residues in the α2I domain structure within 4.2 Å (non-hydrogen atoms) of the bound GFOGER tripeptide and equivalent residues in the other human and lamprey α1 domains, and the fragment from the hagfish.

α2I, 1DZI, 2.10 Å	*S153*	N154	*S155*	Y157	N189	Q215	G217	G218	D219	L220	*T221*	E256	S257	H258	Y285	L286
α1I, 1PT6, 1.87 Å	S152	N153	S154	Y156	E188	Q214	G216	G217	R218	Q219	T220	E255	S256	H257	S284	Y285
α10I	S	N	S	Y	E	R	E	G	R	E	T	E	S	H	H	Y
α11I	S	N	S	Y	E	Q	G	G	T	E	T	E	S	H	Y	Y
Pma_f1	S	N	S	Y	A	R	W	G	M	E	R†	E	S	H	S	Y
Pma_f2	S	N	S	Y	F	S	P	F	V	R	T	E	S	H	Y	Y
Pma_f3	S	N	S	W	E	Q	G	G	K	V	T	E	S	S	Y	Y
Ebu_af	S	R	S	T	S	Q	K	A	*	G	T	E	S	D	?	?
α1I, 3F74, 1.70 Å	S139	M140	S141	Q143	T175	H201	L203	L204	*	L205	T206	E241	A242	T243	*	*
αωI, 1IDO, 1.70 Å	S141	G142	S143	I146	E178	Q204	L206	G207	*	R208	T209	E244	K245	F246	*	*
αχI, 1N3Y, 1.65 Å	S140	G141	S142	S144	N176	Q202	Q204	G205	*	F206	T207	K242	K243	E244	*	*
αβI	S	G	S	D	N	Q	K	G	*	L	T	Q	K	Y	*	*
αεI	S	G	S	D	G	Q	G	S	*	V	T	G	F	*	*	*

Where available, the sequence numbering is from a three-dimensional structure (PDB codes and resolution are indicated). The metal ion at MIDAS is covalently bound to the tripeptide ligand. Residues from MIDAS (S153, S155 and T221 in α2I, 1DZI) are in italics and two residues, D151 and D254 in α2I (not listed), are absolutely conserved across all of the sequences and bind to the metal at MIDAS via a water molecule (WAT2001). *, no equivalent or aligned residue; ?, residue not present in the sequence fragment; †, alignment uncertain at the position - no threonine present nearby in the sequence and replacement of arginine with threonine did not alter binding to collagens of the expressed mutant (data not shown).

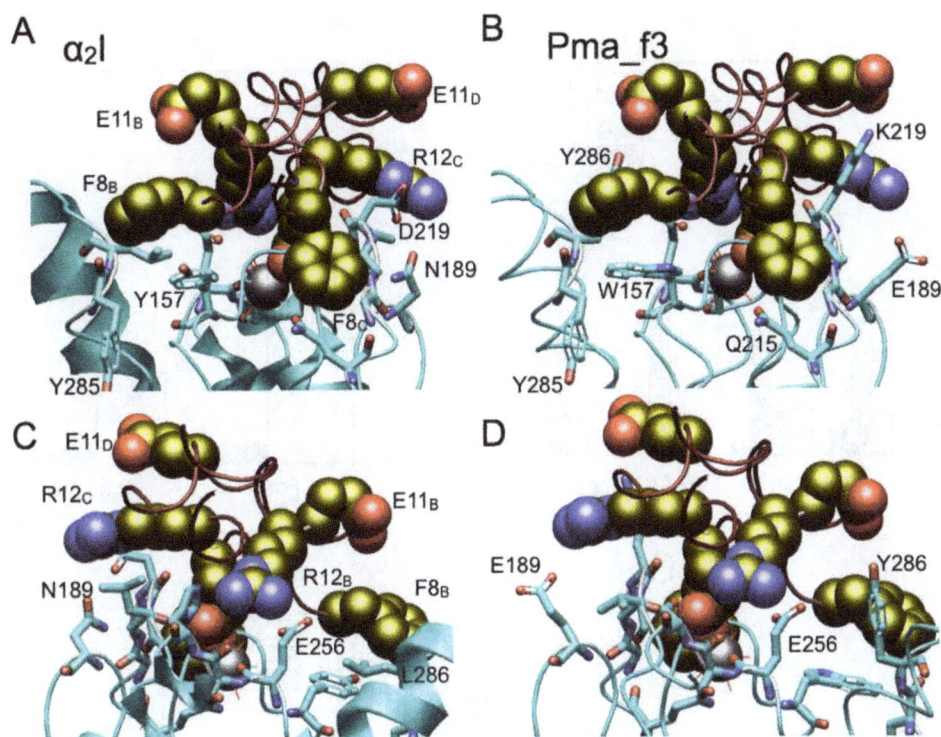

Figure 5. Views of (A, C) the structure of the αI domain with bound GFOGER tripeptide (PDB code: 1DZI) and (B, D) a model constructed for lamprey Pma_f3 αI domain; (C) and (D) are rotated approximately 180° from the view in (A) and (B). The model of Pma_f3 αI was superposed on the αI-peptide complex in order to place the peptide in the same relative position in the model of Pma_f3. Relevant residue side chains of the peptide are shown as CPK models and residues from the αI domains are shown as ball and stick models. For clarity, residues and water molecules binding the metal (grey sphere) at MIDAS are not shown.

[53]. This residue is lysine (K219) in Pma_f3 and could reach E11D and form a strong electrostatic interaction that is seen in models for both human α1I and α10I domains where arginine is present. As positioned in the model, E189 in Pma_f3 would interact strongly with R12$_C$ of the peptide and this residue is also present in α1I, α10I and α11I.

Pma_f2, like Pma_f1, is identical in sequence at 9 of 16 ligand-interacting positions seen for the α2I domain. One key position in α2I, T221, functions to chelate the metal ion at MIDAS and the equivalent residue in the Pma_f1 sequence is uncertain and there is no threonine residue nearby. In Table 2, the alignment of the Pma_f1 sequence ^{219}MER^{221} with ^{219}DLT^{221} in the α2I domain cannot be correct as the large arginine side chain in the Pma_f1 sequence cannot substitute for threonine (the engineered, expressed R221T mutant behaves like the expressed wild-type Pma_f1 αI domain; data not shown) but it may be that the adjacent E220 can substitute for threonine; it remains to be tested.

Sea Lamprey αI Domains Recognize Different Mammalian Collagen Types and GFOGER tripeptide

The three sea lamprey αI domain sequences of Pma_f1, Pma_f2, and Pma_f3 were synthesized and cloned into expression vectors pGEX-2T producing the recombinant GST-fusion proteins. Recombinant proteins were expressed in the *E. coli* strain BL21 tuner. The expressed proteins were sufficiently pure for kinetic experiments to be carried out. A minor amount of GST was observed in each protein preparation and in Pma_f3 preparations a small amount of processed fusion protein was occasionally observed (Fig. S3 in File S1.doc). The ability of recombinant Pma αI domains to recognize and bind to various

collagens was tested with a solid-phase assay as described previously [54]. Binding studies, performed using a fixed concentration of Pma αI domain (400 nM), showed that all recombinant Pma αI domains recognize and bind to several different collagens types: rat collagen I and bovine collagen II (fibrillar collagens), mouse collagen IV (network-forming collagen), and recombinant human collagen IX (FACIT) (Fig. 6A). The highest binding for all Pma αI domains is seen with rat collagen I and generally Pma_f3 αI showed the highest binding with all ligands tested. All Pma αI domains show metal-dependence in binding rat collagen I since when recombinant Pma αI domains were incubated with EDTA in the binding step the observed binding levels were clearly lower (Fig. 6A).

GFOGER is a well-known motif in collagen receptor integrins [55] and one of the most important recognition sequences in, e.g. collagen I. We tested whether triple-helical GFOGER peptide could be recognized by recombinant Pma αI domains. All Pma αI domains bind the GFOGER peptide (Fig. 6B), showing a similar binding profile to the rat collagen I binding profile (Fig. 6A); Pma_f1 and Pma_f3 αI domains show the highest binding and the Pma_f2 αI domain binds to a lesser extent.

In order to compare the binding of Pma αI domains and human collagen receptor integrin αI domains, Pma_f3 αI domain, human wild type α2I wt and human α2I E318W ("open conformation" mutant) were tested for binding to rat collagen I. Recombinant Pma_f3 αI domain shows significantly lower binding levels at a high αI concentration (400 nM) (Fig. 6C), possibly indicating that there is a lower number of binding sites available on rat collagen I for Pma_f3 αI domain than for human α2I wt or human α2I E318W. It is known that for human α2I wt there are at least three

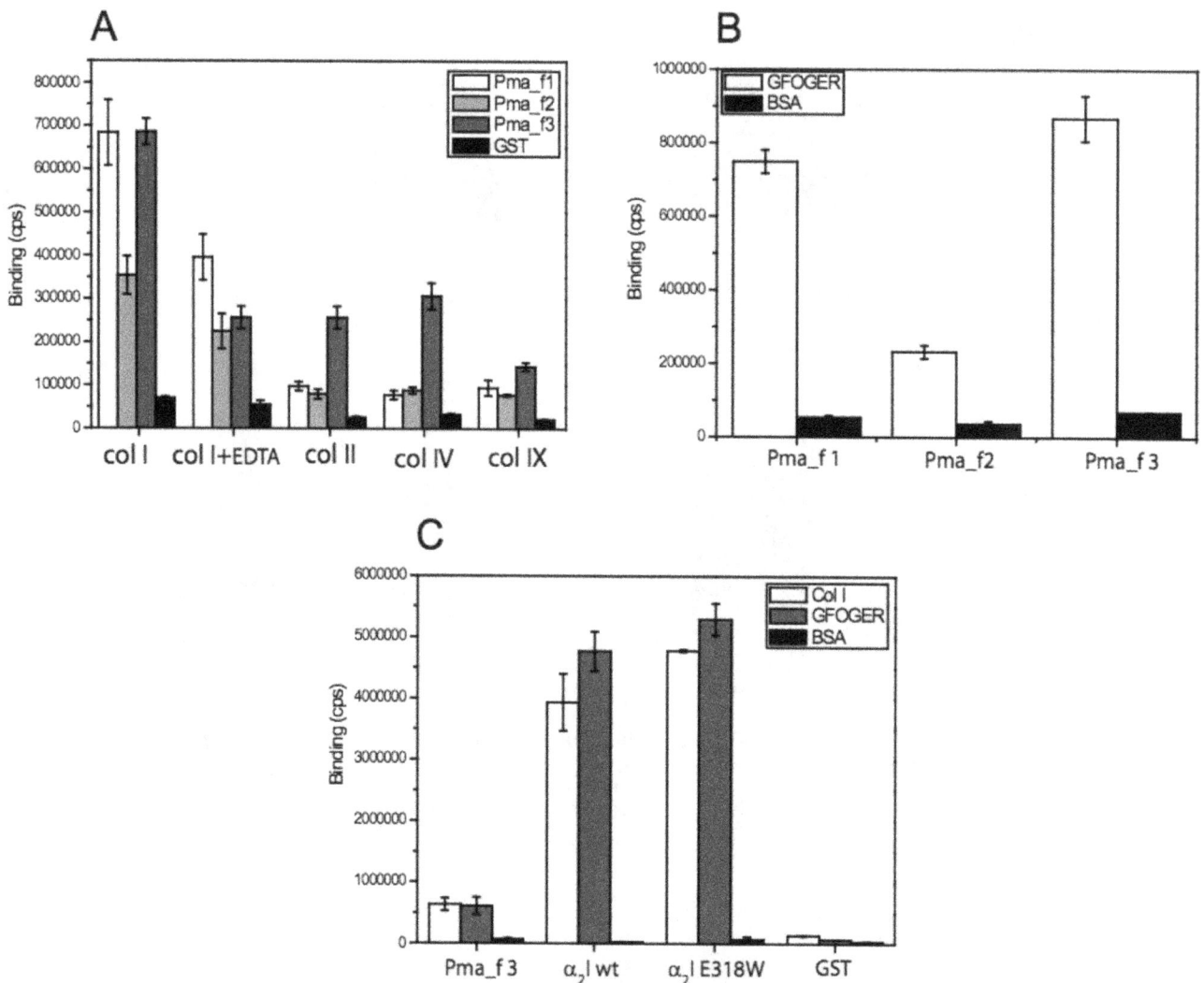

Figure 6. Lamprey αI domains recognize mammalian collagens and the GFOGER-motif in a metal-dependent manner. (A) Binding of Pma_f αI domains to various mammalian collagens at a fixed (400 nM) concentration. The EDTA concentration was 10 mM. GST binding serves as a negative control. (B) Binding of Pma_f αI domains to triple-helical GFOGER-tripeptide at a fixed concentration (400 nM). Binding to BSA serves as a negative control. (C) Binding of Pma_f3 αI, human α2I wt, and human α2I E318W domains to rat collagen I, GFOGER-peptide, and BSA. Binding of GST serves as a control.

high-affinity binding sites on bovine collagen I [56] and a few sites with lower affinity [57].

Pma_f1 and Pma_f3 αI Domains Bind Rat Collagen I at Relatively High Affinity

In order to determine the binding affinity of recombinant Pma αI domains we tested their binding to rat collagen I at various αI domain concentrations and estimated the affinity as described previously [54,58–59] (Fig. 7). Recombinant αI domains of Pma_f1 and Pma_f3 show clear saturation at higher αI domain concentrations and estimates for the apparent affinity constants can be made (the Kd for Pma_f1 αI is 200±35 nM and the Kd for Pma_f3 αI is 195±15 nM). Recombinant Pma_f2 αI does not indicate clear saturation, which leads to a poorer estimate of the Kd (375±120 nM). The Kd values for lamprey Pma_f1 αI and Pma_f3 αI are comparable to the affinities we have measured typically for the binding of human α2I wt to mouse collagen IV [53].

Discussion

The basic integrin heterodimeric structure arose early (Fig. 1A), probably within a single-cell eukaryote [2], thus predating the first metazoans. The integrin was key for recognizing important extracellular matrix proteins e.g. fibronectin, having roles in, for example, cell adhesion, cell migration and tissue remodeling. Ligands with short e.g. RGD and LVD recognition sequences form a direct interaction at the βI-like domain of the β subunit via aspartate with the metal cation at MIDAS, and arginine in RGD cross-links via salt bridges with an aspartate residue in the β-propeller domain of the α subunit (see e.g. the X-ray structure of the αVβ3 ectodomain with bound RGD peptide; PDB code: 1L5G, [60]). Because of the narrow confines at the subunit-subunit interface (in αVβ3 the distance from ligand atom OD1 of aspartate D5003, bound to Mn^{2+}, to the ligand atom NH2 of arginine R5001 is 14.3 Å and the two "walls" of the α subunit, 8.8-9.7 Å between atoms near the aspartate where R5001 binds, restricts the ligand to be an extended chain), the early integrins were limited to

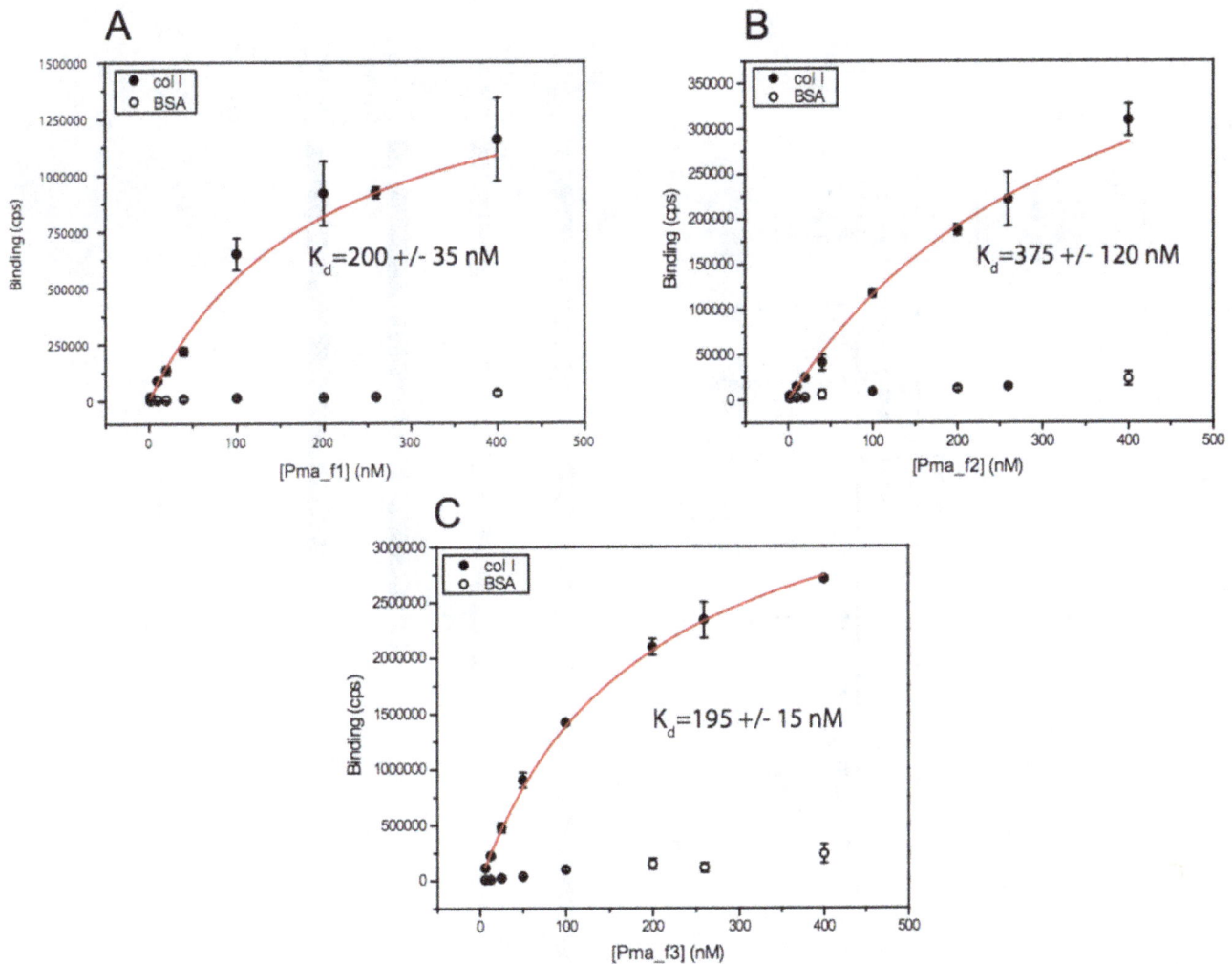

Figure 7. Binding of Pma_f αI domains to rat collagen I as a function of the concentration of Pma_f αI. (A–C) Binding affinities of Pma_f αI domains to rat collagen I were estimated by fitting binding data using a hyperbolic function, which is identical to Hill's equation when h = 1. BSA was used as a control.

the recognition of exposed loop regions of ligands that could occupy the restricted binding cleft and having restricted options for motif specificity. This integrin organization usefully served for the recognition of proteins from the extracellular matrix and cell surfaces with exposed loops but would have been unable to accommodate other, more bulky ligands.

This original organization of the integrin heterodimer is found across the span of metazoan species and is the sole integrin type identified in species diverging prior to the tunicates (Fig. 8). Thus the plan of the α subunit has remained remarkably constant since its inception and half of the integrin α subunits encoded in the human genome abide by this original domain organization.

The insertion of the αI domain into an α subunit occurred approximately 550 MYA, after the deuterostomes first appeared and after the chordate line was established. The αI domain is observed in integrin α subunits from the tunicates but not in the lancelet (Cephalochordata), which is congruent with the lancelet now being acknowledged on the basis of genome comparison studies [61,28] as having diverged before the ascidians as the earliest extant chordate instead of vice versa as previously thought on the basis of phenotypic characteristics. The αI domain bestowed additional flexibility in terms of ligand recognition by integrins, helping to meet the challenges of major cellular and system-wide changes occurring within the chordate lineage.

The αI domain has a highly solvent-exposed ligand binding surface capable of recognizing larger ligands and surfaces, thus the integrin binding site would no longer be limited to external loops that could access the fairly narrow cleft between the β-propeller and βI-like domains. With the αI domain, ligands bind to the metal at MIDAS via a glutamate residue instead of aspartate found in ligands targeting MIDAS of the βI-like domain. The αI domain allows unfettered access to the binding site facilitating recognition of ICAM immunoglobulin-fold domain surfaces and collagen triple helices bundled into large structures could be more easily accessed and recognized. The more exposed binding site also means that the interaction of the αI domain with ligands involves more residues, upwards of 15 residues in collagen-like peptide and ICAM immunoglobulin fold recognition. As a consequence of the relocation of the binding site, a C-terminal glutamate residue of the αI domain acts as an intrinsic ligand binding to MIDAS of the βI-like domain, participating in the dynamic conformational mechanisms associated with the function of integrins with αI domains.

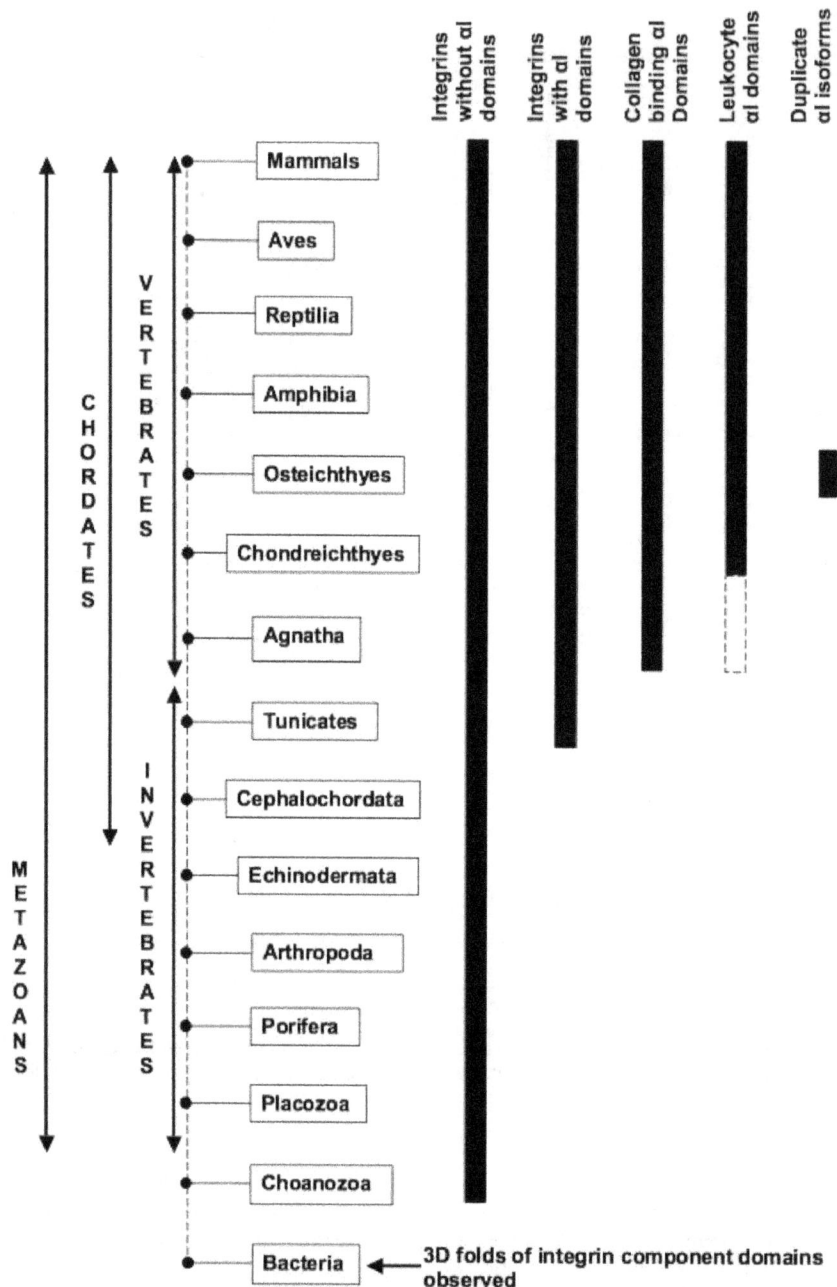

Figure 8. Summary of integrin evolution across a broad range of species: αI domain specialization, as seen in humans, is a vertebrate invention. Individual domains having the same fold class as integrin component domains (i.e. β propeller, immunoglobulin fold, epidermal growth factor fold, vWFA) are observed already in prokaryotes but the earliest diverging sets of identifiable integrin subunits have been observed in the choanozoan *C. owczarzaki*, a single-cell eukaryote. The number of α and β subunits expands with increasing organismal complexity with 18 α and 8 β subunits forming up to 24 heterodimers in humans. Integrins undergo considerable functional diversification with the introduction of the αI domain in some α subunits. Tunicates like *C. intestinalis* and *H. roretzi* are the earliest diverging organisms where integrins with αI domains have been identified, but they are not direct vertebrate orthologues as they form a distinct clade. αI domain containing fragments can be detected in the lamprey *P. marinus* and possibly the hagfish *E. burgeri*; both are extant representatives of the first vertebrates. The lamprey fragments share characteristic features in common with the human collagen-binding αI domain group and they bind different mammalian collagens at MIDAS; four shark sequences are orthologues of the corresponding human α subunits, three collagen binding and one from the leukocyte clade, and duplicate isoforms are observed in observed in bony fish e.g. *D. rerio*, *C. carpio* and *O. niloticus*.

Here, we show that the first appearance of features characteristic of the collagen receptor integrins, and possibly immune system integrins, are found in the agnathastomes, whereas the integrins with αI domains of the earlier diverging tunicates cluster together and have clearly not specialized into the types observed in humans (Fig. 8). This is not to say that the tunicate αI domains cannot bind collagens or have roles in immune function – Miyazawa et al. [30] have reported that *H. roretzi* αII functions in a primitive form of complement recognition and Tulla et al. [62] have shown that the *C. intestinalis* αII domain can bind human recombinant collagen IX that is both metal and MIDAS independent. Orthologues of the human collagen receptor αI domains always have the αC helix and this is found in all three reported lamprey sequences, one of which is a fairly complete α subunit, lacking only the first two repeats of the β-propeller domain. The expressed lamprey αI domains bind mammalian collagens, as shown here but binding, in contrast to *C. intestinalis* α1I, is metal and MIDAS dependent as is the case for mammalian collagen binding with the human αI domains; thus the mechanism of mammalian collagen binding in the tunicate is clearly different from that shared by lamprey and humans. Furthermore, this study shows that the determinants for collagen recognition by integrins with αI domains was established early on in chordate evolution and persists throughout the vertebrates.

I domains in the integrin α subunit have provided a means to diversify chordate integrins to fulfill new tasks associated with the increasing complexity of organs and systems within the chordates, including both complement-based and an adaptive immune system, a circulatory system with the blood clotting, a complex nervous system, cartilaginous and skeletal framework and support system for larger organisms. This study fills in several gaps in our understanding of the evolution of the integrin αI domains, establishing that orthologues of the human integrins with I domains are observed in the agnathostomes, lamprey and perhaps hagfish, present in sharks, but have not been observed in earlier diverging extant chordates or in other invertebrates. The presence of collagen receptor α1, α2 and α11 integrin subunits strongly suggests that integrin α10 must also be present in the cartilaginous fish. The immune system integrins with αI domains appear to diversify fully at a later date than the collagen-recognizing integrins, since only an αE subunit is so far identifiable in the shark *C. milii*. Both αE and αL are present in bony fish but the presence of other bony fish integrins within the leukocyte clade show that the αM/αD/αX specialization had not yet occurred (Fig. 3). It remains uncertain as to the functions of the individual αI integrins in the ascidian *C. intestinalis*, but the function of the integrin fragments with αI domains from the sea lamprey appears clear – they do bind collagens.

Conclusions

The origin and evolution of integrins with inserted I domains in the α subunit has been clarified by the identification of sea lamprey sequences and their comparison with other chordate integrins. Orthologues of human collagen and some leukocyte receptor integrins extend from the cartilaginous fish, being present in the genome of the elephant shark. The lamprey fragments do not cluster with the earlier diverging tunicates. Instead the fragments share key sequence and thus structural similarities of the collagen receptor integrin clade. Moreover, the expressed lamprey sequences recognize different mammalian collagens at MIDAS as do human collagen receptor integrins and the binding is metal dependent unlike that observed for the tunicate *C. intestinalis* α1I. Leukocyte α subunits are present in cartilaginous fish, possibly in the ascidians too, but they do not diversify into the complete set of five subunits see in humans until after the divergence of the bony fish. Thus, integrin α subunits with inserted I domains whose functions are vertebrate specific were established between the divergence of the ascidians and the appearance of the jawless vertebrates.

Materials and Methods

Sequence Searches and Homologue Detection

Searches were made with sequences of human integrin I domain containing α subunits utilizing the BLAST [52] services at the NCBI homepage (http://blast.ncbi.nlm.nih.gov/Blast.cgi) in order to identify potential candidate sequences for this study. Various ongoing and completed genome projects at the Ensembl webpage (http://www.ensembl.org/index.html) were also searched (based on human integrin sequences and on key words like "integrin", "integrin alpha" or "integrin-like") in order to finalize and create a dataset for our analysis (see Table 1 for list of organism and genomes utilized). In addition to the genome assembly searches, we also utilized the tBLASTn [52] to identify any fragments or Expression Sequences Tags (EST's) from organisms that diverged between the appearance of the Ascidians and Osteichthyes (bony fish). These specific searches included the genomes of the green spotted puffer [63], Nile tilapia [64], zebrafish [65], sea lamprey [41] and elephant shark [39]. All identified sequences were also cross-referenced against the conserved domain database (CDD; [18]) and the protein families database (PFAM; [66]) for confirmation.

Sequence Alignment, Phylogenetic Tree Construction and Multivariate Analysis

Sequence alignments were carried out using TCOFFEE [67] and CLUSTALW [68] and examined for obvious errors. Phylogenetic trees were constructed using the Maximum Likelihood (ML) and Neighbor Joining (NJ) implementations in MEGA [69] and Phylip [70]. For the NJ trees, the Jones-Taylor-Thornton (JTT) distance matrix [43] was implemented for each set of alignments. Additionally, for the ML trees, the best-fit evolutionary model for the dataset was assessed using ProtTest [71] and MEGA; both programs reached the same conclusion and suggested the Whelan and Goldman (WAG) [44] substitution matrix with frequency model and gamma distribution with invariant sites (i.e. WAG+I+G+F) to be the best model to explain this dataset based on the Bayesian Information Criteria. Therefore, the WAG substitution matrix was implemented in order to derive the ML trees. For both the NJ and ML trees, the stability of the topology was explored using Felsenstein's bootstrap replication method [72] with 1000 bootstrap replicates. The ML and NJ trees were drawn with MEGA.

Bayesian phylogenetic analysis was performed using MrBayes [73] by implementing the Whelan and Goldman (WAG+I+G+F) model. Monte Carlo Markov Chain analysis was performed for 106 generations with a sampling frequency of 100 generations. The run was halted when the standard deviation of the split frequencies dropped below 0.01. The LnL graph (Log likelihood versus generation plot) was inspected and found to be satisfactory as there was no increasing or decreasing trend observed in the graph. Confidence level for the nodes was assessed with Bayesian posterior probabilities and the consensus tree was redrawn using Treegraph [74].

To complement the clusterings made by the three tree programs, we also supplied the distance data to a C-program program for multivariate analysis (PCA, MS Johnson). The program displays coordinates for each sequence and their locations such that the variance among the data is a maximum, and projections for various numbers of dimensions are possible. The three most informative dimensions, as a pseudo-PDB coordinate file, were visualized using Bodil [75].

Structure Modeling and Identification of Functionally Important Residues

Protein structures were obtained from the Protein Data Bank [10]. The 2.1 Å resolution X-ray structure of the human integrin α2I domain in complex with the GFOGER tripeptide (IDZI; [15])

was used to model the structures of Pma_f1, Pma_f2 and Pma_f3. Structures of human αI domains were aligned using Vertaa in Bodil [75] and used as the basis to optimize the sequence alignments (optimal placement of gaps based on key functional residues and secondary structure) made using Malign [76]. Models were constructed using the Homodge package in Bodil and using Modeller [77] in Discovery studio (http://accelrys.com/products/discovery-studio/). Furthermore, energy minimization was included by using the Charmm force field [78] in Discovery studio.

Bodil [75] was used to visualize the model structures, explore the side-chain conformations using the rotamer utility, and to construct figures from the models. A simple C program, Surf2, was written to identify interactions between the α2I domain and the GFOGER tripeptide and structural water molecules (PDB code: 1DZI), and between the α1I domain NMR structure and GLOGEN tripeptide (PDB code: 3M32) and apo-form of the α1I domain (PDB code: 1PT6; [79]). A 4.2 Å distance cutoff between atoms was used to identify a contact. All contacts were confirmed visually using Bodil.

Cloning and Protein Expression

Lamprey nucleic acid sequences for the predicted αI domain regions were synthesized by Eurofins MWG Operon (Germany) for Pma_f1, Pma_f2, and Pma_f3 and the genes were transferred into the pGEX-2T vector for expression. The expression strain *E. coli* BL21 tuner (Invitrogen, USA) was used for protein production, which was performed as earlier [58]. Human α2I domains (α2I wt and α2I E318W) of α2β1 integrin as well as glutathione S-transferase (GST) were expressed as described earlier [58]. Lamprey αI domains were expressed as either the full sequences below or the sequence minus the N-terminal amino acids that are highlighted in bold.

Pma_f1

SGFNVSESYAPTLQKCGSYMDIVFVLDGSNSIYPWSDVQ-NFLVKTLQSFHIGPDQTQDDVCLPGANVVVVFKLSDTPLY-ERWGVSLVVLWRRWGMERGNDLNVYPSRSEAFSPERGA-RPDAQKVMIVVTDGESHDKYLLPEVIDQCERDGITRYA-IAVLRSYSSNADDVARLINEVRSIASHPVERHFFNVTSEA-TLIDIVGTLGERIFSLEGTR

Pma_f2

ADFQVTSTLTPAAQRCGLFMDIVIVLDGSNSIYPWQEV-QNFVINIVKKFHIGPGQSRNGGGSTRFGVRTIHWHLGIA-RWACEGVQDVENIYSRPFVRTASALCQSLQVVRSEAFS-PLFGAREGASKVMIVVTDGESHDSEDLTEAIAACERDN-ITRYAIAVLGYYKRKNIDPSNFISELKAISSEPEEKHFINV-ADEAALNDIVGTLGERIFSLEGTV

Pma_f3

PNFQQLGSPFAPTMTGCRSFLDIVLVLDGSNSIWPWPSV-LDFLSSILETFSIGPGQTQVGIMQYGETVSNEMNLNQFTN-KAQLKIAASKIPQRGGKVTNTAMGIEAARFFFFENGGRA-EASKVMIVVTDGESSDAYKLPGVIKDCNDDGITRFGIA-VLDYYISSNMNVEKLQAEIRSIASTPTEKYYFDVKSTGA-LVDITKALGERIYSLEGTS

For both the short and long versions of Pma_f1 αI and Pma_f3 αI we did not see any differences in their binding properties (not shown), however the short version of Pma_f2 αI was not expressible.

Binding Studies

The following collagens were used in experiments: rat tail collagen I (Sigma Aldrich, USA), bovine collagen II (Chemicon, USA), mouse collagen IV (EHS mouse tumor; Becton-Dickinson, USA), and recombinant human collagen IX (a kind gift from Dr. Leena Ala-Kokko, University of Oulu). The GFOGER tripeptide

was synthesized by Auspep (Australia). The triple-helical nature of the peptide has been checked with CD-spectroscopy.

Binding studies were performed as earlier [54]. In general, 96-well plates were coated with collagen (16.4 µg/ml) or GFOGER-peptide (5 µg/ml) or BSA (negative control; 1:1 with Diluent II, Perkin-Elmer, USA) overnight at 4°C. Wells were washed once with PBS +2 mM MgCl2 and blocked with 1:1 BSA-Diluent II, incubated for one hour at RT. Wells were washed once with PBS +2 mM MgCl2 and samples (all αI domains were used as a GST-fusion protein) were added to the wells for one hour at RT. Wells were then washed three times with PBS +2 mM MgCl2 and for each well Europium-labeled anti-GST antibody (Perkin Elmer, USA) was added in the Assay buffer (Perkin Elmer, USA) with 2 mM MgCl2. Wells were washed three times with PBS +2 mM MgCl2 and Enhancement Solution (Perkin Elmer, USA) was added to each well. Wells were measured using a Victor3-multilabel counter (Perkin Elmer, USA) using time-resolved fluorescence. Binding affinities of αI domains to rat collagen I were estimated by fitting the binding data using a hyperbolic function, which is identical to Hill's equation when h = 1.

Supporting Information

File S1 Table S1: Sequences utilized in the phylogenetic analysis. Table S2. Residues in the α1I domain structure within 4.2 Å (non-hydrogen atoms) of the bound GLOGEN tripeptide (NMR structure; [21]) and equivalent residues in the human αI domains and the sequence fragments from the lamprey and hagfish. Where available, the sequence numbering is from a three-dimensional structure (PDB codes and resolution are indicated for the known X-ray structures). The metal ion at MIDAS is covalently bound to the tripeptide ligand. Residues from MIDAS (S13, S15, T81 and D114 in α1I, 3M32) are in italics and one residue, D11 in α1I (not listed) is absolutely conserved across all of the sequences. In the X-ray structure of α1I (PDB code: 1PT6; [79]) and this residue (D150 in 1PT6) binds to the metal at MIDAS via an intervening water molecule (WAT603). Table S3. Residues in the αLI domain structure within 4.2 Å (non-hydrogen atoms) of the bound ICAM and equivalent residues in the human αI domains and the sequence fragments from the lamprey and hagfish. Where available, the sequence numbering is from a three-dimensional structure (PDB codes and resolution are indicated for the known X-ray structures). The metal ion at MIDAS is covalently bound to the tripeptide ligand. Residues from MIDAS (S139, S141 and T206 in αLI, 1T0P) are in italics and two residues, D137 and D239 in αLI (not listed), are conserved across all of the sequences and functions to bind the metal at MIDAS via a water molecule (WAT943). Figure S1. Phylogenetic analysis of integrin sequences with the Bayesian method using MrBayes and based on the species and sequences listed in Tables 1 and S1. (A) Full-length sequence alignment of integrin α subunits his dataset contains the nearly full-length integrin α subunit from the sea lamprey Pma_f3 (highlighted in bold). (B) Tree based on the aligned common sequence region in all three lamprey sequence fragments Pma_f1, Pma_f2 and Pma_f3 (highlighted in bold). (C) Tree based on the alignment of the integrin αI domain sequences; this dataset includes the three lamprey αI domain sequences Pma_f1, Pma_f2 and Pma_f3 (highlighted in bold) and the hagfish fragment Ebu_f (highlighted in bold). Bayesian phylogenetic trees were constructed by implementing the Whelan and Goldman substitution matrix with frequency model and gamma distribution with invariant sites (WAG+I+G+F). Statistical support, in the form of the percentage posterior probability, was obtained with a MCMC run of 106 generations and the resulting percentage support value is indicated at each node. Figure S2. Phylogenetic analysis of integrin sequences with the Neighbor joining method using MEGA and based on the species and sequences listed in Tables 1 and S1. (A) Full-length sequence alignment of integrin α subunits his dataset contains the nearly full-length integrin α subunit from the sea lamprey Pma_f3 (highlighted in bold). (B) Tree based on the aligned common sequence region in all three lamprey sequence fragments Pma_f1, Pma_f2 and Pma_f3 (highlighted in bold). (C) Tree based on the alignment of the integrin αI domain sequences; this dataset includes the three lamprey αI domain sequences Pma_f1, Pma_f2 and Pma_f3 (highlighted in bold) and the hagfish fragment Ebu_f (highlighted in bold). Neighbor joining trees were constructed by implementing the Jones and Thornton (JTT) matrix. Statistical support for each phylogenetic tree was obtained with 1000 bootstrap replicates and the percentage bootstrap support value is indicated at each node. Figure S3. SDS PAGE of Pma_f1-3, human wild-type α2I, GST

and molecular weight standards (st). SDS PAGE was run according to manufacturer's instructions using the GE Healthcare PhastSystem (GE, USA) and 8-25% gradient gel. Protein samples were adjusted to 300 ng/ml and the sample size was 1 μl. The gel was stained with Coomassie Brilliant Blue.

Acknowledgments

We gratefully acknowledge the use of bioinformatics infrastructure supported by Biocenter Finland and CSC IT Center for Science.

Author Contributions

Conceived and designed the experiments: BSC JK KD AD JH MSJ. Performed the experiments: BSC JK KD AD. Analyzed the data: BSC JK KD AD JH MSJ. Contributed reagents/materials/analysis tools: JH MSJ. Wrote the paper: BSC JK JH MSJ.

References

1. Eble JA, Kühn K (1997) Integrin-ligand interactions.Chapman and Hall (New York).
2. Sebé-Pedrós A, Roger AJ, Lang FB, King N, Ruiz-Trillo I (2010) Ancient origin of the integrin-mediated adhesion and signaling machinery. Proc Natl Acad Sci USA 107: 10142–10147.
3. Ponting CP, Aravind L, Schultz J, Bork P, Koonin EV (1999) Eukaryotic signalling domain homologues in archaea and bacteria. Ancient ancestry and horizontal gene transfer. J Mol Biol 289: 729–745.
4. Johnson MS, Lu N, Denessiouk K, Heino J, Gullberg D (2009) Integrins during evolution: evolutionary trees and model organisms. BBA 1788: 779–789.
5. Chouhan B, Denesyuk A, Heino J, Johnson MS, Denessiouk K (2011) Conservation of the human integrin-type β-propeller domain in bacteria. PLoS One 6: e25069.
6. Johnson MS, Käpylä J, Denessiouk K, Airenne TA, Chouhan B, et al. (2013) Evolution of cell adhesion to extracellular matrix. In: Keeley W, Mecham RP, editors. Evolution of Extracellular Matrix, Biology of Extracellular Matrix.Springer-Verlag Berlin (Heidelberg). pp. 243–283.
7. Hynes RO (2002) Integrins: bidirectional, allosteric signaling machines. Cell 110: 673–687.
8. Larson RS, Corbi AL, Berman L, Springer T (1989) Primary structure of the leukocyte function-associated molecule-1 α subunit: an integrin with an embedded domain defining a protein superfamily. J Cell Biol 108: 703–712.
9. Arnaout MA (1990) Structure and function of the leukocyte adhesion molecules CD11/CD18. Blood 75: 1037–1050.
10. Berman HM, Westbrook J, Feng Z, Gilliland G, Bhat TN, et al. (2000) The Protein Data Bank. Nucleic Acids Res 28: 235–242.
11. Lee JO, Rieu P, Arnaout MA, Liddington R (1995a) Crystal structure of the A domain from the α subunit of integrin CR3 (CD11b/CD18). Cell 80: 631–638.
12. Lee JO, Bankston LA, Arnaout MA, Liddington RC (1995b) Two conformations of the integrin A-domain (I-domain): a pathway for activation? Structure 3: 1333–1340.
13. Qu A, Leahy DJ (1995) Crystal structure of the I-domain from the CD11a/CD18 (LFA-1, αLβ2) integrin. Proc Nat Acad Sci USA 92: 10277–10281.
14. Emsley J, King SL, Bergelson JM, Liddington RC (1997) Crystal structure of the I domain from integrin α2β1. J Biol Chem 272: 28512–28517.
15. Emsley J, Knight CG, Farndale RW, Barnes MJ, Liddington RC (2000) Structural basis of collagen recognition by integrin α2β1. Cell 101: 47–56.
16. Xie C, Zhu J, Chen X, Mi L, Nishida N, et al. (2010) Structure of an integrin with an αI domain, complement receptor type 4. EMBO J 29: 666–679.
17. Murzin AG, Brenner SE, Hubbard T, Chothia C (1995) SCOP: a structural classification of proteins database for the investigation of sequences and structures. J Mol Biol 247: 536–540.
18. Marchler-Bauer A, Lu S, Anderson JB, Chitsaz F, Derbyshire MK, et al. (2011) CDD: a Conserved Domain Database for the functional annotation of proteins. Nucleic Acids Res D225–229.
19. Ruoslahti E (1996) RGD and other recognition sequences for integrins. Annu Rev Cell Dev Biol 12: 697–715.
20. Xiong J-P, Stehle L, Zhang R, Joachimiak A, Frech M, et al. (2002) Crystal structure of the extracellular segment of integrin αVβ3 in complex with an Arg-Gly-Asp ligand. Science 296: 151–155.
21. Chin YK, Headey SJ, Mohanty B, Patil R, McEwan PA, et al. (2013) The structure of integrin α1I domain in complex with a collagen-mimetic peptide. J Biol Chem 288: 36796–36809.
22. Shimaoka M, Xiao T, Liu JH, Yang Y, Dong Y, et al. (2003) Structures of the αL I domain and its complex with ICAM-1 reveal a shape-shifting pathway for integrin regulation. Cell 112: 99–111.
23. Kang S, Kim CU, Gu X, Owens RM, van Rijn SJ, et al. unpublished.

24. Song G, Yang Y, Liu JH, Casasnovas JM, Shimaoka M, et al. (2005) An atomic resolution view of ICAM recognition in a complex between the binding domains of ICAM-3 and integrin αLβ2. Proc Natl Acad Sci USA 102: 3366–3371.
25. Zhang H, Casasnovas JM, Jin M, Liu JH, Gahmberg CG, et al. (2008) An unusual allosteric mobility of the C-terminal helix of a high-affinity αL integrin I domain variant bound to ICAM-5. Mol Cell 31: 432–437.
26. Chouhan B, Denesyuk A, Heino J, Johnson MS, Denessiouk K (2012) Evolutionary origin of the alpha C helix in integrins. WASET 65: 546–549.
27. Johnson MS, Chouhan BS (2014). Evolution of integrin I domains. In: Gullberg D, editor.I Domain Integrins (Second Edition).Advances in Experimental Medicine and Biology, Springer (Amsterdam). In press.
28. Putnam NH, Butts T, Ferrier DE, Furlong RF, Hellsten U, et al. (2008) The amphioxus genome and the evolution of the chordate karyotype. Nature 45: 1064–1071.
29. Huhtala M, Heino J, Casciari D, Luise AD, Johnson MS (2005) Integrin evolution: insights from ascidian and teleost fish genomes. Matrix Biol 24: 83–95.
30. Miyazawa S, Azumi K, Nonaka M (2001) Cloning and characterization of integrin α subunits from the solitary ascidian Halocynthia roretzi. J Immunol 166: 1710–1715.
31. Ewan R, Huxley-Jones J, Mould AP, Humphries MJ, Robertson DL, et al. (2005) The integrins of the urochordate Ciona intestinalis provide novel insights into the molecular evolution of the vertebrate integrin family. BMC Evol Biol 5: 31.
32. Dehal P, Satou Y, Campbell RK, Chapman J, Degnan B, et al. (2002) The draft genome of Ciona intestinalis: insights into chordate and vertebrate origins. Science 298: 2157–2166.
33. Ivaska J, Käpylä J, Pentikäinen O, Hoffren A-M, Hermonen J, et al. (1999) A peptide inhibiting the collagen binding function of integrin alpha2I domain. J Biol Chem 274: 3513–3521.
34. Pentikäinen O, Hoffren A-M, Ivaska J, Käpylä J, Nyrönen T, et al. (1999) RKKH peptides from the snake venom metalloproteinase of Bothrops jararaca bind near the MIDAS site of the human integrin α2I -domain. J Biol Chem 274: 31493–31505.
35. Xing L, Huhtala M, Pietiäinen V, Käpylä J, Vuorinen K, et al. (2004) Structural and functional analysis of integrin α2I domain interaction with echovirus 1. J Biol Chem 279: 11632–11638.
36. Jokinen J, White DJ, Salmela M, Huhtala M, Käpylä J, et al. (2010) Molecular mechanism of α2β1 integrin interaction with human echovirus 1. EMBO J 29: 196–208.
37. Alonso JL, Essafi M, Xiong JP, Stehle T, Arnaout MA (2002) Does the integrin αA domain act as a ligand for its βA domain? Curr Biol 12: R340–342.
38. Yang W, Shimaoka M, Salas A, Takagi J, Springer TA (2004) Intersubunit signal transmission in integrins by a receptor-like interaction with a pull spring. Proc Natl Acad Sci USA 101: 2906–2911.
39. Venkatesh B, Lee AP, Ravi V, Maurya AK, Lian MM, et al. (2014) Elephant shark genome provides unique insights into gnathostome evolution. Nature 505: 174–179.
40. Johnson MS, Tuckwell D (2003) Evolution of Integrin I-domains. In: Gullberg D, editor.I domains in integrins, Landes Bioscience (Texas, USA). pp. 1–26.
41. Smith JJ, Kuraku S, Holt C, Sauka-Spengler T, Jiang N, et al. (2013) Sequencing of the sea lamprey (Petromyzon marinus) genome provides insights into vertebrate evolution. Nat Genet 45: 415–421.
42. Suzuki T, Shin-IT, Kohara Y, Kasahara M (2004) Transcriptome analysis of hagfish leukocytes: a framework for understanding the immune system of jawless fishes. Develop Comp Immunol 28: 993–1003.
43. Jones DT, Taylor WR, Thornton JM (1992) The rapid generation of mutation data matrices from protein sequences. Comput Applic Biosci 8: 275–282.

44. Whelan S, Goldman N (2001) A general empirical model of protein evolution derived from multiple protein families using a maximum likelihood approach. Mol Biol Evol 18: 691–699.

45. DeSimone DW, Hynes RO (1988) *Xenopus laevis* integrins. Structure and evolutionary divergence of the β subunits. J Biol Chem 163: 5333–5340.

46. Hughes AL (1992) Coevolution of vertebrate integrin α- and β-chain genes. Mol Biol Evol 9: 216–234.

47. Fleming JC, Pahl HL, Gonzalez DA, Smith TF, Tenen DG (1993) Structural analysis of the CD11b gene and phylogenetic analysis of the α-integrin gene family demonstrate remarkable conservation of genomic organization and suggest early diversification during evolution. J Immunol 150: 480–490.

48. Burke RD (1999) Invertebrate integrins: structure, function, and evolution. Int Rev Cytol 191: 257–284.

49. Hynes RO, Zhao Q (2000) The evolution of cell adhesion. J Cell Biol 150: F89–96.

50. Hughes AL (2001) Evolution of the integrin α and β protein families. J Mol Evol 52: 63–72.

51. Takada Y, Ye X, Simon S (2007) The integrins. Genome Biol 8: 215.

52. Altschul SF, Gish W, Miller W, Myers EW, Lipman DJ (1990) Basic local alignment search tool. J Mol Biol 215: 403–410.

53. Tulla M, Pentikäinen OT, Viitasalo T, Käpylä J, Impola U, et al. (2001) Selective binding of collagen subtypes by integrin α1I, α2I, and α10I domains. J Biol Chem 276: 48206–48212.

54. Tulla M, Lahti M, Puranen JS, Brandt AM, Käpylä J, et al. (2008) Effects of conformational activation of integrin α1I and α2I domains on selective recognition of laminin and collagen subtypes. Exp Cell Res 314: 1734–1743.

55. Knight CG, Morton LF, Peachey AR, Tuckwell DS, Farndale RW, et al. (2000) The collagen-binding A-domains of integrins α1β1 and α2β1 recognize the same specific amino acid sequence, GFOGER, in native (triple-helical) collagens. J Biol Chem 275: 35–40.

56. Xu Y, Gurusiddappa S, Rich RL, Owens RT, Keene DR, et al. (2000) Multiple binding sites in collagen type I for the integrins α1β1 and α2β1. J Biol Chem 275: 38981–38989.

57. Farndale RW, Lisman T, Bihan D, Hamaia S, Smerling CS, et al. (2008) Cell-collagen interactions: the use of peptide toolkits to investigate collagen-receptor interactions. Biochem Soc Trans 36: 241–250.

58. Lahti M, Bligt E, Niskanen H, Parkash V, Brandt AM, et al. (2011) Structure of collagen receptor integrin α1I domain carrying the activating mutation E317A. J Biol Chem 286: 43343–43351.

59. Lahti M, Heino J, Käpylä J (2013) Leukocyte integrins αLβ2, αMβ2 and αXβ2 as collagen receptors-receptor activation and recognition of GFOGER motif. Int J Biochem Cell Biol 45: 1204–1211.

60. Xiong JP, Stehle T, Goodman SL, Arnaout MA (2004) A novel adaptation of the integrin PSI domain revealed from its crystal structure. J Biol Chem 279: 40252–40254.

61. Dunn CW, Hejnol A, Matus DQ, Pang K, Browne WE, et al. (2008) Broad phylogenetic sampling improves resolution of the animal tree of life. Nature 452: 745–750.

62. Tulla M, Huhtala M, Jäälinoja J, Käpylä J, Farndale RW, et al. (2007) Analysis of an ascidian integrin provides new insight into early evolution of collagen recognition. FEBS Lett 581: 2434–2440.

63. Jaillon O, Aury JM, Brunet F, Petit JL, Stange-Thomann N, et al. (2004) Genome duplication in the teleost fish *Tetraodon nigroviridis* reveals the early vertebrate proto-karyotype. Nature 431: 946–957.

64. Guyon R, Rakotomanga M, Azzouzi N, Coutanceau JP, Bonillo C, et al. (2012) A high-resolution map of the Nile tilapia genome: a resource for studying cichlids and other percomorphs. BMC Genomics 13: 222.

65. Howe K, Clark MD, Torroja CF, Torrance J, Berthelot C, et al. (2013) The zebrafish reference genome sequence and its relationship to the human genome. Nature 496: 498–503.

66. Finn RD, Mistry J, Tate J, Coggill P, Heger A, et al. (2010) The Pfam protein families database. Nucleic Acids Res D211–222.

67. Notredame C, Higgins DG, Heringa J (2000) T-Coffee: A novel method for multiple sequence alignments. J Mol Biol 302: 205–217.

68. Larkin MA, Blackshields G, Brown NP, Chenna R, McGettigan PA, et al. (2007) ClustalW and ClustalX version 2. Bioinformatics 23: 2947–2948.

69. Tamura K, Peterson D, Peterson N, Stecher G, Nei M, et al. (2011) MEGA5: molecular evolutionary genetics analysis using maximum likelihood, evolutionary distance, and maximum parsimony methods. Mol Biol Evol 28: 2731–2739.

70. Felsenstein J (1989) PHYLIP - Phylogeny Inference Package. Cladistics 5: 164–166.

71. Darriba D, Taboada GL, Doallo R, Posada D (2011) ProtTest 3: fast selection of best fit models of protein evolution. Bioinformatics 27: 1164–1165.

72. Felsenstein J (1985) Confidence limits on phylogenies: An approach using the bootstrap. Evolution 39: 783–791.

73. Huelsenbeck JP, Ronquist F (2001) MRBAYES: Bayesian inference of phylogeny. Bioinformatics 17: 754–755.

74. Stöver BC, Müller KF (2010) TreeGraph 2: Combining and visualizing evidence from different phylogenetic analyses. BMC Bioinformatics 11: 7.

75. Lehtonen JV, Still DJ, Rantanen VV, Ekholm J, Björklund D, et al. (2004) BODIL: a molecular modeling environment for structure-function analysis and drug design. J Comput Aided Mol Des 18: 401–419.

76. Johnson MS, Overington JP (1993) A structural basis for the comparison of sequences: An evaluation of scoring methodologies. J Mol Biol 233: 716–738.

77. Sali A, Blundell TL (1993) Comparative protein modelling by satisfaction of spatial restraints. J Mol Biol 234: 779–815.

78. Brooks BR, Bruccoleri RE, Olafson BD, States DJ, Swaminathan S, et al. (1983) CHARMM: A program for macromolecular energy, minimization, and dynamics calculations. J Comp Chem 4: 187–217.

79. Nymalm Y, Puranen JS, Nyholm TK, Käpylä J, Kidron H, et al. (2004) Jararhagin-derived RKKH peptides induce structural changes in α1I domain of human integrin α1β1. J Biol Chem 279: 7962–7970.

9

Deep Sequencing of HIV-1 near Full-Length Proviral Genomes Identifies High Rates of BF1 Recombinants Including Two Novel Circulating Recombinant Forms (CRF) 70_BF1 and a Disseminating 71_BF1 among Blood Donors in Pernambuco, Brazil

Rodrigo Pessôa[1], Jaqueline Tomoko Watanabe[1], Paula Calabria[1], Alvina Clara Felix[1], Paula Loureiro[2], Ester C. Sabino[3], Michael P. Busch[4], Sabri S. Sanabani[5]* for the International Component of the NHLBI Recipient Epidemiology and Donor Evaluation Study-III (REDS-III)[¶]

1 Virology Department, São Paulo Institute of Tropical Medicine, University of São Paulo, São Paulo, Brazil, 2 Pernambuco State Center of Hematology and Hemotherapy, Recife, Pernambuco, Brazil, 3 Department of Infectious Disease/Institute of Tropical Medicine, University of São Paulo, São Paulo, Brazil, 4 Blood Systems Research Institute, San Francisco, California, United States of America, 5 Clinical Laboratory, Department of Pathology, Hospital das Clínicas, School of Medicine, University of São Paulo, São Paulo, Brazil

Abstract

Background: The findings of frequent circulation of HIV-1 subclade F1 viruses and the scarcity of BF1 recombinant viruses based on *pol* subgenomic fragment sequencing among blood donors in Pernambuco (PE), Northeast of Brazil, were reported recently. Here, we aimed to determine whether the classification of these strains (n = 26) extends to the whole genome sequences.

Methods: Five overlapping amplicons spanning the HIV near full-length genomes (NFLGs) were PCR amplified from peripheral blood mononuclear cells (PBMCs) of 26 blood donors. The amplicons were molecularly bar-coded, pooled, and sequenced by Illumina paired-end protocol. The prevalence of viral variants containing drug resistant mutations (DRMs) was compared between plasma and PBMCs.

Results: Of the 26 samples studied, 20 NFLGs and 4 partial fragments were *de novo* assembled into contiguous sequences and successfully subtyped. Two distinct BF1 recombinant profiles designated CRF70_BF1 and CRF71_BF1, with 4 samples in profile I and 11 in profile II were detected and thus constitute two novel recombinant forms circulating in PE. Evidence of dual infections was detected in four patients co-infected with distinct HIV-1 subtypes. According to our estimate, the new CRF71_BF1 accounts for 10% of the HIV-1 circulating strains among blood donors in PE. Discordant data between the plasma and PBMCs-virus were found in 15 of 24 donors. Six of these strains displayed major DRMs only in PBMCs and four of which had detectable DRMs changes at prevalence between 1-20% of the sequenced population.

Conclusions: The high percentage of the new RF71_BF1 and other BF1 recombinants found among blood donors in Pernambuco, coupled with high rates of transmitted DRMs and dual infections confirm the need for effective surveillance to monitor the prevalence and distribution of HIV variants in a variety of settings in Brazil.

Editor: Lars Kaderali, Technische Universität Dresden, Medical Faculty, Germany

Funding: This work was supported by grants 2011/11090-5 and 2011/12297-2 from the Fundação de Amparo à Pesquisa do Estado de São Paulo. The study was also supported by the Retrovirus Epidemiology Donor Study-II (HHSN268200417175C) and the Recipient Epidemiology and Donor Outcomes Study-III (HHSN268201100007I) contracts from NHLBI. The funders had no role in study design, data collection and analysis, decision to publish, or preparation of the manuscript.

Competing Interests: The authors have declared that no competing interests exist.

* Email: sabyem_63@yahoo.com

¶ Those responsible for the Recipient Epidemiology and Donor Outcomes Study-III are listed in the acknowledgments.

Introduction

One of the most prominent features of HIV-1 is the remarkable accumulation of genetic diversity in its population during the course of infection. This diversity can be attributed to the high mutation rate of reverse transcriptase (3×10^{-5} substitutions per site per generation) [1], rapid viral turnover (10^8 to 10^9 virions per day) [2], large number of infected cells (10^7 to 10^8 cells) [3], and recombination events that are taking place during replication [4]. Consequently, the HIV-1 population is composed of a swarm of highly genetically related variants, i.e. a *quasispecies*, capable of rapidly adapting to various selective pressures. This diversity has been shown to have an impact not only on viral phenotypes at the level of transmission patterns, pathogenicity and immunology but also in responses to antiretroviral therapy and vaccines [5,6]. Nine distinct genetic subtypes, (A–D, F–H, J and K) are joined in the pandemic today by more than 70 major circulating recombinant forms (CRFs) [http://www.hiv.lanl.gov/content/sequence/HIV/CRFs/CRFs.html] and numerous unique recombinant forms (URFs) have been isolated from individual patients [7]. Recombination between the URFs and CRFs and between the existing CRFs (inter-CRF recombinants) results in emergence of novel second and third generation recombinant forms which would further continue to shape the future of HIV epidemic through the generation of other variants with improved fitness to influence viral transmissibility [8,9]. It has been reported that recombinant viruses including the URFs and CRFs may account for at least 20% of all HIV infections [10]. The existence of recombinant viruses is an evidence of simultaneous infection of multiple viruses during a single transmission event (co-infection) or from the sequential infection of viruses during multiple transmission events (superinfection).

Brazil, the most populous country in the Latin America, is home to about one third of the people living with HIV (608,230) in Central and South America (UNAIDS. 2010–2011 Report on the Brazilian response to HIV/AIDS). HIV-1 subtype B is a major genetic clade circulating in the country but the overall prevalence of non-B strains, particularly URF BF1, C and URF BC, is increasing [11,12,13,14]. Data from recent studies of the viral near full-length genomes (NFLGs) have provided evidence of Brazilian CRF strains designated as CRF28_BF, CRF29_BF, CRF39_BF, CRF40_BF, CRF46_BF and CRF31_BC [12,13,15,16]. Recently, Alencar et al [17] performed a molecular epidemiological survey of HIV-1 in Brazil and analyzed the partial *pol* gene of 341 samples from seropositive blood donors collected between 2007 and 2011 from 4 Brazilian blood centers of the REDS-II (Retrovirus Epidemiological Donor Study) international program. The study reported a relatively high prevalence of subclade F1 (26 [24%] of 110), and only one case of BF1 recombinant among blood donors from Recife, Pernambuco (PE). These findings contrast with those from the previous studies of HIV-1 NFLGs in Brazil [13,18,19,20,21]. We therefore undertook this study to determine whether the classification of these strains extends to the whole genome sequences. Additionally, we aimed to compare the rate of drug resistance mutations (DRMs) in plasma bulk RNA and PBMC massively parallel sequencing (MPS) data to elucidate the differences in resistance profile between both compartments. The results revealed a considerable diversity of BF1 mosaic structures and high percentage of the new RF71_BF1 and other BF1 recombinants among blood donors in PE with high rates of transmitted DRMs and dual infections.

Materials and Methods

Study samples

The 24 peripheral blood mononucleated cells (PBMCs) samples reported in this study were from HIV-1 seropositive blood donors in Recife, capital of Pernambuco in the northeast region of Brazil. The rationale for collection of these samples has been previously reported [17]. No evidence of direct epidemiological linkage could be established. All study subjects provided written informed consent. Ethical approval was obtained from the local ethical review committee of the HEMOPE foundation as well as the Recipient Epidemiology and Donor Evaluation Study-II collaborating centers (Blood Systems Research Institute/University of California San Francisco) and Data Coordinating Center (Westat, Inc.) in the US.

DNA extraction and amplification of the NFLGs

The genomic DNA used for the PCR analyses was extracted using the QIAamp blood kit (Qiagen GmbH, Hilden Germany) according to the manufacturer's instructions. The NFLGs from five overlapping fragments were obtained by PCR using the Platinum *Taq* DNA Polymerase High Fidelity (5 U/µl) (Invitrogen, Life Technologies, Carlsbad, CA) and determined by a previously reported method [12,18]. The amplified DNA fragments from the nested PCR products were separated by gel electrophoresis and purified using Freeze 'N Squeeze DNA Gel Extraction Spin Columns (Bio-Rad, Hercules, CA, USA). Each purified amplicon was quantified using Quant-IT HS reagents (Invitrogen, Life Technologies, Carlsbad, CA), and all five amplicons from a single viral genome were pooled together at equimolar ratios.

Whole viral genome library preparation

Sequencing libraries were prepared as described previously [22]. Briefly, one ng of each sample amplicon pool was used in a fragmentation reaction mix using the Nextera XT DNA sample prep kit according to the manufacturer's protocol (Illumina, San Diego, CA). The tagmentation and fragmentation of each pool were simultaneously performed by incubation for 5 min at 55°C followed by incubation in neutralizing tagment buffer for 5 min at room temperature. After neutralization of the fragmented DNA, a light 12-cycle PCR was performed with Illumina Ready Mix to add Illumina flowcell adaptors, indexes and common adapters for subsequent cluster generation and sequencing. Amplified DNA library was then purified using Agencourt AMPure XP beads (Beckman Coulter), which excluded very short library fragments. Following AMPure purification, the quantity of each library was normalized to ensure equally library representation in our pooled samples. Prior to cluster generation, normalized libraries were further quantified by realtime PCR (qPCR) using the SYBR fast Illumina library quantification kit (KAPA Biosystems) following the instructions of the manufacturer. The qPCR was run on the 7500 Fast Real-Time PCR System (Applied Biosystems). The thermocycling conditions consisted of an initial denaturation step at 95°C for 5 min followed by 35 cycles of 30 s at 95°C and 45 s at 60°C. The final libraries were pooled at equimolar concentration and diluted to 4 nM. To denature the indexed DNA, 5 µL of the 4 nM library were mixed with 5 µL of 0.2 N fresh NaOH and incubated for 5 min at room temperature. 990 µL of chilled Illumina HT1 buffer was added to the denatured DNA and mixed to make a 20 pM library. After this step, 360 µL of the 20 pM library was multiplexed with 6 µL of 12.5 pM denatured PhiX control to increase sequence diversity and then mixed with 234 µL of chilled HT1 buffer to make a 12 pM sequenceable library.

Finally, 600 μL of the prepared library was loaded on an Illumina MiSeq clamshell style cartridge for paired end 250 bp sequencing reads.

Data analysis

Fastq files were generated by the Illumina MiSeq reporter for downstream analysis and validated to evaluate the distribution of quality scores and to ensure that quality scores do not drastically drop over each read. To take the sequencing error rate into account, we only considered variants detected at a frequency higher than 1% and Phred quality score of >30%, i.e., a base call accuracy of 99.9%. Validated fastq files from each viral genome were *de novo* assembled into contiguous sequences and annotated with CLC Genomics Workbench version 5.5 (CLC Bio, Aarhus, Denmark) with default parameters and were additionally assembled using Velvet implemented in the Sequencher program 5.2 (Gene Code Corp., Ann Arbor, MI). The contiguous genomic sequence from each virus strain was extracted from the assembly and used for further analysis. The full designation of samples, according to WHO-proposed nomenclature, is YYBR_PEXXX, where YY stands for the year of study, BR for Brazil, PE for Penambuco, and XXX stands for sample number.

Screening for recombination events and identification of breakpoints

The *de novo* assembled NFLGs and partial consensus sequences were aligned with reference sequences representing subtypes A–D, F–H, J and K obtained from the Los Alamos database (http://hiv-web.lanl.gov) using MAFFT version 7 [23]. Aligned sequences were manually edited and trimmed to the minimal shared length in the BioEdit Sequence Alignment Editor Program. The gap-stripped aligned sequences were screened for the presence of recombination by the bootscan methods and similarity plots implemented in the SIMPLOT program v3.5.1 [24,25]. The following parameters were used in bootscan method: window size, 350 bp; step size, 30 bp; the F84 model of evolution (Maximum likelihood (ML)) as a model to estimate nucleotide substitution; transition\transversion ratio, 2.0; and a bootstrap of 100 trees. In addition, the significant threshold for the bootscan was set at 70%. The jumping profile Hidden Markov Model (jpHMM) [26] was also used to confirm the recombination events and to define the recombination breakpoints according to the HXB2 coordinate system. Recombinant regions of the alignment as determined by the crossover points from the jpHMM and bootscan were analyzed separately by phylogenetic analysis. In further analysis, a network reconstruction was performed for the data set with evidence of recombination using SplitsTree4 version 4.3 [27] using the Neighbor-Net method. The NeighborNet method and the GTR+I+G distance model were used to create the network.

Phylogenetic Analysis

ML phylogenies were constructed using the GTR+I+G substitution model and a BIONJ starting tree. Heuristic tree searches under the ML optimality criterion were performed using the NNI branch-swapping algorithm. The approximate likelihood ratio test (aLRT) based on a Shimodaira-Hasegawa-like procedure was used as a statistical test to calculate branch support. The maximum composite likelihood in MEGA version 6 [28] was used to calculate the genetic distances between and within isolates. All trees were displayed using MEGA version 6.0 software.

Genotyping Analysis

For provirus sequences generated in this study, the MPS reads of partial *pol* gene associated with DRMs in the protease and reverse transcriptase regions of the HIV-1 genome of each sample were aligned to their corresponding consensus sequence using the CLC Genomics Workbench version 5.5 (CLC Bio, Aarhus, Denmark). The minority HIV-1 resistant variants were identified using a threshold of >1.0% of the reads sequenced. Reads with < 1% were discarded to account for potential errors due to the error rate of PCR. Amino acid positions including all listed major mutations and minor mutations associated with drug resistance were identified according to the IAS-USA 2011 and Stanford HIV drug resistance database. Due to the polymorphic nature of most minor protease substitutions, we only considered major mutations as evidence of transmission of drug resistance. The list of the NRTIs resistance related sites **41**, 62, **65**, **67**, **69**, **70**, 71, 74, 75, 77, 115, 116, **151**, **184**, **210**, **215** and **219**; NNRTIs resistance related sites 90, 98, **100, 101, 103, 106**, 108, 138, 179, **181, 188, 190**,221, 225, 227, **230**; PIs resistance related sites 10, 16, 11, 20, 24, **30, 32, 33**, 35, 36, 43, **46, 47, 48, 50**, 53, **54, 58**, 60, 62, 63, 64, 69, 71, 73, **74, 76**, 77, **82**, 83, **84**, 85, **88**, 89, **90** and 93. Bolded numbers are major drug resistance mutation sites.

GenBank accession numbers

All consensus genome assemblies generated in this study were submitted to NCBI's GenBank database (KJ849757 - KJ849783). The two distinct profiles of BF1 identical recombinant structure were registered with the Los Alamos National Database as CRF70_BF1 and CRF71_BF1.

Results

A total of 26 strains preliminarily classified as subclade F1 (n = 25) and BF1 recombinant (n = 1) by sequence analysis of a partial *pol* region in a previous study [17] were corroborated by further phylogenetic analysis of the NFLGs and larger partial fragments. Of note, one BF1 sample (10BR_PE059) was erroneously classified as F1 in our previous study. This sequence was thus corrected in this study and used for further analyses. Sequences were obtained for all five overlapped fragments that cover the NFLGs of 20 participants. Partial sequences were obtained from four fragments derived from 4 samples as shown in **Table 1**. Two samples did not amplify for any fragment. This might be a result of technical difficulties in recovering the provirus, but it is also possible that cells other than PBMCs are infected during the early stages of HIV infection. For the purposes of this report, these 2 samples were not considered further, and the analyses include only the 24 samples whose subtypes were successfully determined.

Analysis of the proviral NFLGs and partial consensus sequences revealed all isolates retain intact reading frames for a majority of their genes and no gross deletions or rearrangements were observed. Recombination analysis from each strain shows two distinct recombinant profiles with 4 samples in profile I and 11 in profile II. The recombinant genomes of both profiles essentially consisted of subclades F1 and subtype B as parental sequences and appeared different from all previously documented CRFs in Brazil and South America. Plausible breakpoints identified in each profile using bootscanning analyses were consistent with those identified using jpHMM (schematically illustrated in Figure 1 and 2). Therefore, the new recombinants strains in profile I and II are now designated CRF70_BF1 and CRF71_BF1, respectively. All of the breakpoints in the two CRFs were mapped and compared. The results revealed great similarity in structure between the two

Table 1. The near full-length genomic (NFLG) and partial fragments subtyping of HIV-from plasma and blood samples.

Sample ID	Sequence Fragment					Subtype		Number of Reads	Av. coverage
	A$_{(546-2598)}$	B1$_{(2157-3791)}$	B2$_{(3236-5220)}$	C$_{(4890-7808)}$	D$_{(7719-9537)}$	Plasma	Provirus		
10BR_PE002	-	+	+	+	+	F	BF1	145757	21.299
10BR_PE004	+	+	+	+	+	F	BF1_CRF70	30843	2625
10BR_PE008	+	+	+	+	+	F	BF1_CRF71	111362	2564
10BR_PE009	+	+	+	+	+	F	BF1_CRF71	23923	5624
10BR_PE016	+	+	+	+	+	F	BF1_CRF70	29409	584
10BR_PE025	+	+	+	+	+	F	BF1_CRF70	84718	2173
10BR_PE026	+	+	-	+	+	F	BF1_CRF71	20503	2635
10BR_PE032	+	+	+	-	+	F	BF1	55741	6084
10BR_PE053	+	+	+	+	+	F	B	78936	3697
10BR_PE059[1]	+	+	+	+	+	F	BF1	30166	2619
10BR_PE064	+	+	+	+	+	F	BF1_CRF71	219755	5608
10BR_PE066	+	+	+	+	+	F	BF1_CRF71	32024	6701
10BR_PE071	+	+	+	+	+	F	BF1_CRF71	98156	5578
10BR_PE073	+	+	+	+	-	F	BF1	47041	1365
10BR_PE084	+	+	+	+	+	F	BF1_CRF71	15195	2034
10BR_PE085	-	-	-	-	-	F	-	-	-
10BR_PE086	+	+	+	+	+	F	BF1	15348	1106
10BR_PE087	-	-	+	-	+	F	-	-	-
10BR_PE088	+	+	+	+	+	F	BF1_CRF71	42801	700
10BR_PE090	+	+	+	+	+	F	BF1_CRF71	18052	187
10BR_PE092	+	+	+	+	+	F	BF1_CRF71	4546	201
10BR_PE094	+	+	+	+	+	F	BF1_CRF71	273899	6870
10BR_PE102	+	+	+	+	+	F	BF1	494376	7257
10BR_PE104	+	+	+	+	+	BF1	B/BF1	26901	3324
10BR_PE107	+	+	+	+	+	F	F1	78655	2159
10BR_PE109	+	+	+	+	+	F	BF1_CRF70	165634	6190

[1]Variant was erroneously classified as F1 in our previous study.

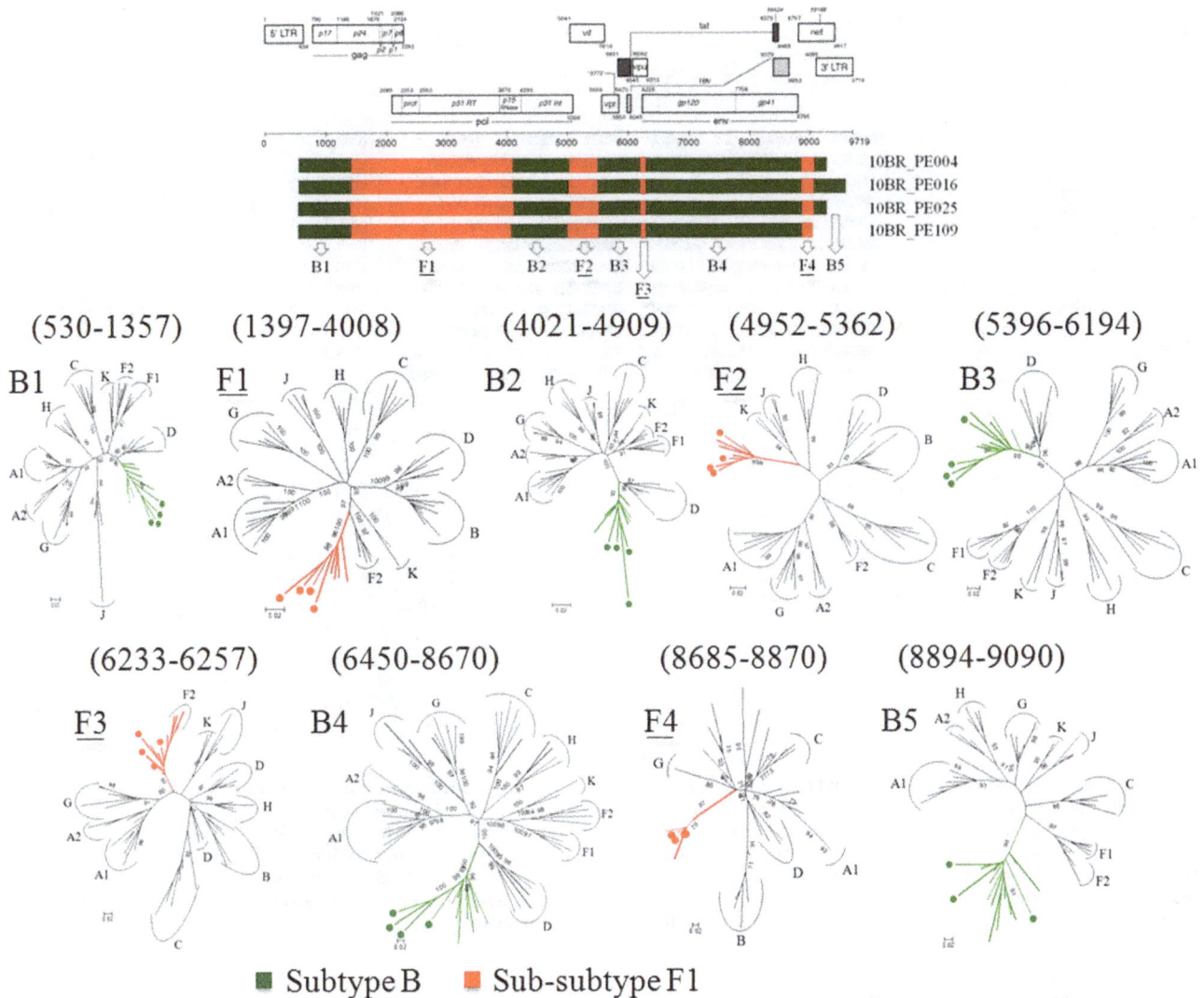

Figure 1. Schematic representation of NFLGs structure and breakpoint profiles with confirmatory phylogenetic trees of the four sequences identified in this study as CRF70_BF1 (colored circles). The phylogenetic trees of the nine mosaic segments defined by jpHMM, similarity plot and bootscan analysis were constructed with PHYML v.3.0. For clarity purposes, the trees were midpoint rooted. The region of subclade F1 and subtypes B are as indicated at the bottom. Positions of breakpoints are numbered relative to the HXB2 numbering system. The approximate likelihood ratio test (aLRT) values of ≥90% are indicated at nodes. The scale bar represents 0.02 nucleotide substitution per site. The results from the ML analysis were sufficiently robust to confirm the structure for the four specimens that were suggested by the recombination analysis.

CRFs. Among the sub-subtype F1 stretches, identified by letter F1 through F4 and F1 through F3 in CRF70_BF1 and CRF71_BF1, respectively (Figure 1 and 2), fragments F1, F2 and F4 appear to be in the same locations. For patient 10BR_PE026, we were unable to amplify fragment designated as B2 in **Table 1** (position 3236–5220) and therefore could not investigate the location of breakpoint at the 5′ of fragment B2 as depicted in Figure 2. Different from CRF71_BF1, CRF70_BF1 appears to contain one too short sub-subtype F1 fragment within the *Vpu* gene (34 bp, nucleotide 6233–6266 according to position in HXB2 GenBank accession no. K03455) interspersed with subtype B. In attempt to find sequences with recombinant structure similar to any of the strains in the two CRFs, multiple sequence alignments including all published BF1 unique NFLGs sequences were built, aligned and subjected to bootscan and automated jpHMM. The results showed that the CRF71_BF1 sequences had a mosaic sequence pattern nearly identical to the previously published Brazilian BF1

isolate 02BR033 (GenBank: DQ358811) in sample collected in 2002 from patient living in São Paulo, Southeast of Brazil [13]. To further test for recombination, ML phylogenetic trees were inferred for the regions of nucleotide sequence on either side of the breakpoints detected by bootscan and jpHMM methods. As shown in Figure 1 and 2, the exploratory tree analysis revealed that each fragment of recombinant virus from each CRF clustered tightly (>90% aLRT) with corresponding fragments of subtype B or F1 reference sequences in agreement with the subtype assigned by recombination analysis. The only disagreement between the recombination and exploratory tree analyses was for a short CRF70_BF1 fragment in the middle of the *Vpu* region (F3 in Figure 2). Despite the fact that F3 fragment in CRF70_BF1 strains have shorter sequences and some group M variants cannot resolve some of the internal nodes, all of the CRF70_BF1 can resolve the terminal nodes and appeared more closely related phylogentically to subclade F1 than the other groups of reference sequences. To

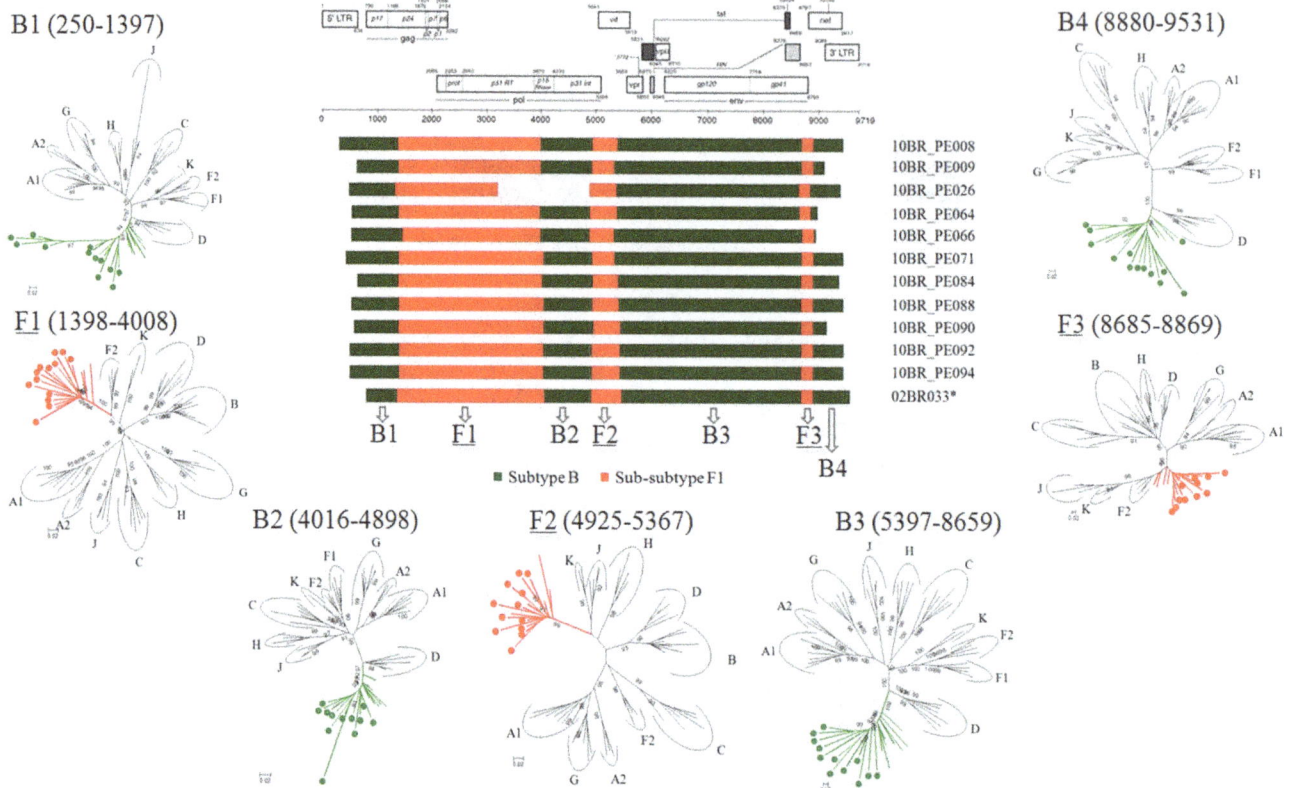

Figure 2. Schematic representation of NFLGs structure and breakpoint profiles with confirmatory phylogenetic trees of the twelve sequences identified in this study as CRF70_BF1 (colored circles). For patient 10BR_PE026, we were unable to amplify fragment designated as B2 and therefore we could not investigate the location of breakpoint at the 5′ of fragment B2. All ML phylogenic trees were constructed using the PHYML v.3.0 package. The region of subclade F1 and subtypes B are indicated at the bottom. Positions of breakpoints are numbered relative to the HXB2 numbering system. For clarity purposes, the trees were midpoint rooted. The approximate likelihood ratio test (aLRT) values of ≥90% are indicated at nodes. The scale bar represents 0.02 nucleotide substitution per site. The results from the ML analysis were sufficiently robust to confirm the structure for the twelve specimens that were suggested by the recombination analysis.

increase the phylogenetic signal, ML trees were performed on concatenated data of discontinuous subtype B and F1 fragment from CRF70_BF1 (Figure 3A and 3B) and CRF71_BF1 (Figure 3A and 3B), which was well confirmed by aLRT values above 90% as shown in Figure 3. Split decomposition was then used to visualize the relationship of the two CRFs (Figure 4). While sequences within each of these two CRF did not group directly with each other, they were present as one group of sequences within a cluster between subtype B and F1. The observed mean intrasubtype genetic distances were 8.2% (range 6.7–10.0%) and 8.7% (range 6.2–9.9%) for CRF70_BF1 and CRF71_BF1 strains, respectively, and closer to the intersubtype distance between both CRFs (8.3%).

Beside the CRFs, in five of the seven strains initially classified as infected with subclade F1, subtype B fragments were detected and appeared to represent URFs strains (Figure 5). The relationships of the viral sequences from patients' PBMCs to the sequences obtained from the corresponding RNA viruses within the same regions were examined for each patient to roughly assess the viral diversity in both compartments (Figure 6). Surprisingly, the intra-individual plasma and proviral sequence variation in four patients (10BR_PE073, 10BR_PE053, 10BR_PE104 and 10BR_PE032) in the partial *pol* regions depicted in Figure 5 (marked with orange boxes) were remarkably high, indicating that the plasma viruses were derived from a population significantly distinct from those of the cellular sources; a result consistent with dual infection with

different subtypes (Figure 5 and 6). Except those with dual infections, all the other sequences from both compartments were located close to one another on the same branch and had plasma RNA and proviral DNA variation only ranging between 0.1–0.8% (Figure 4). In the case of subject 10BR_PE104, MPS data revealed a mixture of two distinct consensus sequences, one NFLGs from strain B and a second recombinant of strains B and F1 (4450 bp) almost identical to the plasma virus in the same region. The BF1 segment was present in 7% of MPS reads with a median coverage of 250 reads in the sample from subject 10BR_PE104, while the NFLGs present in 93% of MPS reads with a median coverage of 7250 reads. Thus, the low BF1 copies would not have been detectable using the provirus traditional bulk sequencing approaches. Again, these data confirms that infection in this subject was founded by two genetic lineages. In Figure 5, there were three samples (10BR_PE002 (pro), 10BR_PE032 [17], and 10BR_PE102 [17], which shared one breakpoint at the integrase gene (position 2393–2462 nt), similar to the third breakpoint of the CRF71_BF variants (5′ of fragment F2, Figure 2). This increased prevalence of genetic breakpoints in the integrase region may indicate a possible preference region for recombination to occur.

Three viruses that were classified as BF1 (n = 2; 10BR_PE059, 10BR_PE086) and F1 (10BR_PE107) on partial *pol* analysis were confirmed by NFLGs analysis (Figure 5).

As stated before, the 26 samples selected to this study are derived from 110 blood donors from Recife PE, and all were

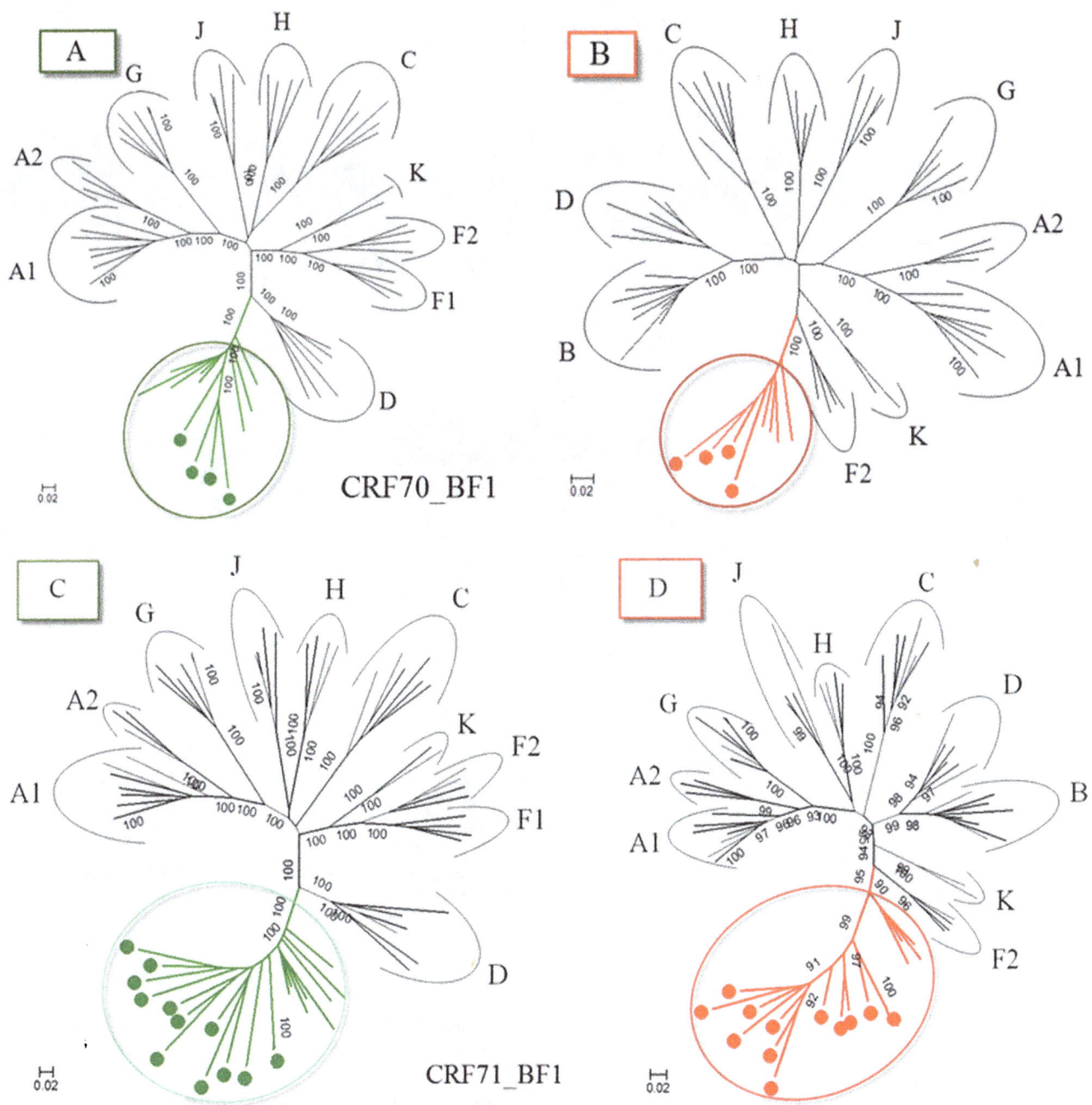

Figure 3. ML phylogenetic tree from concatenated regions assigned as subtype B and F1 from four CRF70_BF1 (3A and 3B) and twelve CRF71_BF1 (3C and 3D) isolates as defined by recombination analysis. Sequences from each region were aligned with reference sequences representing subtypes A–D, F–H, J and K obtained from the Los Alamos database (http://hiv-web.lanl.gov). For clarity purposes, the trees were midpoint rooted. The approximate likelihood ratio test (aLRT) values of ≥90% are indicated at nodes. The scale bar represents 0.02 nucleotide substitution per site. The results from this analysis revealed that each segment of the CRF70_BF1 (3A and 3B) and twelve CRF71_BF1 (3C and 3D) viruses was found to cluster with corresponding segments of subtype B or F1 viruses in agreement with the subtype assigned by recombination analysis.

identified as infected with subclade F1 except two donors who were found to be infected with BF1 recombinant viruses. Genotyping based on NFLGs or long partial fragments of these samples was successful in 24 donors, of whom 11 were infected with CRF71_BF1, 4 with CRF70_BF1, 6 BF1 URF, 1 subtype B, 1 subtype F1, and 1 dually infected with BF1 and B viruses. Based on this analysis, it is evident that non-recombinant subclade F1 accounts for <1% of HIV-1 subtype's circulating in PE and that

CRF71_BF1 is responsible for 50% (11/22) of infections caused by BF1 recombinants.

We also aimed in this study to analyze the emergence of proviral HIV DRMs detected by MPS and compare the results with the plasma HIV DRMs detected by previously described plasma bulk sequencing of the same blood donors group to better understand the resistance profile between the two compartments. A summary of these results are provided in **Table 2**. Discordant data between

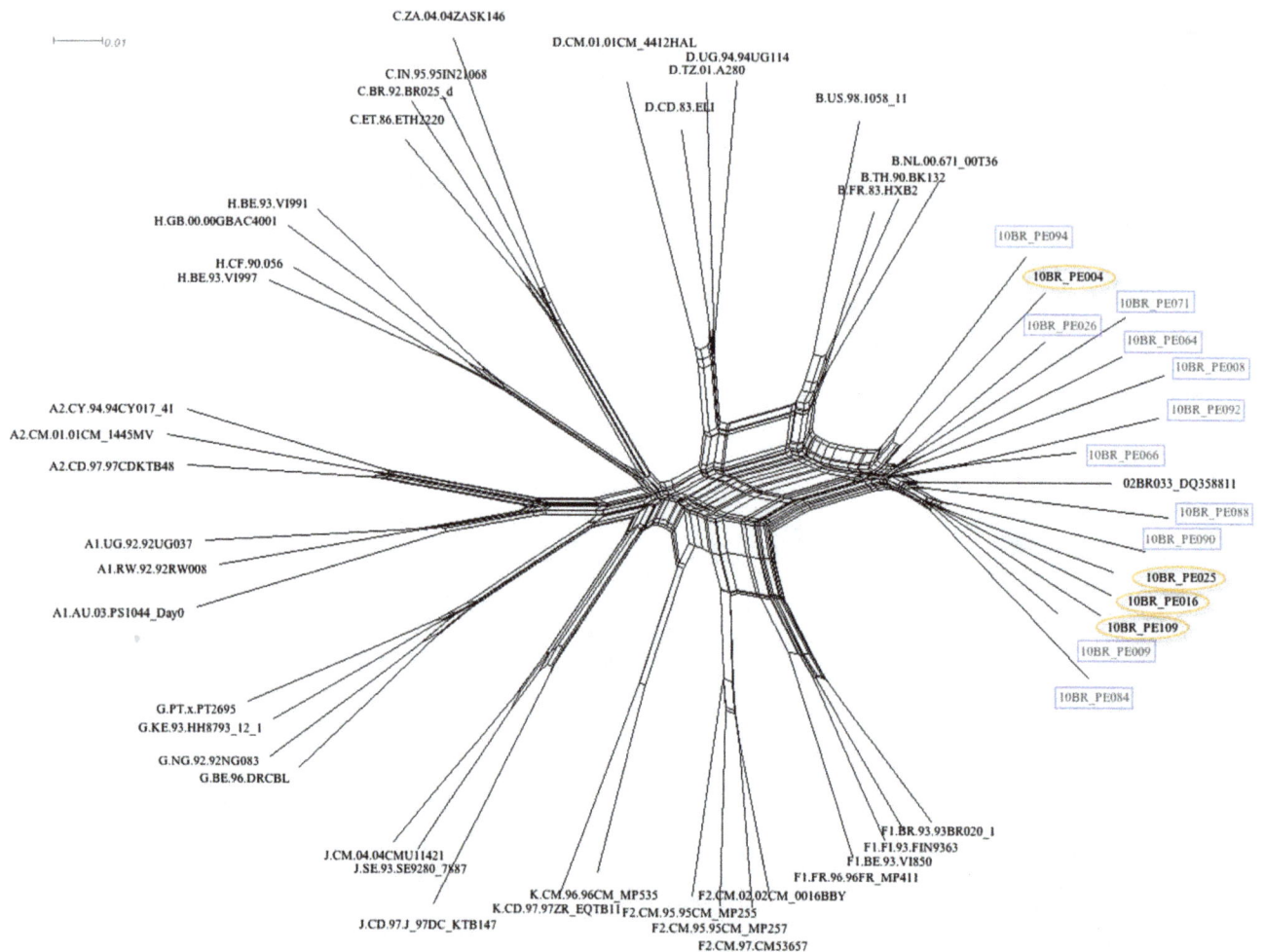

Figure 4. Split networks for four sequences of CRF_70BF1 (marked with orange circles) and eleven CRF_71BF1 (marked with blue boxes) and representatives of HIV subtypes A–K sequences from the Los Alamos database (http://hiv-web.lanl.gov). The splits graphs used distances computed under the GTR+I+G model. The scale bar represents 0.01 nucleotide substitution per site. The results demonstrate that sequences within each of CRF did not group directly with each other, but they were present as one group of sequences within a cluster between subtype B and subclade F1.

the cell free viruses and the PBMC-viruses were found in 15 participants. Six of these strains displayed major amino-acid changes only in the cellular compartment and four of which had detectable major amino-acid changes at prevalence between 1–20% of the sequenced population. Neither major amino-acid changes in the protease nor in the reverse transcriptase coding region were detectable by plasma bulk sequencing. Eleven and two DRMs in the protease and reverse transcriptase regions, respectively, were detected in plasma bulk sequencing but were not observed in MPS data. For the coding region of the protease gene, major PI resistance mutations, namely, M46I, and L33F were detected in four blood donors. As shown in **Table 2**, the overall frequency of minor mutations in both compartments in the protease gene was significantly greater than that detected in reverse transcriptase. Although these mutations were mostly polymorphisms and not directly responsible for drug resistance but they are able to compensate the low protease processivity caused by primary mutations [29]. The proviral DNA MPS analysis in the reverse transcriptase region showed major mutations at the following codons in five patients: M184I, M230I, and E138K. It is well known that HIV-1 harboring the M184I/V mutations has a low viral fitness because of deficient

dNTP usage and that the E138K mutation can compensate for the deficit in dNTP usage associated with the M184I/V mutations [30]. The compensatory substitution at codon 230 in motif E contributes to reduced viral replication and has also been shown to confer resistance to all currently available NNRTIs in both phenotypic and biochemical assays [31].

Discussion

This study describes the MPS of proviral NFLGs and larger fragment from 26 well sampled groups of blood donors from PE who had previously been diagnosed as infected with subclade F1 (n = 25) and BF1 recombinant (n = 1) based on *pol* subgenomic fragment from cell free viruses using conventional bulk sequencing. The most remarkable observations in this study are that 23 of the 24 donors in whom genotyping was successful infected with HIV-1 BF1 recombinant variants. Of these, two novel BF1 CRFs with high genetic diversities that exceed >8% difference at both inter- and intra-host level were identified. These results suggest that both CRFs have been in circulation early in the epidemic and have been evolving independently ever since. Based on the similarity of their recombination profile, it is tempting to speculate

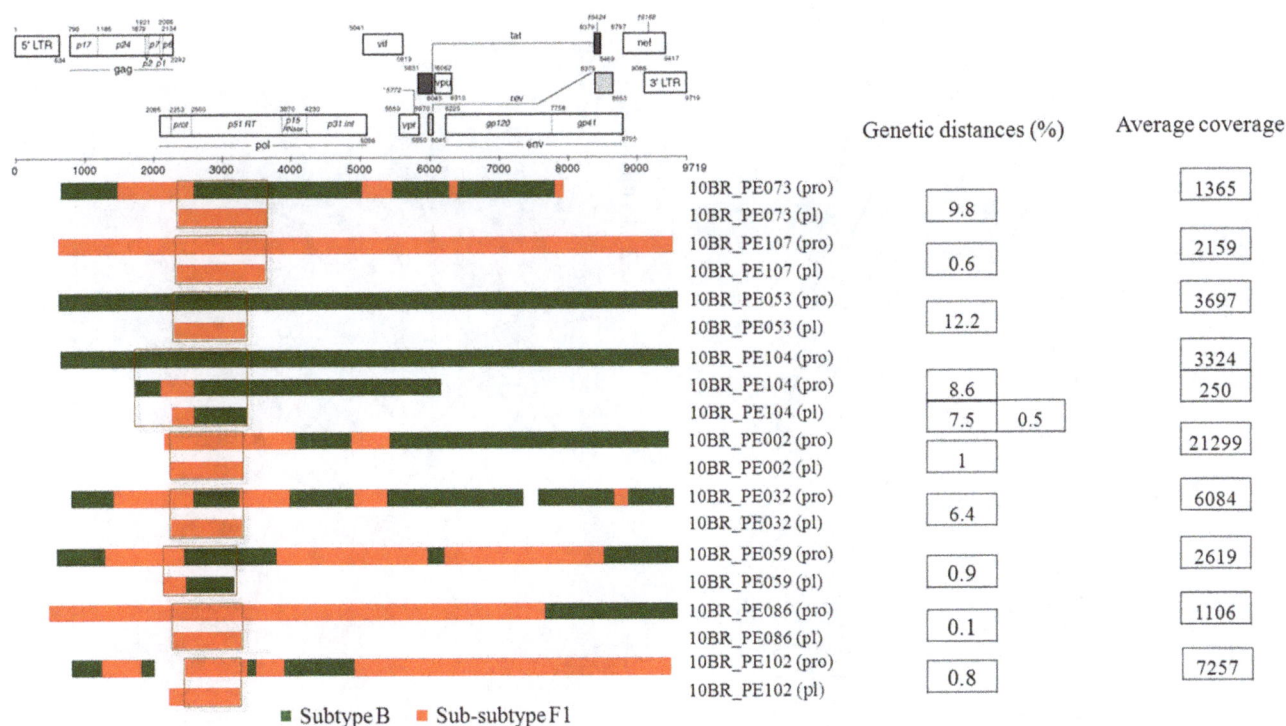

Figure 5. Schematic representation of the NFLGs, partial structure and breakpoint profiles of the BF1 sequences identified in this study from proviral DNA generated by deep sequencing approach and previously published cell free viruses generated by bulk sequencing approach. Sequences were mapped relative to the HXB2 numbering system. Genetic distances of overlapping regions (marked with orange boxes) between sequences from plasma and PBMCs together with the overall mean coverage depth are demonstrated. Distances were computed using the maximum composite likelihood method in MEGA version 6 [28]. As depicted in the figure, the intra-individual plasma and proviral sequence variation in four patients (10BR_PE073, 10BR_PE053, 10BR_PE104 and 10BR_PE032) in the partial *pol* regions (marked with orange boxes) were remarkably high. These results may indicate that the plasma viruses were derived from a population significantly distinct from those of the cellular sources; a result consistent with dual infection with different subtype. In sample 10BR_PE104, MPS data revealed the existence of subtype B NFLGs and a second BF1 recombinant strain (4450 bp) almost identical to the plasma virus in the same region.

that the CRF70_BF1 variants were old "second-generation" recombinants of CRF71_BF1 circulating in PE. Our estimates indicated that the CRF71_BF1 variants are responsible of 50% of infection caused by BF1 recombinants among blood donors in this region of northeast of Brazil. Moreover, if we assume no recombination in the remainder of the genome characterized as subtype B in the previous study, then the prevalence of BF1 recombinant variants is estimated at 20.4% (22/108) and CRF71_BF1 at 10.2% (11/108) of HIV-1 strains circulating among blood donors in PE. Given the high prevalence of CRF71_BF1 observed among the low risk blood donors, it may be suspected that a high prevalence for this variant could be found among other high-risk groups with tight transmission chain and that it has been able to break the transmission barrier from high-risk groups into the general population. Moreover, detection of both CRFs indicates that these variants are actively competing with other BF1 recombinants and other HIV-1 subtypes circulating in this region.

The proportion of BF1 variants described in this study is much greater than the rate reported in 169 individuals recently diagnosed as seropositive for HIV1 (2.8%) in the Metropolitan Region of Recife, Northeastern of Brazil [32] and in other study of 84 patients chronically infected with HIV (3.6%) naïve to antiretroviral treatment [33]. This difference is not surprising, because small fragments from different regions of HIV genomes were characterized in the previous studies while we used larger overlapped fragments to sequence the NFLGs, which is the

preferred method for accurate characterization of HIV-1 isolates. Thus, the previous studies are likely to have missed the recombinants samples. Despite the high rate of recombination as estimated using our fairly conservative assumptions, it is probable that our results have underestimated the true rate of infection with BF1 recombinant viruses; particularly our study was limited to donors infected with F1 and BF1 with partially sequenced viral fragments. Thus, it is possible that the BF1 infection in this group will be higher than what was observed if we had sequenced the virus NFLGs in all the 80 subtype B infected donors described in the previous study.

Additional observations of this study are the description of mixed infections with B, F1 and/or BF1 recombinants. The use of MPS technology enhanced our power to determine the dual infection of larger fragment of BF1 and genuine NFLGs of subtype B in the PBMCs of patient 10BR_PE104. The BF1 recombination profile was almost identical to the plasma virus in the same region demonstrating an intra-individual plasma and proviral sequence variation of only 0.5%. Thus, it is possible to assume that the primary infected PBMCs harboring the BF1 recombinants in this patient were likely the source of the plasma circulating viral sequences. The observation of an intact NFLGs of subtype B in the PBMCs of patient 10BR_PE104 may argue against the inactivity of this provirus but may agree with assumption that subtype B provirus may be integrated into the chromosome at a site at which its expression is prevented, or it may be transcriptionally inactive by virtue of being extrachromosomal [34]. It is also possible that

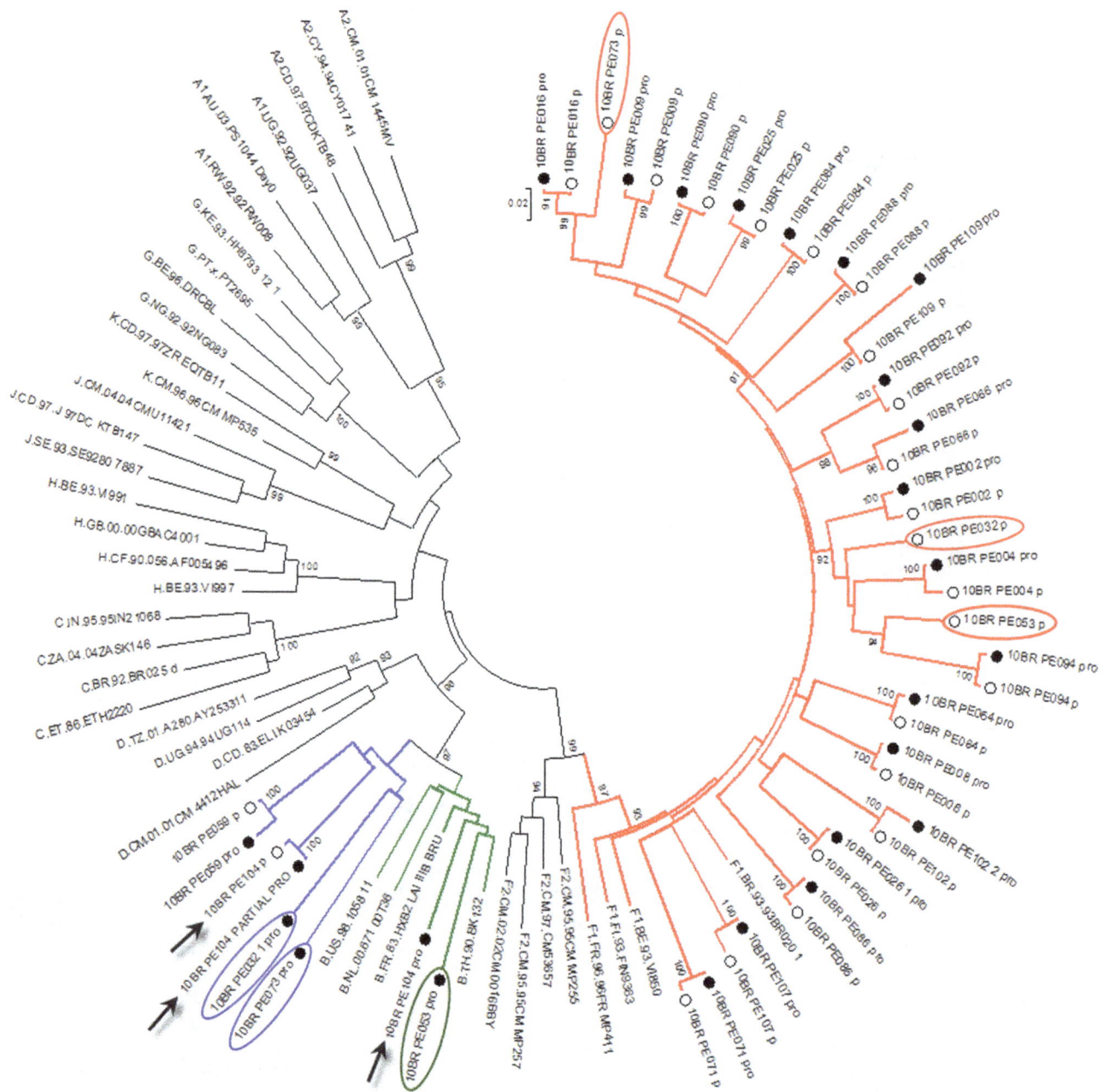

Figure 6. Comparison of phylogenetic clustering profile of sequences from both plasma (empty circles) and provirus (black circles) and other HIV-1 reference sequences from the Los Alamos HIV-1 database representing 11 genetic subtypes. Viral sequences from both compartments were aligned with the complete set of reference sequences obtained from the Los Alamos database (http://hiv-web.lanl.gov). Green, blue and red colored branches represent subtype B, BF1 recombinants, and subclade F1, respectively. Sequence with discordant results between PBMCs and plasma are marked with blue and red oval circles, respectively. For purposes of clarity, the tree was midpoint rooted. The approximate likelihood ratio test (aLRT) values of ≥90% are indicated at nodes. The scale bar represents 0.02 nucleotide substitutions per site. Except subjects 10BR_PE073, 10BR_PE053, 10BR_PE104 and 10BR_PE032, each patient forms a tight cluster and is distinct from other subjects with aLRT SH-like supports >95% for all inter-subject clusters. The results from the ML analysis added further support to the results depicted in figure and were sufficiently robust to confirm the event of dual infection with different subtype. In case 10BR_PE104 (indicated by black arrow), MPS data revealed the existence of subtype B NFLGs and a second BF1 recombinant strain almost identical to the plasma virus in the same region.

the adaptation of the B strains to the PBMC in this patient allows them to avoid competitive exclusion by the dominant strain in plasma. In the other three samples with dual infections, subtype F1 seen in the plasma were completely absent in the PBMCs. Discordances in the HIV subtypes in both compartments may suggest differential sources of infecting viruses. It is also conceivable that the discordances in the HIV subtypes in PBMCs and plasma are likely due to low-level minority strains present as B/BF1 variants that are not detected with bulk plasma sequencing or that the F1 viruses shed in the plasma were more fit. The evidence of dual infections in this study adds support to previous studies [35,36,37,38,39], as this event is far more common in

Table 2. Drug resistance mutations detected with bulk sequencing (plasma) and deep sequencing (PBMCs).

Sample ID	Protease gene coding region		Reverse transcriptase gene coding region(nRTIs)		Reverse transcriptase gene coding region (NNRTIs)	
	PLASMA (Bulk sequencing)	PBMC (Deep sequencing)	PLASMA (Bulk secuencing)	PBMC (Deep sequencing)	PLASMA (Bulk secuencing)	PBMC (Deep sequencing)
10BR_PE002*	M36I, V77I¶, L89M	M36I,*V82I*, L89M		**M184I***		***M230I***
10BR_PE004*	M36I, V77I, L89M	M36I,*M46I*,V77I, L89M			V106I¶	
10BR_PE008*	M36I, D60E, L89M	M36I, D60E, L89M				*V106I*
10BR_PE009*	L10V, M36I, L89M	L10V, M36I, L89M		F77L*¹, **M184I***		***M230I***
10BR_PE016*	L10V, M36I, L89M	L10V, *G16E*, M36I, L89M				
10BR_PE025	L10V, K20R, M36I, L89M	L10V, K20R, M36I, L89M				
10BR_PE026	M36L, L89M	M36I, L89M				
10BR_PE032*	L10V, M36I, L89M	L10V, *L33F**, M36I, L89M		**M184I***		*V108I, M230I**
10BR_PE053*	M36I¶, L89M¶	*V77I*				
10BR_PE059	L10I, M36I, L89M	L10I, M36I, L89M				
10BR_PE064	M36I, L89M	M36I, L89M				
10BR_PE066*	L10V, K20R¶, M36I, D60E¶, I62V, L63P, L89M, I93L	L10V, M36I, D60E, I62V, L89M, I93L				
10BR_PE071*	M36I	M36I		F77L*¹		
10BR_PE073	M36I, I64M	M36I, I64M				
10BR_PE084	L10V, M36I, I64M, L89M	L10V, M36I, I64M, L89M				
10BR_PE086*	L10I¶, K20R, M36I, I62V, L89M, I93L	K20R, M36I, I62V, L89M, I93L				
10BR_PE088	M36I, V82I, L89M	M36I, V82I, L89M			V106I	V106I
10BR_PE090*	L10I, M36I, L89M	L10I,M36I, ***M46I****, L89M		**M184I***		
10BR_PE092	M36I, D60E, I62V, I64M, L89M	M36I, D60E, I62V, I64M, L89M				
10BR_PE094	K20R, M36I, L89M	K20R, M36I, L89M				
10BR_PE102*	L10I¶, M36I¶, I64L¶					
10BR_PE104*	K20R¶, M36I	M36I, *L63P, V77I, I93L*				
10BR_PE107*	M36I, I62V, L63P¶, L89M	M36I, I62V, L89M			V108I¶	
10BR_PE109*	K20M, M36I, D60E, L63P, L89M	K20M, M36I, ***M46I****, D60E, L63P, L89M		F77L*¹		*E138K*

Samples displayed discordant genotypic data between the cell free viruses and the PBMC-viruses are marked by black dots, Major mutations are marked in boldface type, Major drug resistance mutations at prevalence >20% of the sequenced population are underlined, Major drug resistance mutations at prevalence <20% of the sequenced population are marked by star symbol, Mutations detect only in PBMCs are marked in italic face type.
¹Transmitted drug resistance mutation, Mutations detect only in plasma are marked by pilcrow symbol.

Brazil where these subtypes co-circulate. Furthermore, these data agree with the consensus that the presence of two or more HIV-1 subtypes within an infected individual is relatively frequent [40,41]. It is unclear from this study whether the occurrence of HIV multiple distinct strains was the result of superinfection with a second variant at a later time, or whether simultaneous infection with multiple viral strains occurred during a single transmission event. However, the circulation of multiple subtypes in Brazil fortifies the possibility of both scenarios. The overall results indicate that the rate of HIV-1 mixed infections within this Brazilian group is higher than 16%. From these results, we believed that the use of PBMC DNA in addition to plasma RNA is expected to provide highest sensitivity to detect mixed infection via population MPS. Whether dual infection and/or recombination had an impact on the clinical outcomes of the blood donors in this study was unknown, since the available clinical data resulted from one assessment that seek to understand risk exposures and motivations to donate blood.

Other important observation of this study is the underestimation of transmitted resistance obtained by routine plasma analysis that is revealed by the examination of the MPS populations of the archived proviruses in PBMCs. In this study, comparing both sources would have detected 11 DNA provirus disclosed DRMs by MPS not detected by routine plasma bulk analysis. In this study, eleven DNA provirus sequence had detectable DRMs previously missed by plasma bulk analysis. Six of these strains had 17 DRMs only in the PBMCs and four of which had detectable major DRMs at prevalence between 1–20% of the MPS population. These results support the previous observations that standard bulk sequencing cannot fully access the spectrum of viral variants archived in the proviral DNA [42]. The relatively higher proportion of recently infected donors carried low-abundance resistance mutations leads us to believe that the rate of the current transmitted DRMs is underestimated. It is well known that conventional Sanger sequencing of bulk PCR products are limited to the detection of high-frequency variants that present in greater than 20% of the sequenced viral population [43,44]. However, we were able to detect additional DRMs at prevalence higher than 20% of the MPS provirus variants that had not been identified using plasma bulk sequencing. This result probably indicates that MPS approach permit characterization of considerable heterogeneity in the diversity and frequency variations in the proviral DNA [45]. Since the time of infection is unknown, it possible that some of the transmitted mutants in cell free viruses may disappear over time while persisting in cellular HIV-1 DNA [46,47,48]. All together, these results justify the inclusion of proviral DNA from PMBCs as a valuable source for resistance analysis, which is in agreement with previous reports [49,50]. On the other hand, we expected that our MPS approach would detect the entire mutations spectrum from the amplified and sequenced viral population that was displayed by conventional sequencing of the cell free viruses, but this does not hold true for some mutations. Seven patient samples with some detectable DRMs by bulk sequencing approach were not observed in PBMCs reads obtained through the MPS approach. One possible explanation beside the limiting factors intrinsic to PCR is that our MPS approach of the amplified fragments is not sufficiently sensitive to detect all DRMs. It is also possible that the discordant genotypes in these patients are due to dual infection. This was confirmed in patients 10BR_PE053 and 10BR_PE104 who displayed two phylogenetically distinct populations in the sequenced regions.

Apart from small sample size, the present study was limited by the investigation of only the HIV-1 proviruses. Although plasma HIV-1 RNA remains the material of choice for the determination of drug-resistant mutations and guiding therapeutic decisions [51,52] the proviral PBMC DNA sequence can contain a variety of multiple DRMs that are not present in plasma [47,53]. This, combined with the stability of DNA compared with RNA [54], and the fact that HIV DNA recovered from the proviral compartment can reliably be used for the determination of DRMs in treatment naïve patients [55] influenced our decision to use proviral DNA in this study.

Finally, needless to say, HIV phylogenetic analysis based on the complete genome sequences is more reliable than that of the *pol* or other small partial fragment alone. The high percentage of the new CRF71_BF1 and other BF1 recombinants found among blood donors in Pernambuco, coupled with high rates of transmitted DRMs and dual infections confirm the need for effective surveillance to monitor the prevalence and distribution of HIV variants in a variety of settings in Brazil.

Acknowledgments

The Recipient Epidemiology and Donor Outcomes Study-III is the responsibility of the following persons:

Cesar de Almeida Neto[1], Alfredo Mendrone-Junior[1], Anna Bárbara Carneiro-Proietti[2], Divaldo de Almeida Sampaio[3], Paula Loureiro[3], Clarisse Lobo[4] Maria Esther Lopes[4], Ester Cerdeira Sabino[5], Ligia Capuani[5], João Eduardo Ferreira[5], Pedro Losco Takecian[5], Cláudia Di Lorenzo Oliveira[6], Brian Custer[7], Michael P. Busch[7*], Thelma T. Gonçalez[7], Donald brambilla[8], Christopher McClure[8], Simone A. Glynn[9]

[1]Fundação Pró-Sangue Hemocentro de São Paulo, SP, Brazil.
[2]Fundação Hemominas, Belo Horizonte, MG, Brazil.
[3]Fundação Hemope, Recife, PE, Brazil.
[4]Hemorio, Rio de Janeiro, RJ, Brazil.
[5]University of São João Del-Rei, MG, Brazil
[6]University of São Paulo, SP, Brazil.
[7]Blood Systems Research Institute, San Francisco, CA, USA/University of California, San Francisco, San Francisco CA, USA
[8]RTI international
[9]National Heart, Lung, and Blood Institute, NIH
*Group leader:
Michael P. Busch, M.D., Ph.D.
Blood Systems Research Institute
270 Masonic Avenue
San Francisco, CA 94118-4417
mbusch@bloodsystems.org

Author Contributions

Conceived and designed the experiments: SSS. Performed the experiments: RP JTW PC SSS. Analyzed the data: RP JTW PC ACF SSS. Contributed reagents/materials/analysis tools: PL ECS MPB. Contributed to the writing of the manuscript: MPB SSS.

References

1. Mansky LM, Temin HM (1995) Lower in vivo mutation rate of human immunodeficiency virus type 1 than that predicted from the fidelity of purified reverse transcriptase. J Virol 69: 5087–5094.
2. Wei X, Ghosh SK, Taylor ME, Johnson VA, Emini EA, et al. (1995) Viral dynamics in human immunodeficiency virus type 1 infection. Nature 373: 117–122.
3. Chun TW, Carruth L, Finzi D, Shen X, DiGiuseppe JA, et al. (1997) Quantification of latent tissue reservoirs and total body viral load in HIV-1 infection. Nature 387: 183–188.
4. Sabino EC, Shpaer EG, Morgado MG, Korber BT, Diaz RS, et al. (1994) Identification of human immunodeficiency virus type 1 envelope genes recombinant between subtypes B and F in two epidemiologically linked individuals from Brazil. J Virol 68: 6340–6346.
5. Spira S, Wainberg MA, Loemba H, Turner D, Brenner BG (2003) Impact of clade diversity on HIV-1 virulence, antiretroviral drug sensitivity and drug resistance. J Antimicrob Chemother 51: 229–240.
6. Rambaut A, Posada D, Crandall KA, Holmes EC (2004) The causes and consequences of HIV evolution. Nat Rev Genet 5: 52–61.
7. Peeters M, Toure-Kane C, Nkengasong JN (2003) Genetic diversity of HIV in Africa: impact on diagnosis, treatment, vaccine development and trials. AIDS 17: 2547–2560.
8. Thomson MM, Perez-Alvarez L, Najera R (2002) Molecular epidemiology of HIV-1 genetic forms and its significance for vaccine development and therapy. Lancet Infect Dis 2: 461–471.
9. Konings FA, Haman GR, Xue Y, Urbanski MM, Hertzmark K, et al. (2006) Genetic analysis of HIV-1 strains in rural eastern Cameroon indicates the evolution of second-generation recombinants to circulating recombinant forms. J Acquir Immune Defic Syndr 42: 331–341.
10. Arien KK, Vanham G, Arts EJ (2007) Is HIV-1 evolving to a less virulent form in humans? Nat Rev Microbiol 5: 141–151.
11. Barreto CC, Nishyia A, Araujo LV, Ferreira JE, Busch MP, et al. (2006) Trends in antiretroviral drug resistance and clade distributions among HIV-1–infected blood donors in Sao Paulo, Brazil. J Acquir Immune Defic Syndr 41: 338–341.
12. Sanabani S, Neto WK, de Sa Filho DJ, Diaz RS, Munerato P, et al. (2006) Full-length genome analysis of human immunodeficiency virus type 1 subtype C in Brazil. AIDS Res Hum Retroviruses 22: 171–176.
13. Sanabani S, Kleine Neto W, Kalmar EM, Diaz RS, Janini LM, et al. (2006) Analysis of the near full length genomes of HIV-1 subtypes B, F and BF recombinant from a cohort of 14 patients in Sao Paulo, Brazil. Infect Genet Evol 6: 368–377.
14. Passaes CP, Guimaraes ML, Bello G, Morgado MG (2009) Near full-length genome characterization of HIV type 1 unique BC recombinant forms from Southern Brazil. AIDS Res Hum Retroviruses 25: 1339–1344.
15. Guimaraes ML, Eyer-Silva WA, Couto-Fernandez JC, Morgado MG (2008) Identification of two new CRF_BF in Rio de Janeiro State, Brazil. AIDS 22: 433–435.
16. Santos AF, Sousa TM, Soares EA, Sanabani S, Martinez AM, et al. (2006) Characterization of a new circulating recombinant form comprising HIV-1 subtypes C and B in southern Brazil. AIDS 20: 2011–2019.
17. Alencar CS, Sabino EC, Carvalho SM, Leao SC, Carneiro-Proietti AB, et al. (2013) HIV genotypes and primary drug resistance among HIV-seropositive blood donors in Brazil: role of infected blood donors as sentinel populations for molecular surveillance of HIV. J Acquir Immune Defic Syndr 63: 387–392.
18. Sanabani SS, Pastena ER, da Costa AC, Martinez VP, Kleine-Neto W, et al. (2011) Characterization of partial and near full-length genomes of HIV-1 strains sampled from recently infected individuals in Sao Paulo, Brazil. PLoS One 6: e25869.
19. Sanabani SS, Pastena ER, Neto WK, Martinez VP, Sabino EC (2010) Characterization and frequency of a newly identified HIV-1 BF1 intersubtype circulating recombinant form in Sao Paulo, Brazil. Virol J 7: 74.

20. Sanabani SS, Pastena ER, Kleine Neto W, Barreto CC, Ferrari KT, et al. (2009) Near full-length genome analysis of low prevalent human immunodeficiency virus type 1 subclade F1 in Sao Paulo, Brazil. Virol J 6: 78.

21. Thomson MM, Sierra M, Tanuri A, May S, Casado G, et al. (2004) Analysis of near full-length genome sequences of HIV type 1 BF intersubtype recombinant viruses from Brazil reveals their independent origins and their lack of relationship to CRF12_BF. AIDS Res Hum Retroviruses 20: 1126–1133.

22. Pessoa R, Watanabe JT, Nukui Y, Pereira J, Kasseb J, et al. (2014) Molecular characterization of human T-cell lymphotropic virus type 1 full and partial genomes by illumina massively parallel sequencing technology. PLoS One 9: e93374.

23. Katoh K, Misawa K, Kuma K, Miyata T (2002) MAFFT: a novel method for rapid multiple sequence alignment based on fast Fourier transform. Nucleic Acids Res 30: 3059–3066.

24. Salminen MO, Carr JK, Burke DS, McCutchan FE (1995) Identification of breakpoints in intergenotypic recombinants of HIV type 1 by bootscanning. AIDS Res Hum Retroviruses 11: 1423–1425.

25. Lole KS, Bollinger RC, Paranjape RS, Gadkari D, Kulkarni SS, et al. (1999) Full-length human immunodeficiency virus type 1 genomes from subtype C-infected seroconverters in India, with evidence of intersubtype recombination. J Virol 73: 152–160.

26. Schultz AK, Zhang M, Leitner T, Kuiken C, Korber B, et al. (2006) A jumping profile Hidden Markov Model and applications to recombination sites in HIV and HCV genomes. BMC Bioinformatics 7: 265.

27. Huson DH, Bryant D (2006) Application of phylogenetic networks in evolutionary studies. Mol Biol Evol 23: 254–267.

28. Tamura K, Stecher G, Peterson D, Filipski A, Kumar S (2013) MEGA6: Molecular Evolutionary Genetics Analysis version 6.0. Mol Biol Evol 30: 2725–2729.

29. Nijhuis M, Schuurman R, de Jong D, Erickson J, Gustchina E, et al. (1999) Increased fitness of drug resistant HIV-1 protease as a result of acquisition of compensatory mutations during suboptimal therapy. AIDS 13: 2349–2359.

30. Xu HT, Oliveira M, Quashie PK, McCallum M, Han Y, et al. (2012) Subunit-selective mutational analysis and tissue culture evaluations of the interactions of the E138K and M184I mutations in HIV-1 reverse transcriptase. J Virol 86: 8422–8431.

31. Xu HT, Quan Y, Schader SM, Oliveira M, Bar-Magen T, et al. (2010) The M230L nonnucleoside reverse transcriptase inhibitor resistance mutation in HIV-1 reverse transcriptase impairs enzymatic function and viral replicative capacity. Antimicrob Agents Chemother 54: 2401–2408.

32. Cavalcanti AM, Brito AM, Salustiano DM, Lima KO, Silva SP, et al. (2012) Recent HIV infection rates among HIV positive patients seeking voluntary counseling and testing centers in the metropolitan region of Recife - PE, Brazil. Braz J Infect Dis 16: 157–163.

33. de Medeiros LB, Lacerda HR, Cavalcanti AM, de Albuquerque Mde F (2006) Primary resistance of human immunodeficiency virus type 1 in a reference center in Recife, Pernambuco, Brazil. Mem Inst Oswaldo Cruz 101: 845–849.

34. Simmonds P, Balfe P, Peutherer JF, Ludlam CA, Bishop JO, et al. (1990) Human immunodeficiency virus-infected individuals contain provirus in small numbers of peripheral mononuclear cells and at low copy numbers. J Virol 64: 864–872.

35. Soares de Oliveira AC, Pessoa de Farias R, da Costa AC, Sauer MM, Bassichetto KC, et al. (2012) Frequency of subtype B and F1 dual infection in HIV-1 positive, Brazilian men who have sex with men. Virol J 9: 223.

36. Sanabani SS, Pessoa R, Soares de Oliveira AC, Martinez VP, Giret MT, et al. (2013) Variability of HIV-1 genomes among children and adolescents from Sao Paulo, Brazil. PLoS One 8: e62552.

37. Ramos A, Tanuri A, Schechter M, Rayfield MA, Hu DJ, et al. (1999) Dual and recombinant infections: an integral part of the HIV-1 epidemic in Brazil. Emerg Infect Dis 5: 65–74.

38. Diaz RS, Sabino EC, Mayer A, Mosley JW, Busch MP (1995) Dual human immunodeficiency virus type 1 infection and recombination in a dually exposed transfusion recipient. The Transfusion Safety Study Group. J Virol 69: 3273–3281.

39. Blackard JT, Mayer KH (2004) HIV superinfection in the era of increased sexual risk-taking. Sex Transm Dis 31: 201–204.

40. Redd AD, Mullis CE, Serwadda D, Kong X, Martens C, et al. (2012) The rates of HIV superinfection and primary HIV incidence in a general population in Rakai, Uganda. J Infect Dis 206: 267–274.

41. Cornelissen M, Pasternak AO, Grijsen ML, Zorgdrager F, Bakker M, et al. (2012) HIV-1 dual infection is associated with faster CD4+ T-cell decline in a cohort of men with primary HIV infection. Clin Infect Dis 54: 539–547.

42. Wirden M, Soulie C, Valantin MA, Fourati S, Simon A, et al. (2011) Historical HIV-RNA resistance test results are more informative than proviral DNA genotyping in cases of suppressed or residual viraemia. J Antimicrob Chemother 66: 709–712.

43. Van Laethem K, Van Vaerenbergh K, Schmit JC, Sprecher S, Hermans P, et al. (1999) Phenotypic assays and sequencing are less sensitive than point mutation assays for detection of resistance in mixed HIV-1 genotypic populations. J Acquir Immune Defic Syndr 22: 107–118.

44. Gunthard HF, Wong JK, Ignacio CC, Havlir DV, Richman DD (1998) Comparative performance of high-density oligonucleotide sequencing and dideoxynucleotide sequencing of HIV type 1 pol from clinical samples. AIDS Res Hum Retroviruses 14: 869–876.

45. Bansode V, McCormack GP, Crampin AC, Ngwira B, Shrestha RK, et al. (2013) Characterizing the emergence and persistence of drug resistant mutations in HIV-1 subtype C infections using 454 ultra deep pyrosequencing. BMC Infect Dis 13: 52.

46. Geretti AM, Fox Z, Johnson JA, Booth C, Lipscomb J, et al. (2013) Sensitive assessment of the virologic outcomes of stopping and restarting non-nucleoside reverse transcriptase inhibitor-based antiretroviral therapy. PLoS One 8: e69266.

47. Bon I, Gibellini D, Borderi M, Alessandrini F, Vitone F, et al. (2007) Genotypic resistance in plasma and peripheral blood lymphocytes in a group of naive HIV-1 patients. J Clin Virol 38: 313–320.

48. Parisi SG, Boldrin C, Cruciani M, Nicolini G, Cerbaro I, et al. (2007) Both human immunodeficiency virus cellular DNA sequencing and plasma RNA sequencing are useful for detection of drug resistance mutations in blood samples from antiretroviral-drug-naive patients. J Clin Microbiol 45: 1783–1788.

49. Jakobsen MR, Tolstrup M, Sogaard OS, Jorgensen LB, Gorry PR, et al. (2010) Transmission of HIV-1 drug-resistant variants: prevalence and effect on treatment outcome. Clin Infect Dis 50: 566–573.

50. Bon I, Alessandrini F, Borderi M, Gorini R, Re MC (2007) Analysis of HIV-1 drug-resistant variants in plasma and peripheral blood mononuclear cells from untreated individuals: implications for clinical management. New Microbiol 30: 313–317.

51. Bi X, Gatanaga H, Ida S, Tsuchiya K, Matsuoka-Aizawa S, et al. (2003) Emergence of protease inhibitor resistance-associated mutations in plasma HIV-1 precedes that in proviruses of peripheral blood mononuclear cells by more than a year. J Acquir Immune Defic Syndr 34: 1–6.

52. Kaye S, Comber E, Tenant-Flowers M, Loveday C (1995) The appearance of drug resistance-associated point mutations in HIV type 1 plasma RNA precedes their appearance in proviral DNA. AIDS Res Hum Retroviruses 11: 1221–1225.

53. Riva E, Pistello M, Narciso P, D'Offizi G, Isola P, et al. (2001) Decay of HIV type 1 DNA and development of drug-resistant mutants in patients with primary HIV type 1 infection receiving highly active antiretroviral therapy. AIDS Res Hum Retroviruses 17: 1599–1604.

54. Banks L, Gholamin S, White E, Zijenah L, Katzenstein DA (2012) Comparing Peripheral Blood Mononuclear Cell DNA and Circulating Plasma viral RNA pol Genotypes of Subtype C HIV-1. J AIDS Clin Res 3: 141–147.

55. Chew CB, Potter SJ, Wang B, Wang YM, Shaw CO, et al. (2005) Assessment of drug resistance mutations in plasma and peripheral blood mononuclear cells at different plasma viral loads in patients receiving HAART. J Clin Virol 33: 206–216.

Genomic Characterization and Phylogenetic Position of Two New Species in *Rhabdoviridae* Infecting the Parasitic Copepod, Salmon Louse (*Lepeophtheirus salmonis*)

Arnfinn Lodden Økland[1], Are Nylund[1]*, Aina-Cathrine Øvergård[2], Steffen Blindheim[1], Kuninori Watanabe[1], Sindre Grotmol[1,3], Carl-Erik Arnesen[4], Heidrun Plarre[1]

1 Department of Biology, University of Bergen, 5020 Bergen, Norway, **2** SLRC-Sea Lice Research Center, Institute of Marine Research, 5817 Bergen, Norway, **3** SLRC-Sea Lice Research Center, Department of Biology, University of Bergen, 5020 Bergen, Norway, **4** Firda Sjøfarmer AS, 5966 Eivindvik, Norway

Abstract

Several new viruses have emerged during farming of salmonids in the North Atlantic causing large losses to the industry. Still the blood feeding copepod parasite, *Lepeophtheirus salmonis*, remains the major challenge for the industry. Histological examinations of this parasite have revealed the presence of several virus-like particles including some with morphologies similar to rhabdoviruses. This study is the first description of the genome and target tissues of two new species of rhabdoviruses associated with pathology in the salmon louse. Salmon lice were collected at different Atlantic salmon (*Salmo salar*) farming sites on the west coast of Norway and prepared for histology, transmission electron microscopy and Illumina sequencing of the complete RNA extracted from these lice. The nearly complete genomes, around 11 600 nucleotides encoding the five typical rhabdovirus genes N, P, M, G and L, of two new species were obtained. The genome sequences, the putative protein sequences, and predicted transcription strategies for the two viruses are presented. Phylogenetic analyses of the putative N and L proteins indicated closest similarity to the Sigmavirus/Dimarhabdoviruses cluster, however, the genomes of both new viruses are significantly diverged with no close affinity to any of the existing rhabdovirus genera. *In situ* hybridization, targeting the N protein genes, showed that the viruses were present in the same glandular tissues as the observed rhabdovirus-like particles. Both viruses were present in all developmental stages of the salmon louse, and associated with necrosis of glandular tissues in adult lice. As the two viruses were present in eggs and free-living planktonic stages of the salmon louse vertical, transmission of the viruses are suggested. The tissues of the lice host, Atlantic salmon, with the exception of skin at the attachment site for the salmon louse chalimi stages, were negative for these two viruses.

Editor: Amit Kapoor, Columbia University, United States of America

Funding: The authors have no support or funding to report.

Competing Interests: The authors have declared that no competing interests exist.

* Email: are.nylund@bio.uib.no

Introduction

The salmon louse, *Lepeophtheirus salmonis*, feeding on mucus, skin and blood of the host, is a serious problem during farming of the Atlantic salmon, *Salmo salar*, in Norway [1,2]. The life cycle of the salmon louse includes an egg/embryonic stage, two free-living stages, one free-living parasitic stage, and five parasitic stages on the surface of the salmonid host. The salmon louse is attached to the host via a frontal filament during the first two parasitic stages (chalimi stages), while moving freely on the surface of the host during the two preadult and the adult stage [3]. The reproduction of *L. salmonis* in salmon farms and its subsequent release of larvae into the surrounding sea are also recognized as a problem for wild salmonids, *S. salar* and *S. trutta*, along the Norwegian coast [4]. Several control strategies are being used including neurotoxins, hydrogen peroxide, and the use of cleanerfish. The latter method has a limited effect and represents an additional danger of

introducing other fish pathogens (ex. *Paramoeba perurans*) into the salmon cages [5]. The development in the industry is moving towards a critical situation, where the requirements (from the Norwegian Food Authorities, NFA) of a low number of lice on each farmed salmon has led to an increased use of neurotoxins, resulting in the emergence of multiple resistance against these chemicals in the lice populations [6]. Unless new groups of anti-parasitica are developed in the coming years, the aquaculture industry could be facing a critical situation where they are not able to meet the requirements from the NFA and environmental organizations that to a certain degree represent the public opinion on salmon farming.

This development, combined with new advances in biotechnology, may lead to a future use of lice pathogens in the control of this salmonid ectoparasite. One possibility is the use of lice viruses, or their constitutive parts, into novel lice control agents or strategies. There are no published studies of viruses in *L. salmonis*, but several studies

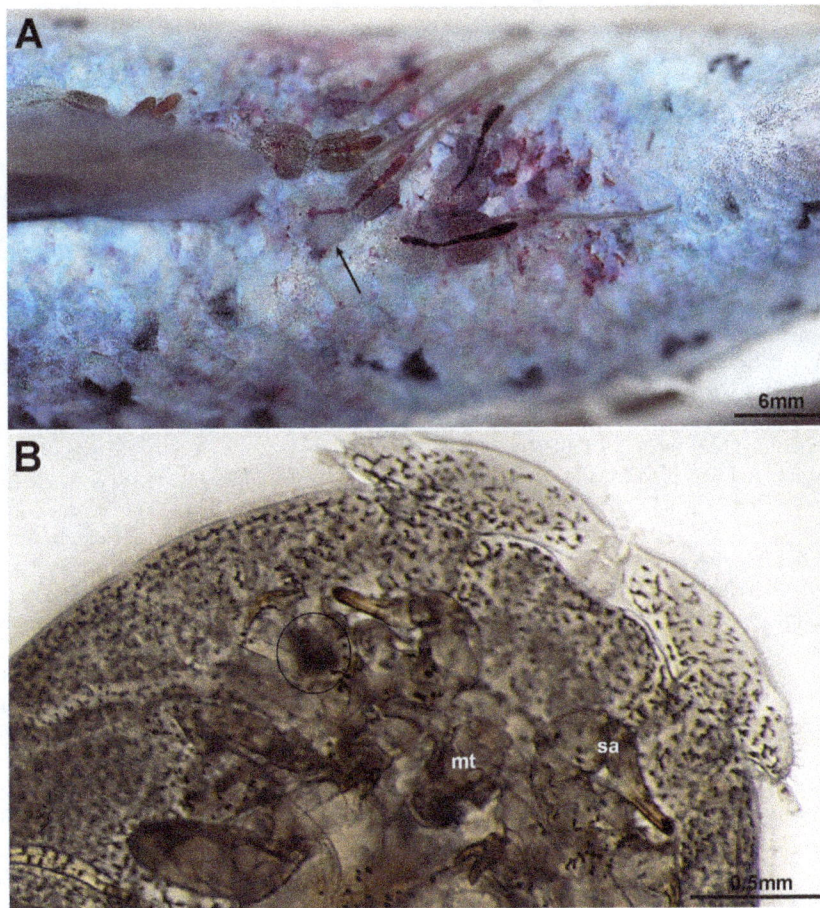

Figure 1. Adult female salmon lice, *Lepeophtheirus salmonis* (arrow), feeding on Atlantic salmon (A). An area of reduced transparency (ring) in the cephalothorax in the vicinity of the second antenna (sa) adult lice (B). Mouth tubule (mt).

have focused on viruses in other crustaceans with a main focus on viruses in commercially important decapods [7,8,9,10, 11,12,13,14,15,16,17,18,19,20,21,22,23,24,25,26,27,28,29]. These studies have shown the presence of members of several different virus families among the crustaceans, including both DNA and RNA viruses.

Studies using transmission electron microscopy on tissues from *L. salmonis* collected from farmed Atlantic salmon in western Norway have shown the presence of different morphs of virus-like particles (A. Nylund, pers. obs.). These viruses, based on the virion morphology and site of assembly, include both DNA and RNA viruses, and the associated histopathology suggests that they may have a significant negative effect on the salmon louse. These viruses, or some of them, could possibly be developed as a tool for future lice control in salmonid aquaculture, but before that can be a reality there are some major problems that have to be resolved. Prior experiences with insect viruses have shown that improvements in the virus efficacy, large scale production and perceived safety will be needed if the lice viruses are to play a major role in the control of this parasite. Knowledge about the genome of these viruses is needed to develop specific and sensitive methods for detection and identification. Fast and safe methods for detection and identification are a necessity for the work towards developing lice viruses as a strategy for control of *L. salmonis*. This study describes the genome of two new species of rhabdoviruses present

in salmon louse, the target tissues and the possible virion morphology.

Materials and methods

Material

Lice (*Lepeophtheirus salmonis*) showing signs of internal changes were collected at five different farming sites on the west coast of Norway in the summer-autumn periods in 2008 – 2013, and transported live to the Fish Disease Research Laboratory at the University of Bergen. A selection of the individuals were sampled both for histology/transmission electron microscopy (TEM) and RNA/DNA extraction, while a large bulk of lice, all the different developmental stages and egg strings, were collected for RNA extraction only. Small subsamples of lice tissues, showing signs of morphological changes, were stored at $-80°C$ for later culture of possible viruses present.

Tissues (gills, skin, heart and kidney) from Atlantic salmon (*Salmo salar*) infected with different stages of *L. salmonis* were collected from a farm in western Norway. The skin tissues were taken from the surface areas where chalimi stages of the lice were attached and from skin areas on the head and behind the dorsal fins, i.e. areas with frequent presence of preadult and adult lice stages. These tissues and different developmental stages of the salmon louse were used for RNA extraction and real time RT PCR.

Histology and TEM

Tissues from lice or one half of the lice cut along the longitudinal axis were fixed in a modified Karnovsky fixative. The fixed tissues were used for histological studies and transmission electron microscopy (TEM). The tissues were processed and sectioned as described in Steigen et al. [30].

RNA extraction

Salmon lice (*L. salmonis*), showing areas of reduced transparency in the cephalothorax in the vicinity of the second antenna (Figure 1), were collected for RNA extraction. The occluded areas, the area from behind the mouth tubule to the anterior of the lice, including the tissues with low light transparency, were used for the RNA extraction. RNA was extracted from individual samples as described by Steigen et al. [30].

The RNA was used for Illumina sequencing, RT PCR and real time RT PCR. The latter method was used for the detection of two rhabdovirus genomes detected in salmon louse after Illumina sequencing.

RNA was also extracted from the collected Atlantic salmon tissues and from the different developmental stages of the salmon louse. The RNA was used for real time RT PCR, PCR and Sanger sequencing.

Illumina sequencing

Total RNA was isolated from the anterior part of the cephalothorax, including the mouth tubule, from five salmon lice collected from five different farms in western Norway. The RNA was pooled and sent to BaseClear (BaseClear Group, Netherlands) for Illumina (Illumina Casava pipeline version 1.8.3) sequencing. At BaseClear a library was created using Illumina TruSeq RNA library preparation kit (Illumina). No polyA capture was used. cDNA synthesis was then performed on fragmented dsRNA, and DNA adapters were ligated to both ends of the DNA fragments before being subjected to PCR-amplification. Prior to sequencing the library was checked on a Bioanalyzer (Agilent) and quantified. The library was sequenced on a full Illumina HiSeq 2500 genome analyzer using a paired-end protocol. The resultant reads were quality checked and low quality reads were removed using the Illumina Chastity filtering. An in-house filtering protocol was used to remove reads containing adapters and/or PhiX control signal. The reads were assembled using the "De novo assembly" option of the CLC Genomics Workbench version 7.0 (CLCbio). This resulted in 10 463 sequences with an average sequence size of 544 bp and a total sum of 5 698 290 bp. Selected sequences were translated using ExPASy's online translation tool (http://web.expasy.org/translate/) and the BLASTP algorithm of the BLAST suite was used to identify the sequences.

Two sequences were identified as possible members of *Rhabdoviridae*. These two sequences, No9 (Accession no: KJ958535) and No127 (Accession no: KJ958536), were used as template for production of primers used to confirm these virus sequences through Sanger sequencing.

Real time RT PCR

Two real time RT PCR assays (Taqman probes) were developed based on the putative nucleoprotein gene sequences of No9 and No127 (Table 1). The assays were optimized for relative quantification. An assay targeting the elongation factor from salmon louse was used as internal control [31]. During real time RT PCR on salmon tissues an assay targeting the elongation factor alpha from Atlantic salmon was used as internal control [32].

Determination of 5′ end terminal sequences of the N protein genes of the two Rhabdovirus from *L. salmonis*

The RNA used in the RNA ligase-mediated amplification of 5′ cDNA ends (GeneRacer Kit version L, Invitrogen) of the two lice rhabdoviruses, No9 and No127, N protein genes were obtained from the anterior part of lice with glandular pathology. The protocol given by the manufacturer was followed using the primers (GeneRacer 5′primer) included in the kit and virus genome specific primers for 5′end race (No9-5′endGSP; CGT TGT TGG GAC CTT CAC GGA CAC A, and No127-5èndGSP; GGC TGG TGT TAC GAG TAT TGA TTT). The final PCR products were cloned into pCR4-TOPO vector (Invitrogen) and sequenced. Sequences were assembled and analysed using the VectorNTI 9.0 software.

Culture system for lice viruses

The only known culture system for these two viruses is the host itself, *L. salmonis*. There are no established cell cultures available from salmon louse or other caligids. Since a range of rhabdoviruses can be cultured in BF2 cells it was decided to test four different cell cultures from fish to see if any of these were susceptible for the two identified rhabdoviruses. In theory, it is possible that these viruses could use the salmon host as a vector for transmission between individual lice, which means that there was a slight possibility that existing cell cultures from salmonids could be susceptible.

The following cell cultures were tested as possible culture systems for these two putative rhabdoviruses; BF-2 (ATCC CCL91), ASK cells [33], CHSE-214 [34], and RT-Gill-W1 cells [35]. The cells were cultured in 25 cm^2 tissue culture flasks at 20°C in Eagle's minimum essential medium (EMEM) supplemented with foetal bovine serum (10%), L-glutamine (4 mM), Non-Essential amino acids and gentamicin (50 µg ml−1). The cells were sub-cultured weekly and formed monolayers within 4–7 days.

For virus propagation, cell culture medium was removed from cell monolayers, and sterile-filtered homogenates from positive salmon lice, diluted 1:10 in serum depleted medium (2% FBS, 4 mM L-glutamine, non-essential amino acids, gentamicin), was added. The cells were incubated at 15°C for 4–5 weeks, or until cytopathic effect (CPE) was observed. The supernatant from the first passage was passed to new cultures, and the cell layers from the first and second inoculation were tested for presence of the two viruses by real time RT PCR.

In situ hybridization

In situ hybridization was performed according to Dalvin et al. [36], with some modifications as described in Tröße et al. [37]. The digoxigenin labelled (DIG-labelled) sense and antisense RNA probes were made with primers listed in Table 2.

Phylogeny

The sequence data were preliminarily identified by GenBank searches done with BLAST (2.0) and the Vector NTI Suite software package was used to obtain multiple alignments of sequences. To perform pairwise comparisons of the two rhabdovirus sequences from the salmon louse, the multiple sequence alignment editor GeneDoc (available at: www.psc.edu/biomed/genedoc) was used for manual adjustment of the sequence alignments. Selected sequences from other members of the family *Rhabdoviridae*, already available on the EMBL nucleotide database, were included in the alignments. Members of the genera *Cytorhabdovirus*, *Novirhabdovirus* and *Nucleorhabdovirus*

Table 1. Primers and probes for Taqman real time RT PCR assays targeting the N protein gene of the two salmon louse rhabdoviruses, LSRV-No9 and LSRV-No127.

Code	Sequence	Position
No9-NF	5'-TCC AAC AGA TCT CCT TAC TCA GTC A -3'	922–946
No9-Nprobe	5'- CGC CAA TCC CTT ATT -3'	948–962
No9-NR	3'- TCC AAT GAT ATG GAC CCA CAT G – 5'	987–966
No127-NF	5- CTA TGG AGC CAT CGG AGG TTA T -3'	873–894
No127-Nprobe	5'- ACC TGG GTG ACT CTT -3'	896–910
No127-NR	5'- CAA GAT CTC AGT CGA GAC GGA AT -3'	934–912

The position of the primer and probes are related to the ORF of the N protein gene of the two viruses.

were excluded because of their large amino acid difference from the two louse viruses. Ambiguously aligned regions were removed using Gblocks [38]. This resulted in sequence alignments of 256 and 1630 amino acids for the N and L proteins, respectively. Phylogenetic relationships were determined using the maximum-likelihood (ML) method available in TREE_PUZZLE 5.2 (available at: http://www.tree-puzzle.de), employing the VT [39] model of amino acid substitution. Quartet puzzling was used to choose from the possible tree topologies and to simultaneously infer support values for internal branches. Quartet trees are based on approximate maximum likelihood values using the selected model of substitution and rate heterogeneity. The robustness of each node was determined using 20 000 puzzling steps. Phylogenetic trees were drawn using TreeView [40].

Protein analysis

The Compute pI/Mw tool in ExPASy was used to calculate the theoretical pI (isoelectric point) and Mw (molecular mass) of the putative proteins coded by the ORFs in the genome of the two rhabdoviruses present in *L. salmonis*. The Phobius server were used to predict N-terminal signal peptide, ectodomain, transmembrane region, and C-terminal cytoplasmic tail in the topology analyses of the glycoprotein genes of the two rhabdoviruses. The ExPASy Bioinformatics Resource Portal (http://www.expasy.org/proteomics) was used for identification of putative glycosylation and phosphorylation sites. Motifscan (http://myhits.isb-sib.ch/cgi-bin/motif_scan) were used on the L protein from the two viruses to predict catalytic domains.

Results

Virus morphology

Salmon lice (*L. salmonis*), showing areas of reduced transparency in the cephalothorax in the vicinity of the second antenna (anterior part of the cephalothorax), were collected from farmed Atlantic salmon (Figure 1). Sectioning of these occluded areas

showed that they consisted of glandular tissues (Figure 2). In some lice the tissues were necrotic or completely disintegrated. One set of glands seems to open in the vicinity of the mouth tubule of the lice. Transmission electron microscopy (TEM) of the glandular tissues showed that they are most likely syncytia or tissue consisting of large multinucleated cells. Large amounts of virus-like particles were seen budding from cellular membranes, surface membranes or membranes of the Golgi/endoplasmatic reticulum system (Figure 3). Modified areas, possibly viroplasm, were observed in the cytoplasm of the glandular cells (Figure 3C). The virus particles were enveloped and rod-shaped or bacilliform with a diameter of 55 nm and a maximum length of 425 nm (Figure 4). The nucleocapsid seemed to exhibit a helical symmetry since in longitudinal tangential sections of the virions they appear as being cross-striated (spacing about 8.5–9.0 nm) (Figure 4B).

Genome

Illumina sequencing of the RNA from lice with glandular pathology and presence of virus-like particles, generated two nearly complete rhabdovirus genomes, *Lepeophtheirus salmonis* rhabdovirus No9 (LSRV-No9) and *L. salmonis* rhabdovirus No127 (LSRV-No127), with lengths of 11 681 and 11 519 nucleotides, respectively. These two sequences were used as template for construction of primers that were used for RT PCR and Sanger sequencing of the two virus genomes. No errors in the two genomes generated by Illumina sequencing were detected. Both viruses (Accession numbers: KJ958535, KJ958536) contain five open reading frames in the negative sense genome in the order '3-N-P-M-G-L-5' also found in other rhabdoviruses.

The 3-leader and 5-trailer regions

The Illumina sequencing generated a leader region of LSRV-No9 and LSRV-No127 consisting of the first 61 and 70 nucleotides, respectively, with trailer regions composed of 122 and 58 nucleotides, respectively. The non-translated 3'-end and

Table 2. The digoxigenin labelled (DIG-labelled) sense and antisense RNA probes were made with primers listed.

Name	Sequence
RhabNt F1(LSRV-No127Npro)	GGAGCCATCGGAGGTTATGACC
RhabNt R1(LSRV-No127Npro)	AAGGGGCCGTGTCAATCCTA
RhabNs F1(LSRV-No9-Norf)	TTCTCCCGAACCGACATGGA
RhabNs R1(LSRV-No9-Norf)	AGGGGATTGGCGGTGACTGA

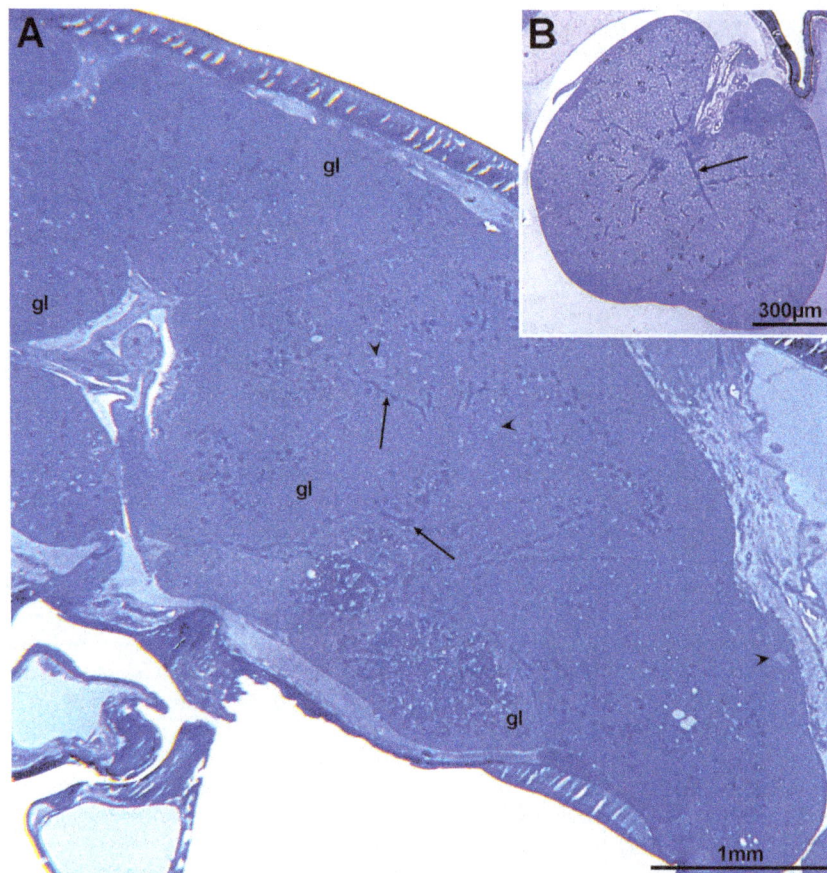

Figure 2. Sections of virus infected glands (gl) situated between the second antenna and the mouth tubule. Accumulation of virions (arrows) and viroplasm-like structures (arrow heads) (A). Virus-infected gland opening in the mouth tubule of the lice (B). Accumulation of virions (arrow).

5′-end regions of the two viruses may not be complete but still exhibit inverse complementarity. The first 27 nt of the leader of LSRV-No9 show 63.0% identity to the last 22 nucleotides of the trailer, while the first 19 nucleotides in the leader of LSRV-No127 show 89.5% identity to the last nucleotides in the trailer (Figure 5). The identities of the leader and trailer region from LSRV-No9 compared to the same regions in LSRV-No127 are 42.6% (61 nt compared) and 47.6% (63 nt compared), respectively. The leader, first 24 nucleotides, of LSRV-No9 and LSRV-No127 show 50.0% and 91.7% identity to Vesicular stomatitis New Jersey virus (NJ89GAS, Accession no: JX121110), respectively.

Gene junctions

The distances between translation stop and start codons in the gene junctions of the two viruses range from 47 (G-L) to 136 (N-P) nucleotides and from 41 (G-L) to 271 (N-P) nucleotides in the genomes of LSRV-No-9 and LSRV-No127, respectively. The untranscribed intergenic regions, the gene junctions between the polyadenylation sequence and the transcription initiation sequence, of the two lice rhabdoviruses vary in length (0 to 6 nt). The nucleotide sequences of the intergenic regions are not conserved between the two viruses and are also different from that of other related rhabdovirus genera (Table 3).

The putative transcription termination and polyadenylation signal, based on its homology to other rhabdoviruses, is conserved in the genomes of the two salmon louse viruses and comprises the motif $TATG(A)^7$ with the exception of the transcription stop/

polyadenylation signal of the G protein gene of LSRV-No127 which is $TAAG(A)^7$ (Table 3).

The potential start sequence in the genome of LSRV-No9 is not conserved and the same sequence, AACAA, can only be found in the start of the P, M and G protein genes (Table 3). The start sequence of the N and L protein genes is AACAG. The start of the N protein gene was determined by 5-end RACE. The junction between the P/M genes in LSRV-No9 differs from the other junctions in these two viruses in that the transcriptional start signal of the M gene seems to start with the last two nucleotides of the transcriptional stop signal of the P gene, or, as an alternative, it precedes the transcriptional stop signal of the P gene resulting in a possible overlap of 27 nt.

The transcription initiation site sequences, expected to occur shortly after the previous transcription termination signal, seem to be TAAGAA in the genome of LSRV-No127 with the exception of the transcription initiation of the L protein gene, which seems to be CAAGAA (Table 3). The start of the N protein gene was determined by 5-end RACE.

Protein genes

To annotate the coding sequences it has been assumed that each open reading frame (ORF) starts at the first AUG occurring after the previous transcription termination sequence, and that it continues to the first stop codon. The G protein gene is in reading frame one, the N, P and L protein genes are in reading frame two and the M protein gene is in reading frame three in the genome of

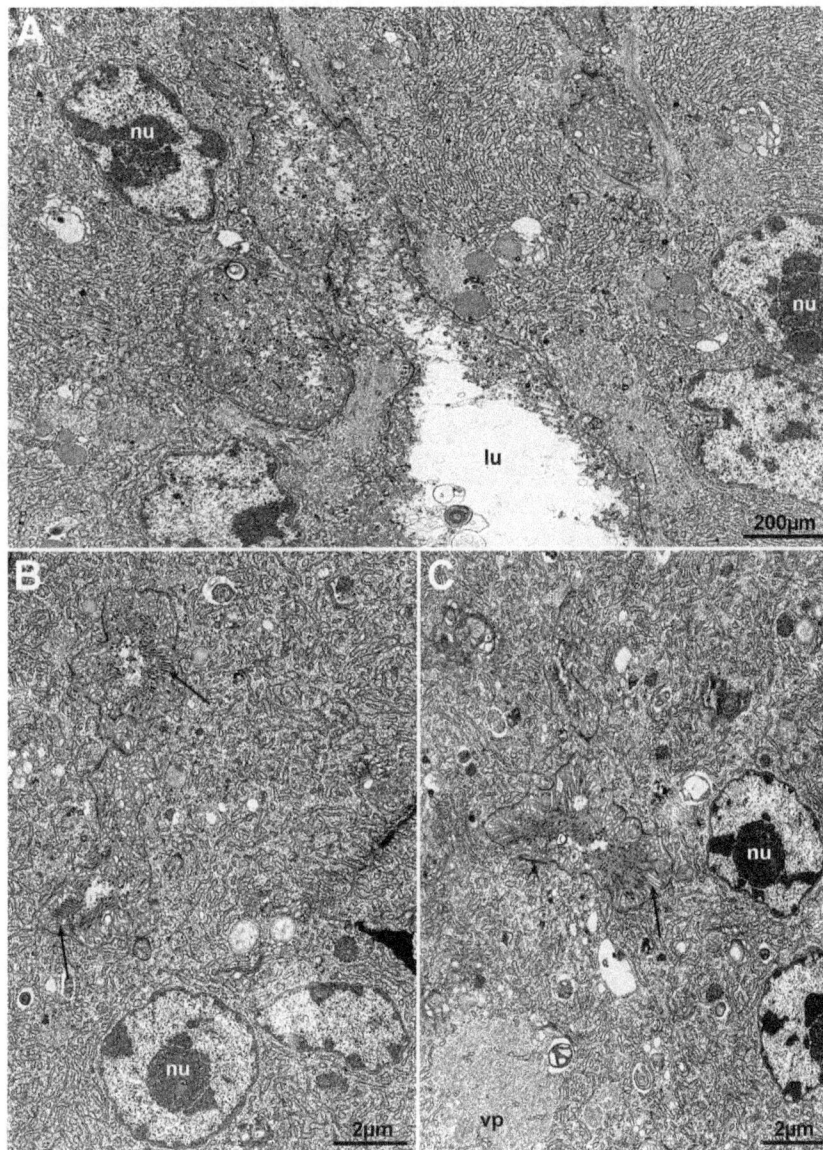

Figure 3 Multinucleated (nu) gland cells with channels containing virus-like particles (arrows) and amorphic material (A and B). C) This figure shows viroplasm (vp) in the vicinity of a channel containing virus-like particles (arrow). Note the accumulation of electron dense material (arrow head) on the inside of the cell membrane. Nucleus (nu).

LSRV-No9, while the N, P, M and G genes are in reading frame three and the L gene in reading frame one in the genome of LSRV-No127.

N gene. The 5′ ends of the N protein gene of the two salmon louse viruses were obtained using the GeneRacer Kit (Invitrogen) for full-length RNA ligase-mediated amplification of 5′ cDNA ends. The N gene in LSRV-No9 is 1691 nt long from the putative transcriptional start signal (AACAG) to the transcription termination signal (TATG(A)^7), and contains a single ORF of 1491 nt encoding a putative protein of 497 amino acids (Table 4). The calculated molecular mass (Mw) of the protein is 56.8 kDa with an isoelectric point (pI) of 5.8. The N gene of LSRV-No9 also contains a 199-nt 3′-UTR of unknown function between the stop codon and the polyadenylation signal. Amino acid sequence comparisons with other rhabdoviruses using BLAST search show that LSRV-No9 shares the highest identity (28.0–33.0%) and similarity with members of viruses belonging to the Dimarhabdo-

virus and Sigma virus groups. However, the N protein of LSRV-No9 shows 89.9% nucleotide identity (97.2% amino acid similarity) to a possible rhabdovirus nucleoprotein (Accession no: BT077702) obtained from salmon lice (*L. salmonis*) in the Pacific Ocean (Canada).

The N gene of LSRV-No127 is 1680 nt long from the putative transcription initiation site (TAAGAA) to the polyadenylation signal (TATG(A)^7) containing a single ORF consisting of 1398 nt encoding a putative protein of 466 aa (Table 4). The calculated Mw of the protein is 52.8 kDa with a pI of 5.9. The identity of the nucleotide and putative amino acid sequences of the N protein of LSRV-No9 compared to LSRV-No127 are 48.7% and 25.6%, respectively.

The N proteins of LSRV-No9 and LSRV-No127 contain 26 and 31 potential phosphorylation sites, and the sequences, $_{306}$GISNRSPYSV$_{315}$ and $_{288}$GISAKSPYSV$_{297}$, respectively. These sequences are relatively conserved with the RNA binding

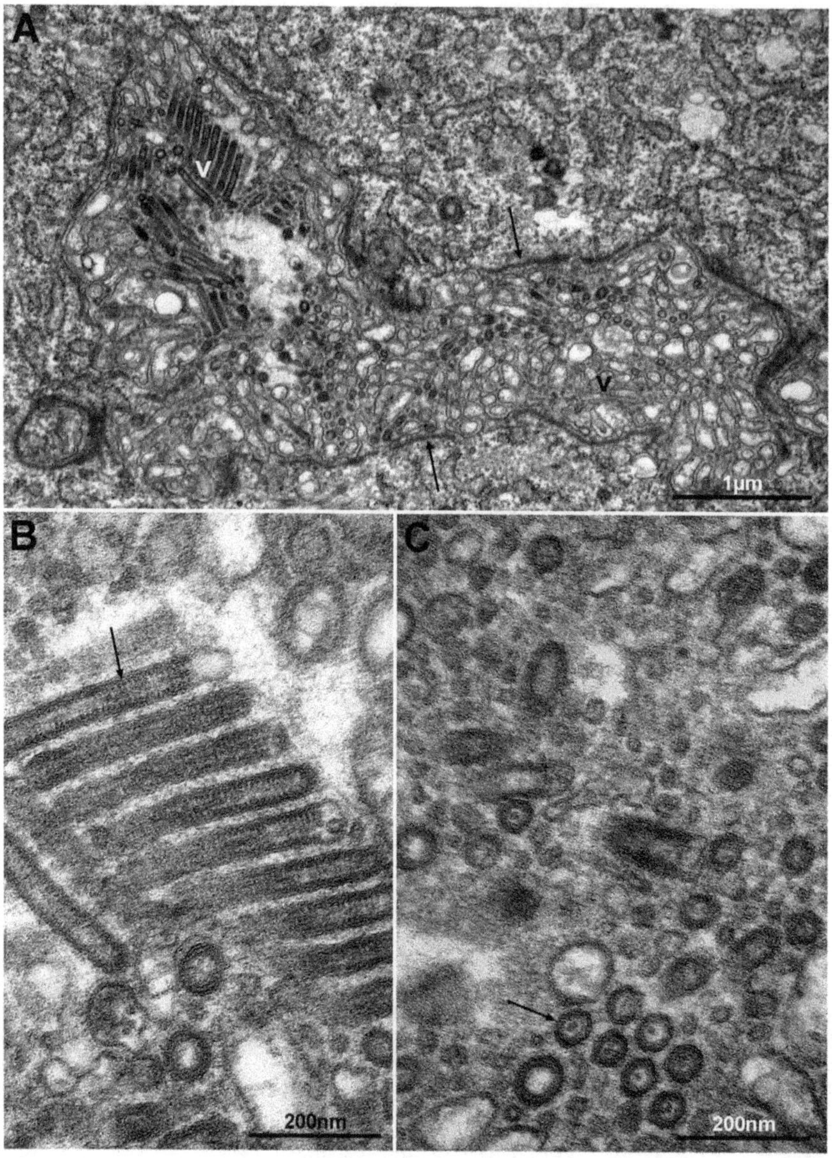

Figure 4. Section through a channel (lu) in a gland cell containing large amounts of virus-like particles (V). Note the accumulation of electron dense material (arrows) on the inside of the cell membrane (A). The virus particles (arrow) are enveloped, rod-shaped or bacilliform, and appear as being cross-striated in tangential longitudinal sections (B). Transverse section through virus particles showing surrounding unit membrane and an electron dense core (C).

motif (G(L/I)SXKSPYSS) present in vesiculoviruses, ephemeroviruses and lyssaviruses.

P gene. The putative LSRV-No9 P gene is 994 nt long and contains a single ORF of 888 nt encoding a putative protein of 296 aa, while the LSRV-No127 P gene is 926 nt long with a single ORF of 789 nt encoding a putative protein of 263 amino acids (Table 4). The calculated Mw of these two proteins are 32.6 kDa and 30.3 kDa with pI of 5.0 and 5.3, respectively. The P proteins of LSRV-No9 and LSRV-No127 contain 19 and 15 potential phosphorylation sites, respectively. Based on the predicted phosphorylation pattern it appears that both LSRV-No9 and LSRV-No127 P proteins contain a non-phosphorylated stretch in the centre, from amino acids 49–161 and 95–142, respectively. The two putative P protein sequences share no clear homology with the P proteins from other rhabdoviruses, while the amino acid similarity between the two viruses is 33.4%.

M gene. The M gene in LSRV-No9 is 763 nt long and contains a single ORF of 675 nt encoding a putative protein of 225 aa with calculated Mw of 25.1 kDa and a pI of 7.8. Amino acid sequence comparison with other rhabdoviruses, BLAST search, shows that it shares 25% identity with *Scophthalmus maximus* rhabdovirus, SMRV (ADU05404), and no significant match with other rhabdoviruses.

The M gene in LSRV-No127 is 736 nt long and with a single ORF of 657 nt encoding a putative protein of 219 amino acids with a calculated Mw of 24.0 kDa and a pI of 8.7. The identity and similarity of the putative amino acid sequences of LSRV-No9 compared to LSRV-No127 are 27.6% and 46.2%, respectively, while a BLAST search using the LSRV-No127 putative M protein gives identities in the range 21–23% with the M protein from Flanders virus (AGV98721), *Anguillid rhabdovirus* (AFJ94645), *Perch rhabdovirus* (YP007641365).

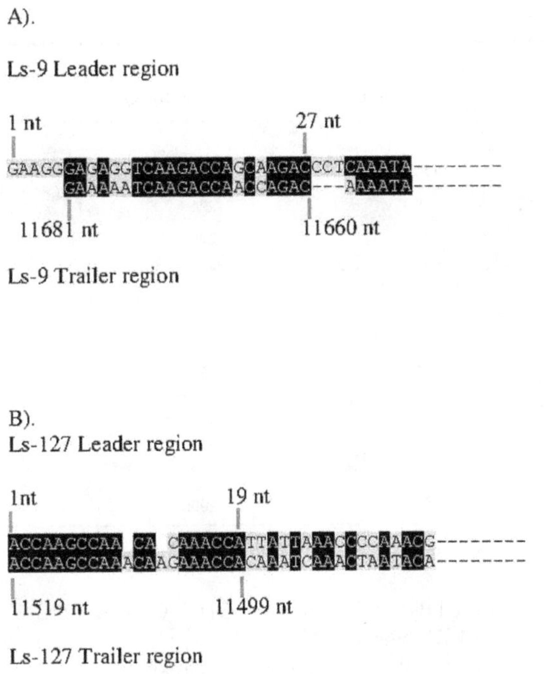

A).

Ls-9 Leader region

Ls-9 Trailer region

B).
Ls-127 Leader region

Ls-127 Trailer region

Figure 5. The non-translated 3'-end and 5'-end regions of the two viruses may not be complete but still exhibit inverse complementarity. The first nucleotides of the leader of LSRV-No9 aligned with the inverse complementary last nucleotides of the trailer (A). The first nucleotides of the leader of LSRV-No127 aligned with the inverse complementary last nucleotides of the trailer (B).

Both the predicted M proteins from LSRV-No9 and LSRV-No127 contain several phosphorylation sites, 14 and 18, respectively.

G gene. The G gene in LSRV-No9 is 1659 nt long and contains a single ORF of 1596 nt encoding a putative protein of 532 amino acids with a calculated Mw of 59.7 kDa and a pI of 6.7 (Table 4). Topology analyses using the Phobius server predict a transmembrane region spanning from amino acid 478–503 and a C-terminal cytoplasmic tail from aa 501–532. The protein is predicted to contain four putative N-glycosylation sites, $_{33}$NGTT, $_{249}$NQSC, $_{350}$NSTL, and $_{445}$NASI, respectively. Amino acid

sequence comparisons with other rhabdoviruses, BLAST search, show that this virus ORF shares the highest identity (22.0–23.0%) and similarity with that of Spring viraemia of carp virus (Genus *Sprivivirus*).

The G gene of LSRV-No27 is 1657 nt long containing a single ORF consisting of 1626 nt encoding a putative protein of 542 aa with a calculated Mw of 62.2 kDa and a pI of 7.3. Topology analyses using the Phobius server predict an N-terminal signal peptide (aa 1–24, N-region aa 1–3, H-region aa 4–15, C-region aa 16–24), an ectodomain from aa 25–486, a transmembrane region spanning from amino acid 487–511, and a C-terminal cytoplasmic tail from aa 512–542. The protein is predicted to contain two putative N-glycosylation sites, $_{16}$NLSI and $_{410}$NSSD, respectively. The identity of the nucleotide sequence and the similarity of the putative amino acid sequences of LSRV-No9 compared to LSRV-No127 are 31.3% and 46.4%, respectively. BLAST searches show that LSRV-No127 shares the highest identity (24.0%) with a virus isolated from tick or bat, Kolente virus (Accession no: AHB08864, unclassified Rhabdovirus) which possibly belongs to the Dimarhabdovirus group. However, the G protein of LSRV-No127 shows 50.9% nucleotide identity (39.8% amino acid similarity) to a possible rhabdovirus glycoprotein (Accession no: BTO75815) obtained from *Caligus rogercresseyi* in the Pacific Ocean (Chile).

L gene. The last gene in the genome of the two salmon louse rhabdoviruses, the L protein gene, shows a clear affinity to other members of *Rhabdoviridae*, with the closest affinity (>35.0% identity) to the Dimarhabdoviruses and members of the genus *Sigmavirus*. The full length L proteins from LSRV-No9 and LSRV-No127 are closest to each other (40.4% identity) and to the L protein from turbot rhabdovirus, SMRV, (>38.9%), and VSV (>38,8%) (Table 5). The *Sigmavirus* (>35.9%) and BEFV (>35.4%) are slightly more distant.

The L gene from LSRV-No9 is 6380 nt long and contains a single ORF of 6351 nt encoding a putative protein of 2117 aa, while the L gene from LSRV-No127 is 6376 nt long with a single ORF of 6288 nt encoding a putative protein of 2096 amino acids (Table 4). The calculated Mw of these two proteins are 241.8 kDa and 240.7 kDa with pI of 8.5 and 8.7, respectively.

The L gene is the most conserved in the family *Rhabdoviridae* and is structured into six conserved blocks that contain motifs for the structure and function of the L protein [41]. Pairwise alignments of the LSRV-No9 and LSRV-No127 L proteins with L proteins of selected members of Dimarhabdovirus and

Table 3. Leader and trailer regions for isolates Ls9 and Ls127.

Isolate	Gene	Leader	Trailer	Intergenic sequence	
Ls9	N	AACAG	TATGAAAAAAA		
Ls9	P	AACAA	TATGAAAAAAA	N-P	CAGT
Ls9	M	AACAA	TATGAAAAAAA	P - M	-
Ls9	G	AACAA	TATGAAAAAAA	M - L	CGGTTT
Ls9	L	AACAG	TATGAAAAAAA	G - L	TCT
Ls127	N	TAAGAA	TATGAAAAAAA	-	
Ls127	P	TAAGAA	TATGAAAAAAA	N - P	CT
Ls127	M	TAAGAA	TATGAAAAAAA	P - M	CCTC
Ls127	G	TAAGAA	TAAGAAAAAAA	M - G	CTAT
Ls127	L	CAAGAA	TATGAAAAAAA	G - L	T

Conserved transcription initiation (TI) and transcription termination/polyadenylation (TTP) sequences flank each gene to direct transcription of capped and polyadenylated mRNAs.

Table 4. Predicted genes and putative proteins of LSRV-No9 and LSRV-No127.

Protein	Gene length (nt)	ORF (nt)	5'-UTR (nt)	3'-UTR (nt)	Protein (aa)
No9					
N	1691	1491	93	106	497
P	994	888	28	78	296
M	763	675	18	72	225
G	1659	1596	28	35	532
L	6380	6351	12	17	2117
No127					
N	1680	1398	69	213	466
P	926	789	57	80	263
M	736	657	23	56	219
G	1657	1626	20	11	542
L	6376	6288	30	58	2096

Sigmavirus show a pattern that conforms to the given conserved blocks. Block II is the most conserved of the major domains and block I is the least conserved showing identities at the same level as seen for the entire L protein (Table 5).

The first conserved amino acid motif, DYxLNSP, in the L proteins of the rhabdoviruses compared is found in position 46–52 and 39–45 of LSRV-No9 and LSRV-No127, respectively. Three amino acid motifs, LMxKD (LSRV-No9 residue 237–241, LSRV-No127 residues 231–235), SFRHxGHP (LSRV-No9 res. 359–366, LSRV-No127 res. 353–360), and LASDLA (LSRV-No9 res. 395–400, LSRV-No127 res. 389–394), are highly conserved among the rhabdoviruses included in the alignment of block I.

Block II is highly conserved among the rhabdoviruses compared and the KERELK motif present in *Vesiculovirus* is found as $_{535}$KEREVK$_{540}$ and $_{529}$KEREMK$_{534}$ in LSRV-No9 and LSRV-No127, respectively. This motif has been shown to be involved in the positioning and binding of RNA template and the KEREMK motif is also present in other rhabdoviruses like Tibrogargan virus, Wongabel virus, and Flanders virus. LSRV-No9 share this motif, $_{535}$KEREVK$_{540}$, with Ngaingan virus.

Within block III the subdomain III-A is the most conserved, while subdomain III-D shows lower amino acid identity than the overall identity for the complete L protein. The GGLEGLR motif and the sequence LAQGDNQVI (with the invariant peptide QGDNQ), the latter in position 715–723 in LSRV-No9 and 709–717 in LSRV-No127, could correspond to motifs B and C, in block III which is important for the polymerase function. Using motifscan (http://myhits.isb-sib.ch/cgi-bin/motif_scan) on the L protein from LSRV-No9 and LSRV-No127 a predicted catalytic domain between amino acids 603–791 and 587–785, respectively, is detected.

The conserved domains, the RNA polymerase domain, mRNA capping-region (block V), and a methyltransferase region, are also present in both the salmon louse viruses. The conserved motif GSxT-(60–70 aa)-HR in block V, which is essential for mRNA capping activity could correspond to the sequences $_{1162}$GSKT-69aa-HR$_{1236}$ and $_{1153}$GSKT-69aa-HR$_{1227}$ in the L protein of LSRV-No9 and LSRV-No127, respectively. The conserved motif, IKRA (present in *Vesiculovirus*) was also present in both the louse viruses as LKRA (position LSRV-No9: 1181–1184, LSRV-No127: 1175–1178).

Block VI showed the GxGxG motif as GDGSG in both LSRV-No9 (res. 1673–1677) and LSRV-No127 (res. 1666–1670) which could play a role of polyadenylation or protein kinase activity.

Phylogeny

To reveal the relationships of the two louse viruses, LSRV-No9 and LSRV-No127, to other members of the family Rhabdoviridae, phylogenetic trees based on the L and N proteins were generated. Members of the genera *Cytorhabdovirus*, *Novirhabdovirus*, and *Nucleorhabdovirus* were excluded due to their large divergence which reduced the phylogenetic resolution, and the lyssaviruses were also removed from the alignment of the N protein due to high divergence. The ambiguously aligned regions in the alignments were removed using Gblocks resulting in sequence alignments of the L and N protein of 1630 and 256 amino acids, respectively.

In the phylogeny based on the L protein the two viruses from salmon louse, LSRV-No9 and LSRV-No127, group in a distinct clade with uncertain affinity to the other rhabdovirus genera included in the study and distant from the lyssaviruses (Figure 6). The phylogeny based on the N protein shows even less affinity between the two salmon louse viruses and no clear affinity to any of the assigned genera included in the study (Figure 7). However, LSRV-No9 groups closely with a rhabdovirus N protein sequence (Accession no: ACO12126) obtained from salmon louse (*L. salmonis*) in the Pacific Ocean (Canada).

Screening

Selected tissues from Atlantic salmon (N = 70) infected with *L. salmonis* and different developmental stages of the salmon louse (N = 165), including egg strings, were tested for presence of both rhabdoviruses, LSRV-No9 and LSRV-No127, using real time RT PCR.

All life stages of the salmon louse tested positive for both rhabdoviruses, but the largest amounts of virus RNA were detected in adult lice (Ct values as low as 12 were obtained for both viruses). Virus RNA were also present in the eggs and embryos. All tissues (skin, gills, heart, kidney) from the Atlantic salmon were negative or only slightly positive (CT values >30) with the exception of skin tissues surrounding the attachment site for the chalimi stages. The Ct values at the attachment site were in

Table 5. Percent amino acid identity of Ls9 and Ls127 L protein domains and subdomains compared with a selection of related rhabdoviruses.

Virus	Entire	Blocks % identity						Subdomains block III (%)			
	L	I	II	III	IV	V	VI	III-A	III-B	III-C	III-D
Ls9											
VSV	38.8	**40.3**	64.4	**56.3**	46.2	53.2	56.5	**92.3**	88.5	**90.0**	**30.8**
SVCV	37.8	33.5	60.6	55.8	45.6	53.7	62.4	**92.3**	92.3	80.0	**30.8**
PFRV	37.6	35.1	59.6	55.8	45.0	53.7	60.0	**92.3**	92.3	**90.0**	**30.8**
PRV	38.0	36.1	59.6	52.7	46.8	56.7	58.3	**92.3**	92.3	80.0	**30.8**
SMRV	**38.9**	35.6	**65.4**	53.6	**52.0**	**61.0**	58.8	**92.3**	88.5	**90.0**	**30.8**
BEFV	36.1	35.1	61.5	50.9	39.2	52.8	50.6	84.6	88.5	80.0	23.1
TIBV	34.7	31.4	58.6	53.1	42.7	51.9	**67.9**	69.2	80.8	80.0	30.8
DURV	36.0	38.2	**65.4**	52.2	46.8	46.3	63.1	76.9	92.3	80.0	30.8
WONV	36.0	38.2	59.6	51.3	48.5	54.1	52.4	**92.3**	**100.0**	80.0	15.4
NGAV	36.5	31.9	62.5	53.6	44.4	55.0	57.1	84.6	88.5	**90.0**	15.4
MOUV	28.2	27.2	52.9	44.6	38.6	38.1	41.7	69.2	69.2	80.0	23.1
KOLEV	35.2	36.1	62.5								
SIGMAV	35.3	34.6	56.7	47.8	41.5	52.8	49.4	76.9	80.8	80.0	**30.8**
Ls127											
VSV	39.0	40.3	63.5	**53.6**	**55.0**	48.9	**60.0**	**100.0**	84.6	**90.0**	30.8
SVCV	38.1	36.6	58.7	51.8	48.5	51.1	54.1	**100.0**	84.6	80.0	30.8
PFRV	37.3	37.2	59.6	52.2	49.7	49.4	54.1	**100.0**	84.6	**90.0**	30.8
PRV	37.7	37.7	61.5	50.9	52.0	54.5	47.6	84.6	84.6	80.0	30.8
SMRV	**39.1**	35.6	62.4	47.3	54.4	**55.0**	56.5	**100.0**	84.6	**90.0**	30.8
BEFV	35.4	34.0	59.6	50.4	44.4	47.2	44.7	84.6	84.6	80.0	30.8
TIBV	34.1	33.0	55.8	47.3	47.4	46.3	58.3	61.5	76.9	80.0	23.1
DURV	35.8	**40.8**	**63.5**	48.7	51.5	46.3	59.5	76.9	84.6	80.0	23.1
WONV	35.1	35.1	62.5	48.2	53.2	51.5	48.8	84.6	92.3	80.0	15.4
NGAV	36.8	35.6	58.7	52.2	52.0	51.9	50.0	84.6	84.6	**90.0**	15.4
MOUV	28.4	28.3	50.0	41.5	43.3	39.0	40.5	61.5	65.4	80.0	23.1
KOLEV	36.1	37.2	58.7								
SIGMAV	35.9	37.7	55.8	46.9	46.8	48.9	50.6	76.9	76.9	80.0	38.5
Ls9	49.4	53.4	68.3	67.9	66.7	62.8	69.4	76.9	92.3	100	69.2

Vesiculovirus (VSV = Vesicular stomatitis virus (ABP01784)), Sprivivirus (SVCV = Spring viraemia of carp virus (ABW24037) and PFRV = Pike fry rhabdovirus (ACP28802)), Perhabdovirus (PRV = Perch rhabdovirus (AFX72892)), Ephemerovirus (BEFV = Bovine ephemeral fever virus (NP065409)), Sigmavirus (Sigma = Sigma virus (AFV52407)), Tibrovirus (TIBV = Tibrogargan virus (YP007641376)), Tupavirus (DURV = Durham virus (ADB88761)). Unassigned (SMRV = Turbot rhabdovirus (ADU05406), WONV = Wongabel virus (YP002333280), NGAV = Ngaingan virus (YP003518294), MOUV = Moussa virus (ACZ81407), KOLEV = Kolente virus (AHB08865)).

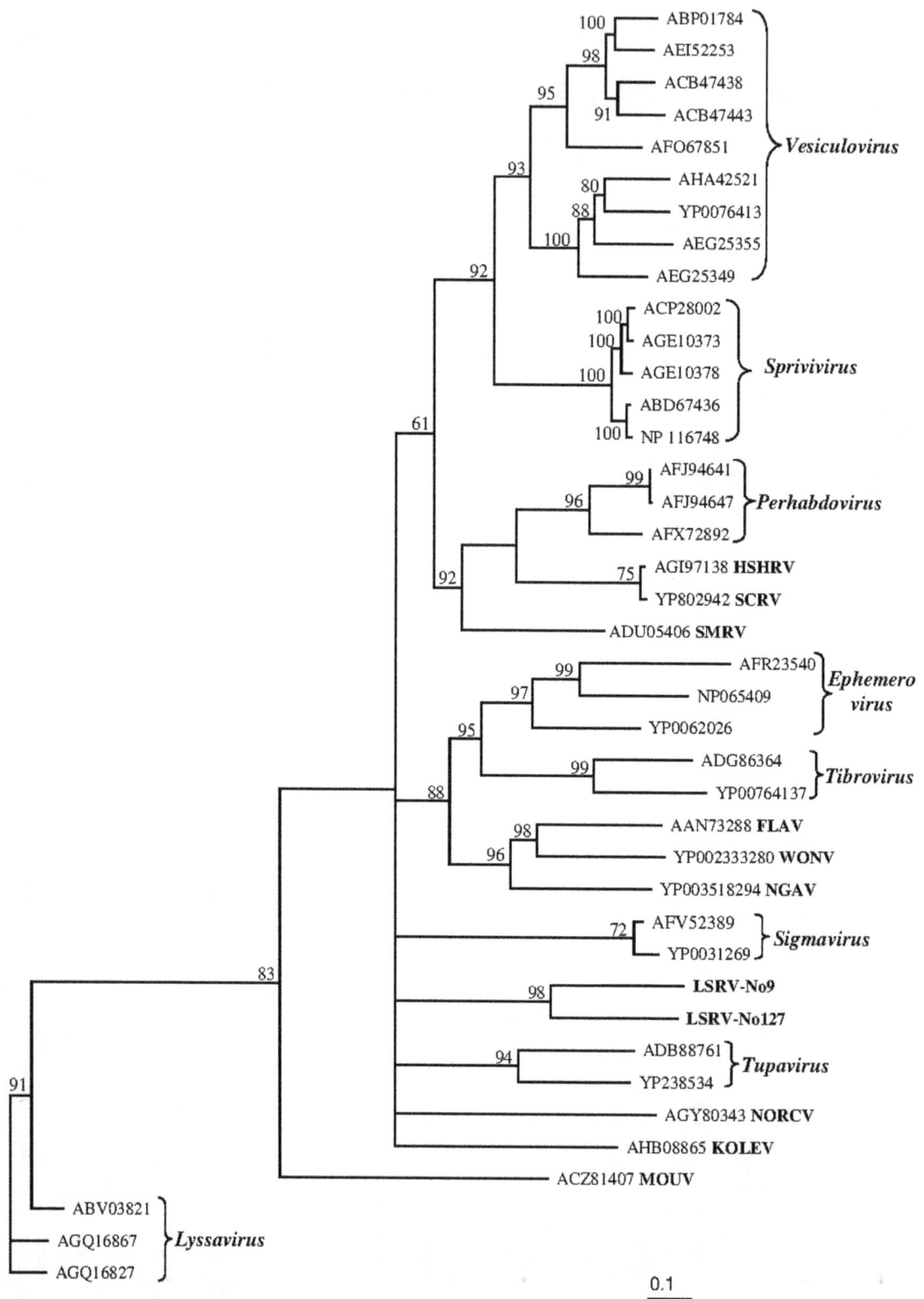

Figure 6. Phylogenetic position of two *Rhabdoviridae*, **LSRV-No9 (Accession no: KJ958535) and LSRV-No127 (Accession no: KJ958536), obtained from salmon louse (***L. salmonis***) in relation to other rhabdoviruses based on analysis of the L protein sequences after removal of ambiguously aligned regions using Gblocks** [38]. The evolutionary relationship is presented as maximum likelihood trees based on 1630 aa from the complete alignment of the L protein amino acid sequences. Branch lengths represent relative phylogenetic distances according to maximum likelihood estimates based on the VT matrix [39]. The scale bar shows the number of amino acid substitutions as a proportion of the branch lengths.

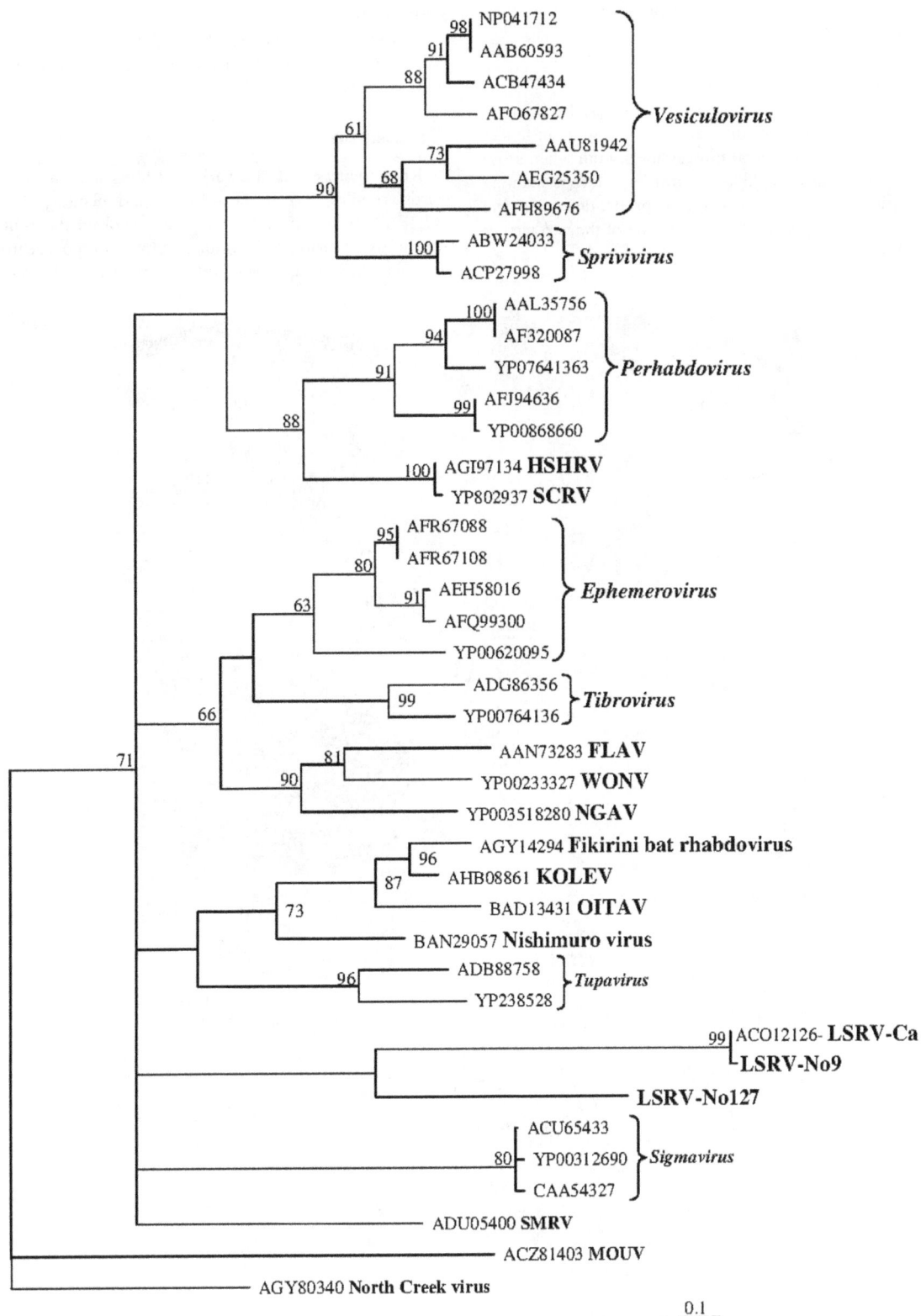

Figure 7. Phylogenetic position of two *Rhabdoviridae*, **LSRV-No9 (Accession no: KJ958535) and LSRV-No127 (Accession no: KJ958536), obtained from salmon louse (***L. salmonis***) in relation to other rhabdoviruses based on analysis of the N protein sequences after removal of ambiguously aligned regions using Gblocks** [38]. The evolutionary relationship is presented as maximum likelihood trees based on 256 aa from the complete alignment of the N protein amino acid sequences. Branch lengths represent relative phylogenetic distances according to maximum likelihood estimates based on the VT matrix [39]. The scale bar shows the number of amino acid substitutions as a proportion of the branch lengths.

the range between 22 and 30 indicating presence of substantial amounts of virus RNA.

In situ hybridization

The two viruses had similar tissue tropism (Figure 8). Staining was observed in glands, subcuticular tissue and, in some instances in peripheral cytoplasm of skeletal muscle fibers, both when sense and antisense probes were employed. In ovaries and eggs, staining was only seen in sections with the antisense probe, detecting viral mRNA. All lice stained positive for at least one of the two viruses (results not shown).

Cell culture

All the tested cell cultures, BF-2, CHSE-214, ASK and RT-Gill-W1, appeared to be refractory to the two rhabdoviruses from salmon louse.

Discussion

Rhabdoviruses infect a variety of hosts such as mammals, fish, birds, reptiles, insects, crustaceans and plants [42,43,44,45,46, 47,48,49,50,51,52,53,54]. They have evolved different modes of transmission including transmission by arthropods, through direct contact, through gametes and through water. Eleven genera of

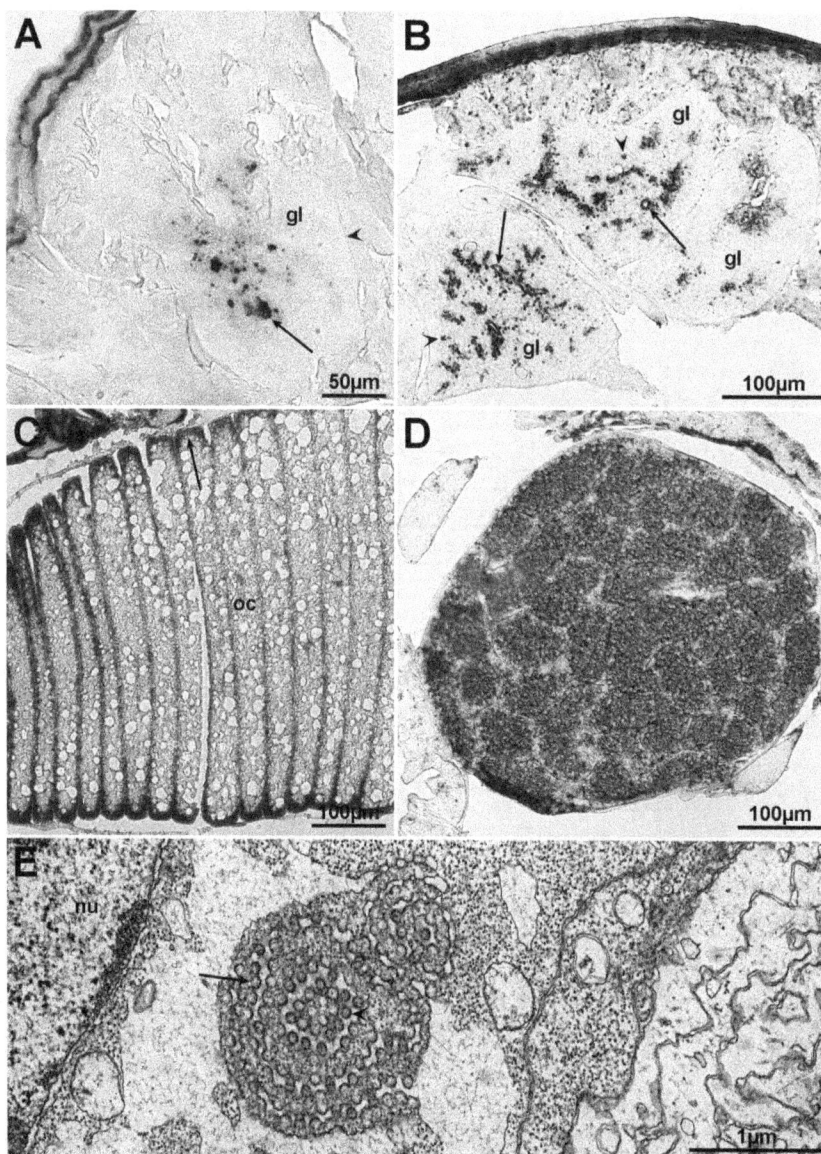

Figure 8. In situ hybridization for localization of LSRV genomes and mRNAs encoding the putative N protein. *In situ* hybridization with an antisense probe targeting mRNA encoding the N protein of LSRV-No9 results in patches of staining (arrow) within an exocrine gland (gl), where the arrowhead is pointed at the gland capsule. These patches may represent viroplasm (A). A sense probe targeted at the LSRV-No127 genome induces coloring in or around gland (gl) secretory ducts, which are indicated by arrows. This may reflect viral budding through the cytoplasmic membrane and the presence of mature virions within the lumen of the duct, as shown by TEM (Figure 3 and 4). Patches of staining (arrowheads) in the cytoplasm may reflect viroplasm (B). Utilization of an antisense probe aimed at LSRV-No127 mRNA encoding the N protein, results in staining (arrow) of cytoplasm at the periphery of oocytes (oc), and within the ovary (figures C and D). TEM picture of putative virions budding (arrow) into the lumen of ER (E). It is not known if these spherical virus-like particles (arrow head) are connected to any of the two rhabdoviruses. Nucleus of the ovary cell (nu).

rhabdoviruses are recognized where viruses associated with arthropods and a wide range of vertebrates, including fish, are found within the genera Vesiculovirus, Ephemerovirus, Sprivivirus, Sigmavirus, Tibrovirus, Tupavirus and some unassigned rhabdoviruses (dimarhabdovirus super group [55]). This is the first study where the nearly complete genomic sequences of new rhabdoviruses obtained from a parasitic copepod, *Lepeophtheirus salmonis*, are presented. Phylogenetic analysis of the two salmon louse viruses, LSRV-No9 and LSRV-No127, based on the L and N protein clearly places them as distinct virus species among these members of *Rhabdoviridae*. The significant divergence of the two lice viruses compared to the closest members of *Rhabdoviridae* suggests that they probably deserve to be recognized as a new genus within this family.

The gene organization, 3'-N-P-M-G-L-5', is the same as for members of *Vesiculovirus* [47]. There are no additional genes interposed between the five structural genes, as found in some genera of the Rhabdoviridae [56,57,58,59,60]. The RNA binding motif (G(L/I)SXKSPYSS) sequences that are relatively conserved among N proteins from vesiculoviruses, ephemeroviruses and lyssaviruses [61,62,63] are also present in a conserved area in the central region of both louse viruses N protein. The P and M proteins of the two salmon louse viruses show little or no similarity to other described rhabdoviruses, while the G protein of the two salmon louse viruses, like that of other rhabdoviruses is predicted to be a class I transmembrane glycoprotein with an N-terminal signal peptide, glycosylated ectodomain, a transmembrane domain and a short C-terminal cytoplasmic domain [47]. The L protein of the two louse viruses have identifiable sequence homology to other rhabdoviruses, containing all six conserved regions, and associated motifs; RNA template binding, RNA-dependent RNA polymerase, mRNA capping, polyribonucleotidyltransferase activity, methyl transferase activity, and polyadenylation/protein kinase activity [41]. The amino acid sequences of the L protein show close to 40% identity to *Vesiculovirus*. Hence the gene organization and the most conserved genes and motifs support that the two louse viruses belong in the family *Rhabdoviridae*. The non-coding gene junctions of the two salmon louse viruses also contained the conserved transcription termination/polyadenylation motif TATG(A)^7 and the relatively conserved transcription initiation motif AAGAA/G found among other related rhabdoviruses [43,45,46,49,50,51,53,57,60,64].

Although arthropods are frequently involved as hosts of rhabdoviruses, only a few have been associated with crustaceans and none characterized from parasitic copepods [7,10,14,16, 44,49,50,51,52,53,56,58]. The salmon louse (*L. salmonis*), parasitizing salmonids in the northern Atlantic and Pacific oceans, is one of several blood feeding fish parasites found among crustacean copepods. Screening of *L. salmonis* collected in Norwegian salmon farms for presence of the two louse viruses, show that all stages including the egg strings of this parasite are positive for both viruses, and *in situ* hybridization and transmission electron microscopy show that the two viruses are present in glandular tissues of adult lice. The ovaries are also positive in the *in situ* hybridization test, but rhabdovirus virions were not observed using TEM on this organ. The host (*S. salar*) for the salmon louse seems to be negative for presence of these two viruses and it has not been possible to culture these viruses in cell cultures obtained from salmonids. The weak positives (Ct values >30) found during screening of skin and gills could possibly be a result of contamination from salmon lice present on the fish. Still, relatively low Ct values were obtained when skin tissues from the Atlantic salmon at the attachment sites for the chalimi stages of the salmon louse were tested. This could suggest that the louse injects the virus

into the host skin during the attachment process which would explain the presence of virions in the mandibular glands of the parasite. It is also tempting to speculate that this could be part of a strategy used by the louse to prevent the rejection of the frontal filament that the louse injects into the host skin during early establishment on the host. It has been shown that bites from arthropods can modulate vertebrate host functions by several mechanisms including modulation of the immune response and vasodilation [65]. If this is the case then this group of viruses could be present in most members of the *Caligidae* (a large group of fish parasites). Sequence comparisons, using the N protein from LSRV-No9 and the G protein from LSRV-No127, indicate that similar viruses are most likely also present in parasitic copepods in the Pacific Ocean. The nucleotide sequence from the N protein ORF of LSRV-No9 shows 89.9% identity to a N protein ORF obtained from subspecies *L. salmonis onchorhynci* [66] in Canadian waters, while the G protein ORF from LSRV-No127 shows 50.9% identity two a sequence obtained from *C. rogercresseyi* (Accession no: BT075815) in Chilean salmon culture. Rhabdoviruses and rhabdovirus-like particles have also been detected in glandular tissues of other arthropods and crustaceans [10,11], however, nothing is known about the genome of viruses from these other crustaceans.

The two rhabdoviruses characterized in this study are the first members of this family that infect copepods, however, there are reports suggesting that spring viraemia of carp virus (SVCV) could be transmitted by a fresh water crustacean, the fish parasite *Argulus foliaceus* [67]. SVCV has also been isolated from crustaceans, *Penaeus stylirostris* and *P. vannamei*, causing mortalities in both fish and penaeid hosts [16]. It has been shown that the salmon louse (*L. salmonis*) may function as a mechanical vector for infectious salmon anaemia virus (ISAV) and infectious haematopoietic necrosis virus (IHNV) [68,69,70], and recently, it was shown that another Caligidae, *Caligus rogercresseyi*, may function as a mechanical vector for ISA virus in the culture of Atlantic salmon in Chile [71]. However none of these viruses have been demonstrated to replicate in these parasitic copepods. Rhabdoviruses have been isolated and detected in several fish species including salmonids like *Salmo trutta* and *S. salar* [64,72,73,74], but these viruses are genetically distant from the two salmon louse rhabdoviruses which are not associated with any disease in Atlantic salmon.

Conclusions

The present study characterize the genome of two new rhabdoviruses obtained from the parasitic copepod *Lepeophtheirus salmonis*, identify their target tissues by *in situ* hybridization, and their putative virion morphology by TEM. Comparison of the genomes show that the two viruses cluster among the Dimarhabdovirus/Sigmavirus groups as two distinct new species that might be classified as distinct from the 11 currently recognized *Rhabdoviridae* genera. The gene organization, 5'-N-P-M-G-L-3', of the two viruses is the same as that described from *Vesiculovirus*.

Detection of substantial amounts of RNA from both lice viruses at the attachment site for the parasite at the salmonid host suggest that the louse injects the viruses into the skin during early establishment on the host. If the salmon louse uses these viruses for modulation of the immune response in the salmonid hosts one can expect that the other fish parasite species in the copepod family *Caligidae* could be using related viruses for the same purpose. This hypothesis is supported by the presence of a G protein gene, showing high similarity to the G protein from the two salmon louse viruses in the parasitic copepod *Caligus rogercresseyi* collected in the South Pacific Ocean. The existing large diversity of the *Rhabdoviridae* is underscored by the uniqueness of these two viruses from the salmon louse and suggests that more studies are needed to map the complexity of this virus family.

References

1. Nylund A, Økland S, Bjørknes B (1992) Anatomy and ultrastructure of the alimentary canal in *Lepeophtheirus salmonis* (Copepoda: Siphonostomatoida). J Crust Biol 3: 423–437.
2. Heuch PA, Bjørn PA, Finstad B, Holst JC, Asplin L, et al. (2005) A review of the Norwegian "National action plan against salmon lice on salmonids": The effect on wild salmonids. Aquaculture 246: 79–92.
3. Hamre L, Eichner C, Caipang CMA, Dalvin ST, Bron JE, et al. (2013) The salmon louse *Lepeophtheirus salmonis* (Copepoda: Caligidae) life cycle has only two Chalimus stages. PLoS ONE 8(9): e73539. doi: 10.1371/journal.pone. 0073539
4. Krkosek M, Revie CW, Gargan PG, Skilbrei OT, Finstad B, et al. (2012) Impact of parasites on salmon recruitment in the Northeast Atlantic Ocean. Proc R Soc B 20122359. Available: http://dx.doi.org/10.1098/rspb.2012. 2359.
5. Karlsbakk E, Olsen AB, Einen AC, Mo TA, Fiksdal IU, et al. (2013) Amoebic gill disease due to *Paramoeba perurans* in ballan wrasse (*Labrus bergylta*). Aquaculture vol 412–413: 41–44.
6. Torrisen O, Jones S, Asche F, Guttormsen A, Skilbrei OT, et al. (2013) Salmon lice – impact on wild salmonids and salmon aquaculture. J Fish Dis 36: 171–194.
7. Jahromi SS (1977) Occurrence of Rhabdovirus-like particles in the Blue crab, *Callinectes sapidus*. J Gen Virol 36: 485–493.
8. Johnson PT (1978) Viral diseases of the Blue crab, Callinectes sapidus. Mar Fish Rev 40 (10): 13–15.
9. Johnson PT (1984) Viral diseases of marine invertebrates. Helgoländer Meeresuntersuchungen 37: 65–98.
10. Yudin AI, Clark WH (1979) A description of Rhabdovirus-like particles in the mandibular gland of the blue crab, *Callinectes sapidus*. J Inv Pathol 33: 133–147.
11. Johnson PT, Farley CA (1980) A new enveloped helical virus from the Blue crab *Callinectes sapidus*. J Invert Pathol 35: 90–92.
12. Pappalardo R, Mari J, Bonami JB (1986) T (tau) virus infection of *Carcinus mediterranus*: Histology, cytopathology, and experimental transmission of the disease. J Invert pathol 47: 361–368.
13. Mari J, Bonami JR (1988) W2 virus infection of the crustacean *Carcinus mediterraneus*: a reovirus disease. J Gen Virol 69: 561–571.
14. Lu Y, Loh PC (1994) Viral structural proteins and genome analysis of the rhabdovirus of penaeid shrimp (RPS). Dis Aquat Org 19: 187–192.
15. Vogt G (1996) Cytopathology of bay of Piran shrimp virus (BPSV), a new crustacean virus from the Mediterranean Sea. J Invert Pathol 68: 239–245.
16. Johnson MC, Maxwell JM, Loh PC, Leong AAC (1999) Molecular characterization of the glycoprotein from two warm water rhabdoviruses; snakehead rhabdovirus (SHRV) and rhabdovirus of penaeid shrimp (RPS)/ spring viraemia of carp virus (SVCV). Virus Res 64: 95–106.
17. Munro J, Owens L (2007) Yellow head-like viruses affecting the penaeid aquaculture industry: a review. Aquaculture Res 38: 893–908.
18. Escobedo-Bonilla CM, Alday-Sanz V, Wille M, Sorgeloos P, Pensaert MB, et al. (2008) A review on the morphology, molecular characterization, morphogenesis and pathogenesis of white spot syndrome virus. J Fish Dis 31: 1–18.
19. Lightner DV (2011) Virus diseases of farmed shrimp in Western Hemisphere (the Americas): A review. J Invert Pathol 106: 110–130.
20. Overstreet RM, Jovonovich J, Ma H (2009) Parasite crustaceans as vectors of viruses, with an emphasis on three penaeid viruses. Int Comp Biol 49 (2): 127–141.
21. Stentiford GD, Bonami JR, Alday-Sanz V (2009) A critical review of susceptibility of crustaceans to Taura syndrome, Yellowhead disease and White spot disease and implications of inclusions of these diseases in European legislation. Aquaculture 29: 1–17.
22. Walker PJ, Winton JR (1010) Emerging viral diseases of fish and shrimp. Vet Res 51: 51.
23. Behringer DC, Butler MJ, Shields JD, Moss J (2011) Review of *Palinurus argus* virus 1 –a decade after its discovery. Dis Aquat Org 94: 153–160.
24. Bonami JR, Widada JS (2011) Viral diseases of the giant fresh water prawn *Macrobrachium rosenbergii*: A review. J Invert Pathol 106: 131–142.
25. Bonami JR, Zhang S (2011) Viral diseases in commercially exploited crabs: A review. J Invert Pathol 106: 6–17.
26. Longshaw M (2011) Diseases of crayfish: A review. J Invert Pathol 106: 54–70.
27. Vega-Heredia, Mendoza-Cano F, Sanchez-Pas A (2011) The infectious hypodermal and hematopoietic necrosis virus: A brief review of what we do and do not know. Transbondary and Emerging Dis 59: 95–105.
28. Flegel TW (2012) Historic emergence, impact and current status of shrimp pathogens in Asia. J Invert Pathol 110: 166–173.
29. Lightner DV, Redman RM, Pantoja CR, Tang KFJ, Noble BL, et al. (2012) Historic emergence, impact and current status of shrimp pathogens in Americas. J Invert Pathol 110: 174–183.
30. Steigen A, Nylund A, Karlsbakk E, Akoll P, Fiksdal IU, et al. (2013) 'Cand. Actinochlamydia clariae' gen. nov., sp. nov., a unique intracellular bacterium causing epitheliocystis in catfish (*Clarias gariepinus*) in Uganda. PLoS ONE 2013 Jun 24; 8(6): e66840. doi: 10.1371/journal.pone.0066840
31. Frost P, Nilsen F (2003) Validation of reference genes for transcription profiling in the salmon louse, *Lepeophtheirus salmonis*, by quantitative real-time PCR. Veterinary Parasitology 118: 169–174.
32. Olsvik PA, Lie KK, Jordal AEO, Nilsen TO, Hordvik I (2005) Evaluation of potential reference genes in real-time RT-PCR studies of Atlantic salmon. BMC Mol. Biol. 6, doi: 10.1186/147 1-2199-6-21
33. Devold M, Krossøy B, Aspehaug V, Nylund A (2000) Use of RT-PCR for diagnosis of infectious salmon anaemia virus (ISAV) in carrier sea trout *Salmo trutta* after experimental infection. Dis Aquat Org 40: 9–18.
34. Lannan CN, Winton JR, Fryer JL (1984) Fish cell lines: Establishment and characterization of nine cell lines from salmonids. In Vitro 20 (9): 671–676.
35. Bols NC, Barlian A, Chirinotrejo M, Caldwell SJ, Goegan P, et al. (1994) Development of a cell-line from primary cultures of rainbow-trout, *Oncorhynrchus mykiss* (Walbaum), gills. J Fish Dis 17 (6), 601–611.
36. Dalvin S, Nilsen F, Skern-Mauritzen R (2013) Localization and transcription patterns of LsVasa, a molecular marker of germ cells in *Lepeophtheirus salmonis* (Krøyer). J Natural History Vol 47, Nos 5–12: 889–900.
37. Tröße C, Nilsen F, Dalvin S (2014) RNA interference mediated knockdown of the KDEL receptor and COPB2 inhibits digestion and reproduction in the parasitic copepod *Lepeophtheirus salmonis*. Comparative biochemistry and physiology. Part B, Biochemistry & molecular biology. doi: 10.1016/j.cbpb.2013.12.006
38. Talavera G, Castresana J (2007) Improvement of phylogenies after removing divergent and ambiguously aligned blocks from protein sequence alignments. Systematic Biology 56: 564–577.
39. Muller T, Vingron M (2000) Modeling amino acid replacement. J Computational Biol 7(6): 761–776.
40. Page RDM (1996) TREEVIEW: an application to display phylogenetic trees on personal computers. Comput Appl Biosci 12: 357–358.
41. Poch O, Blumberg BM, Bougueleret L, Tordo N (1990) Sequence comparison of five polymerases (L proteins) of unsegmented negative-strand RNA viruses: theoretical assignment of functional domains. J Gen Virol 71: 1153–1162.
42. Hoffman B, Schutze H, Mettenleiter TC (2002) Determination of the complete genomic sequence and analysis of the gene products of the virus of Spring viremia of carp, a fish rhabdovirus. Virus Res 84: 89–100.
43. Chen HL, Lui H, Liu ZX, He JQ, Gao LY, et al. (2009) Characterization of the complete genome sequence of pike fry rhabdovirus. Arch Virol 154: 1489–1494.
44. Quan PL, Junglen S, Tashmukhamedova A, Conlan S, Hutchison SK, et al. (2010) Moussa virus: A new member of the Rhabdoviridae family isolated from *Culex decens* mosquitoes in Cote d'Ivoire. Virus Res 147: 17–24.
45. Cherian SS, Gunjikar RS, Banerjee A, Kumar S, Arankalle VA (2012) Whole genomes of Chandipura virus isolates and comparable analysis with other Rhabdovirueses. PLoS ONE 7(1): e30315. doi: 10.1371/journal.pone.0030315
46. Galinier R, van Beurden S, Amilhat E, Castric J, Schoehn G, et al. E (2012) Complete genomic and taxonomic position of eel European X (EVEX), a rhabdovirus of European eel. Virus Res 166: 1–12.
47. King AMQ, Adams MJ, Carstens EB, Lefkowitz EJ (2012) Virus taxonomy. Classification and nomenclature of viruses. Academic Press. London.
48. Rodriguez LL, Pauszek SJ (2012) Genus Vesiculovirus. (In: Rhabdoviruses. Molecular taxonomy, evolution, genomics, ecology, host-vector interactions, cytopathology and control. Eds: Dietzgen RG, Kuzmin IV). Caister Academic Press, Norfolk, UK.
49. Ghedin E, Rogers MW, Widen SG, Guzman H, Travassos de Rosa APA, et al. (2013) Kolente virus, a rhabdovirus species isolated from ticks and bats in the Republic of Guinea. J Gen Virol 94: 2609–2615.
50. Vasilakis N, Widen S, Travassos de Rosa APA, Wood TG, Walker PJ, et al. (2013a) Malpais spring virus is a new species in the genus *Vesiculovirus*. Virology J 10: 69.
51. Vasilakis N, Widen S, Mayer SV, Seymor R, Wood TG, et al. (2013b) Niakh virus: A novel member if the family *Rhabdoviridae* isolated from phlebotomine sandflies in Senegal. Virology 444: 80–89.
52. Coffey LL, Page BL, Greninger AL, Herring BL, Russel RC, et al. (2014) Enhanced arbovirus surveillance with deep sequencing: Identification of novel rhabdoviruses and Bunjaviruses in Australian mosquitoes. Virology 448: 146–158.
53. Torkarz R, Sameroff S, Leon MS, Jain K, Lipkin WI (2014) Genome characxterization of Long island tick rhabdovirus, a new virus identified in *Amblyomma americanum* ticks. Virology J 11: 26.

Author Contributions

Conceived and designed the experiments: ALØ AN SB KW CEA. Performed the experiments: ALØ AN ACØ SB KW SG CEA HP. Analyzed the data: ALØ AN SB SG. Contributed reagents/materials/analysis tools: AN ACØ HP. Wrote the paper: AN ALØ SB SG ACØ HP.

54. Zeng W, Wang Q, Wang Y, Liu C, Liang H, et al. (2014) Genomic characterization and taxonomic position of a rhabdovirus from Hybrid snakehead. Arch Virol. doi: 10.1007/s00705-014-2061-z

55. Bourhy H, Cowley JA, Larrous F, Homes EC, Walker PJ (2005) Phylogenetic relationships among rhabdoviruses inferred using the L polymerase gene. J Gen Virol 86: 2849–2858.

56. Longdon B, Obbard DJ, Jiggins FM (2010) Sigma viruses from three species of *Drosophila* form a major clade in the rhabdovirus phylogeny. Proc R Soc B 277: 35–44.

57. Gubala A, Davis S, Weir R, Melville L, Cowled C, et al. (2010) Ngaingan virus, a macropod-associated rhabdovirus, contains a second glycoprotein gene and seven open reading frames. Virology 399: 98–108.

58. Gubala A, Davis S, Weir R, Melville L, Cowled C, et al. (2011) Tibrogargan and Costal Plains rhabdoviruses: genomic characterization, evolution of novel genes and seroprevalence in Australian livestock. J Gen Virol 92: 2160–2170.

59. Walker PJ, Dietzgen RG, Joubert DA, Blasdell KR (2011) Rhabdovirus accessory genes. Virus Res 162: 110–125.

60. Zhu RL, Lei XY, Ke F, Yuan XP, Zhang QY (2011) Genome of turbot rhabdovirus exhibits unusual non-coding regions and an additional ORF that could be expressed in fish cells. Virus Res 155: 495–505.

61. Tordo N, Poch O, Ermine A, Keith G (1986) Primary structure of leader RNA and nucleoprotein genes of the rabies genome: segmented homology with VSV. Nucl Acid Res 14 (6): 2671–2683.

62. Crysler JG, Lee P, Reinders M, Prevec L (1990) The sequence of the nucleocapsid protein (N) gene of Piry virus: possible domains in the N protein of vesiculoviruses. J Gen Virol 71: 2191–2194.

63. Walker PJ, Wang Y, Cowley JA, McWilliam SM, Prehaud CJN (1994) Structural and antigen analysis of the nucleoprotein of bovine ephemeral fever rhabdovirus. J Gen Virol 75: 1889–1899.

64. Johansson T, Nylund S, Olesen NJ, Bjørklund H (2001) Molecular characterization of the nucleocapsid protein gene, glycoprotein gene and gene junctions of rhabdovirus 903/87, a novel fish pathogenic rhabdovirus. Virus Res 80: 11–22.

65. Schoeler GB, Wikel SK (2001) Modulation of host immunity by haematophagous arthropods. Ann Tropical Med Parasitol vol 95 (8): 755–771.

66. Skern-Mauritzen R, Torrissen O, Glover KA (2014) Pacific and Atlantic *Lepeophtheirus salmonis* (Krøyer, 1838) are allopatric subspecies: *Lepeophtheirus salmonis salmonis* and *L. salmonis onchorhynci* subspecies novo. BMC Genetics 15: 32.

67. Ahne W (1985) *Argulus foliaceus* L. and *Piscicola geometra* L. as mechanical vectors of spring viraemia of carp virus (SVCV). J Fis Dis 8: 241–242.

68. Nylund A, Bjørknes B, Wallace C (1991) *Lepeophtheirus salmonis*, -a possible vector in the spread of diseases on salmonids. Bull Eur Ass Fish Pathol 11 (6): 213–216.

69. Nylund A, Wallace C, Hovland T (1993) The possible role of *Lepeophtheirus salmonis* (Krøyer) in the transmission of infectious salmon anaemia. In: Boxshall GA, Defaye D (eds) Pathogens of Wild and Farmed Fish: Sea lice. Ellis Horwood Limited, Chichester, p 367–373.

70. Jakob E, Barker DE, Garver KA (2011) Vector potential of the salmon louse *Lepeophtheirus salmonis* in the transmission of infectious heamatopoietic necrosis virus (IHNV). Dis Aquat Org 97: 155–165.

71. Oelckers K, Vike S, Duesund H, Gonzales J, Wadsworth S, et al. (2014) *Caligus rogercresseyi* as a potential vector for transmission of infectious salmon anaemia (ISA) virus in Chile. Aquaculture 420–421: 126–132.

72. Adair BM, McLoughlin M (1986) Isolation of pike fry rhabdovirus from brown trout (Salmo trutta). Bull Eur Ass Fish Pathol 6 (3), 85–86.

73. Koski P, Hill BJ, Way K, Neuvonen E, Rintamaki P (1992) A rhabdovirus isolated from brown trout (*Salmo trutta* m. *lacustris* (l.)) with lesions in parenchymatous organs. Bull Eur Ass Fish Pathol 12 (5): 177–180.

74. Borzym E, Matras M, Mai-Paluch J, De Boisseson C, Talbi C, et al. (2014) First isolation of Hirame rhabdovirus from freshwater fish in Europe. J Fish Dis 37: 423–430.

11

Detection Theory in Identification of RNA-DNA Sequence Differences Using RNA-Sequencing

Jonathan M. Toung[1], Nicholas Lahens[1], John B. Hogenesch[2,3,5], Gregory Grant[2,3,4]*

1 Genomics and Computational Biology Graduate Program, University of Pennsylvania School of Medicine, Philadelphia, PA, United States of America, 2 Institute for Biomedical Informatics, University of Pennsylvania School of Medicine, Philadelphia, PA, United States of America, 3 Institute for Translational Medicine and Therapeutics, University of Pennsylvania School of Medicine, Philadelphia, PA, United States of America, 4 Department of Genetics, University of Pennsylvania School of Medicine, Philadelphia, PA, United States of America, 5 Department of Pharmacology, University of Pennsylvania School of Medicine, Philadelphia, PA, United States of America

Abstract

Advances in sequencing technology have allowed for detailed analyses of the transcriptome at single-nucleotide resolution, facilitating the study of RNA editing or sequence differences between RNA and DNA genome-wide. In humans, two types of post-transcriptional RNA editing processes are known to occur: A-to-I deamination by ADAR and C-to-U deamination by APOBEC1. In addition to these sequence differences, researchers have reported the existence of all 12 types of RNA-DNA sequence differences (RDDs); however, the validity of these claims is debated, as many studies claim that technical artifacts account for the majority of these non-canonical sequence differences. In this study, we used a detection theory approach to evaluate the performance of RNA-Sequencing (RNA-Seq) and associated aligners in accurately identifying RNA-DNA sequence differences. By generating simulated RNA-Seq datasets containing RDDs, we assessed the effect of alignment artifacts and sequencing error on the sensitivity and false discovery rate of RDD detection. Overall, we found that even in the presence of sequencing errors, false negative and false discovery rates of RDD detection can be contained below 10% with relatively lenient thresholds. We also assessed the ability of various filters to target false positive RDDs and found them to be effective in discriminating between true and false positives. Lastly, we used the optimal thresholds we identified from our simulated analyses to identify RDDs in a human lymphoblastoid cell line. We found approximately 6,000 RDDs, the majority of which are A-to-G edits and likely to be mediated by ADAR. Moreover, we found the majority of non A-to-G RDDs to be associated with poorer alignments and conclude from these results that the evidence for widespread non-canonical RDDs in humans is weak. Overall, we found RNA-Seq to be a powerful technique for surveying RDDs genome-wide when coupled with the appropriate thresholds and filters.

Editor: Yi Xing, University of California, Los Angeles, United States of America

Funding: This work was funded by Glue Grant: U54HL117798; CSTA Grant: The project described was supported by the National Center for Research Resources and the National Center for Advancing Translational Sciences, National Institutes of Health, through Grant UL1TR000003. The content is solely the responsibility of the authors and does not necessarily represent the official views of the NIH; ITMAT: Supported in part by the Institute for Translational Medicine and Therapeutics of the Perelman School of Medicine at the University of Pennsylvania. The funders had no role in study design, data collection and analysis, decision to publish, or preparation of the manuscript.

Competing Interests: The authors have declared that no competing interests exist.

* Email: greg@grant.org

Introduction

Next-generation sequencing technology provides comprehensive sequence information. The precision afforded by RNA-Seq is useful for studying various aspects of the transcriptome such as alternative splicing [1,2], RNA editing [3,4], and differential allelic expression [5–7]. RNA editing refers to co- or post-transcriptional modification of RNA, resulting in a transcript that is different from the underlying genomic template. In humans, two types of RNA editing processes are known to occur: adenosine deamination by ADAR results in A-to-G edits [8,9] and cytidine deamination by APOBEC1 results in C-to-U changes [10,11].

In recent years, many genome-wide surveys of RNA editing in humans have been performed using next-generation sequencing technology [3,12–15]. In addition to the known A-to-G and C-to-U alterations introduced by RNA editing, researchers have reported the existence of RNA-DNA sequence differences (RDDs) that cannot be explained by known mechanisms [14,16,17]. However, the validity of these results is contested, as many reports cite experimental and technical artifacts as the main determinants of such systematic sequence differences between RNA and DNA [15,18–21].

Current methods for the accurate identification of RDDs mainly involve ad hoc filters aimed at removing false positives [14,15,22]. In this study, we used a detection theory approach to evaluate the relative effect of misalignment and sequencing error on RDD analysis. In particular, we generated simulated RNA-Seq datasets containing simulated RDDs and assessed the performance of various RNA-Seq aligners in accurately identifying RDDs. We also analyzed filtering methods for their efficacy in achieving low false discovery rates of RDD detection and high sensitivity values. Lastly, after determining the optimal thresholds and parameters for sequence difference analysis, we searched for the presence of RDDs in an experimental human RNA-Seq dataset for which deep DNA and RNA sequence information is publicly available.

Overall, our report aims to explore the phenomenon of RDDs in humans as well as provide a framework for those interested in the study of RNA editing, RDDs, or differential allelic expression by elucidating the appropriate thresholds and parameters for accurate detection of allele-specific differences in RNA-Seq data. The simulated datasets generated in this study are publicly available for download (see Methods).

Results

Simulated RNA-Seq datasets

To evaluate the performance of various alignment algorithms and filtering methods in detecting RDDs, we generated simulated RNA-Seq datasets containing RDDs (see Methods). First, we created a "clean" dataset (dataset 1) with no sequencing errors or intronic reads in order to evaluate the degree of bias introduced by alignment error alone. Next, in order to capture the effect of sequencing error on RDD identification, we generated a more realistic RNA-Seq dataset containing substitutional sequencing errors, indel polymorphisms, intronic signal, and lower quality bases at the tail end of reads. We used a simplistic error model in which sequencing errors occur randomly and independently. We considered the effect of non-random sequencing errors and found their presence to be minimal in Illumina Hi-Seq datasets (see Methods). Both datasets contain 50 million pairs of non strand-specific reads of length 100 base pairs (bp) and were generated in triplicates to allow for assessment of variation in our various metrics.

Datasets were aligned using GSNAP, MapSplice, RUM, and Tophat2 (see Methods). For both dataset 1 and 2, GSNAP and MapSplice performed the best in terms of the number of reads mapped in total and uniquely (Table S1 in File S1), aligning approximately 99% of the 50 million read pairs. In contrast, RUM and Tophat2 aligned approximately 98% of the read pairs in dataset 1, but only roughly 95% in dataset 2, which contains sequencing errors. Overall, between 97 to 99% of the read pairs are aligned uniquely with GSNAP, MapSplice, and RUM, whereas only approximately 89% of the read pairs are aligned uniquely with Tophat2.

Simulated RNA-DNA sequence differences

For each of the two datasets, we randomly introduced RDDs throughout the genome (see Methods). Namely, at positions containing RDDs, a percentage of the total reads at the site bear a randomly chosen non-reference allele representing the sequence difference. Furthermore, we define the percentage of reads containing the non-reference base to be the RDD level.

For each dataset, we generated approximately 600,000 total RDDs each in order to obtain reasonable sample sizes for making statistical inferences. Our motivation in choosing this number was not to simulate known frequencies or features of RNA editing events, but rather to accurately probe the ability of next-generation sequencing technology to detect hypothetical sequence differences in the human genome. Each RDD type is equally represented, with sequence differences that originate from cytosine and guanosine (C>A, C>G, C>T, G>A, G>C, G>T) slightly overrepresented than other types. This variation results from differing base compositions throughout the genome, with the effect more pronounced in dataset 2, which contains reads from intronic regions of the genome (Figure S1).

The coverage, or the total number of reads, at a given site is important in the analysis of RDDs as the presence and levels of RDDs at sites that are deeply sequenced are more likely to be robustly assessed (Figure S2). In order to assess the effect of sequencing depth on RDD detection, we stratified sites in the

genome according to coverage and simulated equal numbers of RDDs in each group (see Methods).

For each RDD, we chose the level, or proportion of reads carrying the non-reference base, from a standard uniform random distribution excluding 0. However, because of the discrete nature of coverage, the distribution of RDD levels is not uniform at sites with low coverage; for sites with coverage greater than 100x, the distribution of levels is uniform across all levels with the exception of boundary values (Figure S3).

To understand the effect of hyperediting by ADAR and the observation that non-canonical RDDs often cluster [14], we modeled a subset of RDDs to occur in close proximity of one another (see Methods). Hyperediting refers to a type of editing by members of the ADAR family whereby approximately 50% of the adenosines on each strand of an RNA duplex is edited in a promiscuous fashion [23]. For each dataset, we generated approximately 2,000 clusters of length 100 bp within which approximately 50% of all positions with the same reference base bear the same type of RDD (see Methods).

Overall, in dataset 1, the average distance between neighboring RDDs is 10 bp (median 3 bp) for sites belonging to hyperedited clusters and 815 bp (median of 58 bp) for those that do not. For dataset 2, which contains intronic reads, the average distance between RDDs is also 10 bp (median 3 bp) for sites belonging to hyperedited clusters and 1565 bp (median 225 bp) for those that do not (Table S2 in File S1).

Sensitivity of RNA-DNA sequence difference detection

We began our assessment of the performance of next-generation sequencing technology in identifying RDDs by analyzing the sensitivity or true positive rate of sequence difference identification. We will address the false negative and false positives rates of RDD detection in a subsequent section. We start by defining a simulated RDD as being properly identified by the aligner if at least one read bearing the non-reference base is observed. For dataset 1, we found that overall, GSNAP detected 96.32±6.19E-2% of the simulated RDDs, whereas MapSplice, RUM, and Tophat2 correctly identified 95.30±1.61E-1%, 95.36±1.63E-1%, and 95.04±1.38E-1%, respectively. For dataset 2, which contains sequencing errors and intronic reads, GSNAP identified 93.54±1.04E-1% of all simulated sites, whereas MapSplice, RUM, and Tophat2 found 92.34±1.16E-1%, 91.12±1.33E-1%, and 90.84±1.54E-1%, respectively.

Next, we investigated the effect of sequencing depth or coverage on the detection of RDDs. We observed that for both datasets, the sensitivity of detection increases with higher thresholds on the minimum depth of coverage (Figure 1). For example, the sensitivity of sequence difference detection using GSNAP increases approximately 2 to 4% in datasets 1 and 2 when sites with coverage lower than 10x are removed from consideration. The sensitivity of RDD detection using MapSplice, RUM and Tophat2 increases in a similar fashion with higher coverage (Table S3 in File S1). Given the relatively low true positive rate or high false negative rate of RDD detection for locations with low coverage, we restrict subsequent analyses to sites with a minimum of 10 total reads in the simulated dataset and the corresponding aligned datasets per GSNAP, MapSplice, RUM, or Tophat2.

Next, we analyzed the effect of RDD level on the sensitivity of RDD detection. We binned the simulated sequence differences into 10 groups by RDD levels and evaluated the true positive rate for each group. For both datasets, we found the sensitivity of RDD detection to increase with higher RDD levels (Figure 2; Table S4 in File S1). Furthermore, GSNAP had the highest sensitivity values across all levels among the four aligners. Given the lower recall

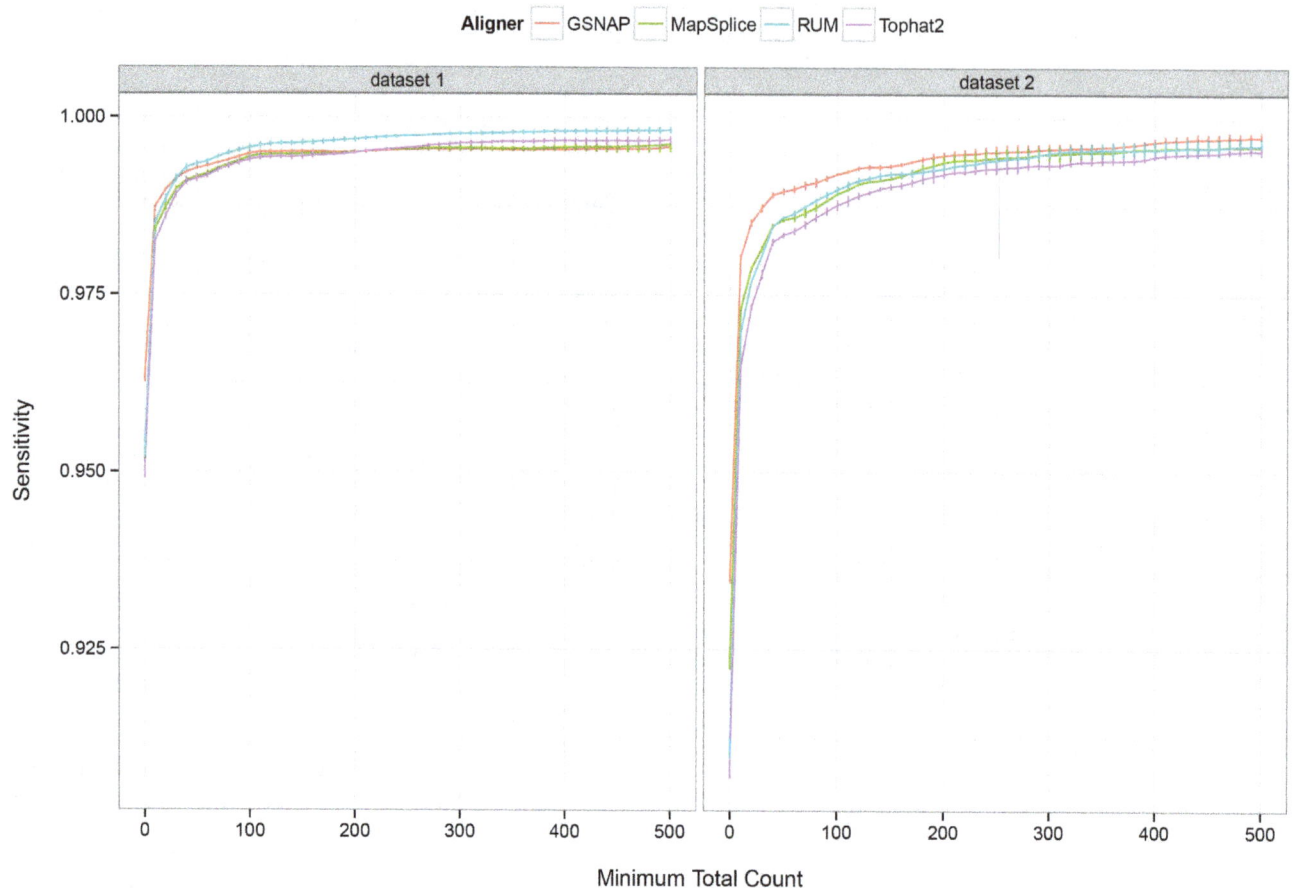

Figure 1. Sensitivity of RNA-DNA sequence difference detection versus coverage. The sensitivity or true positive rate of RNA-DNA sequence difference identification is shown versus various thresholds on the minimum depth of coverage required at the site of the simulated difference. For all four aligners, the true positive rate increases sharply upon raising the minimum depth of coverage required for detection from 0x to approximately 50x, after which it plateaus.

rates for sequence differences with low levels, we restrict our downstream analyses to sites with a minimum level of 10%.

Next, we analyzed whether the repetitive nature of the genomic sequence flanking the RDD site affects the detection of RDDs. We investigated this question by evaluating the sensitivity of detection in regions of the genome that are deemed non-unique by BLAT (see Methods). We observed that the true positive rate of RDD detection using GSNAP in non-unique regions according to BLAT is lower than that in unique regions by approximately 5% (Figure S4). For GSNAP, the average sensitivity in non-unique versus unique regions is 94.98±10.73E-2% versus 99.53±28.42E-4% for dataset 1 and 94.74±14.35E-2% versus 99.25±2.58E-2% for dataset 2. For GSNAP, MapSplice, and Tophat2, the difference in sensitivity of detection between RDDs within non-unique versus unique regions is roughly 4 to 5%, while for RUM, it is interestingly less than 1% (Table S5 in File S1). Upon further investigation, we attributed this difference in sensitivity patterns between the four aligners to different reporting procedures: for reads that align to multiple locations of the genome, GSNAP, MapSplice and Tophat2 will distinguish between primary and secondary alignments, whereas RUM does not. We also examined the sensitivity of RDD identification for sites lying within versus outside of RepeatMasker regions [24] and observed that for all four aligners, the sensitivity of detection is approximately 1 to 2%

higher for sites lying outside of RepeatMasker regions (Table S6 in File S1).

Lastly, we analyzed the effect of proximity to neighboring RDDs on sensitivity of detection. Short-read aligners typically have a limit on the number of mismatches relative to the reference permitted in a reported alignment, and thus sites with many neighboring sequence differences may be harder to identify. We observed that the sensitivity of sequence difference detection for sites that are greater than 10 bp in distance away from a neighboring sequence difference is roughly 1 to 3% higher for dataset 1 and 3 to 6% higher for dataset 2 (Table S7 in File S1).

Correlation between simulated versus observed levels of RNA-DNA sequence differences

In many studies, the mere detection of RDDs is not sufficient. For example, in studies on RNA editing or differential allelic expression, information about the degree or level of difference is important. Here we analyzed the correlation between simulated and observed RDD levels. Based on our previous analyses, we restricted our study to sites with a minimum coverage of 10x, minimum level of 10%, and minimum of 1 read bearing the sequence difference base. Using this threshold, we calculated the correlation between observed and simulated RDD levels to be relatively high, at approximately 98 on average across all three replicates for all four aligners and both datasets (Figure 3; Table

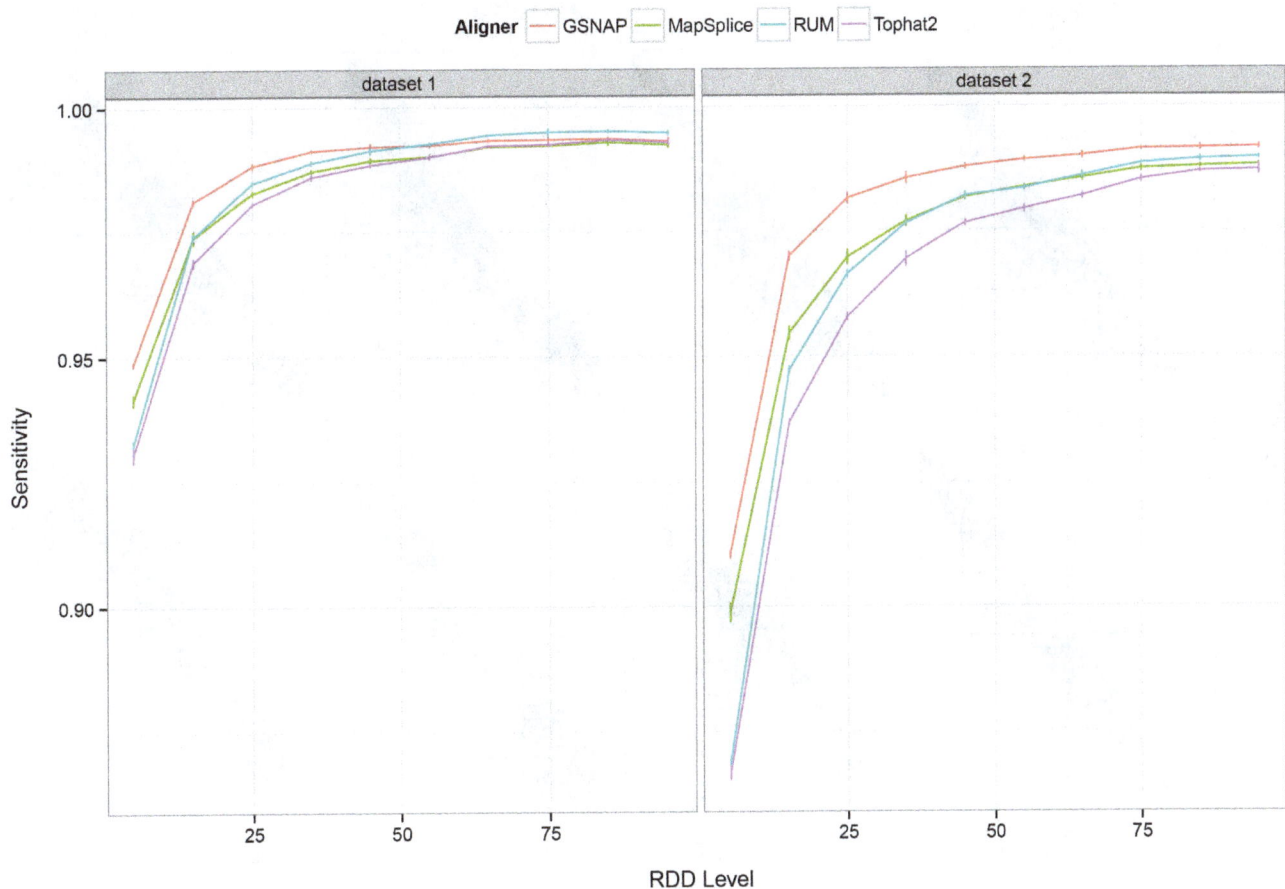

Figure 2. Sensitivity of RDD detection versus the simulated RDD level. Here we depict the true positive rate of RDD detection versus the simulated RDD level, or the percentage of reads at the site bearing the sequence difference allele. A minimum of 1 read bearing the RNA-DNA sequence difference is sufficient for a site to be deemed correctly identified. Sites with coverage less than 10x per the simulated RNA-Seq dataset are removed from consideration.

S8 in File S1). Although the simulated and observed levels correspond well, we found that roughly 20 to 40% of sites in each dataset for any aligner have observed levels that deviate from the simulated values by more than 5% (Figure S5). In particular, we found that in the majority (75 to 90%) of cases in which the observed and simulated levels deviate by at least 10%, the observed level underestimates the simulated level.

We hypothesized that one contributing factor to the discrepancy in RDD levels is the uniqueness or the ability of the region surrounding the site to be aligned accurately. Indeed, we found that approximately 22 to 34% of sites in which the simulated versus observed RDD levels differ by more than 30% are found in non-unique regions of the genome as determined by BLAT versus roughly 7 to 12% for those where the levels do not differ by 30% or more (Table S9 in File S1).

Receiver operating characteristic and false positive analysis of RDD detection

Next, we analyzed the false positive and false discovery rates of RDD detection by evaluating the presence of differences at sites that were not simulated to represent RDDs. Using parameters we identified from our sensitivity analysis, we performed a receiver operating characteristic analysis on RDD detection genome-wide in each of the datasets (Table S10 in File S1). Overall, we observed that using a 'minimum coverage of 10x, minimum level of 10%,

and minimum of 1 read bearing the RDD base' cutoff, the false positive rate of sequence difference detection is low, averaging 3.39E-2% and 6.47E-1% across the different aligners for datasets 1 and 2, respectively. However, these low false positive rates are not unexpected, as the vast majority of sites in the genome do not contain simulated RDDs.

For a better understanding of how false positives affect the analysis of RDDs, we evaluated the false discovery rate (FDR), or the percentage of sites identified as having sequence differences that were not simulated to represent RDDs. For dataset 1, we found the FDR to range from 1.31±4.06E-2% in Tophat2 to 6.24±8.26E-2% in MapSplice when using a 'minimum coverage of 10x, minimum level of 10%, and minimum of 1 one read bearing the RDD base' threshold. These relatively low false discovery rates indicate that in the absence of sequencing error, misalignment issues do not contribute significantly to the incidence of false positives. With the introduction of sequencing error in dataset 2, we found that the false discovery rates are much higher, ranging from approximately 57% in GSNAP to 71% in Tophat2. These results are not surprising, as a threshold requiring only one read to bear the RDD base introduces false positives at sites with sequencing errors. With stricter thresholds on RDD detection, such as requiring a minimum coverage of 20x, a minimum RDD level of 20%, and a minimum of 4 RDD bases observed, we found that the false discovery rate decreases dramatically (Figure 4).

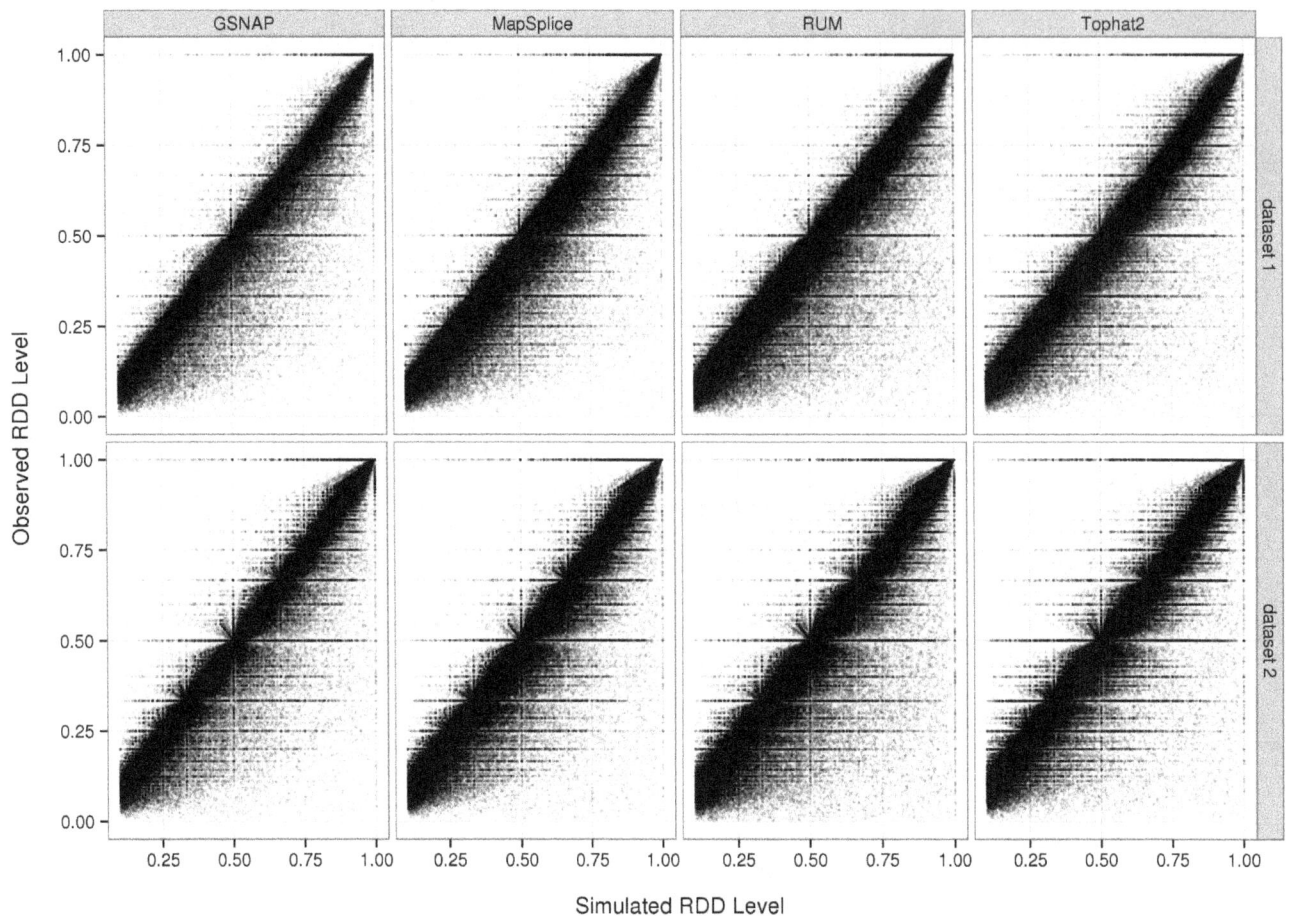

Figure 3. Simulated versus observed levels of RNA-DNA sequence differences. Here we plot the simulated RDD level versus the observed level as determined by GSNAP, MapSplice, RUM, or Tophat for replicate 1. Sites with coverage less than 10x or a RDD level less than 10% per the simulated dataset are removed from consideration. Overall, we observed the correlation between simulated and observed levels to be approximately 98% in both datasets and across the various aligners and replicates.

Evaluation of filters in reducing false positives

Many previous studies on RNA editing and RDDs attempt to remove false positive sites using various filters [3,14,15,17,22]. We investigated the effectiveness of some of these measures in eliminating false positives.

The first filter we analyzed requires RDDs to be identified concordantly by other aligners in order to be considered valid. We hypothesize that this condition will minimize the contribution of aligner-specific biases to the problem of false positive RDDs. For each aligner, we calculated the number of true and false positives remaining after RDDs that were not found by other aligners were removed (Table S11 in File S1). Using a 'minimum coverage of 20x, minimum level of 20%, and a minimum of 4 reads bearing the sequence difference' to identify RDDs, we observed the false discovery rates of RDD detection decreased with increasing numbers of other aligners required to concordantly identify the RDD (Figures S6-S9). In both datasets, requiring concordance among GSNAP, MapSplice, and RUM removed approximately 30% to 50% of false positives in GSNAP to 60 to 90% in MapSplice and RUM, whereas approximately only 2 to 12% of true positives were removed (Table S11 in File S1). We observed that while requiring concordance with Tophat2 led to the largest reductions in the number of false positives, it also led to large (over 50%) decreases in the number of true positives; this is expected as

we previously observed Tophat2 to identify the fewest number of RDDs among the four aligners (Table S10 in File S1).

The second filter we analyzed involves using BLAT to determine whether the sequence surrounding the RDD site can be aligned to other homologous regions of the genome (see Methods). Using a 'minimum coverage of 20x, minimum level of 20%, and minimum of 4 reads bearing the sequence difference' to identify RDDs, we observed that the BLAT method removes approximately 14% and 28% of false positives found by GSNAP in datasets 1 and 2 respectively, but only filters out roughly 1 to 5% of true positives in either datasets (Table S12 in File S1). As expected, we observed that the performance of the BLAT filter varies depending on the repetitive nature of the underlying flanking sequence. For example, within RepeatMasker regions, approximately 22% of false positives and 19% of true positives within the dataset 1 for GSNAP are filtered out, whereas outside of RepeatMasker regions, roughly 14% of false positives are removed compared to less than 1% of true positives (Table S12 in File S1). Interestingly, the difference between the percentage of false versus true positives removed by the BLAT method is largest for RUM, followed by MapSplice, GSNAP, and Tophat2 (Figure S13). Overall, we found that the BLAT filtering approach decreased the FDR of RDD detection for GSNAP by approximately 13% in dataset 1 and 24% in dataset 2 (Figure S14; Table S13 in File S1).

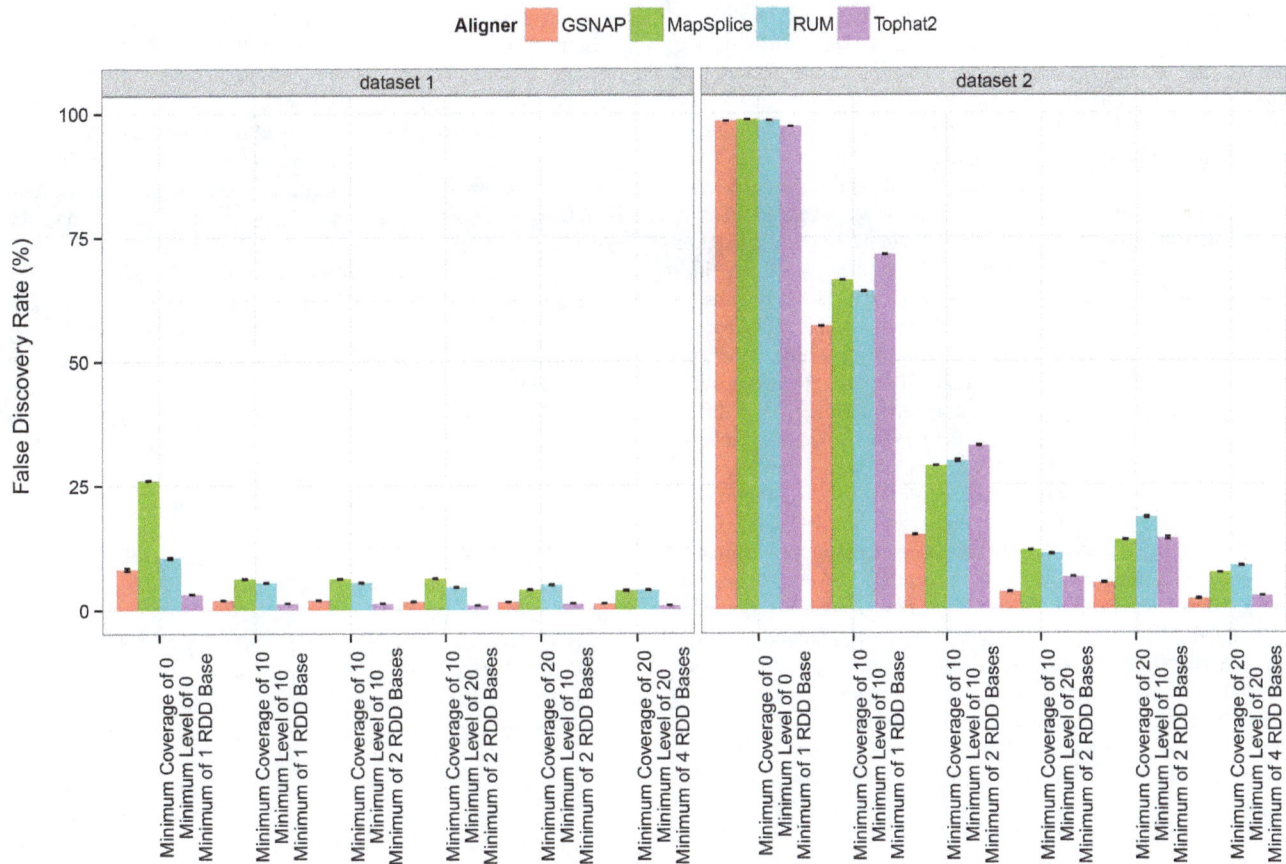

Figure 4. False discovery rate of RNA-DNA sequence difference detection. Here we depict the false discovery rate of RNA-DNA sequence difference detection under various thresholds on the coverage, level of sequence difference, and number of reads bearing the sequence difference base per the aligner. Calculations are averaged across the three replicates and error bars represent standard deviation values.

Pseudogenes are non-functioning homologs of genes that are either not expressed or unable to be translated into protein product, and their high sequence similarity to functioning genes can result in false positive sequence difference calls. We observed that the removal of all sequence differences lying within pseudogenes as annotated by Gencode version 13 [25] decreases the FDR of RDD detection using GSNAP by approximately 45 to 50% in both datasets (Table S14 in File S1).

Misalignments near exon-exon junctions can commonly lead to the identification of false positive sequence differences. We evaluated the effect of such incorrectly spliced alignments on sequence difference detection and found that roughly 2% of the false positives identified by GSNAP in dataset 1 and 5% of those found in dataset 2 are in intronic sequences within 6 bp of exon-exon junctions. Removal of all sites in introns within 6 bp of splice junctions leads to a roughly 2 to 4% decrease in the false discovery rate for GSNAP. RUM and Tophat2 are more robust to misalignments near splice junctions, as less than 1 to 2% of false positives detected by either aligner are in introns near exon-exon junctions (Table S15 in File S1), whereas 40 to 50% of false positives identified by MapSplice are found near junctions.

Finally, we analyzed the effect of implementing the various bioinformatics filters in concert on the false positive rates. In analyzing the RDDs obtained using the 'minimum coverage of 20x, minimum level of 20%, and minimum of 4 reads containing the sequence difference allele', we observed that requiring concordance with at least one other aligner, the BLAT filtering

method, removal of differences in pseudogenes, and elimination of intronic sites within 6 bp of exon junctions in combination removed roughly 50 to 90% of false positives depending on the aligner versus roughly 3 to 11% of true positives (Table S16 in File S1). Across the different aligners, these various filters led to a decrease of approximately 50 to 90% in the FDR of RDD detection (Table S17 in File S1), whereas sensitivity changed by approximately 2 to 11%.

Evaluation of RNA-DNA sequence differences in human lymphoblastoid cell line

Lastly, to evaluate the performance of our pipeline on a real experimental dataset, we analyzed the human lymphoblastoid cell line GM12878, for which deep DNA and RNA sequence is readily available [26]. We used the parameters and thresholds as determined from our previous simulated data analyses to identify RDDs. In particular, we aligned two replicates of RNA-Seq data, each containing approximately 120 million 76 bp paired-end reads, using GSNAP (Table S18 in File S1) and identified RDDs using a 'minimum coverage of 20x, minimum level of 20%, and minimum of 4 reads containing the sequence difference base' threshold. Sequence differences found in dbSNP137 [27] were removed from consideration. Furthermore, to minimize the detection of sequence differences resulting from sequencing error, we focused our analysis on those differences that are observed in both replicates (Table S19 in File S1).

Next, we investigated the percentage of observed RDDs that are removed by filters we previously identified as effective in reducing the amount of false positive RDDs. Other researchers have used these filters in their pipelines to accurately identify RDDs [3,15,22]. The filters we implemented include requiring concordance with at least one other aligner, searching with BLAT for regions homologous to the sequence flanking the sequence difference (see Methods), removing intronic sites near exon-exon junctions, and eliminating differences in annotated pseudogenes or adjacent to homopolymer sequences. We applied the BLAT filter to sequence differences found outside of RepeatMasker regions, as we previously showed that this filter is not as effective in discriminating between true and false positives within repetitive sequences. We separated the differences by type into two groups: A-to-G sequence differences and non-canonical sequence differences, or changes that cannot be explained by known mechanisms. We note that although C-to-T differences can be mediated by APOBEC, APOBEC1 is not expressed in this B-cell cell line, with an FPKM value [28] of 0 in both replicates; thus we classify C-to-T changes as non-canonical. We observed that approximately 72% of non-canonical differences are removed by one or more of these filters, whereas only roughly 36% of A-to-G sites are eliminated (Table 1). The filtering steps that filtered out the greatest percentage of sites are the requirement of concordance with at least one other aligner and the pseudogene and BLAT filters, as nearly 30% to 45% of non-canonical sites are removed by each filter independently. After taking into consideration all of the filters we used, a total of 5,997 sequence differences remained (Figure 5), 75% of which are A-to-G edits and are likely to be mediated by RNA editing via ADAR. Of these 5,997 differences, the majority (78%) are located within RepeatMasker regions. Within RepeatMasker regions, 90% of the differences are A-to-G, as is expected due to the phenomenon of editing in human *Alu* elements [29]. In contrast, the majority (82%) of sites outside of RepeatMasker are noncanonical differences. The distribution of sequence differences we observed are highly concordant with other studies (Table S20 in File S1), and the most common noncanonical RDD types we observed were A-to-C (23%) and its complement T-to-G (54%), as previously seen by others [15,17,22].

For the remaining RDDs that are not removed by our filtering methods, we asked whether features indicative of sequencing error or low-quality mapping are more common in non-canonical versus A-to-G sequence differences. Specifically, we noticed that many non-canonical sequence differences occur within regions where many of the reads overlapping the sequence difference site are either partially mapped via a local alignment with clipped bases or mapped with many mismatches (Figure S9). To investigate the mapping quality near sites of RDDs globally, we calculated for each read that overlaps an RDD the number of bases (out of the total 76 bp sequence) that are neither clipped nor aligned with a mismatch or indel; we refer to this figure as the number of bases aligned properly. We observed that in both replicates, for sequence differences in RepeatMasker, the overall number of bases that are aligned properly is higher for A-to-G changes than for the most of the non-canonical types, particularly the C-to-G type; for sites lying outside of RepeatMasker regions, the number of bases aligned properly for C-to-G and G-to-C are generally much lower than that for other types (Figure S10).

Discussion

RNA-Sequencing is a powerful technology for genome-wide analyses of transcriptome information at the single-nucleotide level. The resolution afforded by next-generation sequencing

technology has allowed for genome-wide studies on RNA editing in humans [3,13] and led to the identification of all 12 types of sequence differences [16]. There are, however, limitations to high-throughput sequencing, as difficulties lie in the alignment of short sequencing reads and errors introduced by sequencing and library preparation among other challenges. The relative effect of these various misalignment and sequencing errors on the identification of RDDs is debated, although many reports assert that the majority of the non-canonical sequence differences observed result from technical artifacts [18–21]. In this study, we dissect the various sources of error leading to false positive RDDs and evaluate their relative contribution. Using a detection theory approach, we generated simulated RNA-Seq datasets containing known RDDs to evaluate the effect of alignment and sequencing error on RDD analysis. In the absence of sequencing error, we found that minimal thresholds are sufficient for sensitivity values above 95% and false discovery rates below 5%. Moreover, we found that the RDD levels reported by the various aligners correlation well ($R \sim 98\%$) with the true levels per our simulation. Upon introduction of sequencing errors following a random and independent distribution, we found that a threshold requiring a 'minimum coverage of 20x, minimum level of 20%, and minimum of 4 reads bearing the RDD base' is necessary for false discovery rates below 10% across the various aligners.

Currently, most pipelines use ad hoc filtering methods to minimize the presence of false positives in sequence difference studies without a full understanding of the efficacy of these methods or the trade-off between sensitivity and false discovery rates. We found that overall while the various filters used in the literature for removal of false positive RDDs are effective in discriminating between true and false positives, a sizeable percentage (roughly 10 to 50% depending on the aligner) of false positives remain even after all filtering methods are implemented.

Lastly, we used our pipeline for identification of RDDs to evaluate the presence of sequence differences in humans. Using parameters and thresholds we deemed as optimal, we identified approximately 6,000 RDDs, the majority (75%) of which are A-to-G changes and likely to be mediated by ADAR. Of the non-canonical RNA-DNA sequence differences that remained after our filtering processes, we found A-to-C and its complement T-to-G to be most common. Notably, A-to-C changes have been found by others to be the most common sequencing error [30,31]. Furthermore, we found that the alignments of reads overlapping non-canonical RNA-DNA sequence differences, with the exception of A-to-C and T-to-G types, contain many more mismatches or clipped bases than those of A-to-G differences. The distribution of sequence differences we observed is highly concordant with previous studies, and like others [15,32], we conclude that there is little evidence for widespread non-canonical editing.

Overall, we observed that next-generation sequencing technology and current bioinformatics tools are a reliable and powerful technique for studying RDDs genome-wide. Furthermore, we found that computational biology methods are an effective means for evaluating the various thresholds and filtering techniques used to accurately identify sequence differences. Our results demonstrate that while RNA-Sequencing allows for precise detection and measurement of RDDs, current bioinformatics filters do not completely remove false positive calls. We aim for this study to provide a general framework for those interested in site-specific allelic differences in humans using RNA-Sequencing, and hope in particular that our work may shed light on the appropriate thresholds and necessary caution to employ for RDD analyses.

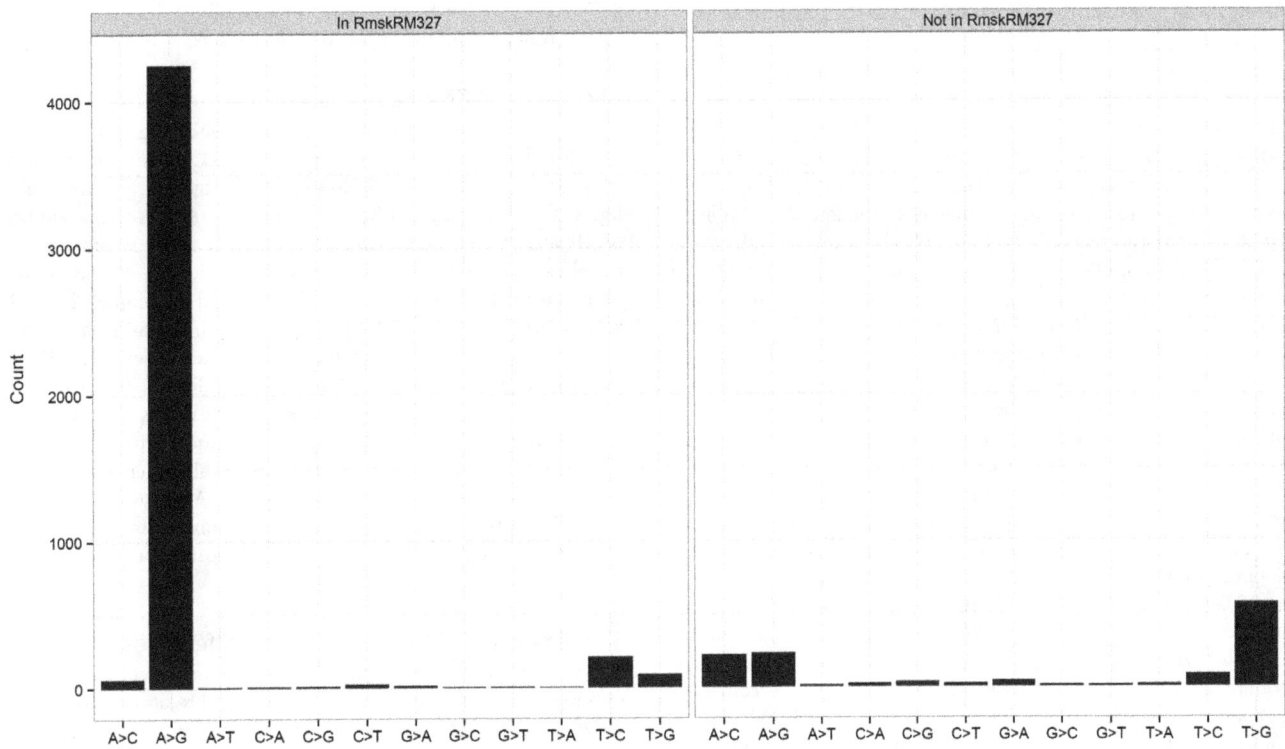

Figure 5. Distribution of RNA-DNA sequence differences in GM12878. Here we depict the distribution of RNA-DNA sequence differences in GM12878 after removing sites using various filters.

Methods

Simulation of RNA-Seq datasets

Simulated datasets were generated using the BEERS simulator [33]. Data are based on human build hg19 and RefSeq transcript models [34], as aligned to the genome by UCSC [35] using BLAT [36]. The expression intensities are Poisson distributed with probabilities estimated from roughly 300 million reads of human retina RNA-Seq data, as described previously [33]. Default settings result in 36,467 transcripts, of which approximately 70% are expressed. We simulated two types of RNA-Seq datasets. Dataset 1 was "clean" and designed to contain no intron signal or sequencing error. Dataset 2 was "realistic" and constructed with intron retention and sequencing error. We used a substitutional error rate of 1 in 200 (0.5%), a value comparable to

sequencing error rates observed in Illumina Genome Analyzer IIx and HiSeq machines [37]. Furthermore, we simulated poorer quality bases at the 3′ ends of reads by increasing the substitutional error rate to 20% in the last 10 bases for 25% of the reads. Approximately 30% of the signal in the dataset originates from introns. These parameters are consistent with real data observations. Lastly, we also included indel polymorphisms at a rate of 1 in 1000 (0.1%). Both datasets 1 and 2 were generated in triplicate, with each replicate containing 50 million pairs of reads of length 100 base pairs (bp). The mean fragment length of each read pair is 330 bp. The simulated datasets are available at http://itmat-public.s3.amazonaws.com/toung_rdd-study_simulated.dataset1.tar and http://itmat-public.s3.amazonaws.com/toung_rdd-study_simulated.dataset2.tar for download.

Table 1. Number of RNA-DNA sequence differences removed by various bioinformatics filters.

	A>G	Non-canonical	Total
Total before filters	**7,036**	**5,444**	**12,480**
Concordance with at least one other aligner filter (removed)	1,937 (27.53%)	2,399 (44.07%)	4,336 (34.74%)
Pseudogene filter (removed)	847 (12.04%)	1,959 (35.98%)	2,806
BLAT filter (removed)	545 (7.75%)	1,722 (31.63%)	2,267
Homopolymer filter (removed)	32 (0.45%)	205 (3.77%)	237
Exon junction filter (removed)	30 (0.43%)	88 (1.62%)	118
Total after filters (remaining)	**4,484 (63.73%)**	**1,513 (27.79%)**	**5,997 (48.05%)**
Total after filters - in RmskRM327	4,251	452	4,703
Total after filters - not in RmskRM327	233	1,061	1,294

Analysis of non-random sequencing errors in Illumina HiSeq RNA-Seq datasets

To evaluate the presence of non-random sequencing errors in Illumina HiSeq datasets, we analyzed two replicates of a dataset (denoted IVT-Seq dataset replicates 1 and 2) comprising 1,062 cDNAs from the Mammalian Genome Collection (MGC) that were expressed *in vitro* and sequenced using Illumina HiSeq 2000 technology [38] to obtain approximately 41 million and 32 million 100-bp paired-end reads. In addition, we also sequenced the same 1,062 cDNA plasmids that served as the template for the IVT-Seq datasets using Illumina HiSeq 2000 technology (denoted plasmid dataset). We aligned all three datasets using GSNAP (see 'Alignment of RNA-Seq datasets' in Materials and Methods) to an index containing the non-spliced reference sequence of the 1,062 cDNAs. For the IVT-Seq datasets, approximately 82% of the total reads were aligned in the correct orientation and with the expected inner distance between read pairs. For the plasmid dataset, approximately 20% of the total reads align properly; this relatively low percentage is expected given the presence of plasmid backbone in the dataset. The coverage distribution in the three datasets is fairly uniform, with an average of approximately 2,600x and 3,000x in the IVT-Seq replicates 1 and 2, respectively (median of 2,300x in replicate 1 and 1,800x in replicate 2) and a mean of roughly 770x in the plasmid dataset (median of 700x). To be confident of the sequencing and alignment results, we restricted our analyses to sites with a minimum coverage of 1,000x in the IVT-Seq datasets and 250x in the plasmid dataset. In total, we obtained 4,923,509,994 and 4,195,153,516 bases of sequence at 1,209,658 and 1,111,552 sites and observed a total of 1,877,330 and 4,902,349 sequencing errors, giving an overall error rate of approximately 3.8×10^{-4} and 1.2×10^{-3} in IVT-Seq replicates 1 and 2, respectively. For the plasmid dataset, we obtained 1,340,601,515 bases of sequence at 1,377,516 sites and found 999,763 sequencing errors, giving an overall error rate of roughly 7.5×10^{-4}. These sequencing error rates are all approximately 4x to 13x smaller than the rate of 0.5% we used in our simulated datasets.

For each site in a particular dataset, we calculated the sequencing error level to be the percentage of total reads at the site bearing an error. To test whether the observed sequencing errors occur randomly, we performed a Kolmogorov-Smirnov test, comparing the observed distribution of sequencing error levels to a null distribution derived from the overall error rate calculated previously. We found that for both of the IVT-Seq replicates as well as the plasmid dataset, the distribution of sequencing error levels deviates from that expected under the null distribution ($P<$ 0.001 for all three datasets). These results indicate that errors are not introduced randomly but occur at error levels that are higher than expected for particular sites. These observations are concordant with previous studies that demonstrate that sequencing errors introduced by Illumina next-generation sequencing platforms may occur in a sequence-specific or non-independent manner [37,39]. These non-random sequencing errors are indistinguishable from RDD events if they occur at high error levels. We found the frequency of errors that (1) occur at levels of 20% or higher and (2) are reproducible across the three datasets to be 4.38×10^{-5} and 4.77×10^{-5} in the IVT-Seq replicates 1 and 2 respectively and 3.85×10^{-5} in the plasmid dataset. However, upon further Sanger sequencing validation, we found that the majority (71%) of these nonrandom sequencing errors were errors in the sequence of the clones and not errors introduced by Illumina sequencing. Correcting for the presence of such errors in the sequence, we estimate that the true frequency of nonrandom errors occurring at levels of 20% or greater to be approximately

4.77×10^{-6}, which is one order of magnitude less than that of RDDs observed in our experimental datasets.

Alignment of RNA-Seq datasets

RNA-Seq datasets were aligned using GSNAP version 2012-07-20 [40], MapSplice version 2.1.5 [41], RUM version 2.0.3-02 [33], or Tophat2 version 2.0.6 [42] to the human genome (build hg19). GSNAP was run with default options. A maximum number of 10 alignments were permitted for each read. Alignments to novel exon-exon junctions (per GSNAP option -N 1) and known junctions as defined by RefSeq (downloaded November 2, 2012) and Gencode version 13 [25] were accepted. Alignments with no more than the default maximum of '(read length +2)/12–2' mismatches were retained. MapSplice and RUM were run with the default command line options. Tophat2 was run with the default options. A maximum edit distance and mismatch count of 6 was allowed for each read. Secondary alignments up to the default maximum of 20 were permitted. After alignment with GSNAP, MapSplice, RUM, or Tophat2, non-primary alignments and alignments placing read pairs in the incorrect orientation were removed.

Simulation of RNA-DNA sequence differences

For each dataset, sites in the genome are first stratified by coverage to ensure the placement of RDDs at locations with varying depths of coverage. The distribution of coverage for dataset 1, which does not contain reads originating from intronic regions of the genome, is fairly uniform, while for dataset 2, the distribution is skewed right (Figure S2); approximately 82% of sites in dataset 2 have coverage of 10x or less compared to approximately 20% in dataset 1. For dataset 1, we grouped sites into quartiles, corresponding to coverage values of approximately 0x to 14x for quartile 1, 15x to 49x for quartile 2, 50x to 133x for quartile 3, and 134x and above for quartile 4. For dataset 2, the presence of introns results in a highly skewed right distribution for coverage. As such, we divided dataset 2 into one group containing sites with coverage below 10x and split the remaining sites into tertiles, corresponding to coverage values of approximately 11x to 19x for tertile 1, 20x to 48x for tertile 2, and 49x and above for tertile 3. After we grouped sites by coverage, we randomly inserted RDDs at different sites such that each coverage group contained approximately the same number of sequence difference sites.

The type of RDD difference (e.g. A-to-C, A-to-G, A-to-T, etc.) was determined randomly and independently for each site. The RDD level, or the proportion of reads containing the sequence difference, was chosen randomly from a random uniform distribution from 0 to 1, excluding 0.

A small subset (5%) of the simulated RDDs was randomly chosen to model hyperediting, or the clustering of many sequence differences in a small window. In particular, we designated all of the sites that are within 100 bp of the chosen site to have a 50% chance of having the same RDD type provided that the coverage belongs to the same coverage group as the initial site.

Repetitive regions of the genome as defined by BLAT

As one measure of the repetitive nature of a region surrounding a sequence difference site, we used BLAT [36] to search for homologous sequences in the genome. In particular, we extracted flanking sequences of length 51 bp, 101 bp, and 151 bp around a given site and queried for alignments in the genome with BLAT (v.35x1). The settings −stepSize = 5 and −repMatch = 2253 were used to increase sensitivity. A maximum of (read length +2)/12 − 2 mismatches per alignment, the same amount permitted by

GSNAP, was tolerated. Sites for which more than one alignment is found for one of the three flanking sequences are deemed "non-unique by BLAT".

Filtering of RNA-DNA sequence differences using BLAT

To ensure that an RDD identified by the various aligners cannot be explained by homologous sequences in the genome, sequences of length 25, 50, and 75 bp upstream and downstream of each sequence difference site were aligned to the genome using BLAT (v. 35x1). The settings -stepSize=5 and -repMatch=2253 were used to increase sensitivity. A maximum of (read length +2)/12 - 2 mismatches per alignment, the same amount allowed by GSNAP, was tolerated. An RDD was filtered out if any of the flanking sequences aligned to a region other than the RDD site and if that alignment explained the sequence difference.

Analysis of BWA performance on RDD detection

We evaluated the performance of the Burrows-Wheeler Aligner (BWA) on RDD detection. We conducted these analyses separately from the other three aligners we used as BWA is not a RNA-Seq aligner capable of mapping across intron-size gaps. In particular, synthetic RNA-Seq dataset 1, which does not contain reads covering intronic sequences, was aligned using BWA version 0.7.9a-r786 [43] to a transcriptome index comprising exonic sequences as defined by a non-redundant union of several annotation efforts as published at the UCSC Genome Browser (RefSeq, UCSC Known, Vega, AceView, ENSEMBL). BWA was run with default options.

We defined a simulated RDD as being properly identified by BWA if at least one read bearing the non-reference base is observed. We found that the sensitivity of RDD detection by BWA is $92.96 \pm 1.94E{-}1\%$ compared to approximately 95 to 96% for GSNAP, MapSplice, RUM and Tophat2. Sensitivity of detection increased by 4% when sites with coverage lower than 10x are removed from consideration. Restricting analysis to sites with a minimum coverage of 10x, minimum level of 10%, and minimum of 1 read bearing the sequence difference base, we found the correlation between observed and simulated RDD levels to be $97 \pm 1.57E{-}2\%$.

Next, we analyzed the false positive and false discovery rates of RDD detection by identifying sites that were not simulated to contain RDDs. Using a 'minimum coverage of 10x, minimum level of 10%, and minimum of 1 read bearing the RDD base' cutoff, we found the false positive rate and false discovery rate of RDD detection to be $1.54E{-}2 \pm 2.49E{-}4\%$ and $1.81 \pm 4.22E{-}2\%$, respectively. Overall, our analyses on dataset 1 which comprises exonic regions of the genome showed that BWA provides comparable results to those reported by the other aligners we analyzed, namely GSNAP, MapSplice, RUM, and Tophat2.

Supporting Information

Figure S1 Total number of simulated RNA-DNA sequence differences. For both datasets, approximately 600,000 RDDs were generated in each replicate. Differences in the number of each type of RDD reflect underlying variation in base composition throughout the genome, as dataset 2 contains reads originating from intronic regions whereas dataset 1 does not.

Figure S2 Distribution of coverage for simulated RNA-Seq datasets. The distribution of coverage, or the total number of reads at a given site, is relatively uniform for dataset 1. In contrast, the distribution of coverage for dataset 2 is skewed right mainly owing to the presence of intronic reads. Approximately 82% of the sites in dataset 2 have a depth of coverage lower or equal to 10x.

Figure S3 Levels of simulated RNA-DNA sequence differences. Here we depict the distribution of RDD levels, or the percentage of reads at the sequence difference site that bear the RNA-DNA sequence difference. Because of the discrete nature of RNA-Seq data, the levels of RDDs at sites with relatively low coverage is not uniform as shown by the blue area, which represents sites with coverage less than 10x. For sites with coverage greater than 100x (red area), the density curve of sequence difference levels is fairly uniform except at boundary conditions.

Figure S4 Sensitivity of RDD detection versus uniqueness of flanking genomic sequence. Here we show the sensitivity or true positive rate of RDD detection for regions in the genome that are unique (in blue) versus not unique (in red) as determined by BLAT (see Materials and Methods). Sites with fewer than 10 total reads per the simulated RNA-Seq dataset or a RDD level less than 10% per the simulated dataset are removed from consideration.

Figure S5 Percentage of sites with observed levels that deviate from simulated RDD levels. Here we calculate the percentage of total sites in each dataset (y-axis) with observed levels that deviate from the simulated RDD level by various degrees (x-axis).

Figure S6 Effect of requiring RDDs to be identified by multiple aligners on FDR of RDD detection for GSNAP. Here we depict the false discovery rate of RDD detection using GSNAP under a 'minimum coverage of 20x, minimum level of 20%, and a minimum of 4 reads bearing the sequence difference' threshold after requiring various numbers of other aligners to concordantly identify the RDD.

Figure S7 Effect of requiring RDDs to be identified by multiple aligners on FDR of RDD detection for MapSplice. Here we depict the false discovery rate of RDD detection using MapSplice under a 'minimum coverage of 20x, minimum level of 20%, and a minimum of 4 reads bearing the sequence difference' threshold after requiring various numbers of other aligners to concordantly identify the RDD.

Figure S8 Effect of requiring RDDs to be identified by multiple aligners on FDR of RDD detection for RUM. Here we depict the false discovery rate of RDD detection using RUM under a 'minimum coverage of 20x, minimum level of 20%, and a minimum of 4 reads bearing the sequence difference' threshold after requiring various numbers of other aligners to concordantly identify the RDD.

Figure S9 Effect of requiring RDDs to be identified by multiple aligners on FDR of RDD detection for Tophat2. Here we depict the false discovery rate of RDD detection using Tophat2 under a 'minimum coverage of 20x, minimum level of 20%, and a minimum of 4 reads bearing the sequence difference' threshold after requiring various numbers of other aligners to concordantly identify the RDD.

Figure S10 Percentage of false versus true positives removed using BLAT filter for dataset 1. Here we depict the percentage of false positives versus true positives that are removed when using the BLAT filter for dataset 1.

Figure S11 Percentage of false versus true positives removed using BLAT filter for dataset 2. Here we depict the percentage of false positives versus true positives that are removed when using the BLAT filter for dataset 2.

Figure S12 Effect of BLAT filter on false discovery rate of RNA-DNA sequence difference detection. Here we depict the effect of the BLAT filter on the FDR for various aligners and thresholds for identification of sequence differences.

Figure S13 T-to-G RNA-DNA sequence difference at chr10:102046378. Here we show an image in the IGV browser [1] of a T-to-G sequence differences at chr10:102046378 in the first replicate of the GM12878 dataset. Each grey bar represents an RNA-Seq read. Mismatches are depicted by colored letters. Black dashes within a read represent a clipped sequence; for reads in the bottom half, the string of colored bases depict clipped portions of the sequence. Clipped portions of alignments represent bases that are not aligned within a local alignment.

Figure S14 Number of properly aligned bases in reads that overlap RNA-DNA sequences. Here we depict the number of bases within each read that overlaps an RNA-DNA sequence difference site that are aligned properly. This number excludes bases that contain mismatches or those that are clipped or part of an insertion deletion (indel).

File S1 Table S1, Alignment statistics of simulated RNA-Seq datasets. **Table S2,** Summary statistics on distance between neighboring RNA-DNA sequence differences. **Table S3,** Sensitivity of RNA-DNA sequence difference detection versus coverage threshold. **Table S4,** Sensitivity of RDD detection versus the level of sequence difference. **Table S5,** Sensitivity of RNA-DNA sequence difference detection in unique versus non-unique regions as determined by BLAT. **Table S6,** Sensitivity of RDD detection within RepeatMasker regions. **Table S7,** Sensitivity of RDD detection versus proximity to nearby RDDs. **Table S8,** Correlation between observed and simulated levels of RDDs. **Table S9,** Percent of sites with levels where the observed and simulated levels deviate by more than 30% versus the uniqueness of the underlying site as determined by BLAT. **Table S10,** Receiver operating characteristic analysis of RNA-DNA sequence difference detection. **Table S11,** Effect of requiring RDDs to be concordantly identified by multiple aligners on FDR of RDD detection. **Table S12,** Percentage of true versus false positives removed by BLAT filter. **Table S13,** Effect of BLAT filter on false discovery rate of RDD detection. **Table S14,** Effect of removing RNA-DNA sequence differences in pseudogenes on the false discovery rate of sequence difference detection. **Table S15,** Effect of removing RDDs near exon junctions on the false discovery rate of sequence difference detection. **Table S16,** Percentage of true versus false positives removed by requiring concordance with at least one other aligner, BLAT filter, pseudogene filter, and removal of intronic sites within 6 bp of exon junctions used in conjunction. **Table S17,** Combined effect of requiring concordance with at least one other aligner, BLAT filter, pseudogene filter, and removal of intronic sites within 6 bp of exon junctions on the false discovery rate of sequence difference detection. **Table S18,** Alignment statistics for GM12878 RNA-Seq dataset. **Table S19,** RNA-DNA sequence differences found in GM12878. **Table S20,** Overlap of RNA-DNA sequence differences found in GM12878 with other published studies.

Author Contributions

Conceived and designed the experiments: GG JT JH. Performed the experiments: GG JT NL. Analyzed the data: GG JT NL. Contributed reagents/materials/analysis tools: GG JT. Wrote the paper: GG JT JH.

References

1. Wang ET, Sandberg R, Luo S, Khrebtukova I, Zhang L, et al. (2008) Alternative isoform regulation in human tissue transcriptomes. Nature 456: 470–476.
2. Pan Q, Shai O, Lee LJ, Frey BJ, Blencowe BJ (2008) Deep surveying of alternative splicing complexity in the human transcriptome by high-throughput sequencing. Nat Genet 40: 1413–1415.
3. Peng Z, Cheng Y, Tan BC, Kang L, Tian Z, et al. (2012) Comprehensive analysis of RNA-Seq data reveals extensive RNA editing in a human transcriptome. Nat Biotechnol 30: 253–260.
4. Park E, Williams B, Wold BJ, Mortazavi A (2012) RNA editing in the human ENCODE RNA-seq data. Genome Res 22: 1626–1633.
5. Heap GA, Yang JH, Downes K, Healy BC, Hunt KA, et al. (2010) Genome-wide analysis of allelic expression imbalance in human primary cells by high-throughput transcriptome resequencing. Hum Mol Genet 19: 122–134.
6. Gregg C, Zhang J, Weissbourd B, Luo S, Schroth GP, et al. (2010) High-resolution analysis of parent-of-origin allelic expression in the mouse brain. Science 329: 643–648.
7. DeVeale B, van der Kooy D, Babak T (2012) Critical evaluation of imprinted gene expression by RNA-Seq: a new perspective. PLoS Genet 8: e1002600.
8. Bass BL, Weintraub H (1988) An unwinding activity that covalently modifies its double-stranded RNA substrate. Cell 55: 1089–1098.
9. Nishikura K (2010) Functions and regulation of RNA editing by ADAR deaminases. Annu Rev Biochem 79: 321–349.
10. Chen SH, Habib G, Yang CY, Gu ZW, Lee BR, et al. (1987) Apolipoprotein B-48 is the product of a messenger RNA with an organ-specific in-frame stop codon. Science 238: 363–366.
11. Smith HC, Bennett RP, Kizilyer A, McDougall WM, Prohaska KM (2012) Functions and regulation of the APOBEC family of proteins. Semin Cell Dev Biol 23: 258–268.
12. Levanon EY, Eisenberg E, Yelin R, Nemzer S, Hallegger M, et al. (2004) Systematic identification of abundant A-to-I editing sites in the human transcriptome. Nat Biotechnol 22: 1001–1005.
13. Li JB, Levanon EY, Yoon JK, Aach J, Xie B, et al. (2009) Genome-wide identification of human RNA editing sites by parallel DNA capturing and sequencing. Science 324: 1210–1213.
14. Bahn JH, Lee JH, Li G, Greer C, Peng G, et al. (2012) Accurate identification of A-to-I RNA editing in human by transcriptome sequencing. Genome Res 22: 142–150.
15. Kleinman CL, Adoue V, Majewski J (2012) RNA editing of protein sequences: A rare event in human transcriptomes. RNA 18: 1586–1596.
16. Li M, Wang IX, Li Y, Bruzel A, Richards AL, et al. (2011) Widespread RNA and DNA sequence differences in the human transcriptome. Science 333: 53–58.
17. Ju YS, Kim JI, Kim S, Hong D, Park H, et al. (2011) Extensive genomic and transcriptional diversity identified through massively parallel DNA and RNA sequencing of eighteen Korean individuals. Nat Genet 43: 745–752.
18. Pickrell JK, Gilad Y, Pritchard JK (2012) Comment on "Widespread RNA and DNA sequence differences in the human transcriptome". Science 335:1302;author reply 1302.
19. Schrider DR, Gout JF, Hahn MW (2011) Very few RNA and DNA sequence differences in the human transcriptome. PLoS One 6: e25842.
20. Kleinman CL, Majewski J (2012) Comment on "Widespread RNA and DNA sequence differences in the human transcriptome". Science 335:1302;author reply 1302.

21. Lin W, Piskol R, Tan MH, Li JB (2012) Comment on "Widespread RNA and DNA sequence differences in the human transcriptome". Science 335: 1302; author reply 1302.

22. Ramaswami G, Lin W, Piskol R, Tan MH, Davis C, et al. (2012) Accurate identification of human Alu and non-Alu RNA editing sites. Nat Methods.

23. Polson AG, Bass BL (1994) Preferential selection of adenosines for modification by double-stranded RNA adenosine deaminase. EMBO J 13: 5701–5711.

24. Smit AF (1996) The origin of interspersed repeats in the human genome. Curr Opin Genet Dev 6: 743–748.

25. Harrow J, Denoeud F, Frankish A, Reymond A, Chen CK, et al. (2006) GENCODE: producing a reference annotation for ENCODE. Genome Biol 7 Suppl 1: S4 1–9.

26. Dunham I, Kundaje A, Aldred SF, Collins PJ, Davis CA, et al. (2012) An integrated encyclopedia of DNA elements in the human genome. Nature 489: 57–74.

27. Sherry ST, Ward MH, Kholodov M, Baker J, Phan L, et al. (2001) dbSNP: the NCBI database of genetic variation. Nucleic Acids Res 29: 308–311.

28. Trapnell C, Williams BA, Pertea G, Mortazavi A, Kwan G, et al. (2010) Transcript assembly and quantification by RNA-Seq reveals unannotated transcripts and isoform switching during cell differentiation. Nat Biotechnol 28: 511–515.

29. Athanasiadis A, Rich A, Maas S (2004) Widespread A-to-I RNA editing of Alu-containing mRNAs in the human transcriptome. PLoS Biol 2: e391.

30. Dohm JC, Lottaz C, Borodina T, Himmelbauer H (2008) Substantial biases in ultra-short read data sets from high-throughput DNA sequencing. Nucleic Acids Res 36: e105.

31. Qu W, Hashimoto S, Morishita S (2009) Efficient frequency-based de novo short-read clustering for error trimming in next-generation sequencing. Genome Res 19: 1309–1315.

32. Ramaswami G, Zhang R, Piskol R, Keegan LP, Deng P, et al. (2013) Identifying RNA editing sites using RNA sequencing data alone. Nat Methods 10: 128–132.

33. Grant GR, Farkas MH, Pizarro AD, Lahens NF, Schug J, et al. (2011) Comparative analysis of RNA-Seq alignment algorithms and the RNA-Seq unified mapper (RUM). Bioinformatics 27: 2518–2528.

34. Pruitt KD, Tatusova T, Brown GR, Maglott DR (2012) NCBI Reference Sequences (RefSeq): current status, new features and genome annotation policy. Nucleic Acids Res 40: D130–135.

35. Kent WJ, Sugnet CW, Furey TS, Roskin KM, Pringle TH, et al. (2002) The human genome browser at UCSC. Genome Res 12: 996–1006.

36. Kent WJ (2002) BLAT—the BLAST-like alignment tool. Genome Res 12: 656–664.

37. Minoche AE, Dohm JC, Himmelbauer H (2011) Evaluation of genomic high-throughput sequencing data generated on Illumina HiSeq and genome analyzer systems. Genome Biol 12: R112.

38. Lahens NF, Kavakli IH, Zhang R, Hayer K, Black MB, et al IVT-seq reveals extreme bias in RNA-sequencing. *Genome Biol* (2014) June 30; 15(6)

39. Nakamura K, Oshima T, Morimoto T, Ikeda S, Yoshikawa H, et al. (2011) Sequence-specific error profile of Illumina sequencers. Nucleic Acids Res 39: e90.

40. Wu TD, Nacu S (2010) Fast and SNP-tolerant detection of complex variants and splicing in short reads. Bioinformatics 26: 873–881.

41. Wang K, Singh D, Zeng Z, Coleman SJ, Huang Y, et al. (2010) MapSplice: accurate mapping of RNA-seq reads for splice junction discovery. Nucleic Acids Res 38: e178.

42. Trapnell C, Pachter L, Salzberg SL (2009) TopHat: discovering splice junctions with RNA-Seq. Bioinformatics 25: 1105–1111.

43. Li H, Durbin R (2009) Fast and accurate short read alignment with Burrows-Wheeler Transform. Bioinformatics, 25: 1754–60.

12

Zinc-Finger Nuclease Knockout of Dual-Specificity Protein Phosphatase-5 Enhances the Myogenic Response and Autoregulation of Cerebral Blood Flow in FHH.1[BN] Rats

Fan Fan[1], Aron M. Geurts[2], Mallikarjuna R. Pabbidi[1], Stanley V. Smith[1], David R. Harder[3], Howard Jacob[2], Richard J. Roman[1]*

1 Department of Pharmacology and Toxicology, University of Mississippi Medical Center, Jackson, Mississippi, United States of America, 2 Human and Molecular Genetics Center, Medical College of Wisconsin, Milwaukee, Wisconsin, United States of America, 3 Department of Physiology and Cardiovascular Research Center, Medical College of Wisconsin, Milwaukee, Wisconsin, United States of America

Abstract

We recently reported that the myogenic responses of the renal afferent arteriole (Af-Art) and middle cerebral artery (MCA) and autoregulation of renal and cerebral blood flow (RBF and CBF) were impaired in Fawn Hooded hypertensive (FHH) rats and were restored in a FHH.1[BN] congenic strain in which a small segment of chromosome 1 from the Brown Norway (BN) containing 15 genes including dual-specificity protein phosphatase-5 (Dusp5) were transferred into the FHH genetic background. We identified 4 single nucleotide polymorphisms in the Dusp5 gene in FHH as compared with BN rats, two of which altered CpG sites and another that caused a G155R mutation. To determine whether Dusp5 contributes to the impaired myogenic response in FHH rats, we created a Dusp5 knockout (KO) rat in the FHH.1[BN] genetic background using a zinc-finger nuclease that introduced an 11 bp frame-shift deletion and a premature stop codon at AA121. The expression of Dusp5 was decreased and the levels of its substrates, phosphorylated ERK1/2 (p-ERK1/2), were enhanced in the KO rats. The diameter of the MCA decreased to a greater extent in Dusp5 KO rats than in FHH.1[BN] and FHH rats when the perfusion pressure was increased from 40 to 140 mmHg. CBF increased markedly in FHH rats when MAP was increased from 100 to 160 mmHg, and CBF was better autoregulated in the Dusp5 KO and FHH.1[BN] rats. The expression of Dusp5 was higher at the mRNA level but not at the protein level and the levels of p-ERK1/2 and p-PKC were lower in cerebral microvessels and brain tissue isolated from FHH than in FHH.1[BN] rats. These results indicate that Dusp5 modulates myogenic reactivity in the cerebral circulation and support the view that a mutation in Dusp5 may enhance Dusp5 activity and contribute to the impaired myogenic response in FHH rats.

Editor: Jaap A. Joles, University Medical Center Utrecht, Netherlands

Funding: This work was funded in part by National Institutes of Health R01 HL36279 and DK104184 (RJR), H105997 (Harder), GO grant HL-101681 (Jacob) and New innovator award OD-8396 (Geurts), VA Research Career Scientist Award (Harder) and Scientist Development Grant 13SDG14000006 (Pabbidi) from American Heart Association. The funders had no role in study design, data collection and analysis, decision to publish, or preparation of the manuscript.

Competing Interests: The authors have declared that no competing interests exist.

* Email: rroman@umc.edu

Introduction

The myogenic response is an intrinsic property of vascular smooth muscle cells (VSMC) that initiates contraction of arterioles in response to elevations in transmural pressure [1,2] and contributes to autoregulation of renal and cerebral blood flow (RBF, CBF). [3–6] We recently reported that the myogenic responses of the renal afferent arteriole (Af-Art) and middle cerebral artery (MCA) and autoregulation of RBF and CBF were impaired in Fawn Hooded hypertensive (FHH) rats and were restored in a FHH.1[BN] congenic strain in which a small segment of chromosome 1 from the Brown Norway (BN) containing 15 genes, including dual-specificity protein phosphatase-5 (Dusp5) were transferred into FHH genetic background. [7–9] However, the

genes that contribute to the impaired myogenic response and the mechanisms involved remain to be determined.

Dusp5 is a serine-threonine phosphatase that inactivates MAPK activity[10–14] by dephosphorylating ERK1/2 MAP kinases [15] which modulate the activities of the large conductance Ca^{2+}-activated K$^+$ channel (BK) and transient receptor potential (TRP) channels. Both of these channels influence vascular reactivity and the myogenic response. [1,16–19] In the present study, we found that there were 17 SNPs in the Dusp5 gene in FHH relative to BN rats. One SNP was in the 5′-UTR and three were in the coding region. Of these, two altered potential CpG methylation sites and one introduced a G155R mutation. To determine whether Dusp5 regulates vascular tone and reactivity and if the sequence variants in this gene contribute to the impaired myogenic response in FHH rats, we created and characterized a Dusp5 Zinc-finger nuclease

(ZFN) knockout (KO) rat in the FHH.1[BN] genetic background since transfer of this region of chromosome 1 containing the Dusp5 gene was shown to restore the myogenic response in cerebral arteries. We first compared the myogenic response of the MCA and autoregulation of CBF in Dusp5 KO, FHH.1[BN] and FHH rats. We then compared the expression of Dusp5 in multiple tissues isolated from Dusp5 ZFN KO, FHH.1[BN] and FHH rats. We also investigated whether there are differences in the expression of p-ERK1/2 in cerebral microvessels isolated from these strains as they are the primary substrates normally dephosphorylated and inactivated by Dusp5 [15,20].

Materials and Methods

General

Experiments were performed on 33 FHH, 68 FHH.1[BN] and 92 Dusp5 KO male rats bred in our in house colonies and 16 age-matched Sprague-Dawley (SD) male rats purchased from Charles River Laboratories (Wilmington, MA). The animal care facility at the University of Mississippi Medical Center is approved by the American Association for the Accreditation of Laboratory Animal Care. The rats had free access to food and water throughout the study and all protocols received prior approval by the Institutional Animal Care and Use Committees (IACUC) of the University of Mississippi Medical Center.

Identification and confirmation of SNPs in Dusp5 in FHH versus FHH.1[BN] rats

We first performed an *in silico* analysis of the sequence of the Dusp5 gene in FHH versus BN rats, which is publically available from the Rat Genome database (RGD, http://rgd.mcw.edu/rgdweb/report/gene/main.html?id=620854). To confirm that the SNPs identified in the database are present in our FHH and FHH.1[BN] colonies at both the DNA and mRNA levels, we isolated genomic DNA from tail biopsies using PureLink Genomic DNA Kits (Life Technologies, Grand Island, NY) and RNA from cerebral arteries using TRIzol solution (Life Technologies, Grand Island, NY) and sequenced across the regions of interest. RNA (1 µg) was reverse transcribed using an iScript cDNA Synthesis Kit (Bio-Rad) to produce cDNA. The regions of interest were amplified in a 25 µl PCR reaction containing 25 ng of genomic DNA or 4 ng of cDNA, 25 ng of each primer, 20 mM Tris-HCl buffer (pH 8.4), 50 mM KCl, 1.5 mM MgCl₂, 200 µM of each dNTP and 0.5 U Taq DNA polymerase (QIAGEN) using several primer pairs to cover the full length sequence. For amplification of exon 1, the following forward and reverse primers were used: 5'-AGCTTTCCGGGGCAGCGAGTG-3' and 5'-TCAGGA-TACTGTGAGTAGAAG-3'. Exons 2, 3 and a portion of exon 4 that is in the codon region were amplified using the following forward and reverse primers: 5'-CGTGCTGGACCAGGG-CAGCCG-3' and 5'-GACAGAGAGAGGTCTTCAGTATTG-3'. Intronic primers were used to amplify the regions of interest from genomic DNA. Amplification of exon 1 or 2 required an additional 5 µl of Q solution since these regions were GC-rich. The PCR products were purified using a PureLink PCR Purification Kit (Life Technologies, Grand Island, NY) and then ligated into a pCR4-TOPO TA vector (Life Technologies, Grand Island, NY) by incubating at room temperature for 20 min. One Shot MAX Efficiency DH5α-T1[R] Competent cells (Life Technologies, Grand Island, NY) were transformed using the ligated vectors according to manufacturer's instructions. The colonies were incubated at 37°C overnight in LB media with 100 µg/ml of Ampicillin. The plasmids were extracted using a QIAprep Spin Miniprep Kit (QIAGEN, Valencia, CA) and sequenced using M13

primers. The data were analyzed using ABI software (Applied Biosystems, Grand Island, NY) and compared to the BN reference sequence available on the NCBI GeneBank and RGD databases.

Comparison of the expressions of Dusp5, p-PKC and p-ERK1/2 in FHH and FHH.1[BN] rats

RT-qPCR. RNA was isolated from microdissected cerebral arteries in FHH and FHH.1[BN] rats that was reverse transcribed as described above. Fast SYBR Green Real-Time PCR Master Mixes (Life Technologies, Grand Island, NY) which contain a blend of dTTP/dUTP that is compatible with Uracil N-Glycosylase (UNG) to eliminate DNA contamination from PCR products synthesized in the presence of dUTP were mixed with 4 ng of cDNA and 25 ng of forward (5'-CTT AAA GGT GGG TAC GAG ACC TTC TAC -3') and reverse (5'-GAG AAT GGG CTT TCC GCA CTG -3') primers and amplified using a real-time PCR system (Mx3000P, Stratagene, La Jolla, CA). The data were analyzed with Mxpro qPCR software (Stratagene, La Jolla, CA) using the $2^{-\Delta\Delta CT}$ Method [21]. The PCR products were also separated on a 1% agarose gel using a Tris-borate-EDTA (TBE) buffer visualized the intensity of the bands with 100 mg/ml of ethidium bromide (Sigma, St. Louis, MO) and analyzed using ChemiDoc MP Imaging System (Bio-Rad, Hercules, CA).

Western Blot. Cerebral microvessels were isolated using the Evans blue sieving procedure as previously described [22,23] . Cerebral microvessels and brain tissue obtained from FHH and FHH.1[BN] rats were homogenized in ice-cold RIPA buffer (R0278, Sigma-Aldrich, St. Louis, MO) in the presence of protease and phosphatase inhibitors (Cat# 88663, Thermo Scientific, Pittsburgh, PA) and the proteasome inhibitor MG 132 (Sigma-Aldrich, St. Louis, MO) [12] using a ground glass homogenizer followed by a FastPrep-24 homogenizer (MP Biomedicals, Santa Ana, CA). The homogenate was centrifuged at 1,000 g for 10 minutes at 4°C. Aliquots of supernatant protein (40 µg for cerebral microvessels and 100 µg for brain tissues) were separated on a 10% SDS-PAGE gel, transferred to nitrocellulose membranes and probed with a pan p-PKC antibody (Cat 9371, Cell signaling, Danvers, MA) at 1:1,000 dilution. Antibodies that against p-ERK1/2 and total ERK1/2 (Cat# 4377 and Cat# 4695, Santa Cruz, Santa Cruz, CA) were used at a 1:1,000 dilution followed by a 1:4.000 dilution of a horseradish peroxidase (HRP)-coupled anti-rabbit secondary antibody. The membranes were then stripped and re-probed with a 1: 8,000 dilution of an anti-beta Actin antibody (ab6276, Abcam, Cambridge, MA) followed by a 1: 20,000 dilution of anti-mouse HRP-coupled secondary antibody as a loading control.

Generation of the Dusp5 ZFN KO rats in the FHH.1[BN] genetic background

ZFNs targeting the following sequence CAGGGCAGCCGC-CACtggcaGAAGCTGCGGGAGGA in exon 1 of the rat Dusp5 gene (NM_133578) were obtained from Sigma-Aldrich (St. Louis, MO) and were used to generate a Dusp5 KO rat in the FHH.1[BN] genetic background as previously described [24–26]. The ZFN mRNA was injected into the pronucleus [25,27] of fertilized FHH.1[BN] embryos and transferred to the oviduct of pseudopregnant females to generate Dusp5 ZFN KO founders. Tail biopsies were obtained and digested with 0.2 mg/ml proteinase K in a direct PCR lysis reagent (102-T, Viagen Biotech) at 85°C with rotation for 45 minutes. Founders were identified using the CEL-1 assay [28] and the mutations were confirmed by Sanger DNA sequencing [29]. Positive founders were backcrossed to parental strain to generate heterozygous F1 rats and the siblings were

SNP	Chr	RGSC Genome Assembly v3.4 start	stop	position on genomic DNA	position on mRNA	reference nuc	variant nuc	AA	reference AA	variant AA	variant ID
1	1	259754340	259754341	107	107 (5'UTR)	C(G)	T(G)				296415523
2	1	259754563	259754564	330	330 (exon 1)	G	T	52	L	L	296415524
3	1	259756067	259756068	1834		T	C				293523829
4	1	259756482	259756483	2249		G	A				296415525
5	1	259757182	259757183	2949		G	A				296415526
6	1	259758392	259758393	4159	(intron 1)	T	C				293523830
7	1	259758588	259758589	4355		A	G				296415527
8	1	259759032	259759033	4799		G	A				296415528
9	1	259759139	259759140	4906	627 (exon 2)	C(G)	T(G)	151	L	L	296415529
10	1	259759148	259759149	4915	637 (exon 2)	G	A	155	G	R	296415530
11	1	259759596	259759597	5363		G	C				261240240
12	1	259759660	259759661	5427	(intron 2)	T	C				296415531
13	1	259762138	259762139	7905		C	T				261240242
14	1	259763857	259763858	9624	(intron 3)	A	G				296415532
15	1	259766107	259766108	11874		T	C				293523831
16	1	259766563	259766564	12330	(intron 4)	G	C				293523832
17	1	259767470	259767471	13237		A	G				296415533

Figure 1. Comparison of sequence variants in the Dusp5 gene in FHH and Brown Norway (BN) rats. Analysis of the NeXT Generation sequence data available on the Rat Genome database (RGD, http://rgd.mcw.edu/rgdweb/report/gene/main.html?id=620854). These results indicate that there are 17 SNPs in the Dusp5 (NM -133578) gene in FHH/EurMcwi(variant nuc) rats as compared to Brown Norway (reference nuc) rats. Most of the SNPs are located in introns. There is a C107T SNP is located in the 5'-UTR, and three are found in the coding region including a G330T SNP in exon 1, a C627T and a G637A SNP in exon 2. The C107T SNP alters a CpG site and the C627T SNP alters one of the six 5 CpG's methylation sites previously identified in exon 2. The G637A SNP causes G155R mutation in the Dusp5 protein in FHH rats.

intercrossed to produce homozygous animals. Thereafter, the rats were genotyped using the following primers: Dusp5-F: 5′-GCT GCA GGA GGG CGG CGG CG -3′, Dusp5 R: 5′-CTT TAA GGA AGT AGA CCC G -3′. These primers amplified a 155 bp band for the wild type allele and a 144 bp band for the knockout allele.

Characterization of the Dusp5 ZFN KO rats

Western Blot. Cerebral microvessels, liver, brain and spleen tissues were isolated from Dusp5 KO and FHH.1[BN] rats as described above. White blood cells were also harvested using Ficoll-Paque Premium 1.084 (GE Healthcare) according to the manufacturer's protocol. A 100 μg aliquot of protein isolated from brain, liver, spleen and white blood cells (WBCs) and a 40 μg aliquot of protein isolated from cerebral microvessels was separated on a 10% SDS-PAGE gel, transferred to nitrocellulose membranes which were probed with antibodies to Dusp5 (H00001847-M04, Abnova, Taiwan; 1:1,000) targeting AA286-384 in the C-terminus of the Dusp5 protein and antibodies raised against p-ERK1/2, total ERK1/2 and beta-Actin as described above.

Myogenic response on MCA. MCAs were microdissected from the brains of 9–12 week old Dusp5 KO, FHH.1[BN], FHH and SD rats and mounted on glass micropipettes in a myograph. The bath solution was equilibrated with O_2 (95%) and CO_2 (5%) to provide adequate oxygenation and to maintain at pH 7.4. The diameters of the vessels were measured using a videomicrometer (VIA-100, Boeckeler Instruments) at intraluminal pressures ranging from 40 to 140 mmHg in steps of 20 mmHg as previously described. [8,9,30]

Autoregulation of CBF. CBF autoregulation were determined on 9–12 week old male Dusp5 KO, FHH.1[BN], FHH and SD rats. The rats were anesthetized with ketamine (30 mg/kg, *i.m.*) and Inactin (50 mg/kg, *i.p.*) and were mechanically ventilated throughout experiment to maintain pO_2 and pCO_2 at 100 and 35 Torr, respectively. Body temperature was maintained

at 37°C during experiment. Catheters (PE-50) were placed in the femoral artery and vein and the rats received an intravenous infusion of 0.9% NaCl solution at a rate of 100 μl/min to replace surgical fluid losses. The scalp was exposed and the cranial bone was thinned 3 mm lateral and 6 mm posterior to the Bregma using a handheld drill until the pial vessels became visible through the thinned window. CBF was monitored bilaterally with a laser-Doppler flow meter (PF5001, Perimed Corp, Jarfalla, Sweden). After surgery and a 30-min equilibration period, mean arterial pressure (MAP) was lowered to 90–100 mmHg by increasing the depth of pentobarbital anesthesia (1–5 mg/kg, *i.v.*). Baseline regional CBF was measured, then systemic pressure was elevated in steps of 10–20 mmHg by graded *i.v.* infusion of phenylephrine (0.5–5 μg/min). [31–33] MAP was maintained for 3–5 min until a new steady-state level of CBF was obtained. CBF was expressed as a percentage of the baseline laser-Doppler flow signal.

Statistics

Mean values ± SEM are presented. The significance of the differences in the expression of various proteins and mRNA in FHH, FHH.1[BN] and Dusp5 KO rats was determined using one-way ANOVA. The significance of the differences in mean values between and within groups in the myogenic responses and autoregulation of CBF was determined using a two-way ANOVA for repeated measures and Holm-Sidak test for preplanned comparisons. A P value <0.05 was considered to be statistically significant.

Results

Sequence analysis and expression of Dusp5, p-ERK1/2 and p-PKC in FHH and FHH.1[BN] rats

The results of the comparative sequence analysis are presented in **Figure 1**. We identified 17 SNPs in the Dusp5 gene in FHH versus the BN reference sequence. Most of the SNPs were in introns, however, there were four SNPs in the Dusp5 mRNA

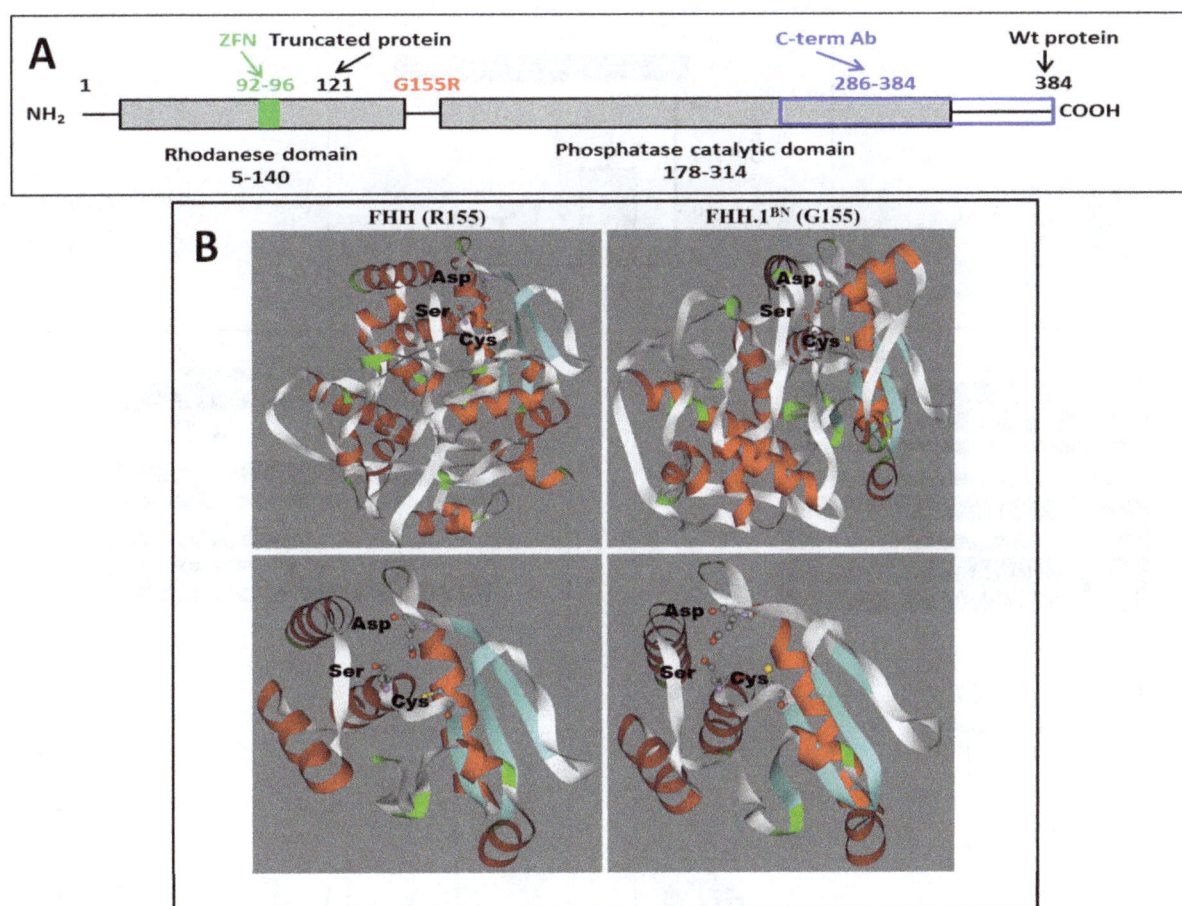

Figure 2. Identification of the Zn-finger target site and deletion in the Dusp5 KO strain. Panel A presents a schematic model of the Dusp5 protein. The Dusp5 Zinc-finger construct targets amino acids (AA) 92–96 in the N-terminal regulatory rhodanese domain (AA5-140) resulting in the introduction of a premature stop codon at AA121 that is predicted to produce a truncated protein. The Dusp5 antibody used in these studies targets AA286-384 in the C-terminal phosphatase catalytic domain (AA178-314). **Panel B** presents a comparison of I-TASSER predicted structure and the folding of the Dusp5 protein in FHH (155R) and FHH.1^{BN} (155G) rats. The upper panels show the predicted structure of the Dusp5 protein in both strains based on the complete AA sequence. The putative catalytic triad (Asp232/Ser268/Cys263) is shown in a "stick figure" form and the 3-letter AA codes are labeled in black. The rest of the protein is represented as ribbon running along the backbone. Secondary structural elements are depicted by color with helices, beta sheets and coils represented in red, cyan and white, respectively. The putative catalytic triad is magnified and shown in "stick figure" form in the lower panel and the 3-letter AA codes are labeled with in black. Only residues 174–320 of Dusp5 protein are presented in order to enhance the view of the putative catalytic triad. There are significant structural differences both in the overall folding of the Dusp5 protein that impact on the structure of the active site/catalytic triad region between the strains. This may account for the observed differences in the activity of the Dusp5 protein in FHH versus FHH.1^{BN} rats.

including a C107T SNP in the 5′-UTR, a G330T SNP in exon 1, a C627T and a G637A in exon 2. All of these SNPs were verified in our FHH and FHH.1^{BN} strains by sequencing cDNAs derived from mRNA extracted from the isolated cerebral vessels. The C107T SNP altered a CpG site and the C627T SNP altered one of six CpGs in exon 2 that were previously identified as methylation sites by bisulfite modification. [34]. Moreover, the G637A SNP caused a G155R mutation (**Figure 2A**) that is predicted using I-TASSER modeling package [35–37] may alter the folding of the protein and the active site conformation of the Dusp5 protein in FHH versus FHH.1^{BN} rats as shown in **Figure 2B.**

To determine if the C107T and C627T SNPs that altered CpG sites and the G637A SNP that caused a G155R mutation might alter the expression of Dusp5 in FHH versus FHH.1^{BN} rats, RT-qPCR and western blot experiments were performed. The results presented in **Figure 3A** indicate that the expression of Dusp5 mRNA is 2-fold higher in cerebral arteries of FHH than in the

FHH.1^{BN} control rats but the expression of protein is not different between these strains (**Figure 3C**). Moreover, as presented in **Figure 3B and 3C**, the expression of p-ERK1/2, the primary substrates for dephosphorylation by Dusp5, is significantly reduced in cerebral microvessels and the brains of FHH relative to FHH.1^{BN} rats, while total ERK1/2 levels are not significantly altered. The expression of p-PKC protein is also significantly elevated in FHH.1^{BN} as compared to FHH rats.

Generation and characterization of the Dusp5 ZFN KO rats

The homozygous Dusp5 knockout (KO) and wild type FHH.1^{BN} control strain were derived from an intercross of the heterozygous Dusp5 founders. Genotyping and sequencing of the Dusp5 KO strain indicated that there is a 14 bp deletion and a 3 bp insertion between nucleotides 449–464 in Dusp5 mRNA that creates a frame shift mutation which is predicted to introduce a premature stop codon at amino acid (AA) 121 (**Figure 2A,**

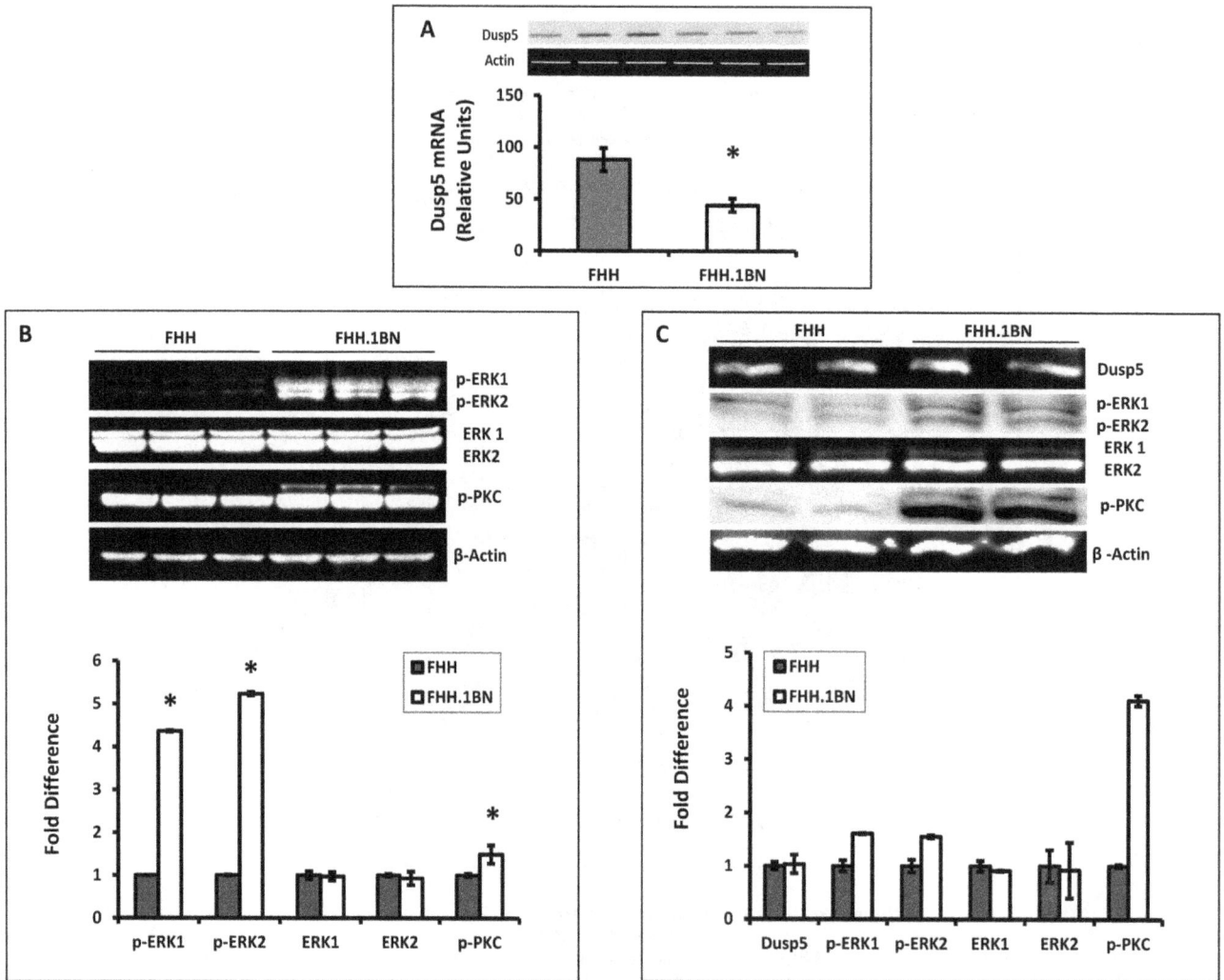

Figure 3. Comparison of the expression and activity of Dusp5 in FHH versus FHH.1^{BN} rats. Panel A presents a comparison of the expression of Dusp5 mRNA in cerebral arteries of FHH versus FHH.1^{BN} rats. The upper portion of the figure presents the representative images of gels showing the qPCR products and the bar graph below compares the expression levels. **Panel B** presents a comparison of the expression of phosphorylated-ERK1/2, total ERK1/2, phosphorylated-PKC and beta-Actin in the brain of FHH and FHH.1^{BN} rats. The upper panel presents the representative images and the lower panel presents the relative quantitation. **Panel C** presents a comparison of expression of these proteins in cerebral microvessels of FHH as compared with FHH.1^{BN} rats. All of the vessels isolated from one strain were pooled into a single sample. The upper panel presents the representative images and the lower panel presents the quantitation of the images. Mean values ± SE from 3 rats per strain are presented in Panel A and Panel B. Panel C represents the results from duplicate aliquots run from a single pooled microvessel sample isolated from 8–10 rats per strain. * indicates a significant difference from the corresponding value in FHH rats.

4A, 4B). The genotypes of the animals were verified using PCR as shown in **Figure 4C**.

Dusp5 is a ubiquitous protein that is abundantly expressed in the brain, spleen and WBCs [11,12,38]. The expression of Dusp5 protein in WBCs isolated from Dusp5 KO versus FHH.1^{BN} control rats was compared using an anti-Dusp5 antibody that targeted AA286-384 (**Figure 2A**) in the C-terminus of the protein to confirm that the introduction of the new stop codon produces a truncated Dusp5 protein in the KO animals. The results presented in **Figure 5** indicate that the expression of Dusp5 protein is markedly reduced in WBCs (**Figure 5A**) and cerebral microvessels (**Figure 5B**) isolated from Dusp5 KO rats relative to the FHH.1^{BN} control strain and in other tissues (brain, liver and spleen, data not shown). We also compared the expression of the primary substrates of Dusp5, p-ERK1/2, in these strains. The results presented in **Figure 5** indicate that the levels of p-ERK1/2

are enhanced, while the expression of total ERK1/2 is not significantly different in WBCs (**Figure 5A**) and cerebral microvessels (**Figure 5B**) in Dusp5 KO rats compared to FHH.1^{BN} control rats.

Comparison of the myogenic response of the MCA of Dusp5 KO and FHH.1^{BN} rats

The luminal diameter of the MCA in FHH.1^{BN} rats (n = 12) decreased by $20 \pm 2\%$ when the perfusion pressure was increased from 40 to 140 mmHg. The myogenic response of MCA isolated from Dusp5 KO rats was significantly greater, and the diameter of these vessels decreased by $34 \pm 7\%$ when the perfusion pressure was increased over the same range. In contrast, the myogenic response of the MCA of FHH rats was markedly impaired as the diameter of these vessels only increased by $10 \pm 4\%$ when pressure was increased from 40 to 140 mmHg (**Figure 6A**). The passive

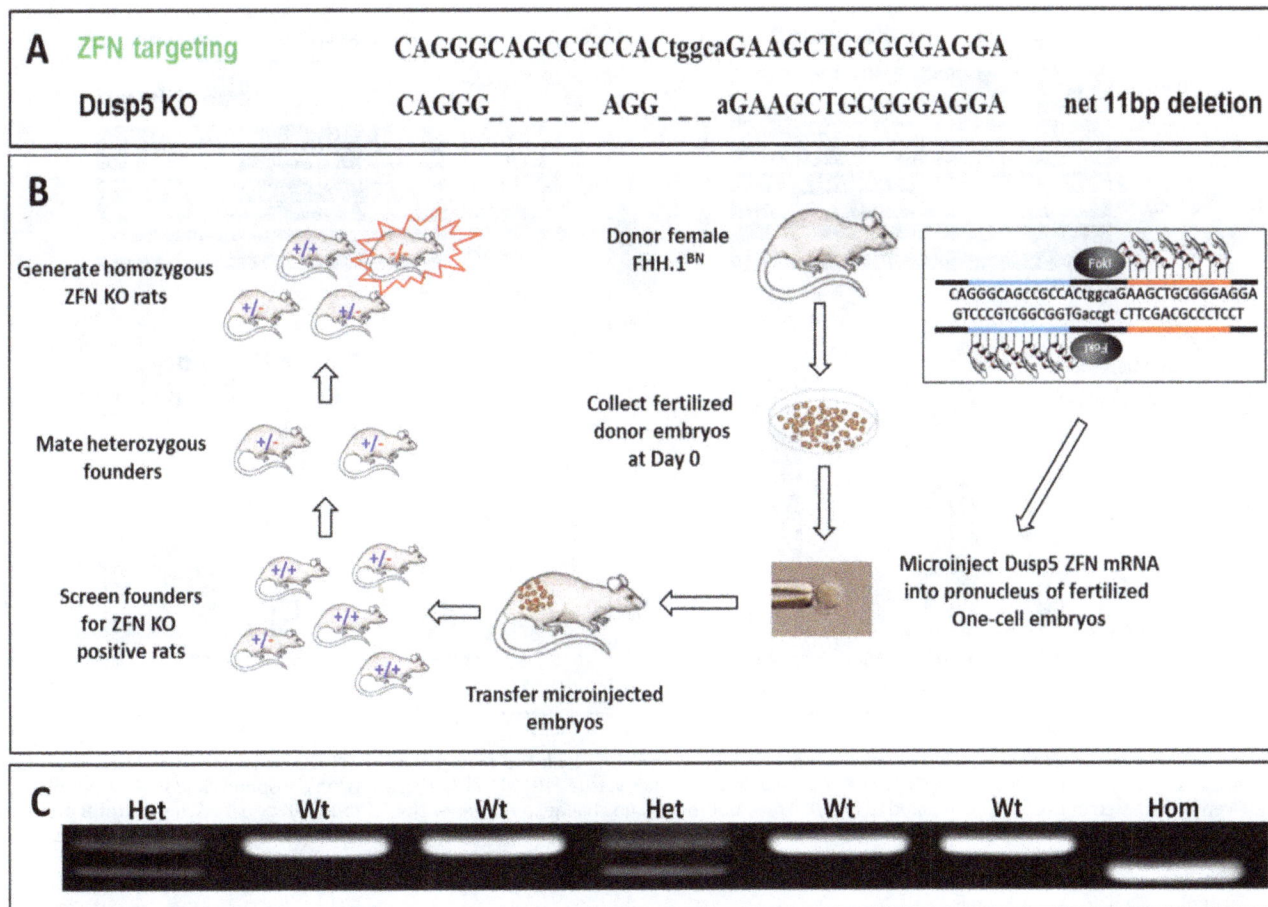

Figure 4. Schematic describing the generation of Dusp5 ZFN KO rats in the FHH.1^{BN} genetic background. Panel A presents the sequence of the Dusp5 ZFN. The Dusp5 specific ZFN introduced a 14 bp deletion and a 3 bp insertion resulting in a net 11 bp deletion between nucleotides 449–464 in the Dusp5 sequence in the KO animals that introduced a frame shift mutation. **Panel B** presents the strategy for the generation of Dusp5 ZFN KO rats in the FHH.1^{BN} genetic background. Fertilized donor embryos from female FHH.1^{BN} rats were collected and the Dusp5 ZFN mRNA was microinjected into pronuclei of the fertilized one-cell embryos. These embryos were transferred back to a foster mother. The heterozygous founders were brother-sister mated to generate the homozygous ZFN KO founders and the FHH.1^{BN} wild type control rats. **Panel C** presents an example of PCR genotyping of the region of interest in FHH.1^{BN} and Dusp5 KO rats. The rats were genotyped using the following primers: Dusp5-F: 5′-GCT GCA GGA GGG CGG CGG CG -3′, Dusp5 R: 5′-CTT TAA GGA AGT AGA CCC G-3′. These primers amplify a 155 bp band for the wild type allele and a 144 bp band for the knockout allele. Both bands are observed in heterozygous rats.

diameter curves generated in Ca^{2+} free solution in all strains were not significantly different (**Figure 6B**).

Comparison of the autoregulation of CBF of Dusp5 KO and FHH.1^{BN} rats

Autoregulation was markedly impaired and CBF increased by 54±6% in FHH rats when MAP was increased from 100 to 160 mmHg. CBF was autoregulated to a greater extent in the FHH.1^{BN} and Dusp5 KO rats were not significant different and only increased by 26±3% and 12±3%, respectively, when MAP was increased over the same range. The range of the autoregulation of CBF since CBF rose by 33±4% when pressure was increased from 100 to 190 mmHg in Dusp5 KO rats versus an increase of 65±5% in the FHH.1^{BN} rats and 99±3% in the FHH animals (**Figure 7**).

Discussion

We recently reported that the myogenic response of the MCA and autoregulation of CBF were markedly impaired in FHH rats

and were restored in a FHH.1^{BN} congenic strain in which Chromosome 1 from the BN rats containing 15 genes was transferred into the FHH genetic background. [7–9] However, the gene or genes that contribute to the impairment of vascular function and the mechanisms involved still remain obscure. In the present study, we identified 17 SNPs in the Dusp5 gene in FHH versus FHH.1^{BN} rats. Most were in the intronic region, but four were in exons including a C107T SNP in the 5′-UTR, a G330T SNP in exon 1, a C627T and a G637A in exon 2. Both of the SNPs, C107T and C627T, altered CpG sites and the C637A SNP caused a G155R mutation. To determine if the altered CpG sites and/or the G155R mutation might underlie the loss of the myogenic response in FHH rats, we created Dusp5 KO rats in the FHH.1^{BN} genetic background using ZFN KO technology [25–27,39]. Site specific Zn-fingers fused to *Fok I* nuclease were introduced into the pronucleus of one cell embryos to induce double-strand DNA breaks at the target site, followed by error-prone non-homologous DNA repair which resulted in a frameshift mutation and formation of a premature stop codon leading to a truncated or non-functional protein. [25–27,40] The ZFNs

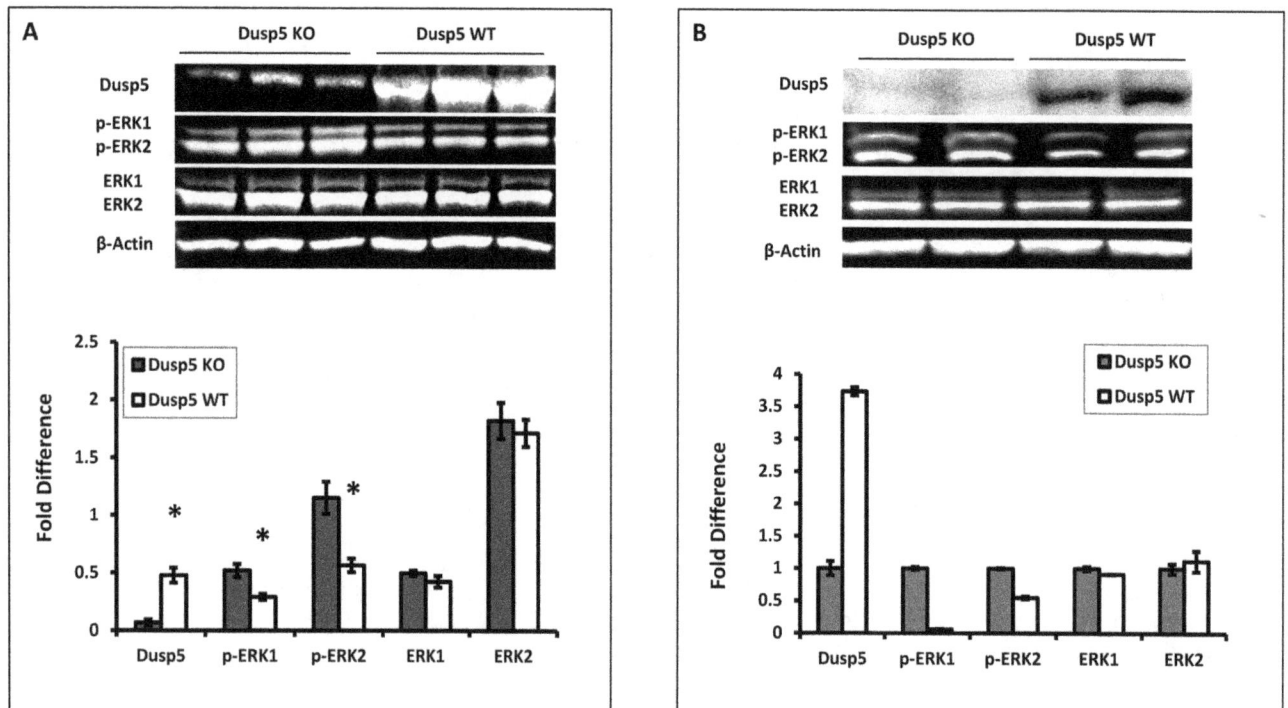

Figure 5. Comparison of the expression and activity of Dusp5 in WBCs isolated from Dusp5 ZFN KO versus FHH.1^BN rats. Panel A: The expression of Dusp5 protein in WBCs is nearly absent in Dusp5 KO versus FHH.1[BN] rats. The levels of phosphorylated-ERK1/2 protein are significantly increased in Dusp5 KO compared to FHH.1[BN] rats, but there is no change in the expression of total ERK or beta-Actin. **Panel B** presents the results of the expression of Dusp5, p-ERK1/2, total ERK1/2 and β-Actin protein from duplicate aliquots of a single pooled microvessel sample isolated from 8–10 Dusp5 KO and FHH.1[BN] rats. The upper panel presents a representative image of the gels and the lower panel presents the quantitation of the images above. Mean values ± SE from 3 rats per strain are presented in Panel A. * indicates a significant difference from the corresponding value in Dusp5 KO rats.

targeted AA92-96 in the N-terminal regulatory rhodanese domain of Dusp5 protein that is the ERK1/2 binding site (**Figure 2A**). [41,42] The Dusp5 ZFN KO strain was successfully generated and we confirmed that this strain had a net 11 bp deletion between nucleotides 449–464 in Dusp5 mRNA that introduced a frame shift mutation and a premature stop codon at AA121 (**Figure 4A, 2A**). We also confirmed that the expression of Dusp5 at the protein level was nearly absent in multiple tissues including cerebral microvessels using an antibody directly targeting AA286-384 in the C-terminus that are beyond the predicted site of the newly introduced stop codon. The loss of Dusp5 protein in the KO animals was associated with an expected increase in p-ERK1/2 levels in various tissues and cerebral microvessels compared to FHH.1[BN] wild type animals because Dusp5 is a serine-threonine phosphatase that inactivates MAPK activity [10–14] by specifically dephosphorylating p-ERK1/2 MAP kinases. [15,38]

An increase in the dephosphorylation of p-ERK1/2 by Dusp5 phosphatase would be expected to modulate BK and TRP channel activities [1,16–19] and downregulate PKC, Rho/ROCK[20,43] and STAT pathways[11] which are regulated by the MAP kinase system. Activation of BK and TRP channel activities alter the myogenic response of small blood vessels by modulating calcium entry in VSMCs. [1,19,22] Discovered over 100 years ago by Bayliss, the myogenic response is an intrinsic property of VSMC that initiates contraction of arterioles in response to elevations in transmural pressure. [1,2] It is impaired following cerebral vasospasm, stroke or traumatic brain injury and the autoregulatory range is shifted to higher pressures in hypertension [3,4,44–46]. Autoregulation of CBF is one of the major mechanisms to protect

the brain from elevations in perfusion pressure that promote vascular leakage and swelling of the brain [9,19,46]_ENREF_64. The myogenic response of MCA plays a major role in autoregulation of CBF and contributes about 50% to overall compensation to elevations in perfusion pressure [47].

In the present study, we found that the myogenic response of the MCA was greater in Dusp5 KO animals than in wild type controls and FHH rats. There was no difference in the passive pressure diameter relationships measured in Ca^{2+} free solution between these strains. The increased myogenic response was associated with enhanced autoregulation of CBF in response to elevations in systemic pressure from 100–160 mmHg in the Dusp5 KO compared to FHH.1[BN] and FHH strains. Moreover, the range of effective autoregulation of CBF was extended to higher pressures in the Dusp5 KO rats versus FHH.1[BN] and FHH animals. Our findings are consistent with the results of a recent study by Wickramasekera, *et al* demonstrating that downregulation of the expression of Dusp5 by siRNA in cultured cerebral arteries enhanced pressure-dependent myogenic constriction [20]. Together, these findings confirm that alteration in the expression or activity of Dusp5 modulates the myogenic response of the MCA *in vitro* and autoregulation of CBF *in vivo*.

We also examined whether the sequence variants we identified in the Dusp5 gene in FHH versus FHH.1[BN] rats might contribute to the impaired myogenic response and autoregulation of CBF in FHH rats by altering the expression of Dusp5 and its phosphatase activity. Our RT-qPCR results indicated that the expression of Dusp5 at the message level in microdissected cerebral arteries was 2-fold higher in FHH relative to FHH.1[BN] rats, but the expression

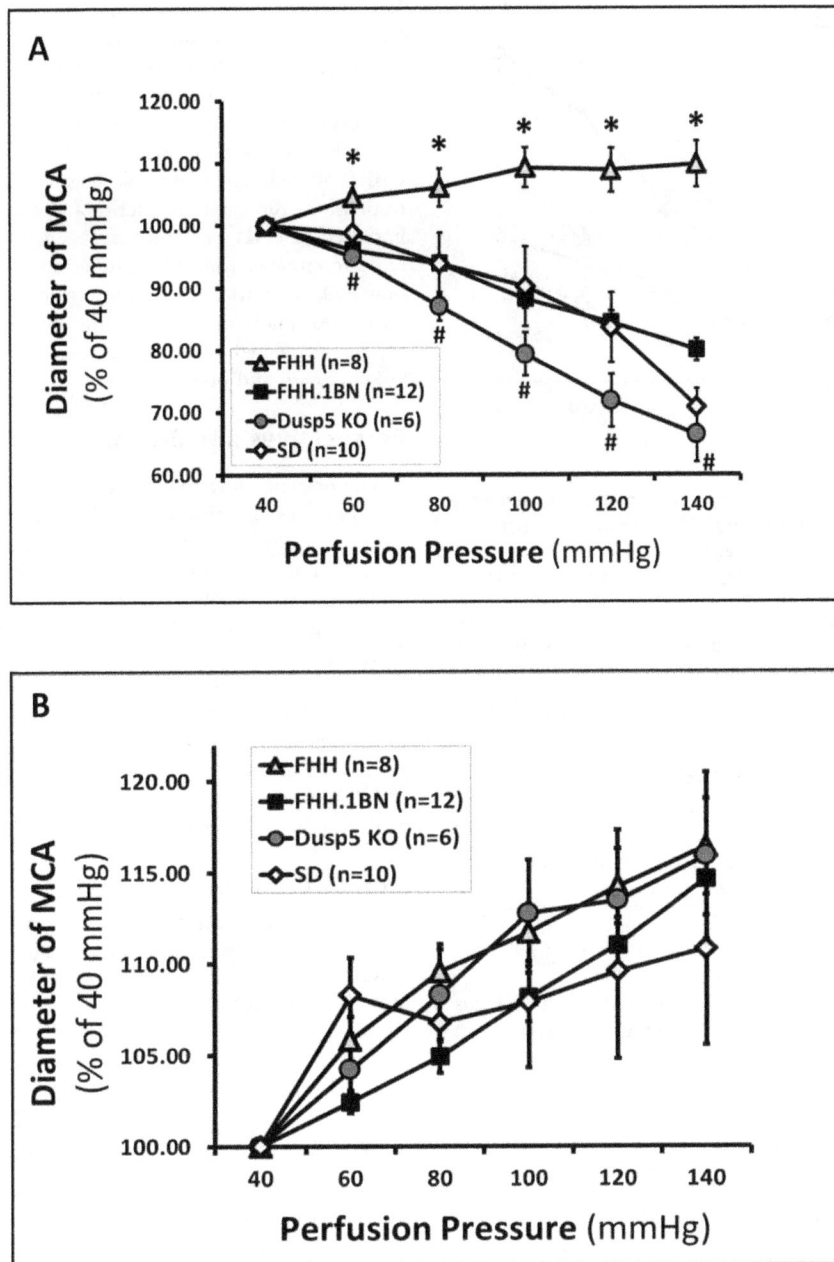

Figure 6. Comparison of the myogenic response in middle cerebral artery (MCA) isolated from Dusp5 ZFN KO versus FHH.1^{BN} and FHH rats. Panel A presents the passive pressure-diameter curves in Ca^{2+} free solution at each pressure in all strains. **Panel B**: The luminal diameter of the MCA decreased from 100 to 66±4% in Dusp5 KO rats and from 100 to 80±2% in FHH.1^{BN} rats when the perfusion pressure was increased from 40 to 140 mmHg, whereas it was dilated in FHH rats (from 100 to 110±4%). The MCA also constricted in Sprague Dawley rats that is widely used as a control strain for the myogenic response. Mean values ± SE are presented. Numbers in parentheses indicate the number of vessels studied per group. * indicates a significant difference in the corresponding value in FHH versus all the other strains. # indicates there is a significant difference between Dusp5 KO and FHH.1^{BN} rats.

at protein level was not significantly different. This suggests that the difference in the myogenic response of MCA between FHH and FHH.1^{BN} rats is not due to changes in the expression of Dusp5 secondary to the two SNPs (C107T and C627T) in the Dusp5 gene in FHH rats that is predicted to alter CpG sites that may possibly cause DNA demethylation and alter transcriptional activity. [34] However, we found that p-ERK1/2 levels were significantly decreased in FHH compared to FHH.1^{BN} rats. Although more work will be needed to rigorously test this

hypothesis, a decrease in p-ERK1/2 levels is entirely consistent with the observed reduction in the myogenic response in the MCA and autoregulation of CBF observed in FHH rats. Moreover, we also found that a G637A SNP causes a G155R mutation in Dusp5 protein in FHH rats. This G155R mutation localized between the N-terminal regulatory rhodanese domain and the C-terminal phosphatase catalytic domain (**Figure 2A**) converts a nonpolar amino acid Glycine (G) to a basic polar Arginine (R) and is predicted by the I-TASSER program [35–37] to affect the folding

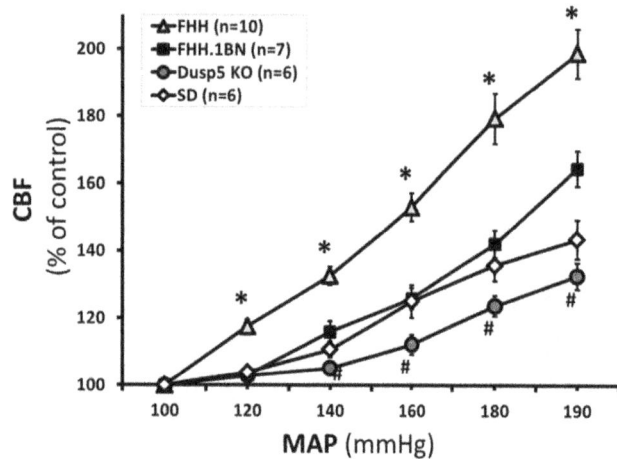

Figure 7. Comparison of autoregulation of CBF in Dusp5 ZFN KO versus FHH.1[BN] and FHH rats. The relationships between cerebral blood flow and mean arterial pressures in 9–12 week old Dusp5 ZFN KO, FHH.1[BN], FHH and Sprague Dawley rats are compared. Mean values ± SE are presented. * indicates significantly difference in the corresponding value in FHH rats versus all the other strains. # indicates a significant difference in the corresponding values in Dusp5 KO and FHH.1[BN] rats. Numbers in parentheses indicate numbers of animal studied per strain.

of the Dusp5 protein and the conformation of the active site. **Figure 2B** illustrates the changes in the folding based on theoretical structural models (upper panel) and the expanded view presents the confirmation of the active/catalytic site (lower panel). These differences in global folding and active site conformation in FHH compared to FHH.1[BN] rats might lead to differences in protein stability, interactions with binding partners, catalytic efficiency or catalytic activity.

An impaired myogenic response and autoregulation of CBF has been reported in various pathological conditions in patients and experimental animals including: in subarachnoid hemorrhage (SAH), [48–53] ischemic stroke [3,54–56] and traumatic brain injury. [57–60] In hypertensive patients, impaired autoregulation of CBF accelerates the development of a cognitive decline. [61] However, the mechanisms involved have been difficult to directly study due to lack of an animal model in which autoregulation of CBF is altered. The present findings indicating that the myogenic response in the cerebral arteries and autoregulation of CBF are impaired in FHH rats and restored in FHH.1[BN] congenic strain and are enhanced in our newly generated Dusp5 ZFN KO rats now fill this knowledge gap and provide an important new model system to study the mechanisms by which genetic defects in myogenic mechanisms contribute to the development of small vessel disease and brain damage.

Perspectives and Significance

The present study reports on the creation of a Dusp5 KO rat and provides the first *in vivo* evidence that Dusp5 plays an important role in modulating the myogenic response of cerebral arteries and autoregulation of CBF. We identified a G155R mutation that might contribute to an increase in Dusp5 phosphatase activity and reduced the phosphorylation of ERK1/2 that is consistent with the impaired myogenic response and autoregulation of CBF in FHH rats. Our newly generated Dusp5 KO rat model also provides the scientific community a new model to investigate the mechanisms of impaired myogenic response in FHH rats, and the essential role of Dusp5 in the regulation of MAP kinase activity in vascular reactivity, immune response, cell proliferation and apoptosis and cancer.

Author Contributions

Conceived and designed the experiments: FF DRH HJ RJR. Performed the experiments: FF AMG MRP. Analyzed the data: FF AMG MRP SVS RJR. Contributed reagents/materials/analysis tools: AMG MRP HJ RJR. Wrote the paper: FF SVS RJR.

References

1. Davis MJ, Hill MA (1999) Signaling mechanisms underlying the vascular myogenic response. Physiol Rev 79: 387–423.
2. Bayliss WM (1902) On the local reactions of the arterial wall to changes of internal pressure. J Physiol 28: 220–231.
3. Paulson OB, Strandgaard S, Edvinsson L (1990) Cerebral autoregulation. Cerebrovasc Brain Metab Rev 2: 161–192.
4. Strandgaard S (1991) Cerebral blood flow in the elderly: impact of hypertension and antihypertensive treatment. Cardiovasc Drugs Ther 4 Suppl 6: 1217–1221.
5. Johansson B (1989) Myogenic tone and reactivity: definitions based on muscle physiology. J Hypertens Suppl 7: S5–8; discussion S9.
6. Mellander S (1989) Functional aspects of myogenic vascular control. J Hypertens Suppl 7: S21–30; discussion S31.
7. Burke M, Pabbidi M, Fan F, Ge Y, Liu R, et al. (2013) Genetic basis of the impaired renal myogenic response in FHH rats. American Journal of Physiology-Renal Physiology 304: F565–F577.
8. Pabbidi MR, Mazur O, Fan F, Farley JM, Gebremedhinm D, et al. (2014) Enhanced large conductance K+ channel (BK) activity contributes to the impaired myogenic response in the cerebral vasculature of Fawn Hooded Hypertensive rats. Am J Physio heart and circulatory phys 2014 Apr 1;306(7): H989–H1000.
9. Pabbidi MR, Juncos J, Juncos L, Renic M, Tullos HJ, et al. (2013) Identification of a region of rat chromosome 1 that impairs the myogenic response and autoregulation of cerebral blood flow in fawn-hooded hypertensive rats. Am J Physiol Heart Circ Physiol 304: H311–317.
10. Owens DM, Keyse SM (2007) Differential regulation of MAP kinase signalling by dual-specificity protein phosphatases. Oncogene 26: 3203–3213.
11. Kovanen PE, Bernard J, Al-Shami A, Liu C, Bollenbacher-Reilley J, et al. (2008) T-cell development and function are modulated by dual specificity phosphatase DUSP5. J Biol Chem 283: 17362–17369.
12. Kucharska A, Rushworth LK, Staples C, Morrice NA, Keyse SM (2009) Regulation of the inducible nuclear dual-specificity phosphatase DUSP5 by ERK MAPK. Cell Signal 21: 1794–1805.
13. Zassadowski F, Rochette-Egly C, Chomienne C, Cassinat B (2012) Regulation of the transcriptional activity of nuclear receptors by the MEK/ERK1/2 pathway. Cell Signal 24: 2369–2377.
14. Patterson KI, Brummer T, O'Brien PM, Daly RJ (2009) Dual-specificity phosphatases: critical regulators with diverse cellular targets. Biochem J 418: 475–489.
15. Mandl M, Slack DN, Keyse SM (2005) Specific inactivation and nuclear anchoring of extracellular signal-regulated kinase 2 by the inducible dual-specificity protein phosphatase DUSP5. Mol Cell Biol 25: 1830–1845.
16. Sun CW, Falck JR, Harder DR, Roman RJ (1999) Role of tyrosine kinase and PKC in the vasoconstrictor response to 20-HETE in renal arterioles. Hypertens 33: 414–418.
17. Murphy TV, Spurrell BE, Hill MA (2002) Cellular signalling in arteriolar myogenic constriction: involvement of tyrosine phosphorylation pathways. Clin Exp Pharmacol Physiol 29: 612–619.
18. Kamkin A, Kiseleva I, Isenberg G (2000) Stretch-activated currents in ventricular myocytes: amplitude and arrhythmogenic effects increase with hypertrophy. Cardiovasc Res 48: 409–420.
19. Toth P, Csiszar A, Tucsek Z, Sosnowska D, Gautam T, et al. (2013) Role of 20-HETE, TRPC channels, and BKCa in dysregulation of pressure-induced Ca2+ signaling and myogenic constriction of cerebral arteries in aged hypertensive mice. Am J Physiol Heart Circ Physiol 305: H1698–1708.
20. Wickramasekera NT, Gebremedhin D, Carver KA, Vakeel P, Ramchandran R, et al. (2013) Role of dual-specificity protein phosphatase-5 in modulating the myogenic response in rat cerebral arteries. J Appl Physiol 114: 252–261.
21. Livak KJ, Schmittgen TD (2001) Analysis of relative gene expression data using real-time quantitative PCR and the 2(-Delta Delta C(T)) Method. Methods 25: 402–408.

22. Fan F, Sun CW, Maier KG, Williams JM, Pabbidi MR, et al. (2013) 20-Hydroxyeicosatetraenoic Acid Contributes to the Inhibition of K+ Channel Activity and Vasoconstrictor Response to Angiotensin II in Rat Renal Microvessels. PLoS One 8: e82482.

23. Dunn KM, Renic M, Flasch AK, Harder DR, Falck J, et al. (2008) Elevated production of 20-HETE in the cerebral vasculature contributes to severity of ischemic stroke and oxidative stress in spontaneously hypertensive rats. Am J Physiol Heart Circ Physiol 295: H2455–2465.

24. Chen CC, Geurts AM, Jacob HJ, Fan F, Roman RJ (2013) Heterozygous knockout of transforming growth factor-beta1 protects Dahl S rats against high salt-induced renal injury. Physiol Genomics 45: 110–118.

25. Geurts AM, Cost GJ, Freyvert Y, Zeitler B, Miller JC, et al. (2009) Knockout rats via embryo microinjection of zinc-finger nucleases. Science 325: 433.

26. Geurts AM, Cost GJ, Remy S, Cui X, Tesson L, et al. (2010) Generation of gene-specific mutated rats using zinc-finger nucleases. Methods Mol Biol 597: 211–225.

27. Rangel-Filho A, Lazar J, Moreno C, Geurts A, Jacob HJ (2013) Rab38 modulates proteinuria in model of hypertension-associated renal disease. J Am Soc Nephrol 24: 283–292.

28. Miller JC, Holmes MC, Wang J, Guschin DY, Lee YL, et al. (2007) An improved zinc-finger nuclease architecture for highly specific genome editing. Nat Biotechnol 25: 778–785.

29. Sanger F, Coulson AR (1975) A rapid method for determining sequences in DNA by primed synthesis with DNA polymerase. J Mol Biol 94: 441–448.

30. Burke M, Pabbidi M, Fan F, Ge Y, Liu R, et al. (2013) Genetic basis of the impaired renal myogenic response in FHH rats. Am J Physiol Renal Physiol 304: F565–577.

31. Bellapart J, Fraser JF (2009) Transcranial Doppler assessment of cerebral autoregulation. Ultrasound Med Biol 35: 883–893.

32. Purkayastha S, Raven PB (2011) The functional role of the alpha-1 adrenergic receptors in cerebral blood flow regulation. Indian J Pharmacol 43: 502–506.

33. Wagner BP, Ammann RA, Bachmann DC, Born S, Schibler A (2011) Rapid assessment of cerebral autoregulation by near-infrared spectroscopy and a single dose of phenylephrine. Pediatr Res 69: 436–441.

34. Fu Q, McKnight RA, Yu X, Callaway CW, Lane RH (2006) Growth retardation alters the epigenetic characteristics of hepatic dual specificity phosphatase 5. FASEB J 20: 2127–2129.

35. Roy A, Kucukural A, Zhang Y (2010) I-TASSER: a unified platform for automated protein structure and function prediction. Nat Protoc 5: 725–738.

36. Zhang Y (2008) I-TASSER server for protein 3D structure prediction. BMC Bioinformatics 9: 40.

37. Roy A, Yang J, Zhang Y (2012) COFACTOR: an accurate comparative algorithm for structure-based protein function annotation. Nucleic Acids Res 40: W471–477.

38. Huang CY, Tan TH (2012) DUSPs, to MAP kinases and beyond. Cell Biosci 2: 24.

39. Rangel-Filho A, Sharma M, Datta YH, Moreno C, Roman RJ, et al. (2005) RF-2 gene modulates proteinuria and albuminuria independently of changes in glomerular permeability in the fawn-hooded hypertensive rat. Journal of the American Society of Nephrology 16: 852–856.

40. Geurts AM, Moreno C (2010) Zinc-finger nucleases: new strategies to target the rat genome. Clin Sci (Lond) 119: 303–311.

41. Caunt CJ, Keyse SM (2013) Dual-specificity MAP kinase phosphatases (MKPs): shaping the outcome of MAP kinase signalling. FEBS J 280: 489–504.

42. (2003) USRDS: the United States Renal Data System. Am J Kidney Dis 42: 1–230.

43. Zhao Y, Zhang L, Longo LD (2005) PKC-induced ERK1/2 interactions and downstream effectors in ovine cerebral arteries. Am J Physiol Regul Integr Comp Physiol 289: R164–171.

44. Faraci FM, Baumbach GL, Heistad DD (1990) Cerebral circulation: humoral regulation and effects of chronic hypertension. J Am Soc Nephrol 1: 53–57.

45. Faraci FM, Mayhan WG, Heistad DD (1987) Segmental vascular responses to acute hypertension in cerebrum and brain stem. Am J Physiol 252: H738–742.

46. Walsh MP, Cole WC (2013) The role of actin filament dynamics in the myogenic response of cerebral resistance arteries. J Cereb Blood Flow Metab 33: 1–12.

47. Faraci FM, Heistad DD (1998) Regulation of the cerebral circulation: role of endothelium and potassium channels. Physiol Rev 78: 53–97.

48. Ishii R (1979) Regional cerebral blood flow in patients with ruptured intracranial aneurysms. J Neurosurg 50: 587–594.

49. Voldby B, Enevoldsen EM, Jensen FT (1985) Regional CBF, intraventricular pressure, and cerebral metabolism in patients with ruptured intracranial aneurysms. J Neurosurg 62: 48–58.

50. Voldby B, Enevoldsen EM, Jensen FT (1985) Cerebrovascular reactivity in patients with ruptured intracranial aneurysms. J Neurosurg 62: 59–67.

51. Dernbach PD, Little JR, Jones SC, Ebrahim ZY (1988) Altered cerebral autoregulation and CO2 reactivity after aneurysmal subarachnoid hemorrhage. Neurosurgery 22: 822–826.

52. Lang EW, Diehl RR, Mehdorn HM (2001) Cerebral autoregulation testing after aneurysmal subarachnoid hemorrhage: the phase relationship between arterial blood pressure and cerebral blood flow velocity. Crit Care Med 29: 158–163.

53. Roman RJ, Renic M, Dunn KM, Takeuchi K, Hacein-Bey L (2006) Evidence that 20-HETE contributes to the development of acute and delayed cerebral vasospasm. Neurological research 28: 738–749.

54. Agnoli A, Fieschi C, Bozzao L, Battistini N, Prencipe M (1968) Autoregulation of cerebral blood flow. Studies during drug-induced hypertension in normal subjects and in patients with cerebral vascular diseases. Circulation 38: 800–812.

55. Hoedt-Rasmussen K, Skinhoj E, Paulson O, Ewald J, Bjerrum JK, et al. (1967) Regional cerebral blood flow in acute apoplexy. The "luxury perfusion syndrome" of brain tissue. Arch Neurol 17: 271–281.

56. Olsen TS, Larsen B, Herning M, Skriver EB, Lassen NA (1983) Blood flow and vascular reactivity in collaterally perfused brain tissue. Evidence of an ischemic penumbra in patients with acute stroke. Stroke 14: 332–341.

57. Cold GE (1981) Cerebral blood flow in the acute phase after head injury. Part 2: Correlation to intraventricular pressure (IVP), cerebral perfusion pressure (CPP), PaCO2, ventricular fluid lactate, lactate/pyruvate ratio and pH. Acta Anaesthesiol Scand 25: 332–335.

58. Cold GE, Christensen MS, Schmidt K (1981) Effect of two levels of induced hypocapnia on cerebral autoregulation in the acute phase of head injury coma. Acta Anaesthesiol Scand 25: 397–401.

59. Cold GE, Jensen FT (1978) Cerebral autoregulation in unconscious patients with brain injury. Acta Anaesthesiol Scand 22: 270–280.

60. Overgaard J, Tweed WA (1974) Cerebral circulation after head injury. 1. Cerebral blood flow and its regulation after closed head injury with emphasis on clinical correlations. J Neurosurg 41: 531–541.

61. Lammie GA (2002) Hypertensive cerebral small vessel disease and stroke. Brain Pathol 12: 358–370.

De novo Assembly of the Grass Carp *Ctenopharyngodon idella* Transcriptome to Identify miRNA Targets Associated with Motile Aeromonad Septicemia

Xiaoyan Xu[1ɕ], Yubang Shen[1ɕ], Jianjun Fu[1], Liqun Lu[3], Jiale Li[1,2]*

1 Key Laboratory of Exploration and Utilization of Aquatic Genetic Resources, Shanghai Ocean University, Ministry of Education, Shanghai 201306, PR China, **2** E-Institute of Shanghai Universities, Shanghai Ocean University, 999 Huchenghuan Road, 201306 Shanghai, PR China, **3** National Pathogen Collection Center for Aquatic Animals, College of Fisheries and Life Science, Shanghai Ocean University, 999 Huchenghuan Road, 201306 Shanghai, PR China

Abstract

Background: *De novo* transcriptome sequencing is a robust method of predicting miRNA target genes, especially for organisms without reference genomes. Differentially expressed miRNAs had been identified previously in kidney samples collected from susceptible and resistant grass carp (*Ctenopharyngodon idella*) affected by *Aeromonas hydrophila*. Target identification for these differentially expressed miRNAs poses a major challenge in this non-model organism.

Results: Two cDNA libraries constructed from mRNAs of susceptible and resistant *C. idella* were sequenced by Illumina Hiseq 2000 technology. A total of more than 100 million reads were generated and *de novo* assembled into 199,593 transcripts which were further extensively annotated by comparing their sequences to different protein databases. Biochemical pathways were predicted from these transcript sequences. A BLASTx analysis against a non-redundant protein database revealed that 61,373 unigenes coded for 28,311 annotated proteins. Two cDNA libraries from susceptible and resistant samples showed that 721 unigenes were expressed at significantly different levels; 475 were significantly up-regulated and 246 were significantly down-regulated in the SG samples compared to the RG samples. The computational prediction of miRNA targets from these differentially expressed genes identified 188 unigenes as the targets of 5 conserved and 4 putative novel miRNA families.

Conclusion: This study demonstrates the feasibility of identifying miRNA targets by transcriptome analysis. The transcriptome assembly data represent a substantial increase in the genomic resources available for *C. idella* and will provide insights into the gene expression profile analysis and the miRNA function annotations in further studies.

Editor: Daniel Doucet, Natural Resources Canada, Canada

Funding: This work was supported by grants from National Key Technology R&D Program of China (2012BAD26B02), the China's Agricultural Research System (CARS-46-04), the Agricultural Seed Development Program of Shanghai City (2012NY10), Shanghai Ocean University Doctoral Research Foundation, Shanghai Universities First-class Disciplines Project of Fisheries and the Innovation Plan of Shanghai Graduate Education. The funders had no role in study design, data collection and analysis, decision to publish, or preparation of the manuscript.

Competing Interests: The authors have declared that no competing interests exist.

* Email: jlli2009@126.com

ɕ These authors contributed equally to this work.

Background

Next-generation sequencing (NGS) -based RNA sequencing for transcriptome methods (RNA-seq) allow simultaneous acquisition of sequences for gene discovery as well as identification of transcripts involved in specific biological processes. This is especially suitable for non-model organisms whose genomic sequences are unknown [1,2]. In addition, the dynamic range, sensitivity and specificity of RNA-seq also make it ideal for quantitatively analyzing various aspects of gene regulation [3]. These techniques do not require prior knowledge of genomic sequence and are much advanced in terms of time, cost, labor,

amount of data produced, data coverage, sensitivity, and accuracy compared to traditional sequencing methods [4,5].

The grass carp (*Ctenopharyngodon idella*) is one of the most important farmed fish species in China, with a cultural history dating back to the 7th century CE (Tang Dynasty) [6]. According to the FAO, the value of farmed *C. idella* reached more than 6.46 billion USD for a production of 5.03 billion tons in 2012, thus accounting for the highest production and third highest value of major cultured fish species worldwide at single species level [7]. Despite favorable growth traits, farmed *C. idella* are rather susceptible to various disease. Outbreaks of disease associated with bacteria such as *Aeromonas hydrophila* have caused high mortality, resulting in reduced production and considerable economic losses [8].

A. hydrophila is a causative agent of a wide spectrum of diseases in humans and animals [9]. While originally thought to be an opportunistic pathogen in immunocompromized humans, an increasing number of intestinal and extraintestinal disease cases suggest that it is an emerging human pathogen irrespective of the immune status of the host [10]. The pathogenesis, pathogenic mechanism, and virulence factors responsible for selected *A. hydrophila* infections in different species are not well understood [11]. *A. hydrophila* is a Gram-negative motile bacillus widely distributed in aquatic environments. It causes motile aeromonad septicemia (MAS), which results in great economic losses in worldwide freshwater fish farming [12]. Thus, more effective measures against *A. hydrophila* infection in fish are needed. Identification of differentially expressed genes (DGEs) following *A. hydrophila* infection is important for an improved understanding of fish MAS.

MicroRNAs (miRNAs) are 20–22 nt non-coding RNAs that play important roles in post-transcriptional gene regulation. In animal cells, miRNAs regulate their targets by translational inhibition and mRNA destabilization [13]. MicroRNAs (miRNAs) are key effectors in mediating host-pathogen interactions and constitute a family of small RNA species; they are considered a promising candidate for regulating the interaction between host and pathogen [14,15]. Therefore, dissecting the biological functions of miRNAs may help us understand the pathogenic mechanism of motile aeromonad septicemia in *C. idella*. Many studies have identified miRNAs and mRNA transcriptome in fish species, like common carp [16,17], nile tilapia [18,19] rainbow trout [20,21] channel catfish [22,23] and silver carp [24,25]. To thoroughly interpret the biological functions of these miRNAs, a first step is to predict their targets. Therefore, establishing a more powerful transcriptome data for target identification is preferred.

Although two parallel *C. idella* expressed sequence tag analyses have already been conducted using head kidney tissue [26,27], the data presented here represent the first effort to analyze the transcriptome of *C. idella* affected by *A. hydrophila*. Two cDNA libraries from SG and RG *C. idella* used for our miRNA analysis were constructed and sequenced with Illumina Hiseq 2000. The obtained reads were assembled into transcripts and annotated by BLAST analysis against various databases before screening the results for differentially expressed genes and the prediction of miRNA targets. Our work will provide an approach to identify the target genes of miRNAs and to characterize their functional/regulatory network to increase our understanding of hemorrhagic septicemia outbreaks in *C. idella*.

Materials and Methods

Ethics statement

All handling of fishes was conducted in accordance with the guidelines on the care and use of animals for scientific purposes set up by the Institutional Animal Care and Use Committee (IACUC) of the Shanghai Ocean University, Shanghai, China. The IACUC has specially approved this study within the project "Breeding of Grass Carp" (approval number is SHOU-09-007).

Sample collection

C. idella with an average weight of 50 g were cultured individually at the Wujiang National Farm of Chinese Four Family Carps (Jiangsu Province, China). Animals were raised at 28°C in 400-L aerated tanks for one week before the experiment and fed twice daily (in the morning and late in the afternoon) at a

ratio of 5% of total biomass. Two groups (30 animals in each group) were maintained in two aquariums and intraperitoneally injected with *A. hydrophila* AH10 (Aquatic Pathogen Collection Center of Ministry of Agriculture, China) at a dose of 7.0×10^6 cells suspended in 100 μl PBS per fish. All fish were observed every 4 h for any mortality and samples were taken until the termination of the experiment at 240 h post-challenge. *C. idella* died in the first 72 h post-challenge were classified as susceptible group (SG) for their high sensitivity to *A. hydrophila*, while the animals that survived over 240 h post-challenge were considered as resistant group (RG) [28]. Spleen and kidney samples were immediately snap-frozen in liquid nitrogen and stored at −80°C until further use.

cDNA library construction and sequencing

Experimental protocols for the cDNA sequence were performed according to the manufacturer's technical instructions. The spleen and kidney tissues of randomly-selected three fish from both the susceptible and resistant groups were collected and, labeled as SG and RG, respectively. The RNA from the same tissue of three fishes of SG and RG *C. idella* was pooled with equal quantity for the construction of SG and RG cDNA libraries. The pooled total RNA was isolated from each spleen and kidney samples with TRIZOL reagent (Invitrogen, Grand Island, NY, USA). RNA integrity was confirmed using a 2100 Bioanalyzer (Agilent Technologies, Inc.) by RNA 6000 nano with a minimum RNA integrity number (RIN) value of 7.0. Poly (A) mRNA was purified from the total RNA using oligo (dT) magnetic beads. Equal amounts of the high-quality mRNA samples were obtained from each group for cDNA library preparation using the NEBNext Ultra RNA Library Prep Kit for Illumina (New England Biolabs, Ipswich, MA, USA) and purified using Agencount AMPure XP beads (Beckman Coulter, Krefeld, Germany) according to the manufacturer's recommendations. The concentration of the cDNA library was determined on an Agilent Technologies 2100 Bioanalyzer by Agilent DNA-1000. Libraries were sequenced, at the Novogene Bioinformatics Institute (Beijing, China) on an Illumina HiSeq 2000 instrument (Illumina, San Diego CA, USA) that generated paired-end reads of 101 nucleotides.

Data processing, assembly, and functional annotation

Raw reads generated by Illumina Hiseq 2000 were then cleaned by removing the adaptor containing sequences, any ambiguous base > 10% reads and low quality reads (1/2 reads with Q-value ≤ 5) to get clean reads. Then, all clean reads were assembled using the *de novo* assembly program Trinity [29]: first, short reads were assembled into high-coverage contigs that could not be extended farther in either direction in a k-mer-based approach for fast and efficient transcript assembly. Then, the related contigs were clustered and a de Bruijn graph for each cluster was constructed. Finally, in the context of the corresponding de Bruijn graph and all plausible transcripts, alternatively spliced isoforms and transcripts were derived.

All assembled transcripts were compared with publicly available databases including Nr (NCBI non-redundant protein sequences), Nt (NCBI non-redundant nucleotide sequences) [30], KOG/COG (Clusters of Orthologous Groups of proteins) [31], Swiss-Prot (a manually annotated and reviewed protein sequence database) [32], KO (KEGG ortholog database) [33], Pfam (protein family) [34] and GO (Gene Ontology) (http://www.geneontology.org/). Nr, Nt, KOG/COG, Swiss-Prot, and KO used the BLASTx analysis with a cut-off E-value of 10^{-5}, Pfam used Hmmerscan and GO used Blast2GO [35]. The best Blast hits from all Blast results were parsed for a homology-based functional

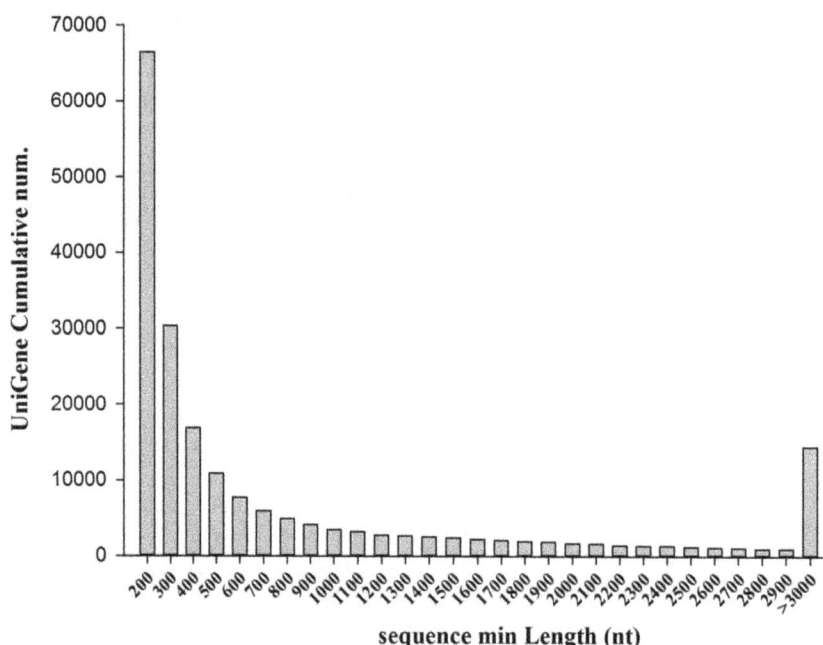

Figure 1. Length distribution of assembled unigenes in the sequenced cDNA library.

annotation. For the nr annotations, the Blast2GO program was used to obtain GO annotations of unique assembled transcripts to describe biological processes, molecular functions, and cellular components.

Differentially expressed genes between the SG/RG libraries

High-quality reads were mapped to reference sequences (unigenes from the transcriptome data of the cDNA library) using RSEM [36]. Gene expression levels were calculated using the fragments per kilo bases per million mapped reads (FPKM) method [37]. The calculation of unigene expression levels and the identification of unigenes that were differentially expressed between the libraries were performed by DEGseq [38] based on TMM normalized counts. The settings "q.value <0.005" and "|log2.Fold change.normalized|>2" were used as thresholds for judging significant differences in transcript expression. Differen-

tially expressed genes across the samples were further annotated by GO and KEGG pathway analysis

MiRNA target prediction

The SG and RG kidney miRNA-seq analysis were conducted in the same biological samples as mRNA-seq. Small RNA libraries were constructed using a Small RNA Cloning Kit (Takara). RNA was purified by polyacrylamide gel electrophoresis (PAGE) to enrich for the molecules in the range of 17–27 nt, then was ligated with 5′ and 3′ adapters. The resulting samples were used as templates for cDNA synthesis followed by PCR amplification. The obtained sequencing libraries were subjected to Solexa sequencing-by-synthesis method. After the run, image analysis, sequencing quality evaluation and data production summarization were performed with Illumina/Solexa pipeline. The sequencing data was pretreated to discard low quality reads, no 3′-adaptor reads, 5′-adaptor contaminants and sequences shorter than 18 nucleo-

Table 1. Summaries of sequencing cDNA library.

Sample name	SG	RG
Total reads	73,063,654	62,737,669
Clean reads	70,210,307	60,668,815
Total mapped to unigenes readcounts	61,396,052.97	51,785,549.97
Reads length (bp)	101	
GC content (%)	48.43	47.13
Number of unigenes	199,554	
Total length of unigenes (bp)	195,075,872	
Mean length of unigenes (bp)	977	
N50 of unigenes (bp)	2,117	
Maximal length of unigenes (bp)	27,185	

Table 2. Statistics of the annotation results for the *C. idella* unigenes.

	All	Nr	Nt	Pfam	KOG	Swiss-Prot	KO	GO
Number of unigenes	61373	28311	46653	33013	16980	22674	27775	34207
% of unigenes	100	46.1	76.0	53.8	27.7	36.9	45.3	55.7

Nr: NCBI non-redundant protein sequences, Nt: NCBI non-redundant nucleotide sequences, Pfam: Protein family, KOG: Clusters of Orthologous Groups of proteins, Swiss-Prot: A manually annotated and reviewed protein sequence database, KO: KEGG Ortholog database and GO: Gene Ontology.

tides. After trimming the 3′ adaptor sequence, sequence tags were mapped onto the transcriptome of *C. idella* using bowtie. Any small RNAs having exact matches to transcriptome of *C. idella* were used from further analysis. The mapped reads were compared to the miRBase (19.0) to annotate conserved miRNAs. To predict novel miRNAs, the miREvo [39] and mirdeep2 [40] were used.

Computational identification of differentially expressed miRNA targets was performed using the miRanda toolbox [41], using the complementary region between miRNAs and mRNAs and the thermodynamic stability of the miRNA-mRNA duplex. All mRNAs used for target prediction came from the differentially expressed unigenes obtained as described above. The miRanda toolbox employed a dynamic programming algorithm to search the complementary regions between the miRNA and the 3′-UTR of the mRNA, and the scores were based on sequence complementary as well as minimum free energy of RNA duplexes, and were calculated with the Vienna RNA package [42]. All detected targets with scores and energies less than the threshold parameters of S>90 (single-residue pair scores) and $\Delta G < -17$ kcal/mol (minimum free energy) were selected as potential targets.

Real time PCR validation

The sequencing results were validated by real time PCR using One Step PrimeScript miRNA cDNA Synthesis Kit for miRNA (TaKaRa) reversely transcribed, PrimeScript RT reagent Kit with gDNA Eraser (TaKaRa) for mRNA and SYBR *Premix Ex Taq* II (2x) (Takara) for qPCR according to the manufacturer protocols. Specific primer assays for *miR-21* [F: 5′-TAGCTTATCA-GACTGGTGTTGGC-3′, R: Uni-miR qPCR Primer (TaR-aKa)], *JNK1* (F: 5′- TGGTCAGAGGTAGTGTGTTG-3′, R: 5′-AGTTTGTTGTGGTCCGAGTC-3′) and *ccr7* (F: 5′-CAAGCCAAGAACTTTGAGAGG-3′,   R: 5′-GGCA-TAAAGGCGAATGTTGTC-3′) were purchased from sangon biotech and real time PCR quantification was carried out in CFX96 Real-Time PCR System (Bio-Rad, CA, USA). To normalize the expression values, *miR-22a* for miRNA and *18s* for mRNA were used as housekeeping control [43,44]. Expression levels were quantitatively analyzed using the $2^{-\Delta\Delta CT}$ method. One-way ANOVA tests were performed using SPSS 17.0 to determine significant differences. Each experiment was repeated in triplicates.

Results and Discussion

Antagonistic bacteria, such as *A. hydrophila*, enhance non-specifically immune-related enzyme activities and disease resistance in *C. idella* and provide a theoretical basis for disease prevention in aquaculture. However, the molecular mechanisms of this disease are still far from fully understood. The identification and characterization of candidate genes involved in MAS would represent the first step in understanding the genetic basis of this process in *C. idella*.

De novo assemblies and unigenes annotation

The sequencing generated 135.80 million raw reads. After trimming, 130.88 million clean reads remained, corresponding to 13.22 GB clean bases. The dataset of each sample, SG and RG, was represented by over 60 million clean reads, a read density sufficient for the subsequent quantitative analysis of genes. The raw sequencing reads have been submitted to the NCBI Short Read Archive under the accession number of SRR1124206 and SRR1125014. Then, all clean reads were assembled using a *de*

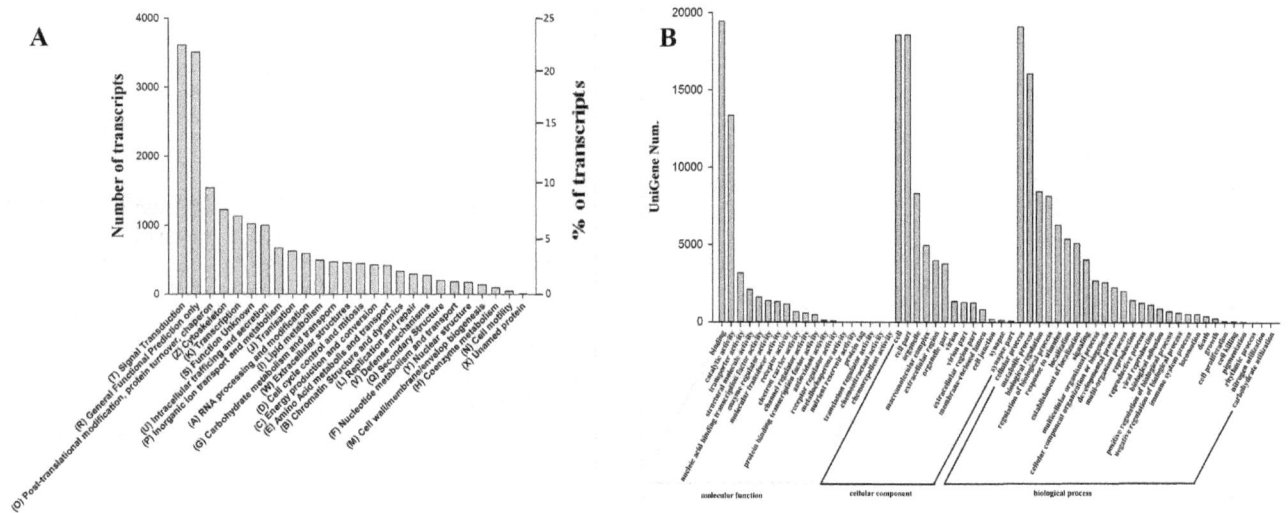

Figure 2. Functional annotations of the unigenes of *C. idella*. (A) KOG annotation. (B) Level 2 GO term distribution for the biological process, cellular component and molecular function categories.

novo assembly program Trinity [45]. These short reads were further assembled into 199,554 transcripts with an average length of 977 bp (Table 1). The size distribution of these transcripts ranged from 201 to 27,185 bp, of which 27,334 were larger than 2,000 bp (Figure 1). The assembled transcriptome data were deposited in NCBI's Transcriptome Shotgun Assembly (TSA) database under the accession numbers from SUB583458.

Annotation of predicted proteins

A total of 61,373 distinct sequences (30.75% of the transcripts) matched known genes corresponding to 28,311 of the annotated proteins (Table 2, Table S1). An additional functional annotation of the unigenes of *C. idella* was performed searching for putative orthologs and paralogs within the KOG database [31]. A total of 16980 unigenes (27.7%) were assigned to 26 eukaryotic orthologous groups (Figure 2A). The category "signal transduction", which contained 3,611 unigenes (21.27% of 16980 unigenes), was the largest, followed by the categories "general functional prediction only" (3506, 20.65%), "post-translational modification, protein turnover, chaperone" (1539, 9.06%) and "cytoskeleton" (1226, 7.22%).

GO annotation and KEGG pathway analyses

After GO annotation, *C. idella* transcripts could be assigned to three categories: biological processes, molecular functions and cellular components. Within the various biological processes, cellular processes (19,121 unigenes) metabolic processes (16,091) and biological regulation (8,463) were the most highly represented members (Figure 2B). Important functions, such as cell death (389) and immune system processes (540), were also identified in this category. Similarly, cell (18,586) as well as cell part (18,586) and binding (19,460) were the most represented sub-categories in the cellular component and molecular function categories, respectively.

Searching against the Kyoto Encyclopedia of Genes and Genomes Pathway database (KEGG) [33] revealed that 10,561 unigenes could be matched to 298 KEGG pathways. The most-represented pathways hierarchy 2 were the "infectious diseases" pathway (3210 unigenes) and the "signal transduction" pathway (2316) (Table 3). Some pathways related to immune system were also identified, such as the "Toll-like receptor signaling" pathway (110) [46] and the "chemokine signaling" pathway (224) [47] (Table 4).

Table 3. Top 10 list of the gene number of Pathway Hierarchy 2.

Pathway Hierarchy 2	Unigene Number
Infectious Diseases	3210
Signal Transduction	2316
Cancers	2235
Nervous System	1648
Immune System	1554
Neurodegenerative Diseases	932
Digestive System	875
Endocrine System	861
Cell Communication	859
Signaling Molecules and Interaction	763

Table 4. Top 10 list of pathways related to immune system.

Pathway Hierarchy 2	KEGG Pathway	Unigene Numbers
Immune System	Chemokine signaling pathway	224
Immune System	Leukocyte transendothelial migration	199
Immune System	T cell receptor signaling pathway	156
Immune System	Fc gamma R-mediated phagocytosis	155
Immune System	Natural killer cell mediated cytotoxicity	129
Immune System	B cell receptor signaling pathway	110
Immune System	Toll-like receptor signaling pathway	110
Immune System	Antigen processing and presentation	86
Immune System	Fc epsilon RI signaling pathway	84
Immune System	RIG-I-like receptor signaling pathway	67

Digital gene expression library sequencing

Based on the transcriptome sequence data, two DGE libraries were constructed to identify the differentially expressed unigenes between the SG and RG samples. After removing low-quality reads, 70,210,307 and 60,668,815 clean reads were generated from the SG and RG libraries, respectively (Table 1). Among these clean reads, 61,396,052.97 of the SG and 51,785,549.97 of the RG readcounts were mapped to unigenes.

Differential gene expression between the SG and RG libraries

The results suggest that the expression of 721 genes differed significantly between the SG and RG groups of *C. idella*. Of these genes, 475 were up-regulated and 246 were down-regulated in the SG samples compared to the RG samples (Figure 3 and Table S2). GO enrichment analysis of DEGs indicated that these genes were significantly enriched in oxidation-reduction processes (biological process), integral to membrane (cellular components), and protein binding (molecular function) (Table S3). Pathway enrichment analysis found the DEGs to be mainly enriched in complement and coagulation cascades, *Staphylococcus aureus* infection and porphyrin and chlorophy II metabolism (Table S4). Notably, several genes involved in the immune and inflammatory response were also identified, such as C-type lectin [48] and matrix metalloproteinase-9 [49].

MiRNA target prediction

The identification of miRNAs and their targets is important for understanding the physiological functions of miRNAs and the functional roles of differentially expressed miRNAs between healthy and diseased fish. We were thus interested in predicting miRNA target genes involved in the immune response or immune system, according to the KEGG analysis. In a previous study, small RNA deep-sequencing data were aligned with miRBase 18.0 to search for known miRNAs with complete matches, namely,

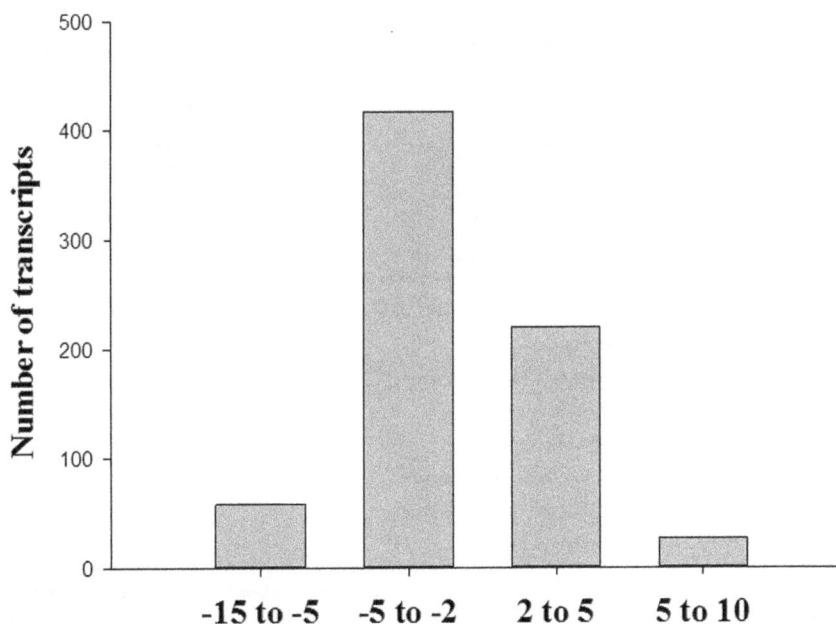

Figure 3. Number and Fold change distribution of differentially expressed genes between the SG/RG libraries.

Table 5. Differentially expressed miRNAs in *C. idella* kidney between SG and RG.

miRNA	Sequence (5'-3')	SG (TMP)	RG (TMP)	log2 (Fold_change) normalized	p-value
let-7i	UGAGGUAGUAGUUUGUGCUGUU	3419.67473	1475.401006	1.212751983	4.62E-174
miR-142a-3p	UGUAGUGUUUCCUACUUUAUGGA	4770.81118	9798.344345	−1.038303405	0
miR-21	UAGCUUAUCAGACUGGUGUUGGC	3431.54242	1640.502276	1.064719594	4.12E-142
miR-217	UACUGCAUCAGGAACUGAUUGG	240.9140969	2983.647091	−3.630486183	0
miR-223	UGUCAGUUUGUCAAAUACCCC	1460.912578	6800.683353	−2.218809871	0
novel_115	UGAAGGCCGAAGUGGAGA	3.560306851	17.95531055	−2.334337113	0.001253505
novel_131	UGCCCGCAUUCUCCACCA	7.713998177	40.28996513	−2.384869847	9.85E-07
novel_154	CCCAGCCAUAUUUGUUUGAAC	16.02138083	0	4.768565529	4.44E-05
novel_3	UGUUUCUGGCUCUGAUAUUUGCU	32.04276166	71.82124218	−1.164412111	8.07E-05

conserved miRNAs (data unpublished). Meanwhile, miRNAs predicted by miRDeep 2.0 that could form stable secondary structures, were identified as novel miRNAs.

We have used a single algorithm, miRanda [50] to predict miRNA targets. As miRNA binding to target 3' UTR generally results in mRNA destabilization and degradation [51], we chose to

Table 6. 26 pathways were related to immune and diseases in all pathways.

KEGG Pathway	Pathway Hierarchy 2	Target gene of different expression miRNA
Toll-like receptor signaling pathway	Immune system	JNK, TLR5, TBK1
Hepatitis C	Infectious diseases: Viral	JNK, TBK1, LDLR
Salmonella infection	Infectious diseases: Bacterial	JNK, TLR5
Fc epsilon RI signaling pathway	Immune system	JNK, FYN
Tuberculosis	Infectious diseases: Bacterial	JNK, PLK3, CTSS
Influenza A	Infectious diseases: Viral	JNK, DNAJB1, TBK1
Measles	Infectious diseases: Viral	FYN, TBK1
MAPK signaling pathway	Signal transduction	JNK, MKP
Toxoplasmosis	Infectious diseases: Parasitic	JNK, LDLR
Chemokine signaling pathway	Immune system	CCR7, ADCY7, DOCK2
HTLV-I infection	Infectious diseases: Viral	JNK, ADCY7, SLC2A1
Herpes simplex infection	Infectious diseases: Viral	JNK, TBK1
Fc gamma R-mediated phagocytosis	Immune system	DOCK2
Prion diseases	Neurodegenerative diseases	FYN
Viral myocarditis	Cardiovascular diseases	FYN
Pathways in cancer	Cancers: Overview	JNK, SLC2A1
Transcriptional misregulation in cancer	Cancers: Overview	CCR7
Type II diabetes mellitus	Endocrine and metabolic diseases	JNK
Cytokine-cytokine receptor interaction	Signaling molecules and interaction	CCR7, TNFSF12
Antigen processing and presentation	Immune system	CTSS
Pertussis	Infectious diseases: Bacterial	JNK
Natural killer cell mediated cytotoxicity	Immune system	FYN
T cell receptor signaling pathway	Immune system	FYN
Chagas disease	Infectious diseases: Parasitic	JNK
Staphylococcus aureus infection	Infectious diseases: Bacterial	CFB
Complement and coagulation cascades	Immune system	CFB

List of gene abbreviations: JNK: c-Jun N-terminal kinase, TLR5: toll-like receptor 5, TBK1: TANK-binding kinase 1, LDLR: low-density lipoprotein receptor, FYN: tyrosine-protein kinase Fyn, CTSS: cathepsin S, PLK3: polo-like kinase 3, DNAJB1: DnaJ homolog subfamily B member 1, MKP: dual specificity MAP kinase phosphatase, DOCK2: dedicator of cytokinesis protein 2, ADCY7: adenylate cyclase 7, SLC2A1: MFS transporter, SP family, solute carrier family 2 (facilitated glucose transporter), member 1, CCR7: C-C chemokine receptor type 7, TNFSF12: tumor necrosis factor ligand superfamily member 12, CFB: component factor B.

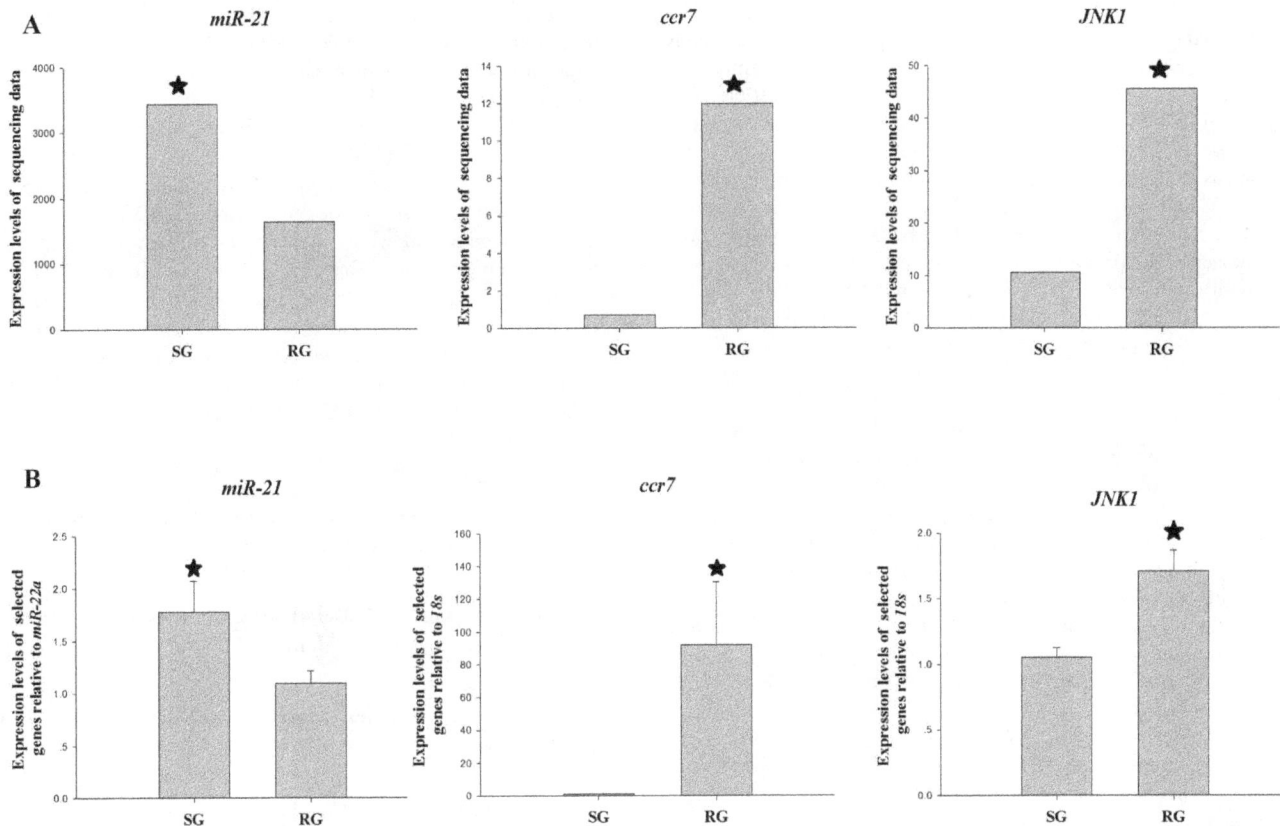

Figure 4. The expression analysis of selected genes from the expression profile by relative quantitative real-time PCR. A Transcriptome sequencing data, B Real-time PCR data. Increases and decreases in relative levels of transcripts with respect to the control 18s for mRNA and *miR-22a* for miRNA are shown. The settings "q.value <0.005" and "|log2.Fold change.normalized|>2" were used as thresholds for judging significant differences in transcript expression. One-way ANOVA tests were performed using SPSS 17.0 to determine significant differences for Real-time PCR data. Statistical significance of the relative expression ratio is indicated *.

narrow down potential targets to those showing differential expression in the opposite direction as the mRNA. This approach increases the strength to discover true target genes and functions affected by miRNA dysregulation. In total, 188 of the target genes predicted by miRanda were differentially expressed in the opposite direction in the target tissue. The identification of 188 unigenes (Table S5) as the predicted target genes of 5 conserved and 4 putative novel miRNA families (Table 5). The identified target genes involved in biological processes, molecular functions, and cellular components were defined using GO annotations. GO analysis demonstrated that these targets were involved in a broad range of physiological processes, including gene expression, transcription regulation, immune system processes, and responses to stress or stimuli (Table S6).

Searching against the KEGG indicated that 188 unigenes mapped to 48 KEGG pathways. 26 pathways were related to immune and diseases in all pathways (Table 6). These included the categories "Toll-like receptor signaling pathway", "Fc epsilon RI signaling pathway", "Chemokine signaling pathway", "Fc gamma R-mediated phagocytosis", "Antigen processing and presentation", "Natural killer cell mediated cytotoxicity", "T cell receptor signaling pathway" and "Complement and coagulation cascades" related immune functions. Toll-like receptor signaling pathway induce the expression of a variety of host defense genes. These include chemokine signaling pathway and other effectors necessary to arm the host cell against the invading pathogen [52]. Cytokine-cytokine receptor interaction play a pivotal role in the generation

of immunological responses during bacterial infection [53]. This implicated functions that are likely regulated by miRNAs and suggests regulation of different pathways during immune activation in susceptible and resistant *C. idella*.

Of particular interest to our study is the fact that several of the most highly expressed miRNAs in SG have been shown to have a role in immunity. let-7i and associated TLR4 expression are involved in cholangiocyte immune responses against *C. parvum* infection [54]. miR-21 targets multiple genes associated with the immuno-logically localized disease [55]. *miR-21* which is up-regulated in SG is predicted to target 28 differentially expressed genes (Table S7) in *C. idella*. Frequently represented in the top immune functions were protein kinase JNK1 (*JNK1*) and chemokine (C-C motif) receptor 7 (*ccr7*) (Table 6). *JNK1* and *ccr7* showed clearly down-regulated expression profiles in the SG samples compared to the RG samples (Figure 4A   http://www.plosone.org/article/info%3Adoi%2F10. 1371%2Fjournal.pone.0073506-pone.0073506.s003   ). The decreased expression profile of these targets in the susceptible samples supported our previous finding that the expression of *miR-21* was significantly up-regulated, with TMP of 3,432 and 1,642 in the SG and RG groups, respectively (Table 5). We validated the *miR-21*, *JNK1* and *ccr7* which had expression change in the sequencing data by performing real time PCR in the same samples used for the sequencing (Figure 4B).

JNK1 is involved in apoptosis, neurodegeneration, cell differentiation and proliferation, inflammatory conditions and cytokine production mediated by AP-1 (activation protein 1), such as

Regulated upon activation normal T cell expressed and presumably secreted (*RANTES*), Interleukin 8 (*IL-8*), and Granulocyte-macrophage colony-stimulating factor (*GM-CSF*) [56]. It has been reported that JNK plays an important role in the innate immune response to microbial challenge [57,58]. Most importantly, JNK1 serves as a negative regulatory factor for MAP kinase phosphatase 5 (*MKP5*) that plays an essential role in innate immune responses [59]. The chemokine receptor CCR7 acts as an important organizer of the primary immune response [60]. A previous study demonstrated a discrete CCR7 requirement in the activation of different T cell subsets during bacterial infection [61]. CCR7 is differentially regulated by macrophages in exposure to bacteria, as it is triggered by exposure to both Gram-negative and Gram-positive bacteria [62].

The discovery of microRNAs dramatically changed our perspective on eukaryotic gene expression regulation [63]. MicroRNAs play important gene-regulatory roles in animals and plants by pairing to the mRNAs of protein-coding genes to direct their posttranscriptional repression [64,65]. The identification of miRNA target genes is an important step in understanding their role in gene regulatory networks. Most miRNA-associated computational methods comprise the prediction of miRNA genes and their targets, and an increasing number of computational algorithms and web-based resources are being developed to fulfill the needs of scientists performing miRNA research, like miRanda [42], TargetScan [66], RNAhybrid [67] and PicTar [68]. However, animal miRNA targets are difficult to predict since miRNA: mRNA duplexes often contain several mismatches, gaps, and G:U base pairs in many positions, thus limiting the maximum length of contiguous sequences of matched nucleotides [69]. The predicted interactions using these computational methods are inconsistent and the expected false positive rates are still high. Recently, several authors suggested integrating expression profiles from both miRNA and mRNA with *in silico* target predictions to reduce the number of false positives and increase the number of biologically relevant targets [70]. These methods have been shown to be effective in identifying the most prominent interactions from the databases of putative targets [71]. To minimize false positive rates in our study, the RNA-seq and miRNA-seq analysis were conducted in the same biological samples. Likewise, to reduce the number of putative target genes, the miRNA targets were predicted from differentially expressed genes. However, some false positive predictions proved inevitable. Further studies will focus on the experimental validation of the differentially expressed mRNA and miRNAs identified in this study. They are likely to be central regulators of the innate immune response to *A. hydrophila* and thus represent potential therapeutic targets or novel biomarkers of infection and inflammation.

Conclusions

In this study, we used high-throughput sequencing data to characterize the transcriptome of *C. idella*, a species for which little genomic data are available. Further, DGE tags were mapped to the assembled transcriptome for further gene expression analysis. A large number of candidate genes involved in MAS were identified. This represents a fully characterized transcriptome, and provides a valuable resource for genetic and genomic studies in *C. idella*. Additionally, DGE profiling provides new leads for functionally studies of genes involved in MAS.

Finally, comparison with our previous miRNA profiling, this study strongly indicates that miRNA is a critical factor in determining mRNA abundance and regulation during MAS. Our on-going effort using experimental approach such as knock-down or over-express candidate miRNAs and mRNAs in vitro is expected to provide new evidence in understanding these regulatory mechanisms of MAS in *C. idella*.

Supporting Information

Table S1   Summary of unigene annotation against Nr, Nt, Pfam, KOG, Swiss-Prot, KO and GO database.

Table S2   721 genes differed significantly between the SG and RG groups of *C. idella*.

Table S3   GO enrichment analysis of 721 genes differed significantly.

Table S4   Pathway enrichment analysis of 721 genes differed significantly.

Table S5   188 genes as the predicted target genes of 9 different expression miRNA between SG and RG groups of *C. idella*.

Table S6   GO enrichment analysis of 188 genes as the predicted target genes of different expression miRNA.

Table S7   28 target gene information of *miR-21*.

Acknowledgments

We thank Zhiwei Wang and Liang Zhang (Novogene Bioinformatics Technology Co. Ltd) for their help in sequencing and data analysis.

Author Contributions

Conceived and designed the experiments: JL LL. Performed the experiments: XX YS. Analyzed the data: XX. Contributed reagents/materials/analysis tools: XX YS JF. Wrote the paper: XX JL.

References

1. Huang da W, Sherman BT, Lempicki RA (2009) Bioinformatics enrichment tools: paths toward the comprehensive functional analysis of large gene lists. Nucleic Acids Res 37: 1–13.

2. Wang Y, Pan Y, Liu Z, Zhu X, Zhai L, et al. (2013) De novo transcriptome sequencing of radish (Raphanus sativus L.) and analysis of major genes involved in glucosinolate metabolism. BMC Genomics 14: 836.

3. Graveley BR (2008) Molecular biology: power sequencing. Nature 453: 1197–1198.

4. Ozsolak F, Kapranov P, Foissac S, Kim SW, Fishilevich E, et al. (2010) Comprehensive polyadenylation site maps in yeast and human reveal pervasive alternative polyadenylation. Cell 143: 1018–1029.

5. Bhardwaj J, Chauhan R, Swarnkar MK, Chahota RK, Singh AK, et al. (2013) Comprehensive transcriptomic study on horse gram (Macrotyloma uniflorum): De novo assembly, functional characterization and comparative analysis in relation to drought stress. BMC Genomics 14: 647.

6. Renkui C (1991) Development History of Freshwater Culture in China. China Press of Science & Technology, Beijing, China.

7. FAO (2014) FAO Yearbook of Fishery and Aquaculture Statistics Summary tables. Rome: FAO.

8. Huang Q, Tang S, Zhang J (1983) Ichthyopathology. Shanghai, China: Shanghai Press of Science & Technology.

9. Igbinosa I, Igumbor E, Aghdasi F, Tom M, Okoh A (2012) Emerging Aeromonas species infections and their significance in public health. The Scientific World Journal 2012: 625023.
10. Figueras MJ (2005) Clinical relevance of Aeromonas sM503. Reviews in Medical Microbiology 16: 145–153.
11. Rahman M, Colque-Navarro P, Kühn I, Huys G, Swings J, et al. (2002) Identification and characterization of pathogenic Aeromonas veronii biovar sobria associated with epizootic ulcerative syndrome in fish in Bangladesh. Applied and environmental microbiology 68: 650–655.
12. Xu X-Y, Shen Y-B, Fu J-J, Liu F, Guo S-Z, et al. (2012) Matrix metalloproteinase 2 of grass carp Ctenopharyngodon idella (CiMMP2) is involved in the immune response against bacterial infection. Fish & shellfish immunology 33: 251–257.
13. Bushati N, Cohen SM (2007) MicroRNA functions. Annual Review of Cell and Developmental Biology. pp.175–205.
14. Zhang P, Li C, Zhu L, Su X, Li Y, et al. (2013) De novo assembly of the sea cucumber Apostichopus japonicus hemocytes transcriptome to identify miRNA targets associated with skin ulceration syndrome. PLoS One 8: e73506.
15. Bartel DP (2004) MicroRNAs: genomics, biogenesis, mechanism, and function. Cell 116: 281–297.
16. Zhu YP, Xue W, Wang JT, Wan YM, Wang SL, et al. (2012) Identification of common carp (Cyprinus carpio) microRNAs and microRNA-related SNPs. BMC Genomics 13: 413.
17. Ji P, Liu G, Xu J, Wang X, Li J, et al. (2012) Characterization of common carp transcriptome: sequencing, de novo assembly, annotation and comparative genomics. PloS one 7: e35152.
18. Yan B, Guo J-T, Zhao L-H, Zhao J-L (2012) microRNA expression signature in skeletal muscle of Nile tilapia. Aquaculture 364–365: 240–246.
19. Tao W, Yuan J, Zhou L, Sun L, Sun Y, et al. (2013) Characterization of gonadal transcriptomes from Nile tilapia (Oreochromis niloticus) reveals differentially expressed genes. PLoS One 8: e63604.
20. Mennigen JA, Panserat S, Larquier M, Plagnes-Juan E, Medale F, et al. (2012) Postprandial regulation of hepatic microRNAs predicted to target the insulin pathway in rainbow trout. PLoS One 7: e38604.
21. Salem M, Rexroad CE 3rd, Wang J, Thorgaard GH, Yao J (2010) Characterization of the rainbow trout transcriptome using Sanger and 454-pyrosequencing approaches. BMC Genomics 11: 564.
22. Mu X, Pridgeon JW, Klesius PH (2011) Transcriptional profiles of multiple genes in the anterior kidney of channel catfish vaccinated with an attenuated Aeromonas hydrophila. Fish Shellfish Immunol 31: 1162–1172.
23. Barozai MY (2012) The MicroRNAs and their targets in the channel catfish (Ictalurus punctatus). Mol Biol Rep 39: 8867–8872.
24. Fu B, He S (2012) Transcriptome analysis of silver carp (Hypophthalmichthys molitrix) by paired-end RNA sequencing. DNA Res 19: 131–142.
25. Chi W, Tong C, Gan X, He S (2011) Characterization and comparative profiling of MiRNA transcriptomes in bighead carp and silver carp. PLoS One 6: e23549.
26. Chen J, Li C, Huang R, Du F, Liao L, et al. (2012) Transcriptome analysis of head kidney in grass carp and discovery of immune-related genes. BMC Vet Res 8: 108.
27. Liu F, Wang D, Fu J, Sun G, Shen Y, et al. (2010) Identification of immune-relevant genes by expressed sequence tag analysis of head kidney from grass carp (Ctenopharyngodon idella). Comp Biochem Physiol Part D Genomics Proteomics 5: 116–123.
28. Heng JF, Su JG, Huang T, Dong J, Chen LJ (2011) The polymorphism and haplotype of TLR3 gene in grass carp (Ctenopharyngodon idella) and their associations with susceptibility/resistance to grass carp reovirus. Fish & Shellfish Immunology 30: 45–50.
29. Grabherr MG, Haas BJ, Yassour M, Levin JZ, Thompson DA, et al. (2011) Full-length transcriptome assembly from RNA-Seq data without a reference genome. Nat Biotechnol 29: 644–652.
30. Pruitt KD, Tatusova T, Maglott DR (2007) NCBI reference sequences (RefSeq): a curated non-redundant sequence database of genomes, transcripts and proteins. Nucleic Acids Res 35: D61–65.
31. Koonin EV, Fedorova ND, Jackson JD, Jacobs AR, Krylov DM, et al. (2004) A comprehensive evolutionary classification of proteins encoded in complete eukaryotic genomes. Genome Biol 5: R7.
32. Boeckmann B, Bairoch A, Apweiler R, Blatter MC, Estreicher A, et al. (2003) The SWISS-PROT protein knowledgebase and its supplement TrEMBL in 2003. Nucleic Acids Res 31: 365–370.
33. Kanehisa M (2000) KEGG: Kyoto Encyclopedia of Genes and Genomes. Nucleic Acids Research 28: 27–30.
34. Bateman A, Birney E, Cerruti L, Durbin R, Etwiller L, et al. (2002) The Pfam protein families database. Nucleic Acids Res 30: 276–280.
35. Götz S, García-Gómez JM, Terol J, Williams TD, Nagaraj SH, et al. (2008) High-throughput functional annotation and data mining with the Blast2GO suite. Nucleic acids research 36: 3420–3435.
36. Li B, Dewey CN (2011) RSEM: accurate transcript quantification from RNA-Seq data with or without a reference genome. BMC Bioinformatics 12: 323.
37. Mortazavi A, Williams BA, McCue K, Schaeffer L, Wold B (2008) Mapping and quantifying mammalian transcriptomes by RNA-Seq. Nat Methods 5: 621–628.
38. Wang L, Feng Z, Wang X, Wang X, Zhang X (2010) DEGseq: an R package for identifying differentially expressed genes from RNA-seq data. Bioinformatics 26: 136–138.
39. Wen M, Shen Y, Shi S, Tang T (2012) miREvo: an integrative microRNA evolutionary analysis platform for next-generation sequencing experiments. BMC Bioinformatics 13: 140.
40. Friedlander MR, Mackowiak SD, Li N, Chen W, Rajewsky N (2012) miRDeep2 accurately identifies known and hundreds of novel microRNA genes in seven animal clades. Nucleic Acids Res 40: 37–52.
41. John B, Enright AJ, Aravin A, Tuschl T, Sander C, et al. (2004) Human MicroRNA targets. PLoS Biol 2: e363.
42. Enright AJ, John B, Gaul U, Tuschl T, Sander C, et al. (2003) MicroRNA targets in Drosophila. Genome Biol 5: R1.
43. Xu XY, Shen YB, Fu JJ, Lu LQ, Li JL (2014) Determination of reference microRNAs for relative quantification in grass carp (Ctenopharyngodon idella). Fish & Shellfish Immunology 36: 374–382.
44. Su J, Zhang R, Dong J, Yang C (2011) Evaluation of internal control genes for qRT-PCR normalization in tissues and cell culture for antiviral studies of grass carp (Ctenopharyngodon idella). Fish Shellfish Immunol 30: 830–835.
45. Grabherr MG, Haas BJ, Yassour M, Levin JZ, Thompson DA, et al. (2011) Full-length transcriptome assembly from RNA-Seq data without a reference genome. Nature Biotechnology 29: 644–U130.
46. Akira S, Takeda K (2004) Toll-like receptor signalling. Nature Reviews Immunology 4: 499–511.
47. Rot A, von Andrian UH (2004) Chemokines in innate and adaptive host defense: basic chemokinese grammar for immune cells. Annu Rev Immunol 22: 891–928.
48. Liu F, Li J, Fu J, Shen Y, Xu X (2011) Two novel homologs of simple C-type lectin in grass carp (Ctenopharyngodon idellus): potential role in immune response to bacteria. Fish Shellfish Immunol 31: 765–773.
49. Xu XY, Shen YB, Fu JJ, Liu F, Guo SZ, et al. (2013) Characterization of MMP-9 gene from grass carp (Ctenopharyngodon idella): An Aeromonas hydrophila-inducible factor in grass carp immune system. Fish Shellfish Immunol 35: 801–807.
50. Enright AJ, John B, Gaul U, Tuschl T, Sander C, et al. (2004) MicroRNA targets in Drosophila. Genome Biology 5: R1–R1.
51. Guo H, Ingolia NT, Weissman JS, Bartel DP (2010) Mammalian microRNAs predominantly act to decrease target mRNA levels. Nature 466: 835–840.
52. Janeway CA, Medzhitov R (2002) Innate immune recognition. Annual Review of Immunology 20: 197–216.
53. Plata-Salamán C, Ilyin S, Gayle D, Flynn MC (1998) Gram-negative and gram-positive bacterial products induce differential cytokine profiles in the brain: Analysis using an integrarive molecular-behavioral in vivo model. International journal of molecular medicine 1: 387–398.
54. Chen XM, Splinter PL, O'Hara SP, LaRusso NF (2007) A cellular micro-RNA, let-7i, regulates Toll-like receptor 4 expression and contributes to cholangiocyte immune responses against Cryptosporidium parvum infection. J Biol Chem 282: 28929–28938.
55. Liu PT, Wheelwright M, Teles R, Komisopoulou E, Edfeldt K, et al. (2012) MicroRNA-21 targets the vitamin D-dependent antimicrobial pathway in leprosy. Nat Med 18: 267–273.
56. Oltmanns U, Issa R, Sukkar MB, John M, Chung KF (2003) Role of c-jun N-terminal kinase in the induced release of GM-CSF, RANTES and IL-8 from human airway smooth muscle cells. Br J Pharmacol 139: 1228–1234.
57. Boutros M, Agaisse H, Perrimon N (2002) Sequential Activation of Signaling Pathways during Innate Immune Responses in Drosophila. Developmental Cell 3: 711–722.
58. Lee J, Mira-Arbibe L, Ulevitch RJ (2000) TAK1 regulates multiple protein kinase cascades activated by bacterial lipopolysaccharide. J Leukoc Biol 68: 909–915.
59. Zhang Y, Dong C (2005) MAP kinases in immune responses. Cell Mol Immunol 2: 20–27.
60. Förster R, Schubel A, Breitfeld D, Kremmer E, Renner-Müller I, et al. (1999) CCR7 Coordinates the Primary Immune Response by Establishing Functional Microenvironments in Secondary Lymphoid Organs. Cell 99: 23–33.
61. Kursar M, Hopken UE, Koch M, Kohler A, Lipp M, et al. (2005) Differential requirements for the chemokine receptor CCR7 in T cell activation during Listeria monocytogenes infection. J Exp Med 201: 1447–1457.
62. Nau GJ, Richmond JF, Schlesinger A, Jennings EG, Lander ES, et al. (2002) Human macrophage activation programs induced by bacterial pathogens. Proc Natl Acad Sci U S A 99: 1503–1508.
63. Mendes ND, Freitas AT, Sagot MF (2009) Current tools for the identification of miRNA genes and their targets. Nucleic Acids Res 37: 2419–2433.
64. Krol J, Loedige I, Filipowicz W (2010) The widespread regulation of microRNA biogenesis, function and decay. Nature Reviews Genetics 11: 597–610.
65. Bartel DP (2009) MicroRNAs: target recognition and regulatory functions. Cell 136: 215–233.
66. Lewis BP, Shih IH, Jones-Rhoades MW, Bartel DP, Burge CB (2003) Prediction of mammalian microRNA targets. Cell 115: 787–798.
67. Rehmsmeier M, Steffen P, Hochsmann M, Giegerich R (2004) Fast and effective prediction of microRNA/target duplexes. RNA 10: 1507–1517.
68. Krek A, Grun D, Poy MN, Wolf R, Rosenberg L, et al. (2005) Combinatorial microRNA target predictions. Nat Genet 37: 495–500.
69. Stark A, Brennecke J, Russell RB, Cohen SM (2003) Identification of Drosophila microRNA targets. PLoS biology 1: e60.
70. Nazarov PV, Reinsbach SE, Muller A, Nicot N, Philippidou D, et al. (2013) Interplay of microRNAs, transcription factors and target genes: linking dynamic expression changes to function. Nucleic Acids Res 41: 2817–2831.
71. Muniategui A, Pey J, Planes FJ, Rubio A (2013) Joint analysis of miRNA and mRNA expression data. Brief Bioinform 14: 263–278.

Exome Sequencing Identifies Three Novel Candidate Genes Implicated in Intellectual Disability

Zehra Agha[1,2,3¶], **Zafar Iqbal**[2¶], **Maleeha Azam**[1], **Humaira Ayub**[1], **Lisenka E. L. M. Vissers**[2], **Christian Gilissen**[2], **Syeda Hafiza Benish Ali**[1], **Moeen Riaz**[1], **Joris A. Veltman**[2], **Rolph Pfundt**[2], **Hans van Bokhoven**[2,4]*, **Raheel Qamar**[1,5]*

1 Department of Biosciences, Faculty of Science, COMSATS Institute of Information Technology, Islamabad, Pakistan, **2** Department of Human Genetics, Nijmegen Centre for Molecular Life Sciences, Radboud University Medical Centre, Nijmegen, the Netherlands, **3** Department of Bioinformatics and Biotechnology, International Islamic University, Islamabad, Pakistan, **4** Department of Cognitive Neurosciences, Donders Institute for Brain, Cognition and Behaviour, Nijmegen, The Netherlands, **5** Department of Biochemistry, Al-Nafees Medical College & Hospital, Isra University, Islamabad, Pakistan

Abstract

Intellectual disability (ID) is a major health problem mostly with an unknown etiology. Recently exome sequencing of individuals with ID identified novel genes implicated in the disease. Therefore the purpose of the present study was to identify the genetic cause of ID in one syndromic and two non-syndromic Pakistani families. Whole exome of three ID probands was sequenced. Missense variations in two plausible novel genes implicated in autosomal recessive ID were identified: lysine (K)-specific methyltransferase 2B (*KMT2B*), zinc finger protein 589 (*ZNF589*), as well as hedgehog acyltransferase (*HHAT*) with a *de novo* mutation with autosomal dominant mode of inheritance. The *KMT2B* recessive variant is the first report of recessive Kleefstra syndrome-like phenotype. Identification of plausible causative mutations for two recessive and a dominant type of ID, in genes not previously implicated in disease, underscores the large genetic heterogeneity of ID. These results also support the viewpoint that large number of ID genes converge on limited number of common networks i.e. ZNF589 belongs to KRAB-domain zinc-finger proteins previously implicated in ID, HHAT is predicted to affect sonic hedgehog, which is involved in several disorders with ID, *KMT2B* associated with syndromic ID fits the epigenetic module underlying the Kleefstra syndromic spectrum. The association of these novel genes in three different Pakistani ID families highlights the importance of screening these genes in more families with similar phenotypes from different populations to confirm the involvement of these genes in pathogenesis of ID.

Editor: Obul Reddy Bandapalli, University of Heidelberg, Germany

Funding: This work was supported by the COMSATS Institute of Information Technology to RQ. This study was also supported by the European Union's Seventh Framework Program [grant agreement number 241995 project GENCODYS and grant agreement number 223143, project TECHGENE]. None of the funding agencies had any role in study design, data collection and analysis, decision to publish or preparation of the manuscript.

Competing Interests: The authors have declared no competing interests exist.

* Email: raheelqamar@hotmail.com (RQ); h.vanbokhoven@radboudumc.nl (HvB)

❧ These authors contributed equally to this work.

¶These authors are co-first authors on this work.

Introduction

Intellectual disability (ID), a neurocognitive disorder, is characterized by substantial limitations both in intellectual functioning and in adaptive behavior. In patients diagnosed before the age of 18 years, it has a prevalence of 2–3% in the general population [1]. Stein et al. [2] reported the prevalence of ID among 3–9 years old children in different populous countries including Pakistan, Brazil, India, Bangladesh and Philippines, which varied from 9/1000 to 156/1000. ID is an unsolved healthcare problem, which creates an enormous socioeconomic burden on the society, especially in the underdeveloped countries where there is a high rate of consanguinity, resulting in further aggravating the genetically inherited disease prevalence [3]. Chromosomal abnormalities and single gene disruptions contribute significantly to all forms of ID, including severe, moderate and mild phenotype [4].

Investigations aiming to unravel the genetic defects initially focused mainly on X-linked ID since male ID patients are overrepresented as compared to females with a ratio of 1:1.3 to 1:1.9 [5]. However, these investigations have revealed that only 10% of ID cases are due to X chromosomal defects while the remaining cases are expected to be caused by genetic defects in the autosomes and equally due to adverse environmental effects such as poor mother health, social deprivation, infections and injuries during prenatal life and hypoxia [6]. Chromosomal aberrations and mutations in more than 450 genes can explain the disorder in about half of all ID patients [5]. The large number of ID genes present a challenge for the identification of the genetic defect in individual families and isolated cases, however, only a limited number of pathways are emerging whose disruption appears to be shared by groups of ID genes [7].

Despite being clinically heterogeneous, syndromic ID (sID) as well as non-syndromic ID (nsID) share common neurological features, such as autism, epilepsy, ADHD (attention deficit hyperactivity disorder) and behavioral anomalies [7]. In addition, syndromic forms of ID are characterized by a pattern of congenital anomalies that can be seen in addition to ID and other neurological features. The latter can help to establish clinical diagnosis, which can subsequently be validated by direct molecular diagnostic testing of targeted gene(s). However, the number of syndromes where similar phenotypes can be caused by mutations in a variety of different genes is increasing. Examples of these include Bardet-Biedl syndrome (MIM 209900; 17 genes), Sotos syndrome (MIM 117550; 2 genes) and Kleefstra syndrome (MIM 610253; 5 genes) [7–10]. Typically, the underlying genes for each of these syndromes have a functional relationship to each other and mutations lead to disruption of the same molecular pathways. As in the case of Kleefstra syndrome (KS), which is characterized by severe to moderate ID, speech impairment, congenital hypotonia, specific distinguishing facial features and complex pattern of other anomalies can be caused by *de novo* mutations affecting epigenetic regulators such as euchromatic histone-lysine N-methyltransferase 1 (*EHMT1*), lysine (K)-specific methyltransferase 2C (*KMT2C*), SWI/SNF related, matrix associated, actin dependent regulator of chromatin, subfamily b, member 1 (*SMARCB1*), nuclear receptor subfamily 1, group I, member 3 (*NR1I3*) and methyl-CpG binding domain protein 5 (*MBD5*) [11–13].

Till now, small families with ID have not been studied extensively due to technical difficulties including non-suitability of homozygosity mapping and use of simple traditional linkage analysis with insufficient power to resolve the genetic cause in such families. The genetic analyses combined with whole exome sequencing have recently enabled the systematic identification of pathogenic mutations in small families with recessive nsID and sID, including mutations in 13 novel nsID genes [13]. Najamabadi et al. [4] have recently proposed 29 candidate genes for autosomal recessive nsID. In the current study missense homozygous mutations in two novel genes including lysine (K)-specific methyltransferase 2B (*KMT2B*), a zinc finger gene *ZNF589* and a heterozygous *de novo* mutation in hedgehog acetyltransferase (*HHAT*) were identified.

Methods

Ethics statement

This study was approved by the Department of Biosciences Ethics Review Board of the COMSATS Institute of Information Technology, Islamabad, Pakistan, and the local Ethics Committee of the Radboud University Medical Centre, Nijmegen, The Netherlands. All family members and 200 ethnically matched control individuals were informed about the purpose of the study and written consent in their local language was taken before recruitment and sampling. The parents or guardians of the individuals in this manuscript have given written consent (as outlined in PLOS consent form) to publish the patients case details.

Clinical features

A consanguineous family MRQ14 (Figure 1A) with three children with severe sID was sampled from central Punjab, Pakistan, including three affected sons (IV:1, IV:2, IV:3) unaffected daughter and son (IV:4 and IV:5), the unaffected mother (III:1), unaffected father (III:2), unaffected paternal grandfather (II:4) and unaffected paternal grandmother (II:5). The three affected

brothers had a similar sID phenotype and were each born after about 39 weeks of uneventful pregnancies with normal labor. At the age of 16 years, the proband (IV:1) had short stature, dysmorphic facial features including a large head, flattened nasal bridge, apparent hypertelorism synophrys, midface hypoplasia, thick eyebrows, everted lower lip, dental anomalies and prognathism. Upon physical examination, he was found to suffer from musculoskeletal anomalies as well as cryptorchidism and micropenis. Being floppy during childhood he was found to be hypotonic on reaching teenage (Figure 2). In addition to speech and motor delay, he was underweight at the time of assessment (23kg, 2nd centile) and was short for his age (81cm, 2nd centile). All the three affected boys lacked language development, had profound ID (IQ<20, IQ was tested using the Wechsler Intelligence Scale for Children (WISC-III)), and they were not able to perform essential activities of daily life. Computed Tomography (CT) scan did not reveal any anomaly of the brain and the biochemical tests including liver transaminases, serum lactate as well as serum electrolyte and complete blood count were all in the normal ranges (Table 1).

The consanguineous family MRQ11 with two affected children (Figure 1B) was sampled from Northern Pakistan. The sampled members were: unaffected mother (III:2), unaffected daughters (IV:1, IV:4, IV:5), affected son (IV:2), and affected daughter (IV:3). The two affected siblings had a similar nsID phenotype. The affected members, a son (IV:2), and a daughter (IV:3) were born after an uneventful 39 weeks pregnancy. Labor was normal and they did not undergo postnatal hypoxia. Both of them had delayed milestones including speech and motor development. At the time of examination, at the ages of 14 years (IV:2) and 10 years (IV:3), they were both below average weight and height: the affected son was 23 kg and 114 cm (both 2nd centile), whereas the affected daughter weighed 22 kg and was 88 cm in height (both being 2nd centile). Both siblings had strabismus and had moderate ID (IQ: 36–51). CT scan did not show any anomaly of the brain, in addition, metabolic and biochemical testing revealed no abnormalities.

Family MRQ15 was a non-consanguineous family (Figure 1C), from Punjab, Pakistan. Five members of the family were sampled that included an unaffected father (I:1), unaffected mother (I:2), affected daughter (II:2), unaffected son (II:4) and affected son (II:7). The two affected siblings had profound ID (IQ<20) and could speak only a few meaningful words and could recognize only their parents and siblings. They were both not able to perform daily activities of life independently. No other syndromic features were present and their CT scan did not show any brain malformation.

Homozygosity mapping and CNV analysis

The pedigrees are concordant with the recessive inheritance in all the three families, therefore in order to obtain copy number variation (CNV) as well as homozygosity mapping data, Affymetrix 250K NSPI SNP (Affymetrix, Santa Clara, CA) array analysis was performed for MRQ14 family members IV:1, IV:2, IV:3 and IV:5, and all sampled affected and unaffected members of MRQ11 and MRQ15 (Figure 1). For CNV determination, the data were scrutinized using Copy Number Analyzer for GeneChip [14]. Affymetrix Genotyping Console (version 2.0) was used to obtain the genotype data and online software Homozygosity Mapper [15] was used to obtain the homozygous regions. The homozygous intervals of at least 1 Mb were also verified visually. All data were mapped using the Human Genome Build hg19.

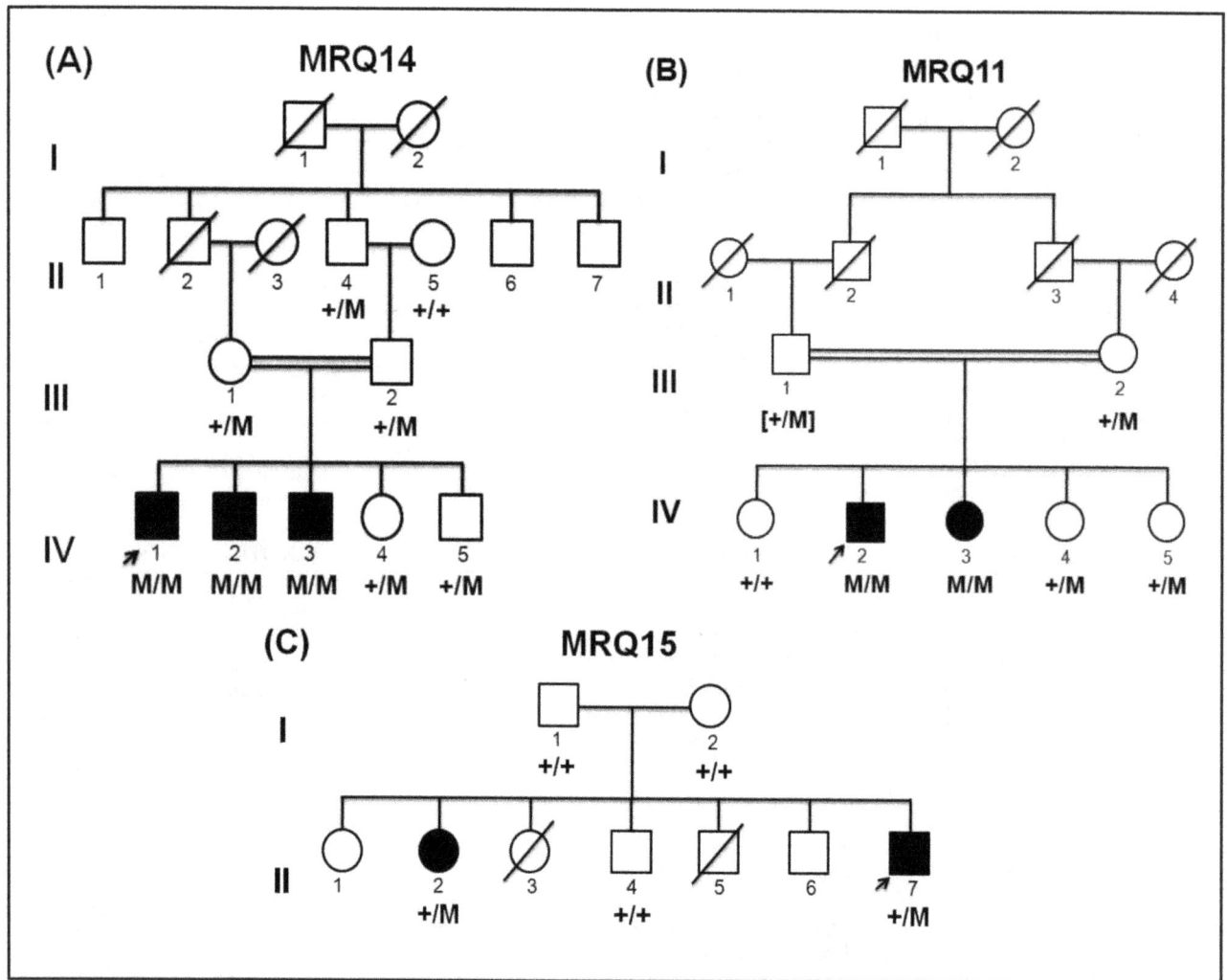

Figure 1. Pedigree of families (A) MRQ14, (B) MRQ11 and (C) MRQ15. The segregation of mutation of *KMT2B*, *ZNF589* and *HHAT* are also in the pedigree. The symbol +/+ represents homozygous ancestral alleles, M/M is for homozygous variant alleles and +/M is for heterozygous carriers. In the panel B, the genotype of the father (III:1) has been deduced.

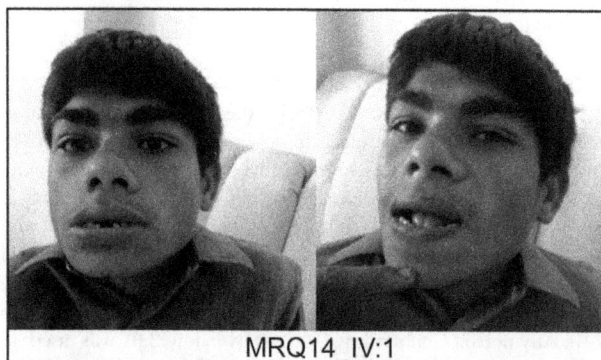

Figure 2. Photographs of MRQ14 proband. The photographs demonstrate the classic facial features representative of Kleefstra syndrome.

Exome sequencing

Homozygosity mapping and CNV analysis did not reveal the genetic cause of the disease in any of the family therefore exome sequencing was performed for the proband of each family. The exomes of the probands were enriched and sequenced as described previously by Vissers et al. [16]. In brief, using an Agilent SureSelect Human All Exon Kit (50 Mb, ~21,000 genes; Agilent Technologies, Santa Clara, CA) exome libraries were prepared as described by the manufacturer and pooled for bead amplification using the emulsion-based clonal PCR (emPCR) of EZbead system (Life Technologies, Santa Clara, CA) and were subsequently sequenced using the SOLiD 4 system (Life Technologies, Santa Clara, CA). As described by Vissers et al. [16] the variants and indels were only selected for further analysis when the overall variant reads were at least 15% of the total number of reads, with a minimum of 5 reads [17]. For each proband, 36,001 to 44,200 variants were annotated. All nongenic, non-splice site, intronic and synonymous variants were excluded from further analysis. Furthermore, the low frequency variants (less than 0.5%) in known ID genes present in dbSNP were also checked, such variants were further filtered on the basis of pathogenicity scores

Table 1. Clinical features of Kleefstra syndrome shared by the three affected brothers of family MRQ14.

Clinical features	Patient-IV:1 (Proband)	Patient-IV:2	Patient-IV:3
High birth weight	no	no	no
Microcephaly	no	no	no
Synophrys	yes	yes	yes
Unusual shape of the eyebrows	yes	yes	yes
Midface hypoplasia	yes	yes	yes
Full everted lower lip	yes	yes	yes
Cupid bowed upper lip	yes	yes	yes
Protruding tongue	yes	yes	yes
Prognathism	yes	yes	yes
Short stature	yes	yes	yes
Overweight (BMI>25)	no	no	no
DD/ID	yes (severe)	yes (severe)	yes (severe)
Heart defect	no	no	no
Genital anomaly	yes	yes	yes
Renal anomaly (including VUR)	no	no	no
Recurrent infections	yes	yes	yes
Hearing deficit	no	no	no
Gastro-esophageal reflux	no	no	no
Epilepsy	yes	yes	yes
Behavioral/psychiatric problems	yes	yes	yes
Anomalies on brain imaging	no	not performed	not performed
Tracheomalacia	no	no	no
Umbilical/inguinal hernia	no	no	no
Anal atresia	no	no	no
Musculoskeletal anomaly	yes	yes	yes
Respiratory complications	no	no	no
Hypertelorism	yes	yes	yes

BMI, body mass index; DD, developmental disability; ID, intellectual disability, VUR, vesico-ureteric reflux

and tested for segregation among the respective family members while all the other variants were excluded (data not shown), which were found in dbSNP132 or those in an in-house database. To predict the pathogenicity of the variants, data were evaluated using *in silico* analysis, including Poly Phen-2 (genetics.bwh.harvard.edu/pph2) and SIFT (sift.jcvi.org/).

Sanger sequencing

Sanger sequencing for confirmation of the candidate gene variants and their segregation in the families was performed (Tables 2, 3 and 4) by designing PCR primers (Table S1, S2 and S3) using the Primer3 program (http://frodo.wi.mit.edu/) and amplifying the regions of interest. PCR amplification was conducted using 0.25 mM dNTPs, 1X PCR buffer (100 mM Tris-HCl, pH 8.3, 500 mM KCl), 2.5 mM Mg^{+2}, 0.5 µM of each primer, 2.5 U *Taq* polymerase (Fermentas Life Sciences, Ontario, Canada) and 50 ng gDNA. The thermal profile consisted of initial denaturation at 95°C for 5 min followed by 30 cycles of amplification at 95°C for 1 min, 57°C for 30 sec and 72°C for 45 sec, a final extension was carried out at 72°C for 5 min. Purified PCR amplicons were then sequenced using the ABI PRISM Big Dye Terminator Cycle Sequencing V3.1 ready reaction kit and the ABI PRISM 3730 DNA analyzer (Applera Corp, Foster City, CA).

Results

After filtering the exome data of the proband (IV:1) of family MRQ14 (Figure 1A), assuming a recessive inheritance model, 53 homozygous and 11 compound heterozygous variants in different genes were obtained. The data were further prioritized to get the most relevant changes using a phyloP score>2.0. This resulted in 11 homozygous and 6 compound heterozygous variants in 14 different genes (Table 2). A variant in *KMT2B* (chr19.hg19:g.36,208,921_36,229,779), c.2456C>T, p.(Pro819-Leu), was the only variant in the shared homozygous region that segregated with the disease in the family after Sanger sequencing (Figure 1A and 3A). In addition, the variant was absent in 200 age and ethnicity-matched control samples (n = 400 alleles) as well as in families MRQ11 and MRQ15. The *KMT2B* c.2456C>T mutation segregated with the disease in the family with the change inherited by descent from the heterozygous paternal grandfather (III:4), the maternal grandparents were not available for screening of the change, but it is likely that the other mutant allele was inherited from the maternal grandfather as the parents are first cousins.

Exome sequencing of the proband of family MRQ11 resulted in the identification of 38 homozygous and 30 compound heterozygous changes in 38 and 15 genes. Homozygosity mapping revealed

Table 2. Family MRQ14 homozygous and compound heterozygous variant validation using Sanger sequencing and *in silico* pathogenecity predictions.

Gene (NM ID)	Protein function	cDNA change	Amino acid change	PhyloP score	Grantham distance	SIFT	Mutation taster	Polyphen2	Zygosity	Segregation in family
NME7 (NM_013330.4)	Unknown	c.38C>T	p.(Arg13Gln)	5.53	43	Deleterious	Disease causing	Probably damaging	Homozygous	No
PPP1R9A (NM_001166160.1)	Nervous system development	c.1387C>T	p.(Pro463Ser)	5.29	74	Deleterious	Disease causing	Probably damaging	Homozygous	No
DYM (NM_017653.3)	Dyggve-melchior-clausen disease, 223800 (3); smith-mc-Cort dysplasia,	c.1205A>T	p.(Leu402*)	4.571	1000	Unknown	Unknown	Unknown	Homozygous	No
DLG1 (NM_004087.2)	Cell to cell adhesion, nervous system involvement	c.574T>C	p.(Ile192Val)	4.518	29	Deleterious	Disease causing	Probably damaging	Homozygous	No
KMT2B (MLL4) (NM_014727.1)	Unknown	c.2456C>T	p.(Pro819Leu)	4.429	98	Tolerated	Disease causing/ polymorphism	Probably damaging	Homozygous	Yes
EHMT2 (NM_006709.3)	Chromatin modification, biological process and nervous system involvement	c.1151C>T	p.(Arg384Gln)	3.503	43	Deleterious	Disease causing	Probably damaging	Homozygous	No
VAV2 (NM_0011343398.1)	Nervous system phenotype	c.2495A>G	p.(Met832Thr)	3.138	81	Deleterious	Disease causing	Probably damaging	Homozygous	No
PLCD4 (NM_032726.2)	Intercellular signaling cascade	c.1885C>T	p.(Leu629Phe)	2.9	22	Deleterious	Disease causing	May be damaging	Homozygous	No
ZNF227 (NM_182490.2)	Regulation of transcription, DNA-dependent	c.956C>T	p.(Thr319Ile)	2.715	89	Deleterious	Polymorphism	May be damaging	Homozygous	No
SMEK1 (NM_032560.4)	Unknown	c.2239A>C	p.(Ser747Ala)	3.03	98	Tolerated	Disease causing	Not damaging	Homozygous	No
UGT8 (NM_001128174.1)	CNS development	c.359A>G	p.(Asn120Ser)	2.47	46	Tolerated	Disease causing	Not damaging	Homozygous	No
DNAH17 (NM_173628.3)	Microtubule-based movement, ciliary or flagellar motility	c.12599A>C	p.(Val4200Gly)	4.821	109	Deleterious	Unknown	Probably damaging	Heterozygous	No
DNAH17 (NM_173628.3)	Microtubule-based movement, ciliary or flagellar motility	c.12267C>T	p.(Met4089Ile)	5.974	10	Deleterious	Unknown	Probably damaging	Heterozygous	No
SACS (NM_014363.5)	Protein folding	c.10291C>G	p.(Val3431Leu)	4.266	32	Deleterious	Disease causing	Probably damaging	Heterozygous	No
SACS (NM_014636.2)	Protein folding	c.5461A>G	p.(Cys1821Arg)	2.631	180	Deleterious	Disease causing	Probably damaging	Heterozygous	No
TEP1 (NM_007110.4)	Telomere maintenance via recombination	c.3519>C	p.(Lys1174Glnfs*16)	2.109	1000	Unknown	Unknown	Unknown	Heterozygous	No
TEP1 (NM_007110.4)	Telomere maintenance via recombination	c.1817G>A	p.(Pro606Leu)	3.26	98	Tolerated	Polymorphism	Not damaging	Heterozygous	No

NM, mRNA accession number; PhyloP, Phylogenetic P-values; Polyphen, Polymorphism phenotyping; SIFT, Sorting intolerance from tolerance, Mutation taster (http://www.mutationtaster.org).

Table 3. Family MRQ11 homozygous and compound heterozygous variant validation using Sanger sequencing and *in silico* prediction.

Gene (NM ID)	Protein function	cDNA change	Amino acid change	PhyloP	Grantham distance	SIFT	Mutation taster	Polyphen2	Zygosity	Segregation in family
SMOX (NM_175839.2)	Oxidation reduction	c.1604C>A	p.(Ser535Tyr)	5.409	144	Deleterious	Disease causing	Probably damaging	Homozygous	No
TAS1R2 (NM_152232.2)	Signal transduction	c.971C>T	p.(Gly324Asp)	5.23	94	Deleterious	Disease causing	Probably damaging	Homozygous	No
ATP11A (NM_015205.2)	ATP biosynthetic process	c.64G>T	p.(Asp22Tyr)	5.14	160	Deleterious	Disease causing	Probably damaging	Homozygous	No
ADORA2B (NM_000676.2)	Positive regulation of cAMP biosynthetic process	c.590T>C	p.(Ile197Thr)	4.87	89	Deleterious	Disease causing	Probably damaging	Homozygous	No
ZNF589 (NM_016089.2)	Regulation of transcription, DNA-dependent	c.956T>A	p.(Leu319His)	2.965	100	Deleterious	Polymorphism	Probably damaging	Homozygous	Yes
ZNF502 (NM_033210.4)	Unknown	c.746G>A	p.(Arg249His)	1.425	29	Tolerated	Not disease causing	Probably not damaging	Homozygous	No
ADHFE1 (NM_144650.2)	Oxidation reduction	c.955A>G	p.(Ile319Val)	3.51	29	Tolerated	Disease causing	Probably not damaging	Heterozygous	No
ADHFE1 (NM_144650.2)	Oxidation reduction	c.1172C>T	p.(Thr391Ile)	1.26	89	Tolerated	Disease causing	Probably damaging	Heterozygous	No
CMYA5 (NM_153610.3)	Unknown	c.7472T>C	p.(Ile2491Thr)	2	89	Deleterious	Disease causing	Probably damaging	Heterozygous	No
CMYA5 (NM_153610.3)	Unknown	c.8534G>A	p.(Arg2845Lys)	0.51	26	Tolerated	Not disease causing	Probably not damaging	Heterozygous	No
DCHS1 (NM_003737.2)	Calcium-dependent cell-cell adhesion	c.3017C>T	p.(Arg1006His)	2.55	29	Deleterious	Disease causing	Probably damaging	Heterozygous	No
DCHS1 (NM_003737.2)	Calcium-dependent cell-cell adhesion	c.1265T>A	p.(Tyr422Phe)	4.91	22	Deleterious	Disease causing	Probably damaging	Heterozygous	No
DPAGT1 (NM_001382.3)	UDP-N-acetylglucosamine metabolic process	c.951G>C	p.(Ser317Arg)	4.39	110	Deleterious	Disease causing	Probably damaging	Heterozygous	No
DPAGT1 (NM_001382.3)	UDP-N-acetylglucosamine metabolic process	c.38A>T	p.Ile13Asn)	4.86	149	Deleterious	Disease causing	Probably damaging	Heterozygous	No
DENND2C (NM_001256404.1)	Unknown	c.1129T>C	p.(Lys377Glu)	3.608	56	Unknown	Unknown	Unknown	Heterozygous	No

NM, mRNA accession number; PhyloP, Phylogenetic P-values; Polyphen, Polymorphism phenotyping; SIFT, Sorting intolerance from tolerance, Mutation transfer (http://www.mutationtaster.org)

Table 4. Family MRQ15 homozygous and compound heterozygous variants validation using Sanger sequencing and *in silico* prediction.

Gene (NM ID)	Protein function	cDNA change	Amino acid change	PhyloP	Grantham distance	SIFT	Mutation taster	Polyphen2	Zygosity	Segregation in family
SF3B3 (NM_012426.2)	RNA splicing	c.82C>A	p.(Gln28Lys)	6.119	53	Deleterious	Disease causing	Probably damaging	Homozygous	No
LARGE (NM_004737.4)	N-acetylglucosamine metabolic process	c.251C>G	p.(Ser84Thr)	5.72	58	Deleterious	Disease causing	Probably damaging	Homozygous	No
HHAT (NM_0011222834.2)	Multicellular organism development	c.1158G>C	p.(Trp386Cys)	5.433	215	Deleterious	Disease causing	Probably damaging	Heterozygous	Present in affected only
NEDD4 (NM_1984400.1)	Protein ubiquitination during ubiquitin-dependent protein catabolic process	c.872C>T	p.(Gly291Glu)	2.915	98	Deleterious	Polymorphism	Probably damaging	Homozygous	No
PLCH1 (NM_0011130960.1)	Cell division	c.13+1C>T	No	4.407	0	Unknown	Unknown	Unknown	Homozygous	No
PDSS8 (NM_015032.3)	Intercellular signaling cascade	c.25-1G>A	No	6.172	0	Unknown	Unknown	Unknown	Homozygous	No
WASF1 (NM_003931.2)	Protein complex assembly	c.4+1C>T	No	5.63	0	Unknown	Unknown	Unknown	Homozygous	No
DENND2A (NM_015689.3)	Unknown	c.54G>C	p.(Pro297Arg)	2.342	110	Deleterious	Disease causing	Probably damaging	Homozygous	No
HMCN1 (NM_031935.2)	Response to stimulus	c.7163G>A	p.(Gly2388Glu)	4.013	98	Deleterious	Disease causing	Probably damaging	Heterozygous	No
HMCN1 (NM_031935.2)	Response to stimulus	c.13190G>A	p.(Arg4397Gln)	0.754	43	Tolerated	Disease causing	Probably not damaging	Heterozygous	No
MED13L (NM_015335.4)	Regulation of transcription from RNA polymerase II promoter	c.1447G>T	p.(Pro483Thr)	3.577	38	Tolerated	Disease causing	Probably not damaging	Heterozygous	No
MED13L (NM_015335.4)	Regulation of transcription from RNA polymerase II promoter	c.740A>G	p.(Leu247Pro)	2.644	98	Tolerated	Disease causing	Probably not damaging	Heterozygous	No
ZNF772 (NM_0010024596.2)	Regulation of transcription, DNA-dependent	c.1145T>G	p.(Glu382Ala)	2.468	107	Tolerated	Disease causing	Probably not damaging	Heterozygous	No
ZNF772 (NM_015335.4)	Regulation of transcription, DNA-dependent	c.878G>T	p.(Pro293His)	1.828	77	Deleterious	Polymorphism	Probably damaging	Heterozygous	No

NM, mRNA accession number; PhyloP, Phylogenetic P-values; Polyphen, Polymorphism phenotyping; SIFT, Sorting intolerance from tolerance, Mutation taster (http://www.mutationtaster.org).

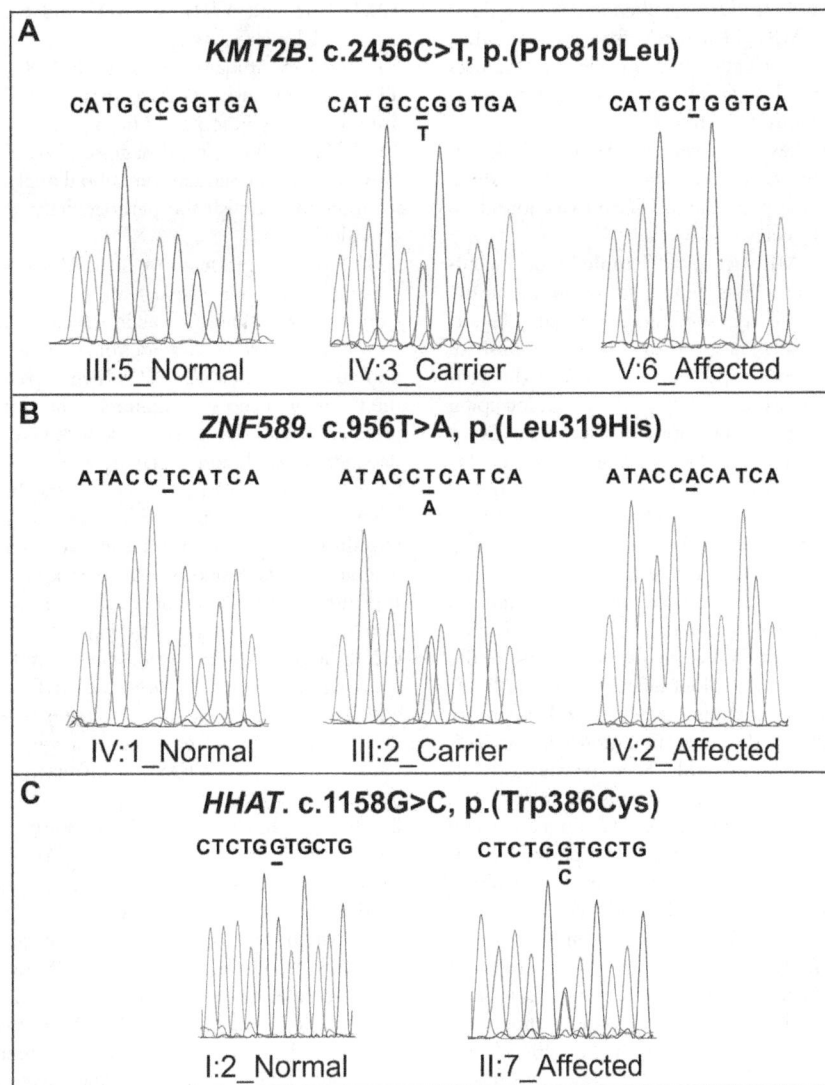

Figure 3. The sequencing chromatograms of the families MRQ14, MRQ11 and MRQ15. (A) Shows the panels containing the region with the identified *KMT2B* mutation in family MRQ14: ancestral (left panel), heterozygous (middle panel) and variant (right panel) (B) shows the region containing the identified *ZNF589* mutation in family MRQ11: ancestral (left panel), heterozygous (middle panel) and variant (right panel). (C) shows the *de novo* variant of *HHAT* in family MRQ15: ancestral (left panel), heterozygous (right panel).

only four common homozygous regions between the 2 affected members of the family (Table S4). The data were further prioritized (as described above), which revealed a total of 15 variants in 11 different genes (7 homozygous and 8 compound heterozygous; Table 3). These variants were further validated by Sanger sequencing for segregation in the respective family. Variant c.956T>A, p.(Leu319His) in *ZNF589* (chr3.hg19:g.48,282,596_48,329,115) segregated with the phenotype in the family (Figure 1B and 3B) and was absent in the control population as well as families MRQ14 and MRQ15.

Genotyping of family MRQ15 by microarray analysis did not reveal any homozygous region shared by the affected members, suggesting that compound heterozygosity as well as *de novo* dominant mutation may cause the disease in this family. Whole exome sequencing was carried out to identify the genetic cause of ID in this family using a trio-based approach [18,19]. After variant filtration as described above, 8 homozygous and 6 compound heterozygous variants were identified in 11 different genes, which

were further screened in the family by Sanger sequencing (Table 4). However, none of these selected variants segregated with the disease. In-line with a hypothesized dominant *de novo* mutation model, a heterozygous change c.1158G>C, p.(Trp386Cys) in the *HHAT* was present in both the affected members (Figure 1C and 3C) and absent in the parents and the unaffected sibling as well as healthy controls, and families MRQ14 and MRQ11. The occurrence of the mutation in both affected members is consistent with a germline mosaicism in one of the unaffected parents. Of note, non-paternity was excluded by segregation analysis of rare paternal variants. It was not possible to obtain other tissues of the parents therefore presence of variants could not be checked in those tissues.

Discussion

Genetic screening of three Pakistani ID families in the current study resulted in the identification of three novel plausible ID genes, *KMT2B*, *ZNF589* and *HHAT*. The phenotype of the three affected members of family MRQ14 had Kleefstra syndrome-like phenotype (clinical details are described in clinical features section), which is characterized by facial dysmorphism, hypotonia and mild to severe ID (Figure 2). This syndrome is rare with unknown prevalence and has not been reported in Pakistan previously. The homozygous variant c.2456C>T; p.(Pro819Leu) identified in the current study in the *KMT2B* was found to segregate in the family in a recessive pattern.

KMT2B belongs to the MLL (myeloid/lymphoid or mixed-lineage leukemia) family, and was found to be ubiquitously expressed in adult tissues as well as in solid tumor cell lines [19], its involvement in human cancer has already been established. Furthermore, *KMT2B* has been reported to express in different parts of human brain such as medial frontal cortex, occipital cortex, hippocampal cortex, amyloid complex and basal ganglia at different ages ranging from 0 to 48 months (Allen Institute for Brain Science. Allen Human Brain Atlas (http://human.brain-map.org/)). The protein encoded by *KMT2B/MLL4* has multiple domains such as a CXXC zinc finger, three PHD zinc fingers, a SET (suppressor of variegation, enhancer of zeste, and trithorax) and two FY domains. Of all the domains, the SET domain is highly conserved in KMT2B and is a hallmark of the KMT gene family. Kleefstra et al. [12] reported that the disease in 25% of the patients with KS was caused by haploinsufficiency of *EHMT1*, while the disease in few other patients diagnosed with a similar phenotype was explained by *de novo* mutations in *MBD5*, *SMARCB1*, *NR1I3* and *KMT2C* [8]. Kleefstra et al. [8], reported a *de novo* mutation in *KMT2C* to cause dominant ID, similarly 41 likely pathogenic mutations in *KMT2D* gene causing Kabuki syndrome (MIM 147920) have been reported in 86 patients, having dysmorphic facial features, bone deformities, hypotonia, congenital heart defects, ID, urinary tract and respiratory tract infections [20]. Kerimogulo et al. [21] have recently reported that *KMT2B/MLL2* belongs to myeloid leukemia gene family, which mediates hippocampal histone 3 lysine 4 di- and trimethylation in memory formation, thus their data supports the *KMT2B/MLL4* involvement along with *KMT2B/MLL2* in cognition [22]. These findings support the current results that mutations in *KMT2B/MLL4* could possibly lead to a Kleefstra syndrome-like phenotype. The current work is the first report of a recessive KS finding, involving the KMT gene family, which has already been implicated in dominant forms of KS and Kabuki syndrome. The identified variant was excluded in 200 age and ethnicity matched controls. KMT genes (*KMT2A-KMT2E*) as well as their *Drosophila* orthologs; trithorax (trx) and trithorax related (trr), express protein products, which are capable of methylating histone H3 on lysine 4 (H3K4). Of note, the *Drosophila trr* gene is the single ortholog of mammalian *KMT2C* and *KMT2B*, and has been shown to be involved in cell proliferation but its mechanism of action is not yet known [23]. However, its ablation results in restricted tissue growth, which explains the growth retardation in mouse model with low levels of *KMT2B* and could also, be the reason for growth delays in affected members in the current study. KMT2B, like KMT2C is a part of the ASCOM (activating signal cointegrator-2) co-activator complex, which has an important role in epigenetic regulation together with the nuclear-receptor transactivation and forms a connection between the two complexes [12]. Kim et al. [23] have previously reported that ASCOM-KMT2B plays an essential role in Farnesoid X receptor trans-activation through their H3K4 trimethylation activity [23]. Hence, it can be proposed that KMT2B could also be a part of the chromatin modification module proposed by Kleefstra et al. [12], along with *KMT2C*,

SMARCB1 and *NR1I3, MBD5* and *EHMT1*. It is remarkable that the Kleefstra syndrome-like phenotype could be the result of recessive variant identified in family MRQ14 (Figure 1A), whereas all other previously reported gene mutations associated with KS are dominantly inherited. This may reflect an intrinsic property of *KMT2B*, making it less dosage-sensitive than the other KS genes. However, the subsequent functional analyses are highly warranted in order to establish the pathogenicity of the identified recessive mutation.

The phenotypic features of MRQ11 were nsID, however, no striking dysmorphic facial features or structural brain anomalies were observed. The IQ was in the range of moderate ID (IQ: 36-51). Exome sequencing identified a substitution at c.956T>A (p.(Leu319His)) in *ZNF589*, which segregated with the disease in the family in a recessive manner. The gene has not been reported previously for ID and is predicted to be involved in DNA dependent regulation of transcription (Gene ontology database: www.geneontology.org). Based on the brain atlas, *ZNF589* has been shown to be expressed at different ages ranging from 0 to 48 months in different parts of brain such as amyloid complex, medial frontal cortex, hippocampal cortex and basal ganglia. (Allen Institute for Brain Science. Allen Human Brain Atlas; http://human.brain-map.org/). The variant is located in the zinc finger C2H2 domain and *in silico* analysis predicted that the two amino acids differ from each other with a Grantham distance of 100, which depicts a moderate physiochemical difference. The SIFT and polyphen2 also describe this variant to be pathogenic in nature (Table 4). *ZNF589* is localized at the cytogenetic band 3p21.31, it belongs to the kruppel C2H2-type zinc-finger protein family. *ZNF589* consists of a conserved KRAB domain at the amino terminus and four zinc fingers of the C2H2 type at the carboxy terminus. Upon alternative splicing of *ZNF589*, two products are obtained that encode a protein of 361 and 421 amino acids, which differ from each other at the carboxy terminus [24]. The missense change c.956T>A is located in the domain C2H2-type. To date, limited functional data of *ZNF589* is available. Liu et al. [24] have shown that *ZNF589* is involved in hematopoiesis, because of its localization in the bone marrow derived stem cells. The zinc finger genes are housekeeping genes such as the *ZNF589*, which is expressed in the bone marrow stem cell and is involved in DNA dependent transcription repression. The missense change identified in the current study is localized in the C2H2 type domain, which is a highly conserved motif in the ZNF589 protein involved in transcriptional regulation by interacting with different cellular molecules [25], therefore it is predicted that this variant would likely disrupt the function of the protein at the cellular level.

Notably, mutations in other C2H2-type zinc finger proteins have been reported before in relation to ID, including *ZNF526* (one patient), *ZNF41* (4 patients) [26] and *ZNF674* (1 family) [27], which supports the pathogenic role of the variant in *ZNF589* encountered in this study.

In the family MRQ15, the *de novo* variant which was found in the two affected members was c.1158G>C; p.(Trp386Cys); NM_001122834.2 in *HHAT* (hedgehog acyltransferase), which was located in MBOAT domain of the HHAT protein. The change has a high phyloP score of 5.433 as well as a high Grantham distance of 215. The substitution was not found in any of the unaffected members but it was present only in the two affected children as a heterozygous change. It is proposed that this variation among the affected members of MRQ15 has occurred as a result of a *de novo* germline change, which is not uncommon. Previously Rauch et al. [28] and de Light et al. [29], have reported a number of *de novo* variants in known and novel ID genes.

HHAT, belongs to the hedgehog family of gene, which is also referred to as skinny hedgehog. It is the precursor of an enzyme that acts within the secretory pathway to catalyze amino-terminal palmitoylation of hedgehog. It encodes a glycoprotein that undergoes autoproteolytic cleavage to generate its active form, the lipid modification is required for multimerization and distribution of hedgehog proteins. Defects in this protein could lead to improper signaling of shh and can lead to multiple defects including neural tube defects [30] as hedgehog proteins are involved in cell growth, survival and pattern of almost every plan of vertebrate body and has a major role in development of forebrain and midbrain [31]. Dennis et al. [32] have shown *HHAT* to be the candidate gene for congenital human holoprosencephaly, functional assays demonstrated that defects in *HHAT* could diminish secretion of hedgehog proteins. This defect in secretion can lead to abnormal patterning and extensive apoptosis within the craniofacial primordial leading to the structural defects in holoprosencephaly [32]. The role of *HHAT* in cognition is supported by the findings of Das et al. [33], where they treated trisomy 21 mice with Sonic hedgehog agonist and the treatment resulted in behavioral improvements and normalized performance in the Morris Water Maze task for learning and memory, the effect was due to improvement in cerebellar development and hippocampal function. Pan et al. [34] and Kyttala et al. [35] reported the role of shh signaling in primary cilia function. Ciliary defects have already been reported to be causative of many human syndromes involving ID as a clinical feature, such as bardet-biedl syndrome, holoprocencephaly, Kartagener syndromes, polycystic kidney disease, and retinal degeneration. These studies could possibly support the current findings in which only a heterozygous *de novo* change in *HHAT* was identified in the two affected siblings, the other allele being normal, hence the variation did not cause any structural malformation in both the siblings but the brain cognitive function was severely affected, which resulted in speech impairment in them.

Conclusion

In conclusion, the identification of probable pathogenic variations in *KMT2B*, *ZNF589* and *HHAT* in the current study suggests potential importance of the particular pathways associated with these genes involved in cognitive dysfunctioning. In the absence of functional data the definitive role of variations in these novel genes cannot be ascertained without any doubt. However, after exome sequencing and segregation analysis of all the filtered variants, the currently reported were the only variants that segregated with the phenotype in the families. Therefore, it is proposed that these variations could be the most likely cause of ID in the studied families. The finding of additional variations in these genes in other ID families could validate the current results. In addition, functional studies could also define the role of these mutations. The current findings point to the possibility that there are many more unknown pathogenic genes in the known pathways for ID, which are yet to be identified.

Supporting Information

Table S1 Selected homozygous and compound heterozygous variants for the family MRQ14 and polymerase chain reaction conditions.

Table S2 Family MRQ11 selected homozygous and compound heterozygous variants and polymerase chain reaction conditions.

Table S3 Selected homozygous and compound heterozygous variants of family MRQ15 and polymerase chain reaction conditions.

Table S4 Homozygous regions obtained among the affected members of family MRQ14 and MRQ11.

Acknowledgments

We are grateful to all the family members for their participation in this study.

Author Contributions

Conceived and designed the experiments: ZA ZI MA HvB RQ. Performed the experiments: ZA ZI MA HA LELMV CG SHBA MR JAV RP. Analyzed the data: ZA ZI MA HvB RQ. Contributed reagents/materials/analysis tools: HvB RQ. Wrote the paper: ZA ZI MA HvB RQ. Read and approved the final version of the manuscript: ZA ZI MA HA LELMV CG SHBA MR JAV RP HvB RQ.

References

1. Gecz J, Shoubridge C, Corbett M (2009) The genetic landscape of intellectual disability arising from chromosome X. Trends Genet 25: 308–316.
2. Stein Z, Belmont L, Durkin M (1987) Mild mental retardation and severe mental retardation compared: Experiences in eight less developed countries. Ups J Med Sci Suppl 44: 89–96.
3. Iqbal Z, van Bokhoven H (2014) Identifying genes responsible for intellectual disability in consanguineous families. Hum Hered 77: 150–160.
4. Najmabadi H, Hu H, Garshasbi M, Zemojtel T, Abedini SS, et al. (2011) Deep sequencing reveals 50 novel genes for recessive cognitive disorders. Nature 478: 57–63.
5. van Bokhoven H (2011) Genetic and epigenetic networks in intellectual disabilities. Annu Rev Genet 45: 81–104.
6. Ropers HH (2010) Genetics of early onset cognitive impairment. Annu Rev Genomics Hum Genet 11: 161–187.
7. van Bokhoven H, Kramer JM (2010) Disruption of the epigenetic code: an emerging mechanism in mental retardation. Neurobiol Dis 39: 3–12.
8. Kleefstra T, Brunner HG, Amiel J, Oudakker AR, Nillesen WM, et al. (2006) Loss-of-function mutations in euchromatin histone methyl transferase 1 (EHMT1) cause the 9q34 subtelomeric deletion syndrome. Am J Hum Genet 79: 370–377.
9. Fickie MR, Lapunzina P, Gentile JK, Tolkoff-Rubin N, Kroshinsky D, et al. (2011) Adults with Sotos syndrome: review of 21 adults with molecularly confirmed NSD1 alterations, including a detailed case report of the oldest person. Am J Med Genet A 155A: 2105–2111.
10. Agha Z, Iqbal Z, Azam M, Hoefsloot LH, van Bokhoven H, et al. (2013) A novel homozygous 10 nucleotide deletion in BBS10 causes Bardet-Biedl syndrome in a Pakistani family. Gene 519: 177–181.
11. Kleefstra T, Nillesen WM, Yntema HG (1993) Kleefstra Syndrome. Available http://www.ncbi.nlm.nih.gov/books/NBK1116. Accessed 30 October 2014.
12. Kleefstra T, Kramer JM, Neveling K, Willemsen MH, Koemans TS, et al. (2012) Disruption of an EHMT1-associated chromatin-modification module causes intellectual disability. Am J Hum Genet 91: 73–82.
13. Puffenberger EG, Jinks RN, Sougnez C, Cibulskis K, Willert RA, et al. (2012) Genetic mapping and exome sequencing identify variants associated with five novel diseases. PLoS One 7: e28936.
14. Nannya Y, Sanada M, Nakazaki K, Hosoya N, Wang L, et al. (2005) A robust algorithm for copy number detection using high-density oligonucleotide single nucleotide polymorphism genotyping arrays. Cancer Res 65: 6071–6079.
15. Seelow D, Schuelke M, Hildebrandt F, Nurnberg P (2009) HomozygosityMapper-an interactive approach to homozygosity mapping. Nucleic Acids Res 37: W593–599.
16. Vissers LE, Stankiewicz P (2012) Microdeletion and microduplication syndromes. Methods Mol Biol 838: 29–75.
17. Veeramah KR, Johnstone L, Karafet TM, Wolf D, Sprissler R, et al. (2013) Exome sequencing reveals new causal mutations in children with epileptic encephalopathies. Epilepsia 54: 1270–1281.
18. Barbaro V, Nardiello P, Castaldo G, Willoughby CE, Ferrari S, et al. (2012) A novel de novo missense mutation in TP63 underlying germline mosaicism in

AEC syndrome: implications for recurrence risk and prenatal diagnosis. Am J Med Genet A 158A: 1957–1961.

19. Varier RA, Timmers HT (2011) Histone lysine methylation and demethylation pathways in cancer. Biochim Biophys Acta 1815: 75–89.

20. Makrythanasis P, van Bon BW, Steehouwer M, Rodriguez-Santiago B, Simpson M, et al. (2013) MLL2 mutation detection in 86 patients with Kabuki syndrome: a genotype-phenotype study. Clin Genet 84: 539–545.

21. Kerimoglu C, Agis-Balboa RC, Kranz A Stilling R, Bahari-Javan S, et al. (2013) Histone-methyltransferase MLL2 (KMT2B) is required for memory formation in mice. J Neurosci 8: 3452–3464.

22. Kanda H, Nguyen A, Chen L, Okano H, Hariharan IK (2013) The Drosophila ortholog of MLL3 and MLL4, trithorax related, functions as a negative regulator of tissue growth. Mol Cell Biol 33: 1702–1710.

23. Kim DH, Lee J, Lee B, Lee JW (2009) ASCOM controls farnesoid X receptor transactivation through its associated histone H3 lysine 4 methyltransferase activity. Mol Endocrinol 23: 1556–1562.

24. Liu C, Levenstein M, Chen J, Tsifrina E, Yonescu R (1999) SZF1: a novel KRAB-zinc finger gene expressed in CD34+ stem/progenitor cells. Exp Hematol 27: 313–325.

25. Ding G, Lorenz P, Kreutzer M, Li Y, Thiesen HJ (2009) SysZNF: the C2H2 zinc finger gene database. Nucleic Acids Res 37: D267–273.

26. Shoichet SA, Hoffmann K, Menzel C, Trautmann U, Moser B, et al. (2003) Mutations in the ZNF41 gene are associated with cognitive deficits: identification of a new candidate for X-linked mental retardation. Am J Hum Genet 73: 1341–1354.

27. Lugtenberg D, Yntema HG, Banning MJ, Oudakker AR, Firth HV, et al. (2006) ZNF674: a new kruppel-associated box-containing zinc-finger gene involved in nonsyndromic X-linked mental retardation. Am J Hum Genet 78: 265–278.

28. Rauch A, Wieczorek D, Graf E, Wieland T, Endele S, et al. (2012) Range of genetic mutations associated with severe non-syndromic sporadic intellectual disability: an exome sequencing study. Lancet 380: 1674–1682.

29. de Ligt J, Willemsen MH, van Bon BW, Kleefstra T, Yntema HG, et al. (2012) Diagnostic exome sequencing in persons with severe intellectual disability. N Engl J Med 367: 1921–1929.

30. Murdoch JN, Copp AJ (2010) The relationship between sonic Hedgehog signaling, cilia, and neural tube defects. Birth Defects Res A Clin Mol Teratol 88: 633–652.

31. Britto J, Tannahill D, Keynes R (2002) A critical role for sonic hedgehog signaling in the early expansion of the developing brain. Nat Neurosci 5: 103–110.

32. Dennis JF, Kurosaka H, Iulianella A, Pace J, Thomas N, et al. (2012) Mutations in Hedgehog acyltransferase (Hhat) perturb Hedgehog signaling, resulting in severe acrania-holoprosencephaly-agnathia craniofacial defects. PLoS Genet 8: doi: 10.1371/journal.pgen.1002927.

33. Das I, Park JM, Shin JH, Jeon SK, Lorenzi H, et al. (2013) Hedgehog agonist therapy corrects structural and cognitive deficits in a Down syndrome mouse model. Sci Transl Med 5: 201ra120.

34. Pan J, Wang Q, Snell WJ (2005) Cilium-generated signaling and cilia-related disorders. Lab Invest 85: 452–463.

35. Kyttala M, Tallila J, Salonen R, Kopra O, Kohlschmidt N, et al. (2006) MKS1, encoding a component of the flagellar apparatus basal body proteome, is mutated in Meckel syndrome. Nat Genet 38: 155–157.

High Diversity and Low Specificity of Chaetothyrialean Fungi in Carton Galleries in a Neotropical Ant–Plant Association

Maximilian Nepel[1], Hermann Voglmayr[2,3], Jürg Schönenberger[1], Veronika E. Mayer[1]*

1 Division of Structural and Functional Botany, Department of Botany and Biodiversity Research, University of Vienna, Vienna, Austria, **2** Division of Systematic and Evolutionary Botany, Department of Botany and Biodiversity Research, University of Vienna, Vienna, Austria, **3** Institute of Forest Entomology, Forest Pathology and Forest Protection, Department of Forest and Soil Sciences, BOKU-University of Natural Resources and Life Sciences, Vienna, Austria

Abstract

New associations have recently been discovered between arboreal ants that live on myrmecophytic plants, and different groups of fungi. Most of the – usually undescribed – fungi cultured by the ants belong to the order Chaetothyriales (Ascomycetes). Chaetothyriales occur in the nesting spaces provided by the host plant, and form a major part of the cardboard-like material produced by the ants for constructing nests and runway galleries. Until now, the fungi have been considered specific to each ant species. We focus on the three-way association between the plant *Tetrathylacium macrophyllum* (Salicaceae), the ant *Azteca brevis* (Formicidae: Dolichoderinae) and various chaetothyrialean fungi. *Azteca brevis* builds extensive runway galleries along branches of *T. macrophyllum*. The carton of the gallery walls consists of masticated plant material densely pervaded by chaetothyrialean hyphae. In order to characterise the specificity of the ant–fungus association, fungi from the runway galleries of 19 ant colonies were grown as pure cultures and analyzed using partial SSU, complete ITS, 5.8S and partial LSU rDNA sequences. This gave 128 different fungal genotypes, 78% of which were clustered into three monophyletic groups. The most common fungus (either genotype or approximate species-level OTU) was found in the runway galleries of 63% of the investigated ant colonies. This indicates that there can be a dominant fungus but, in general, a wider guild of chaetothyrialean fungi share the same ant mutualist in *Azteca brevis*.

Editor: Petr Karlovsky, Georg-August-University Göttingen, Germany

Funding: This work was supported by a University of Vienna, Austria, KWA fellowship to MN; https://international.univie.ac.at/graduate-students/kurzfristige-auslandsstipendien-kwa. The funders had no role in study design, data collection and analysis, decision to publish, or preparation of the manuscript.

Competing Interests: The authors have declared that no competing interests exist.

* Email: veronika.mayer@univie.ac.at

Introduction

It is now clear that microorganisms are major partners in obligate interactions between ants and plants. Ant–fungus associations have been recognised since the mid-19th century (e.g. [1]), and the best-studied examples are the fungal gardens of the leaf-cutter ants in the tribe Attini. Leaf-cutter ants grow monocultures of basidiomycetes on shredded leaf material and feed on the nutrient-rich tips of the fungal hyphae [2,3]. Other examples of ant–plant–fungus interactions have also been found recently in different groups of non-attine ants, where ascomycete fungi are cultivated in domatia (nesting spaces provided by host plants) or on a cardboard-like construction material (named "carton" in ant-plant literature) [4–7]. Such ant–plant–fungus associations have been described from Africa, America and Asia and involve a wide range of plant lineages associated with an equally wide range of ant groups [7].

There is evidence that the fungi cultivated within the domatia are used as a food source [8], whereas those in the carton-like material do not appear to be consumed. Rather, they seem to serve to stabilise the carton mechanically. Carton structures with fungi were first documented in nest walls of the European ant *Lasius fuliginosus* inside hollow tree-trunks [1,9,10]. They have since been found in the walls of free-hanging canopy ant nests in the Palaeotropics [11,12] and in the Neotropics, where ants use fungus-infused, carton-like material to construct tunnel systems called "runway galleries" along branches of their host trees [6,13,14] (Figure 1A–D).

In the tripartite ant–plant–fungus interactions involving non-attine ants studied so far, the vast majority of the fungi have belonged to the ascomycete order Chaetothyriales, the so-called "black yeasts" [7]. These are usually dark, melanised, slow-growing fungi that often colonise extreme environments [15–18], but little is known about the order's ecology and diversity.

A recent survey based on molecular phylogenetics showed that ant-associated chaetothyrialean fungi belong to four clades within the order: a domatia-symbiont clade, two clades with carton fungi, and a mixed clade containing both domatia symbionts and carton fungi [7]. Only a few isolates were placed outside these four clades, and ant-fungi cultivated in the domatia seemed to be specific to each ant species [7]. Carton structures have been less well investigated, and studies to date have produced disparate results: in the *Hirtella* (Chrysobalanaceae)/*Allomerus* (Formicidae) asso-

Figure 1. Carton runway galleries built by *Azteca brevis* **ants, consisting of mycelia and various organic particles.** (A, B) Galleries on the lower side of branches of *Tetrathylacium macrophyllum*: note the scattered circular openings in the gallery walls. (B) Alarmed workers wait with open mandibles below the holes for prey or intruders. (C, D) Scanning electron microscope images of the gallery walls infused with different types of hyphae. (E, F) Light-microscope images of both hyphal types: (E) thin-walled hyaline hyphae typical for carton clades 2 and 3; (F) pigmented thick-walled hyphae typical for carton clade 1 (see Figure 2–3). Bars: (A, B) 1 cm; (C, D) 100 μm; (E, F) 20 μm.

ciation, it was reported that a specific fungus is cultivated by the ants in the wall material of their galleries [14]; in contrast, a wider guild of fungi seems to be involved in the structurally analogous galleries of the *Tetrathylacium/Azteca* association [6]. The aim of the present investigation was (1) to unravel the diversity and geographical pattern of the carton fungi found in the carton galleries, and (2) to investigate the hyphal morphology of the relevant fungal strains with respect to the proposed function of the galleries.

Material and Methods

Species and study site

Azteca brevis Forel, 1899 (Formicidae, Dolichoderinae) is a reddish-brown ant, c. 4 mm long, known from wet forests of the southern Pacific lowlands of Costa Rica [19]. Colonies have been found on *Tetrathylacium macrophyllum* (Salicaceae), *Licania* sp. (Chrysobalanaceae), *Grias* sp. (Lecythidaceae), *Myriocarpa* sp. (Urticaceae), *Ocotea nicaraguensis* (Lauraceae) [19] and *Lonchocarpus* sp. (Fabaceae). The nesting chambers inside the stems are connected externally by runway galleries dotted with small, circular holes (Figure 1A, B).

The most common host plant for *Azteca brevis* is *Tetrathylacium macrophyllum* Poepp. (Salicaceae), a small tree (c. 8 m) that grows on the Pacific slopes of Central and South America in areas characterised by high annual rainfall (>5000 mm). It is found chiefly on steep slopes near rivers and streams in primary forest [20]. About 30% of *T. macrophyllum* trees are occupied by *Azteca brevis* [21,22]. The ants start by colonising hollow chambers in the branches that the plant forms through pith degeneration. As the colony grows, the ants excavate the remaining pith between adjacent naturally formed chambers, and build large nest sites inside the branches.

We sampled within a 5-km circle around the Tropical Research Station La Gamba, Costa Rica (8° 42′ 03″ N, 83° 12′ 06″ W) along the Waterfall Trail, Bird Trail, Río Gamba, Río Bolsa and Río Sardinal. Carton samples, ants and plant parts for herbarium specimens were collected from 18 *T. macrophyllum* trees and one *Lonchocarpus* tree colonised by *Azteca brevis*, under permission from SINAC – Sistema Nacional de Areas de Conservación de Costa Rica of the Ministry of Environment and Energy (MINAE) to M.N. and V.E.M. (No. 182-2010-SINAC). In recent years, trees colonised by *Azteca brevis* have become inexplicably rare at the study site (VEM, pers. obs.), limiting the sample size to 19 trees colonised by *Azteca brevis*. At least three carton pieces per tree and colony, each c. 1 cm long, were taken from runway galleries and stored in 1.5-mL reaction tubes sealed with air-permeable cotton wool. The reaction tubes were kept in a sealed plastic bag with

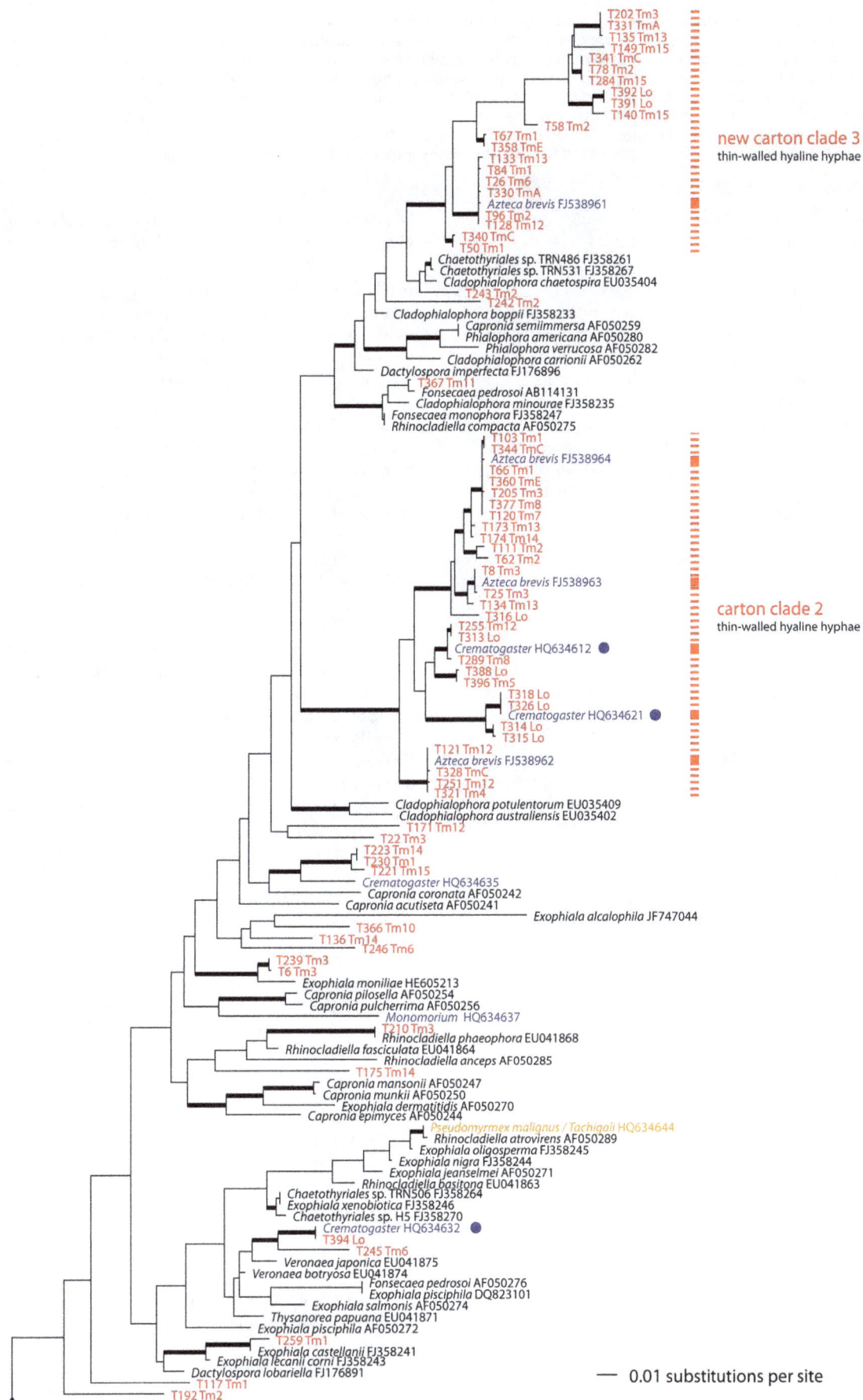

Figure 2. Phylogram of Chaetothyriales, top part. The maximum-likelihood tree is shown, based on partial SSU, complete ITS and 5.8S, and partial LSU rDNA regions. Bold branches are supported in all three analyses: BA probabilities higher than 0.9, ML and MP bootstrap support above 70%. Red labels denote fungal genotypes isolated in this study from ant-built carton structures on *Tetrathylacium macrophyllum* and *Lonchocarpus* sp. trees; orange and blue mark domatia fungi and carton fungi, respectively, from Voglmayr *et al.* [7]; GenBank accession numbers follow taxon names; solid red, violet and orange vertical lines indicate clade definitions and captions from Voglmayr *et al.* [7]; dotted lines mark clade extensions from this study. Blue dots point out three sequences from other continents (2× Cameroon; 1× Thailand) differing by only three mutations from our Costa Rican genotypes. Note the high diversity of isolated genotypes (the large clade extensions compared to Voglmayr *et al.* [7] are due to a greater number of samples and the new monophyletic carton-fungi cluster (new carton clade 3).The tree is continued in Figure 3.

silica gel for three weeks before culturing the fungi; the bags were transported and stored at room temperature.

Fungal cultures and DNA-extraction

At the University of Vienna, a c. 10-mm^2 piece of carton was placed into a droplet (c. 20 µL) of sterile water and fragmented with sterile forceps to make a mycelial suspension. An aliquot of mycelial suspension was then diluted with 1 mL sterile water and spread over each of two 2% malt extract agar plates (MEA) containing 0.5% penicillin and 0.5% streptomycin. This was carried out on average for three samples per colonised tree. The plates were stored at room temperature and visually checked under a dissecting microscope at least once a day for 11 days. Fast-growing "weeds" (*Aspergillus, Cladosporium, Fusarium*) were excised to prevent overgrowth of the slower-growing carton fungi. The thick, darkly pigmented hyphae that are typical for the carton usually started to grow after 2–4 days and were then transferred to new 2% malt extract agar plates. Several carton samples from each ant colony and tree were processed in this way to minimise any cultivation bias.

Sections of approximately 25 mm^2 were cut out from mycelia on pure-culture agar plates and stored in 2-mL reaction tubes at −20°C. The frozen samples were subsequently freeze-dried overnight and ground with five glass beads (3 mm diameter) for 10 min at 30 Hz in an MM 400 mixer mill (Retsch, Germany), after which DNA was extracted using NucleoSpin 96 Plant II kits (Macherey-Nagel, Düren, Germany).

PCR and cleanup

A 1.5–3.5-kb nuclear ribosomal DNA (rDNA) fragment comprising partial small subunit (SSU), complete ITS1–5.8S–ITS2 (ITS) and partial long subunit (LSU) sequences was amplified with the fungal primers V9G [23] and LR5 [24] using Thermo Scientific 2.0× ReddyMix Extensor PCR Master Mix and 1.1× ReddyMix PCR Master Mix (ABgene, Epsom, UK) (for primer sequences and detailed PCR protocol, see Table S1 in Appendix S1). The PCR products were purified with 6 U exonuclease I and 0.6 U FastAP thermosensitive alkaline phosphatase (Fermentas, St. Leon-Rot, Germany) [25]; the PCR product was then incubated for 30 min at 37°C, followed by enzyme deactivation for 15 min at 85°C.

Sequencing

DNA was cycle-sequenced with ABI PRISM BigDye Terminator Cycle Sequencing Ready Reaction v. 3.1 (Applied Biosystems, Warrington, UK) using the PCR primers and primers LR3 [24] and ITS4 [26]. For sequences with large indels, the additional primers LR2R-A, LR2-A [27], F5.8Sr, F5.8Sf [28] and LR3-CH (5′-GGT ATA GGG GCG AAA GAC TAA TC-3′) were necessary to obtain full-length sequences (see Appendix S1 for detailed sequencing protocol). Sequencing was performed on an ABI 3730xl Genetic Analyzer automated DNA sequencer (Applied Biosystems). One sequence of each genotype was deposited in

GenBank. The complete list of accession numbers for the SSU–ITS–LSU locus can be found in Table S2 in Appendix S1.

Analysis of sequence data

After a BLAST search (Basic Local Alignment Search Tool in GenBank, http://blast.ncbi.nlm.nih.gov/Blast.cgi) of the nuITS1–5.8S–ITS2-LSUr DNA sequences obtained from the ant carton (423 in total), 381 sequences were identified as Chaetothyriales and used for further analyses. For phylogenetic analyses, identical sequences from carton samples were reduced to a single sequence per genotype. Cases where two sequences differed only in homopolymer regions were also merged to a single genotype.

For alignment, the chaetothyrialean sequences used in Voglmayr *et al.* [7] and the closest sequences to our isolates from GenBank were added. *Verrucaria denudata, V. csernaensis* and *V. andesiatica* (Verrucariales) were included as outgroups. Ambiguously aligned regions in ITS1/ITS2 and leading gap regions were excluded. The matrix of 258 sequences contained 7767 alignment positions, with the longest sequence comprising 3425 nucleotides. Alignments were produced with Muscle 3.8.31 [29] and revised in BioEdit 7.1.3.0 [30]. (GenBank accession numbers of the sequences included in the phylogenetic analyses are listed in Table S2 in Appendix S1.)

For Bayesian analyses, MrBayes 3.2.1 [31] was run through the Bioportal web service of the University of Oslo [32]. The six-parameter general time-reversible substitution model was used, with a proportion of invariant sites and a gamma distribution for the remaining sites (GTR + I + G), as determined by Modeltest 3.7 [33]. Three parallel runs of four chains were performed over 30 million generations, sampling 30 000 trees in each run. For 90% majority-rule consensus trees, the first 2000 trees of each run were discarded as burn-in.

Maximum-parsimony (MP) bootstrap analyses were performed with PAUP* 4.0b10 [34], using 1000 replicates of a heuristic search with 10 rounds of random sequence addition during each bootstrap replicate and a limit of 100 000 rearrangements per replicate. TBR branch swapping was used, allowing multitrees, and steepest descent was set to 'no'. Gaps were treated as missing data, and no weighting of nucleotides was applied.

The maximum-likelihood (ML) analyses used RAxML [35], as implemented in the programme raxmlGUI 0.95 [36]. The GTRGAMMAI nucleotide substitution model was applied for the ML heuristic search and the ML rapid bootstrap analysis.

Fungal distribution at the genotype and species levels

To evaluate the specificity of the fungi to *Azteca brevis*, the frequency of occurrence on the sampled trees was analyzed. For each tree, a matrix containing genotypes and sampled trees was compiled (Table S3 in Appendix S1) and Bray–Curtis similarity indices among sampled trees were plotted by non-metric multidimensional scaling (NMDS) with 1000 restarts. In addition, an ANOVA of similarity (ANOSIM) was conducted with up to 999 permutations, based on different geographical study sites

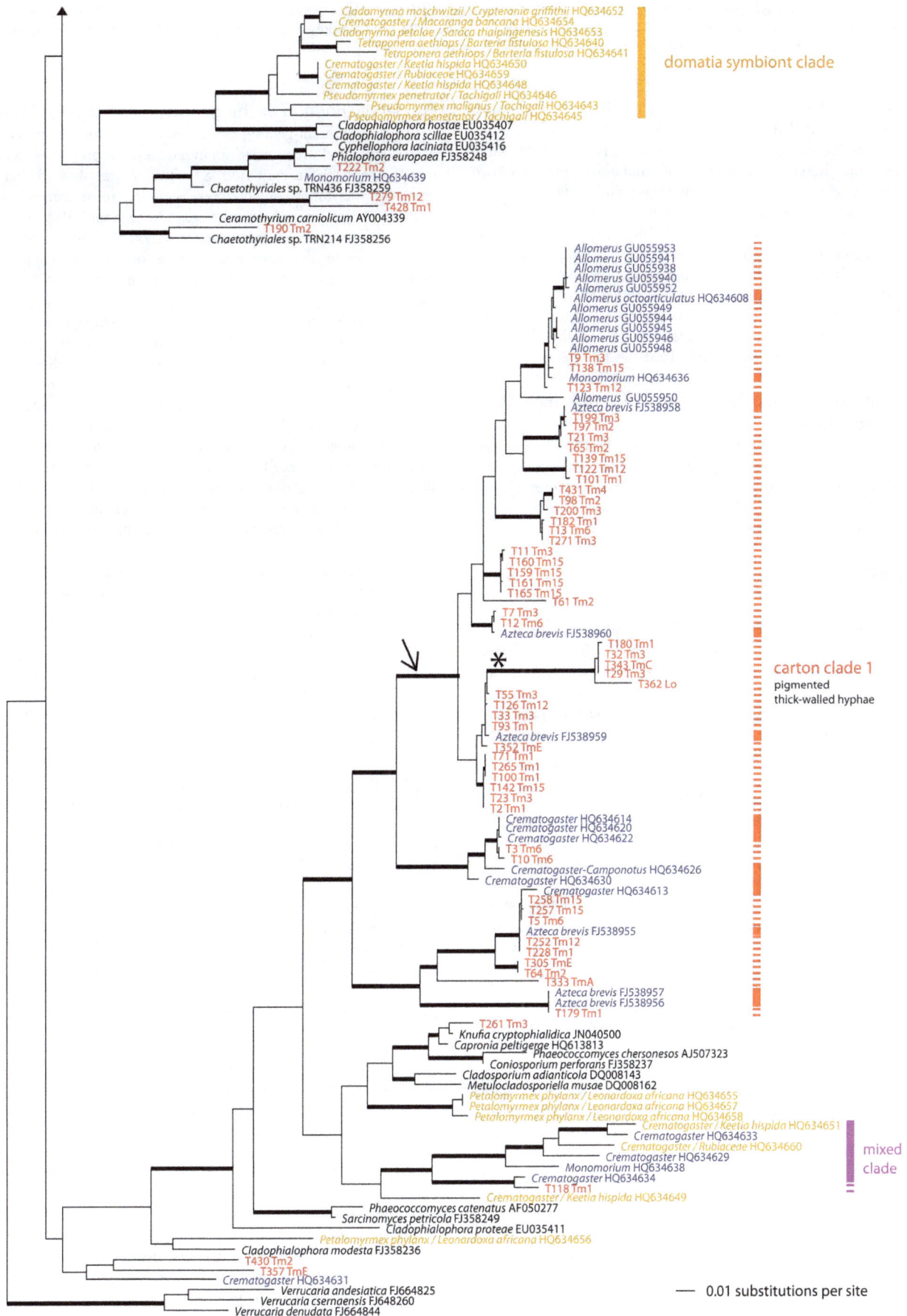

Figure 3. Phylogram of Chaetothyriales, bottom part. Continuation of Figure 2 with arrowheads indicating the connection. For label and colour descriptions see legend to Figure 2. The clade labelled with an asterisk (*) is placed more basally (arrow) in the MP analysis than in ML and BA analyses. Note the domatia-symbiont clade, which remains distinct from carton fungi.

(Table S2 in Appendix S1). All analyses were carried out with Primer 5 5.2.9 (PRIMER-E, 2002).

Slightly different genotypes can, however, represent the same species: the maximum number of mutations between two individuals of the same species (mutation limit) varies with the DNA region and type of organism. For fungi, the ITS region is more variable than SSU or LSU [37]. Because ITS is the main DNA fragment in the present study, the mutation limit was determined by multiplying the maximum intraspecific variation of ITS (0.58%) [37] by the average sequence length. We define an OTU as a species with a maximum genotype variation of 12.02 mutations. The abundance of OTUs was also analyzed using a modified presence–absence matrix (Table S4 in Appendix S1).

Light microscopy and SEM analysis of carton material and fungal hyphae

Pieces of carton from 16 trees and aerial hyphae of the pure cultures were investigated using a Zeiss AxioImager A1 compound microscope with a Zeiss AxioCam ICc3 digital camera. SEM investigations were made with a Jeol JSM-T 300 scanning electron microscope (SEM) at 10 kV.

Results

Cultures

Pure cultures of carton fungi were obtained from carton material of host trees colonised by *Azteca brevis*. Because different species could not be distinguished morphologically, all the pigmented hyphae that germinated were transferred to MEA plates to obtain pure cultures. In total, 423 pure cultures were sequenced, resulting in 128 different genotypes after identical sequences were removed.

Molecular phylogenetic analyses

After adding relevant sequences from GenBank, the final alignment, including outgroups, consisted of 258 sequences and 7767 alignment positions. In the best tree from the ML search (shown as a phylogram in Figure 2–3), backbone support is mostly low or absent, but most of the subclades are well-supported. Consensus trees across the three analyses differed topologically only in one point: one clade of five sequences containing a large indel of c. 1000 bp within the LSU region is located more basally in the topology resulting from the MP analysis than in the topologies from BA and ML analyses (marked with an asterisk and an arrow in Figure 3). Bootstrap support for this node in the MP analysis was only 73%, so this difference was not considered further.

The phylogenetic reconstruction revealed three main clades containing 73% (100 out of 128) of the isolated genotypes (Figure 2–3). Carton clades 1 and 2 were described by Voglmayr *et al.* [7], but carton clade 3, which was represented by a single sequence in previous analyses, is new. The remaining 28 genotypes are distributed across the phylogenetic tree of Chaetothyriales, except for the "domatia-symbiont clade". Three sequences from carton material from Cameroon and Thailand are nearly identical to some fungi sequenced in this study. Each of those three fungal genotypes (marked by blue dots in Figure 2) differs by only three mutations from fungi grown from carton structures collected in Costa Rica.

Fungal distribution at the genotype and species levels

The 128 different genotypes isolated from carton material of 19 *Azteca*-inhabited trees were analyzed with a presence–absence matrix (Table S3 in Appendix S1). The matrix showed that no genotype was found on all trees, with the most common one isolated from nine out of 19 trees. On average, 10.5 different genotypes occurred on each tree, of which 46% were unique to single trees. In the carton sample of one *Lonchocarpus* tree inhabited by *Azteca brevis*, 11 out of 12 genotypes (92%) were unique.

Bray–Curtis similarity indices between sampled trees were calculated to investigate the correlation between genotype composition and collection site. The non-metric multidimensional scaling (NMDS) plot showed no clustering of trees from the same collection site (Figure 4), and the analysis of similarity (ANOSIM) showed a significance level of only $P = 0.32$. A correlation between genotype composition and collection site can therefore be ruled out. One sampled tree (Tm10) had to be excluded because only a single, unique fungus could be isolated, and the Bray–Curtis distance to the other sampled trees was too great for Tm10 to be displayed without clustering all remaining trees too tightly together.

The results were similar at the approximated species level. The 128 genotypes were reduced to 62 OTUs, and the most common OTU in the modified presence–absence matrix, represented by nine genotypes, was found on 12 out of 19 trees (63%) (Table S4; Figure S1 in Appendix S1). Three other OTUs were found on a total of 9 out of 19 trees (47%). The correlation of fungal community and collection site at the species level is weak (NMDS plot, Figure S2 in Appendix S1) and not significant (ANOSIM: $P = 0.16$).

Light microscopy of carton material

Dark, melanised moniliform hyphae with thick cell walls (Figure 1F) appeared to be dominant in each sample. The cell width of these thick-walled hyphae ranged from about 6 to 9 μm. Hyaline hyphae with a cell width less than 5 μm were also present, but appeared less abundant (Figure 1E). Aerial hyphae of pure cultures representing the three carton clades were also examined and, surprisingly, darkly pigmented, thick-walled hyphae are largely restricted to carton clade 1, whereas thin-walled hyaline hyphae are found in carton clades 2 and 3 (Figure 2–3).

Discussion

There is growing evidence that multicellular organisms are shaped by symbioses with smaller partners – often microbial – that contribute to their host's nutrition, protection and even to their normal development [38]. In obligate interactions between ants and plants, for example, it has only recently become apparent that micro-organisms are major partners in interactions that go far beyond the relationship between the ant and the plant. In ant-plant symbioses from Africa, America and Asia, ascomycete fungi are cultivated in domatia and on ant-built carton structures, involving a wide range of distantly related ants and plant families [5,7]. This study is, however, the first dealing with fungi in ant-built cardboard-like carton material in which pure cultures of several samples were made per colony. This resulted in 381 pure cultures and 128 chaetothyrialean genotypes, the highest number

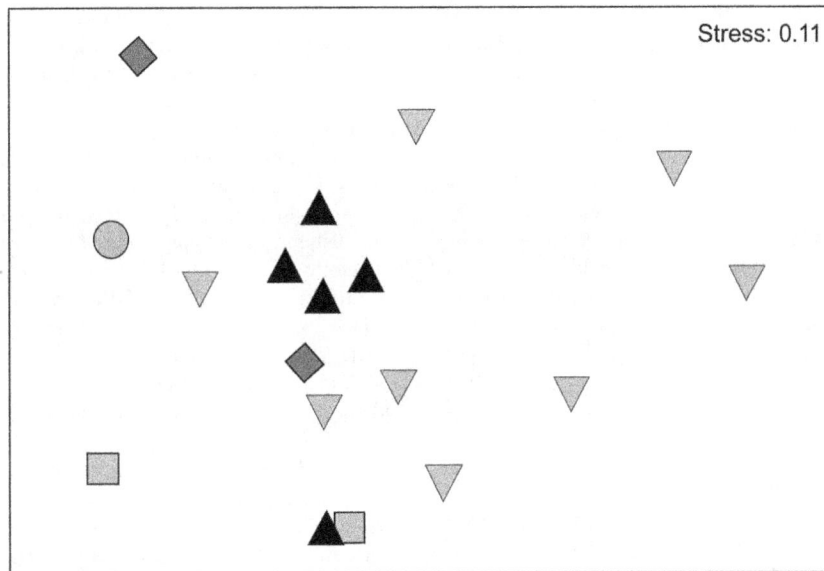

Figure 4. Correlation analysis of genotype sets and collection sites. Non-metric multidimensional scaling (NMDS) plot based on Bray–Curtis similarities between fungal genotype sets occurring on carton material of 18 sampled *Tetrathylacium macrophyllum* trees, and one *Lonchocarpus* sp., colonised by *Azteca brevis*. Different symbols represent trees from different collection sites (squares: Waterfall Trail; triangles: Bird Trail; circle: Río Gamba; diamonds: Río Bolsa; inverted triangle: Río Sardinal).

of carton-associated chaetothyrialean symbionts ever found associated with a single ant species.

This high number of fungus genotypes of the ascomycete order Chaetothyriales ("black yeasts") is astonishing, because these fungi generally seem to have weak competitive abilities. They are slow-growing and often extremophilic. They occur on nutrient-poor substrates, such as leaf or rock surfaces [15,39], or in toxic environments [17,18], and quickly disappear under less extreme conditions [7]. It is not yet known why "black yeasts" are so dominant on the carton material of ants. The frequent germination of fast-growing "weeds" on isolation plates indicates that the spores of moulds probably occur on the carton surface, but that their growth is inhibited. It may be the gallery substrate, the weeding and grooming behaviour of *Azteca brevis*, some ant-specific compounds or antifungal substances released from the fungi themselves that cause this inhibition. One indication that the construction material also shapes the fungal community on the galleries is the fact that 11 of 12 genotypes (92%) isolated from the samples collected from carton on a *Lonchocarpus* tree were unique and not present on the *Tetrathylacium* trees. *Azteca brevis* workers use particles of bark, excavated pith tissue and epiphylls from the host trees as materials for gallery construction, and plant secondary compounds may disfavour fungi other than Chaetothyriales. Furthermore, *Azteca brevis* was observed to groom the carton galleries constantly, and even to nourish them (M. Nepel & V. Mayer, unpubl.). Antibacterial and antifungal compounds produced by ants' exocrine glands, such as the metapleural gland, may play an important role in preventing other moulds from growing [40,41]. In contrast, Chaetothyriales are able to tolerate and even to metabolise aromatic hydrocarbons [18] and can therefore cope and may even use ant-produced antifungal compounds metabolically. Finally, Chaetothyriales themselves might produce bioactive substances against competing fungi [42]. The combination of these factors may account for the relationship between ants and those fungi.

Molecular phylogenetic analyses and fungal diversity

There was no ubiquitous fungus (genotype or OTU) among the 128 genotypes found in this study; 78% of the sequenced carton fungi clustered into three clades. Two of those clades were already established [7], and we have discovered a third (Figure 2). The 28 genotypes that were not assigned to any of those clades were scattered across the whole phylogenetic tree (Figure 2–3), but none of the 128 fungal genotypes isolated from the carton samples arose in the "domatia-symbiont clade" [7]. The fungi belonging to this special clade are highly distinct from carton fungi in terms of their hyphal morphology and growth form (hyaline or less pigmented and tending to produce spores in domatia fungi) [7] probably due to their different functions. Domatia fungi are used as food for the larvae [8], whereas carton fungi probably improve the stability of carton walls in ant nests or runway galleries. The specificity and coevolutionary dynamics between domatia and carton symbionts may differ.

Fungus specificity at the genotype and species level

In most known insect–fungus symbioses (e.g. termites [43,44] or leaf-cutter ants [45]), the associated fungus is cultivated for food, whereas *Azteca brevis* is not likely to eat the fungi, but cultivates them for their nest architecture. In this association, the ants as a single host use a group of multiple fungus species for this purpose. The interpretation of the degree of specificity depends on the taxonomic level and differs between the fungi involved. At higher taxonomic levels, the interaction specificity between *Azteca brevis* and Chaetothyriales is high: chaetothyrialean fungi were found in every carton sample analyzed. At the genotype level, a much more modest degree of interaction specificity was seen (Table S3 in Appendix S1): the most common genotype (T121_Tm12) was associated with 9 out of 19 ant colonies (c. 47%). No correlation between fungal community and collection site was found, and the fungal community seems not to be habitat-specific (Figure 4). Merging genotypes into OTUs based on the maximum intraspecific variation of ITS [37] increased the occurrence of the most common fungus (represented by T66 in Table S4 in Appendix S1) to 63% of the ant colonies and the three next most common OTUs to 47% (Table S4 in Appendix S1). Surprisingly, a mean of eight different OTUs were found per carton sample, but no specialist fungal partner obligate to all *Azteca brevis/Tetrathyla-cium macrophyllum* associations was found. Environmental samples for the most common OTU could not be analyzed, because no primers could be developed which were specific enough for reliable separation of the OTUs, and only culturing was possible. It might be argued that culturing introduces a bias, in that a genotype could have been missed. Because we sampled the carton galleries of every colony and tree several times each, and also performed the isolation procedure on average three times per sample, it is unlikely that we missed any genotype. A heterogeneous spatial distribution of the genotypes along the branches can also be excluded, because we always took samples at positions from the branching point to the tip of a branch, and would therefore have included any heterogeneity.

Moreover, some fungal sequences from *Azteca brevis* carton are nearly identical at the SSU–ITS–LSU region with those of fungi from cartons of other ant species and from samples collected on other continents. Two sequences from *Crematogaster* carton nests from Cameroon (*Crematogaster* HQ634612 and *Crematogaster* HQ634621), and one found in Thailand (*Crematogaster* HQ634632) [7] differ by only three mutations from the Costa Rican sequences (blue dots in Figure 2). This indicates that at least some ant-associated chaetothyrialean fungi have a transcontinental distribution and are associated with more than one ant genus rather than being specific to *Azteca brevis* hosts.

Fungus selection and origin of the fungi

The carton runway galleries of *Azteca brevis* are used as a defence against intruders and as ambush traps for capturing prey (V. Mayer, pers. observ.), analogous to the carton galleries described for *Allomerus decemarticulatus* [13]. The fungal symbionts may play a particularly important functional role by increasing the stability of the carton produced by the ants [6,13]. The hypothesis that fungi are used for reinforcement was first raised by Lagerheim for the carton constructions of *Lasius fuliginosus* [10] and it has been accepted by many authors [6,11,13,46]. Trimming and grooming of the carton fungi has been observed and, due to the reinforcement demand, it may be expected that *Azteca brevis* would select fungal symbionts with a particular phenotype. Because we conducted a pure-culture method, we were able to examine the hyphae microscopically and correlate genotype or OTU with hyphal type. Two types of hyphae were regularly found: thin-walled, hyaline hyphae (Figure 1E), which represent two of our three major carton clades. The other type, thick-walled, melanised hyphae (Figure 1F), are less abundant in the species in our phylogenetic tree. When investigating pieces of carton, however, the thick-walled hyphae visibly dominated the biomass. This indicates that the second hyphae type is favoured on the carton. This may be either due to the substrate or due to the ants' preference. Fungi with thick-walled, melanised hyphae are likely to be better for the carton's stability than thin-walled hyphae.

Unfortunately, no alate queens were found, and interpretations of the transmission of the fungi must therefore be made without experimental evidence. The lack of any strong specificity indicates that the fungi growing on the *Azteca brevis* carton are either transmitted horizontally or environmentally acquired, although the degree of host–symbiont specificity is not always correlated with the transmission mode [47]. *De novo* acquisition of fungal symbionts from the environment in each ant generation would,

however, explain the high number of genotypes and OTUs in the carton samples. *Azteca brevis* workers may collect spores or hyphal fragments from the environment, as termites do [48], but detailed field observations of the worker ants' behaviour are needed to prove this. Also, the plant material used for construction (typically bark and epiphylls) may already be infected with chaetothyrialean spores. The branches and stems of trees inhabited by *Azteca brevis* appear to be completely cleaned of epiphylls (algae, bryophytes and lichens) and epiphytes. In fact, the newly described Chaetothyriales family Trichomeriaceae was found to grow on the surface of living leaves [49]. Investigation of the host plants' surface (stem, branches and leaves) and the epiphylls are needed to clarify whether Chaetothyriales are already present. The chaetothyrialean fungus community of the carton galleries might be a subset of the fungal community found on the host plant's surface.

Two analogous systems: *Allomerus* sp. and *Azteca brevis*

A plant–ant–fungus association with runway galleries that are structurally similar to those in the present study is seen in Amazonian *Allomerus* ants living on *Hirtella physophora* (Chrysobalanaceae) and *Cordia nodosa* (Boraginaceae) [14]. In contrast to the *Azteca brevis* carton, with its guild of numerous species of Chaetothyriales, the *Allomerus* ants are described as cultivating one specific fungal symbiont on the carton galleries (see upper part of carton clade 1; Figure 3). *Allomerus* foundress queens apparently store a pellet with the specific carton fungus represented by a monophyletic group of haplotypes on the domatium wall. While building their runway galleries, *Allomerus* workers were observed to glue pellets from scraped epidermis and mesophyll of the inner domatia walls onto the gallery frame built from trichomes of the host plant [14]. In *Azteca brevis* colonies, no inoculation pellet was found, an observation that supports the hypothesis that fungal symbionts are acquired *de novo* from the environment.

Although the sampling methods were different (only newly produced carton material was collected from the *Allomerus* galleries, whereas it was mainly "mature" black material that was sampled from the *Azteca brevis* galleries), the frequency of the most common fungus or OTU was more or less the same: on 57% (138 out of 240) of the *Allomerus* galleries, and on 63% (12 out of 19) of the *Azteca brevis* galleries. A fungal species found in a little over half of the analyzed carton samples should not, however, be regarded as specific.

Conclusions

We give an insight into the diversity of Chaetothyriales present in the carton galleries of *Azteca brevis*. Our results refute the initial hypothesis that *Azteca brevis* forms a symbiosis with a specific fungus. At the genotype level as well as that of approximated species (OTUs), the fungi we isolated appear to be a guild of different Chaetothyriales. An obligate mutualism with the fungi found in carton galleries of *Azteca brevis* is found for the host-ant with the ascomycete order Chaetothyriales; on the level of fungal species, no obligate mutualism is found. Moreover, *Azteca brevis* does not seem to strongly select for a particular morphological type, as both hyaline, thin-walled hyphae and pigmented, thick-walled hyphae are present in the carton. *Azteca brevis* cultivates and uses many different kinds of Chaetothyriales, and future research is needed to clarify the origins of these fungi. The reasons for the general preference of black yeasts in such ant–plant–fungus associations are still unclear. Knowledge of the diversity, coevolutionary processes and functional role of fungi in ant–plant symbioses is currently very fragmentary and further investigation is needed.

Acknowledgments

We would like to thank the staff of the Tropical Research Station La Gamba, Costa Rica. Research permission (No 182-2010-SINAC) was kindly granted by Javier Guevara of SINAC – Sistema Nacional de Areas de Conservación de Costa Rica of the Ministry of Environment and Energy (MINAE). Many thanks are due to Tamara Bernscherer, Florian Etl, Katharina Kneissl and Rafael Ramskogler for helping us collect samples. We thank Jack Longino for identifying
anonymous reviewers are cordially thanked for their great input, which improved the paper substantially. Chris Dixon eliminated the Germanisms, and made the syntax clearer and less painful to read for the native English-speaking community.

Author Contributions

Azteca brevis. Two

Conceived and designed the experiments: VEM. Performed the experiments: MN VEM. Analyzed the data: MN HV VEM. Contributed reagents/materials/analysis tools: VEM MN JS HV. Wrote the paper: MN VM JS. Other: Received research permission from SINAC – Sistema Nacional de Areas de Conservación de Costa Rica of the Ministry of Environment and Energy (MINAE) (No. 182-2010-SINAC): MN VEM.

References

1. Fresenius G (1852) Beiträge zur Mykologie. vol. 2, Heinrich Ludwig Bönner, Frankfurt am Main.
2. Mueller UG, Rehner SA, Schultz TR (1998) The evolution of agriculture in ants. Science 281: 2034–2038.
3. Mueller UG, Scott JJ, Ishak HD, Cooper M, Rodrigues A (2010) Monoculture of leafcutter ant gardens. PLoS One 5.
4. Schlick-Steiner BC, Steiner FM, Konrad H, Seifert B, Christian E, et al. (2008) Specificity and transmission mosaic of ant nest-wall fungi. Proc Natl Acad Sci 105: 940–943.
5. Defossez E, Selosse MA, Dubois MP, Mondolot L, Faccio A, et al. (2009) Ant-plants and fungi: a new threeway symbiosis. New Phytol 182: 942–949.
6. Mayer VE, Voglmayr H (2009) Mycelial carton galleries of *Azteca brevis* (Formicidae) as a multi-species network. Proc R Soc B 276: 3265–3273.
7. Voglmayr H, Mayer V, Maschwitz U, Moog J, Djieto-Lordon C, et al. (2011) The diversity of ant-associated fungi: insights into a newly discovered world of symbiotic interactions. Fungal Biol 115: 1077–1091.
8. Blatrix R, Djiéto-Lordon C, Mondolot L, La Fisca P, Voglmayr H, et al. (2012) Plant-ants use symbiotic fungi as a food source: new insight into the nutritional ecology of ant-plant interactions. Proc R Soc B 279: 3940–3947.
9. Elliott JSB (1915) Fungi in the nests of ants. Trans Br Mycol Soc 5: 138–142.
10. Lagerheim G (1900) Über *Lasius fuliginosus* (Latr.) und seine Pilzzucht. Entomol Tidskr 21: 17–29.
11. Weissflog A (2001) Freinestbau von Ameisen (Hymenoptera, Formicidae) in der Kronenregion feuchttropischer Wälder Südostasiens. Bestandsaufnahme und Phänologie, Ethoökologie und funktionelle Analyse des Nestbaus. PhD thesis, J W Goethe Univ Frankfurt am Main, Germany.
12. Kaufmann E, Maschwitz U (2006) Ant-gardens of tropical Asian rainforests. Naturwissenschaften 93: 216–227.
13. Dejean A, Solano PJ, Ayroles J, Corbara B, Orivel J (2005) Arboreal ants build traps to capture prey. Nature 434: 973.
14. Ruiz-González MX, Malé P-JG, Leroy C, Dejean A, Gryta H, et al. (2011) Specific, non-nutritional association between an ascomycete fungus and Allomerus plant-ants. Biol Lett 7: 475–479.
15. Selbmann L, Isola D, Zucconi L, Onofri S (2011) Resistance to UV-B induced DNA damage in extreme-tolerant cryptoendolithic Antarctic fungi: detection by PCR assays. Fungal Biol 115: 937–944.
16. Gueidan C, Villaseñor CR, de Hoog GS, Gorbushina AA, Untereiner WA, et al. (2008) A rock-inhabiting ancestor for mutualistic and pathogen-rich fungal lineages. Stud Mycol 61: 111–119.
17. Seyedmousavi S, Badali H, Chlebicki A, Zhao J, Prenafeta-Boldú FX, et al. (2011) Exophiala sideris, a novel black yeast isolated from environments polluted with toxic alkyl benzenes and arsenic. Fungal Biol 115: 1030–1037.
18. Zhao J, Zeng J, de Hoog GS, Attili-Angelis D, Prenafeta-Boldú FX (2010) Isolation and identification of black yeasts by enrichment on atmospheres of monoaromatic hydrocarbons. Microb Ecol 60: 149–156.
19. Longino JT (2007) A taxonomic review of the genus *Azteca* (Hymenoptera: Formicidae) in Costa Rica and a global revision of the *aurita* group. Zootaxa 1491: 1–63.
20. Janzen DH, editor (1983) Costa Rican natural history. Chicago, IL: The University of Chicago Press.
21. Tennant LE (1989) A new ant-plant, *Tetrathylacium costaricense*. Symposium: Interaction between Ants Plants. Oxford, Abstracts vol.: Linnean Society of London. p. 27.
22. Schmidt M (2001) Interactions between Tetrathylacium macrophyllum (Flacourtiaceae) and its live-stem inhabiting ants. Master thesis, Univ Vienna, Austria.
23. De Hoog GS, Gerrits van den Ende AHG (1998) Molecular diagnostics of clinical strains of filamentous Basidiomycetes. Mycoses 41: 183–189.
24. Vilgalys R, Hester M (1990) Rapid genetic identification and mapping of enzymatically amplified ribosomal DNA from several *Cryptococcus* species. J Bacteriol 172: 4238–4246.
25. Werle E, Schneider C, Renner M, Völker M, Fiehn W (1994) Convenient single-step, one tube purification of PCR products for direct sequencing. Nucleic Acids Res 22: 4354–4355.
26. White TJ, Bruns T, Lee S, Taylor J (1990) Amplification and direct sequencing of fungal ribosomal RNA genes for phylogenetics. PCR Protocols: a guide to methods and applications. San Diego, USA: Academic Press. pp. 315–322.
27. Voglmayr H, Rossman AY, Castlebury LA, Jaklitsch WM (2012) Multigene phylogeny and taxonomy of the genus *Melanconiella* (Diaporthales). Fungal Divers 57: 1–44.
28. Jaklitsch WM, Voglmayr H (2011) *Nectria eustromatica* sp. nov., an exceptional species with a hypocreaceous stroma. Mycologia 103: 209–218.
29. Edgar RC (2004) MUSCLE: multiple sequence alignment with high accuracy and high throughput. Nucleic Acids Res 32: 1792–1797.
30. Hall TA (1999) BioEdit: a user-friendly biological sequence alignment editor and analysis program for Windows 95/98/NT. Nucleic Acids Symp 41: 95–98.
31. Huelsenbeck JP, Ronquist F (2001) MRBAYES: Bayesian inference of phylogenetic trees. Bioinformatics 17: 754–755.
32. Kumar S, Skjaeveland A, Orr RJS, Enger P, Ruden T, et al. (2009) AIR: A batch-oriented web program package for construction of supermatrices ready for phylogenomic analyses. BMC Bioinformatics 10: 357.
33. Posada D, Crandall KA (1998) MODELTEST: testing the model of DNA substitution. Bioinformatics 14: 817–818.
34. Swofford DL (2003) PAUP*. Phylogenetic Analysis Using Parsimony (*and Other Methods). Sunderland, Massachusetts, USA: Version 4. Sinauer Associates.
35. Stamatakis A (2006) RAxML-VI-HPC: maximum likelihood-based phylogenetic analyses with thousands of taxa and mixed models. Bioinformatics 22: 2688–2690.
36. Silvestro D, Michalak I (2012) raxmlGUI: a graphical front-end for RAxML. Org Divers Evol 12: 335–337.
37. Schoch CL, Seifert K a, Huhndorf S, Robert V, Spouge JL, et al. (2012) Nuclear ribosomal internal transcribed spacer (ITS) region as a universal DNA barcode marker for fungi. Proc Natl Acad Sci 109: 6241–6246.
38. McFall-Ngai M, Hadfield MG, Bosch TCG, Carey HV, Domazet-Lošo T, et al. (2013) Animals in a bacterial world, a new imperative for the life sciences. Proc Natl Acad Sci U S A 110: 3229–3236. doi:10.1073/pnas.1218525110.
39. Cannon PF, Kirk PM (2007) Fungal families of the world. Wallingford, UK: CABI Publishing.
40. Schlüns H, Crozier RH (2009) Molecular and chemical immune defenses in ants (Hymenoptera: Formicidae). Myrmecological News 12: 237–249.
41. Yek SH, Nash DR, Jensen AB, Boomsma JJ (2012) Regulation and specificity of antifungal metapleural gland secretion in leaf-cutting ants. Proc R Soc B 279: 4215–4222.
42. El-Elimat T, Figueroa M, Raja HA, Graf TN, Adcock AF, et al. (2012) Benzoquinones and Terphenyl compounds as Phosphodiesterase- 4B inhibitors from a fungus of the order Chaetothyriales (MSX 47445). J Nat Prod 76: 382–387.
43. Aanen DK, Ros VID, de Fine Licht HH, Mitchell J, de Beer ZW, et al. (2007) Patterns of interaction specificity of fungus-growing termites and *Termitomyces* symbionts in South Africa. BMC Evol Biol 7: 115. doi:10.1186/1471-2148-7-115.
44. Nobre T, Koné NA, Konaté S, Linsenmair KE, Aanen DK (2011) Dating the fungus-growing termites' mutualism shows a mixture between ancient codiversification and recent symbiont dispersal across divergent hosts. Mol Ecol 20: 2619–2627. doi:10.1111/j.1365-294X.2011.05090.x.
45. Mikheyev a S, Mueller UG, Boomsma JJ (2007) Population genetic signatures of diffuse co-evolution between leaf-cutting ants and their cultivar fungi. Mol Ecol 16: 209–216. doi:10.1111/j.1365-294X.2006.03134.x.
46. Maschwitz U, Hölldobler B (1970) Der Kartonnestbau bei *Lasius fuliginosus* Latr. (Hym. Formicidae). Z Vgl Physiol 66: 176–189.

47. Fabina NS, Putnam HM, Franklin EC, Stat M, Gates RD (2012) Transmission mode predicts specificity and interaction patterns in coral-*Symbiodinium* networks. PLoS One 7: 1–9. doi:10.1371/journal.pone.0044970.

48. Korb J, Aanen DK (2003) The evolution of uniparental transmission in fungus-growing termites (Macrotermitinae). Behav Ecol Sociobiol 53: 65–71. doi:10.1007/sOQ265-002-0559-y.

49. Chomnunti P, Bhat DJ, Jones EBG, Chukeatirote E, Bahkali AH, et al. (2012) Trichomeriaceae, a new sooty mould family of Chaetothyriales. Fungal Divers 56: 63–76.

On the Importance of the Distance Measures Used to Train and Test Knowledge-Based Potentials for Proteins

Martin Carlsen[1], Patrice Koehl[2], Peter Røgen[1]*

1 Department of Applied Mathematics and Computer Science, Technical University of Denmark, Kongens Lyngby, Denmark, **2** Department of Computer Science and Genome Center, University of California Davis, Davis, CA, United States of America

Abstract

Knowledge-based potentials are energy functions derived from the analysis of databases of protein structures and sequences. They can be divided into two classes. Potentials from the first class are based on a direct conversion of the distributions of some geometric properties observed in native protein structures into energy values, while potentials from the second class are trained to mimic quantitatively the geometric differences between incorrectly folded models and native structures. In this paper, we focus on the relationship between energy and geometry when training the second class of knowledge-based potentials. We assume that the difference in energy between a decoy structure and the corresponding native structure is linearly related to the distance between the two structures. We trained two distance-based knowledge-based potentials accordingly, one based on all inter-residue distances (PPD), while the other had the set of all distances filtered to reflect consistency in an ensemble of decoys (PPE). We tested four types of metric to characterize the distance between the decoy and the native structure, two based on extrinsic geometry (RMSD and GTD-TS*), and two based on intrinsic geometry (Q* and MT). The corresponding eight potentials were tested on a large collection of decoy sets. We found that it is usually better to train a potential using an intrinsic distance measure. We also found that PPE outperforms PPD, emphasizing the benefits of capturing consistent information in an ensemble. The relevance of these results for the design of knowledge-based potentials is discussed.

Editor: Yang Zhang, University of Michigan, United States of America

Funding: The first author is financed by internal funding (public, no grant number) from the Technical University of Denmark. PK acknowledges support from the NIH. The funders had no role in study design, data collection and analysis, decision to publish, or preparation of the manuscript.

Competing Interests: The authors have declared that no competing interests exist.

* Email: prog@dtu.dk

Introduction

Proteins are the essential macromolecules inside cells that perform nearly all cellular functions. Just like macroscopic tools, their shapes is a key feature for defining their functions. Structural biologists have embarked upon the challenge of finding the structures of all proteins, in hopes of unraveling this relationship between geometry and biological activity and learn in the process how cells function. Determining experimentally the structure of a protein at the atomic level however is not yet an easy task: this can be indirectly deduced from the fact that we currently know millions of protein sequences but less than hundred thousand protein structures. Predicting the structure of a protein from first principles is not much easier: direct applications of the ideas that have been used for modeling small molecules have not yet been successful on these much larger molecules. Recent reports on the advancements of *ab initio* techniques clearly show that the protein structure prediction community is making progress, but that the quality of the models they generate do not meet yet the stringent accuracy requirements to become useful to the biologists [1]. Interestingly, the series of Critical Assessment of protein Structure Prediction (CASP) meetings have highlighted that while the methods for generating models of protein structures have improved significantly [2], identifying the native-like conforma-

tions among the large collections of model structures (also called decoys) remains a significant challenge [3,4]. In this paper we focus on this problem.

Anfinsen's thermodynamics hypothesis states that the native structure of a protein is determined only by its amino acid sequence [5]. Structural and computational biologists translate this postulate into the statement, that under physiological conditions, the native state of a protein is a unique, stable minimum of the free energy. The key to solving the protein structure prediction problem amounts therefore to finding an accurate representation of this free energy function and several methods have been proposed to construct reasonable approximations of it. The two most common approaches rely on semiempirical and statistical potentials, respectively. Semiempirical methods are derived from knowledge of the basic physical principles whereas statistical potentials are based on the nonrandom statistics of known protein structures [6]. Statistical energy functions are either residue based or atom based and the most recent statistical potentials include pairwise interactions, orientations of side-chains [7], secondary structural preferences, solvent-exposure, and other geometric properties of proteins [8]. We note that there have been attempts to combine physics-based and statistics-based potentials to improve protein structure refinement [9–13].

Current protein structure prediction methods require potentials that ideally should assign "scores" to a protein structure model such that the higher the score, the less native-like the model is, where native-like is measured in terms of a distance d from the model to the native structure. If this condition is satisfied then the potential is expected to detect near native conformations even when the native conformation is not present; in addition, such an ideal potential could then be used for model refinement. In mathematical terms this can be expressed as the score function f satisfying

$$f(seq_i, r_i + dr) = f(seq_i, r_i) + d(r_i, r_i + dr), \qquad (1)$$

for any sequence seq_i and all deformations dr of its native structure r_i.

Several methods have been developed to optimize potentials towards this goal [14–17]. The choice of the distance measure d is critical to the success of these methods. The standard distance measure when comparing protein structural models is RMSD, i.e. the root mean square distance between the two models after optimal translation and rotation. RMSD however has been replaced in recent CASP experiments by the global distance test (GDT-TS [18]) due to its undesirable sensitivity towards local changes in a protein structure; GDT-TS has become one of the most commonly used distance measures in protein structure prediction. A less commonly used distance measure is the fraction of known native contacts, Q. Q quantifies the changes in the number of "contacts" found in the native structure compared to the model structure that is evaluated, where a contact corresponds to two residues being within a given threshold distance from each other. All the distance measures mentioned above identify geometric differences between two structural models but do not attempt to assess if these differences could be assigned to fluctuations due to the dynamics of the protein. Such differences would be less of a concern if they were related to geometric differences that can be explained by dynamics. As an attempt to identify the role of dynamics, Perez et al. recently introduced FlexE, a method based on a simple elastic network model that uses the deformation energy as a measure of the similarity between two structures [19]. As such, FlexE is expected to distinguish biologically relevant conformational changes from random changes.

In this work, we investigate the importance of the distance function d when optimizing an energy function f towards satisfying equation 1. We train two new $C\alpha$-based pairwise potentials, PPD and PPE, to mimic the distance between the model structure considered and its corresponding native structure, using four different definitions of the distance measure, namely RMSD, GDT-TS, Q, and MT, where MT is an anharmonic version of FlexE. These energy functions are trained and tested on sets extracted from the high resolution decoy dataset Titan-HRD [20], as well as on well known decoy datasets from DecoysRUs [21] and Rosetta [22]. We have also analyzed the performance of our potentials on the server generated Stage_1 and Stage_2 decoy sets from CASP 10 [48].

The paper is organized as follows. The next section introduces the different distance measures and describes our procedures for training and testing the potentials PPD and PPE. The following section shows the results on different decoy sets as well as a comparison between PPD, PPE, two statistical knowledge-based potentials and a semi-empirical physical potential. We conclude with a discussion of the importance of the choice of the distance measure and describe potential future work.

Materials and Methods

Geometrical distances between two structural models of the same protein

Let us consider two structural models A and B of the same protein P with N amino acids. We represent the two models as discrete sets of N points, $A = (a_1, a_2, \ldots, a_N)$ and $B = (b_1, b_2, \ldots, b_N)$ where the points a_i and b_i correspond to the positions of the $C\alpha$ atoms i in the two structures. We assume that the correspondence table between A and B is known and set such that a_i corresponds to b_i for all $i \in [1, N]$. We measure the distance between the two models either based on the Euclidean distance between the two sets of points (RMSD and GDT-TS), on differences between contact maps within each set (Q), or on an elastic network (MT).

RMSD, i.e. root mean square deviation, is the Euclidean distance between the corresponding points a_i and b_i after one of the two sets of points (usually set B) has been optimally transformed by a rigid body transformation G:

$$RMSD = \min_G \sqrt{\frac{\sum_{i=1}^{N} \|a_i - G(b_i)\|^2}{N}}. \qquad (2)$$

The rigid body transformation G is a transformation that does not produce changes in the size, shape, or topology of the protein. Such transformations are compositions of rotations and translations. Many closed-form solutions to the problem of finding the optimal G have been derived [23–25]. We note that RMSD as defined above is a metric [26].

RMSD is a distance measure based on the L_2 norm; as such, it is highly sensitive to outliers, for example due to the presence of large albeit local differences between the two structures. The global distance test (GDT) was developed to decrease this sensitivity [18]. GDT focuses on the regions of the structures that can be correctly aligned by counting the number of residues that can be superimposed within a given cutoff distance. GDT-TS (where TS stands for Total Score), combines this information for multiple cutoffs:

$$GDT - TS = \frac{n_1 + n_2 + n_4 + n_8}{4n}, \qquad (3)$$

where n_1, n_2, n_4, and n_8 are the numbers of aligned residues within 1, 2, 4, and 8 Ångströms, respectively, and n is the total aligned length. Note that GDT-TS is a quantity between 0 and 1 that represents similarity, with low values corresponding to bad correspondences, and high values (close to or equal to 1) indicating that the two models are highly similar. We have converted this similarity measure into a distance by considering GDT-TS* = 1- GDT-TS.

RMSD and GDT-TS* are computed after the two model structures have been optimally superposed. An alternative approach is to consider the intrinsic geometry of the two structures, as captured for example by a distance matrix that contains all $C\alpha - C\alpha$ distances internal to one structure. Q and MT are two examples of distance measures that use this alternate approach.

The fraction of native contacts, Q, is a distance measure that quantifies the changes of a contact map between two models for the same structure. A contact map is usually defined as

$$S_{i,j} = \begin{cases} 1 & \text{if residues } i \text{ and } j \text{ are in contact} \\ 0 & \text{otherwise,} \end{cases}$$

where two residues are in contact if they are within a given distance threshold. In this paper, we set this threshold to 9 Å. Q is then defined by

$$Q = \frac{sc}{sc + lc},$$

where sc is the number of shared contacts and lc is the number of lost contacts. Just like GDT-TS, Q is a measure of similarity. We convert it into a distance measure by defining $Q^* = 1-Q$.

Q^* quantifies changes in the contact map of a structure with no consideration of what could have been the reasons for these changes. FlexE is a new measure of similarity between protein structures that was introduced as an attempt to distinguish those changes that are biologically relevant [19]. It is based on the concept of elastic network that assigns virtual isotropic springs between pairs of residues. Elastic network models are used in normal mode analysis [27,28] for example to reconstruct proteins [29], to generate decoy sets [30], or to investigate thermal fluctuations about the native or equilibrium structure [31,32]. In the formalism introduced by Perez et al [19], the distance measure FlexE between two structures N and D is assimilated to the energetic cost of deforming one of the structures into the other:

$$FlexE(N,D) = \frac{1}{N_{\text{res}}} \sum_{i,j=1}^{N_{\text{res}}} S_{i,j}^N k_{ij} \left(r_{ij}^N - r_{ij}^D \right)^2, \qquad (4)$$

where N_{res} is the number of residues in N and D, $S_{i,j}^N$ is a contact map for structure N, r_{ij}^N and r_{ij}^D are the distances between the $C\alpha$ atoms of residues i and j in structures N and D, respectively, and k_{ij} is a force constant associated to the link between i and j. In our implementation of FlexE, we set all force constants to 1. We modify the quadratic term in equation 4 with a term congruent to the potential introduced by Toda [33] to study chains of particles interacting with non-linear forces.

The corresponding variant of FlexE, which we name MT, is defined as:

$$MT(N,D) = \frac{1}{N_{\text{res}}} \sum_{i,j=1}^{N_{\text{res}}} \frac{S_{i,j}^N}{b^2} \left(e^{-(r_{ij}^D - r_{ij}^N)b} + \left(r_{ij}^D - r_{ij}^N \right)b - 1 \right), \qquad (5)$$

where b is a parameter which we set to 0.5. We note that MT is equal to FlexE for small perturbations of the distances between residues; for large perturbations however, it penalizes compression more than extension. Finally the use of the fixed native contact map for all native-decoy comparisons ensures that both Flex-E(N,D) and MT(N,D) are well-defined.

Two new parametric potentials

A smooth, pairwise potential, PPD. We design a smooth knowledge based residue pair potential as done in [34]. For each of the 210 pairs of amino acids types we assume a potential that is determined by the corresponding $C\alpha$-$C\alpha$ distance. We model the interaction as a uniform cubic b-spline with compact support within 1 Å to 12 Å and 8 degrees of freedom, see e.g. [35]. With this model an interaction tends smoothly to zero energy at distances greater than 12 Å and is modeled freely within 4 Å–9 Å. The pair potential has $8\times210 = 1680$ parameters in total. The corresponding potential, PPD, is defined as

$$PPD = \sum_{i<j} \sum_p C_p^{aa(i)aa(j)} B_p(r_{i,j}), \qquad (6)$$

where $aa(i) \in \{1, \ldots, 20\}$ is the amino acid type of the i-th residue and $B_p(r_{i,j})$ is the p-th b-spline basis function evaluated on the distance between the i-th and j-th residues. $C_p^{aa(i)aa(j)}$ are the model parameters determined by the optimization procedure described below.

A consensus potential, PPE. We introduce a novel smooth ensemble based pair potential (PPE) that forms an artificial funnel relative to a pre-calculated contact map:

$$PPE = \sum_{i<j} S_{i,j} \sum_p C_p^{aa(i)aa(j)} B_p(r_{i,j}), \qquad (7)$$

where $S_{i,j}$ is an consensus contact map. The method to calculate the consensus contact map is described below. It is based on a similar consensus method that constructs the reference contact map from an ensemble of decoys [36].

A consensus contact map. We introduce an iterative method to compute a consensus contact map of an ensemble of decoys. The first step is to construct a contact map from the most common contacts in the ensemble. Let $M_{i,j}$ be the fraction of contacts in the ensemble for the i,j-th residue pair. The contact map is then calculated as

$$S_{i,j} = \begin{cases} 1 & \text{if } M_{i,j} > \mu \\ 0 & \text{otherwise} \end{cases} \qquad (8)$$

where μ is a cut-off fixed at 0.25. At each step, we select the 25% closest decoys to this contact map, where "closest" refers to the Hamming-distance to the contact map. This leads to a reduced ensemble from which a new contact map is computed, and the procedure is iterated. The algorithm usually converges in a few steps.

Optimizing the potentials

We design an energy landscape using a sculpting procedure. We assume that we possess a set of natives structures $\{N_i\}$ and that a set $\{D_{i,j}\}$ of decoy structures is known for each of these native structures. Let $\Delta E_{i,j}$ be the energy difference between the i-th native structure, N_i, and its j-th decoy, $D_{i,j}$, and let $d(N_i, D_{i,j})$ be the corresponding distance between N_i and $D_{i,j}$. Our method for optimizing a statistical potential [34] attempts to establish a funnel-shaped energy function by calculating the parameters that minimizes the sum of squared errors between $\Delta E_{i,j}$ and $\alpha_{N_i} d(N_i, D_{i,j})$ where α_{N_i} is a constant of proportionality. The problem can be stated as a quadratic programming (QP) problem with affine constraints,

$$\begin{aligned}\underset{X,\alpha_1\ldots\alpha_M}{\text{minimize}} \quad & \sum_{i,j}\|\Delta E_{i,j}(X)-\alpha_{N_i}d(N_i,D_{i,j})\|^2+\beta\|X\|^2\\[4pt]\text{subject to} \quad & 0.25\leq\alpha_{N_i}\leq4,\quad\text{for}\quad i=1\ldots M \qquad (9)\\[4pt]& \sum_i\alpha_{N_i}=M,\end{aligned}$$

where β is a fixed parameter used for regularization. The variables in this QP problem are X, i.e. the vector of coefficients $C^{i,j}$ introduced above, and the constants of proportionality $\alpha_{N_1}\ldots\alpha_{N_M}$, where M is the number of proteins in the training set. The last term $\beta\|X\|^2$ is a regularization term that adds a penalty onto the modulus of X. The preprocessing is trivially parallelizable since each of the terms, $\|\Delta E_{i,j}(X)-\alpha_i d(N_i,D_{i,j})\|^2$, can be calculated individually. As a consequence, the QP requires little memory and is fast to compute. We use the optimization package cplex to solve it.

Training and test sets

It is a nontrivial task to construct a "good" set of decoy structures. Any such decoy set relies on a sampling of the conformational space accessible to the protein structure of interest. The specific techniques used to generate such sampling are prone to biases [37], leading to poor sampling of the corresponding free energy surfaces. These approximate energy surfaces may not adopt a funnel like geometry in the neighborhood of the native structure and may contain many artificial potential energy barriers. To avoid the risk of learning from a specific bias introduced by one sampling technique, we have considered a variety of test sets to train and measure the performances of our energy functions. Of particular interest to us are near-native test sets since we design energy functions to mimic the neighborhoods of native structures.

We have chosen part of the Titan High Resolution Decoy set [20] as our training set. The list of proteins included in this set was originally proposed by Zhou and Skolnik [17]; it was selected on the basis that it is composed of a representative set of nonhomologous single domain proteins with maximum pairwise sequence similarity reported to be 35%. The models included in the decoy sets were generated using the torsion angle dynamics program DYANA [38] subject to distance constraints that are set to preserve the hydrophobic core of a protein. It is assumed that the hydrophobic core includes all residues within a β strand as well as all hydrophobic residues within an α-helix. The set includes 1400 proteins in total (compared to 1489 proteins in the original set of Zhou and Skolnik [17]). We eliminated all short proteins with a large radius of gyration as these proteins are overfitted by the optimization and are usually separate stretched secondary structures. We divided the remaining proteins into a training set of 1155 proteins with an average of 994 decoys per native structure (Titan-HRD*) and a test set of 142 proteins with an average of 854 decoys per native structure (Titan-HRD). The average GDT-TS distances between native and decoys over the training and test sets are 0.75 and 0.76 with a mean absolute deviation of 0.1, respectively. Note that we will use the mean absolute deviation (the l_1-norm) instead of the standard deviation (the l_2-norm) as it puts less weight on outliers.

Apart from the Titan-HRD set we use 10 freely available decoy sets that were generated using different procedures. These include 6 sets taken from DecoysRUs [21] (4 state reduced [39], hg structal [21], fisa [40], fisa casp3 [40], lmds [41] and lattice ssfit [42,43]). We also included two older versions of the Rosetta decoy

sets (Rosetta-All [44], Rosetta-Tsai [22]), the newest version Rosetta-Baker available at http://depts.washington.edu/bakerpg/decoys/ and the I-Tasser Set II [45].

The different CASP meetings have highlighted successes and failures in generating model structures that resemble the native structures of proteins. A repository of all models that have been proposed as answers to the prediction challenges that were part of these meetings is available on the CASP web page (http://predictioncenter.org). This repository provides a wealth of information on protein structure modeling, as well as useful test cases to assess the quality of new potential energy functions. We have therefore considered five CASP sets each containing models predicted by a variety of methods from the different CASP meetings (302 ensembles in total). We also generated CASP-HRD, a high resolution decoy subset of CASP 5–9, which includes models that have a TM score [46] larger than 0.5 and a RMSD less than 4 Å to the native structures. This cutoff was chosen based on the observation made by Xu and Zhang, which states that two decoys belong to the same fold when their TM-score to a native structure is higher than 0.5 [47]. CASP-HRD is constructed to have nearly the same average distance measure value as Titan-HRD but we find smaller variations of the distance measures for CASP-HRD. In that sense, it does include variations with different structural characteristics compared to Titan-HRD as it is generated by many different methods, while Titan-HRD is more homogeneous.

The total number of ensembles excluding Titan-HRD, Titan-HRD*, and CASP-HRD is 546 with an average GDT-TS between its decoys and their corresponding native structures of 0.47 with a average mean absolute deviation of 0.16. We refer to this set as "Test Set All" (TSA).

Finally, we include decoys from the latest CASP experiment, CASP10. A critical component of the CASP experiment is the assessment of the predictions that are submitted as putative models for the target proteins considered. This assessment is performed by the CASP assessors but also by the CASP community, with considerable enthusiasm, as observed in CASP10 [48]. The procedure for assessing the predictions in CASP10 differed from that of previous CASPs. The main difference was the introduction of two stages, labeled Stage_1 and Stage_2. For the former, twenty of the supposedly best predictions for each CASP target were released for assessment. Subsequently, hundred and fifty decoys were released for each target, defining Stage_2. Stage_1 ensembles are designed to survey single model assessment methods, while stage_2 allows for the survey of methods that rely on ensembles for the assessment of models. We have considered 93 targets from CASP10 for which both Stage_1 and Stage_2 test sets are available from the CASP web site (http://www.predictioncenter.org/casp10/). Compared to the other decoy sets described above, these sets contain longer protein chains. The models they include are usually as distant from their native counterparts as observed for the datasets from the previous CASP meetings. These sets however are more compact, i.e. with less diversity in distances, especially for the Stage_2 sets that resemble the CASP-HRD sets in that respect.

In table 1, we report the mean characteristics of these decoy sets (size, diversity, …) as well as information about their availability.

Preprocessing the decoy sets. To guarantee that the decoys included in a set are consistent in length with their corresponding native structure, we performed the following two-step preprocessing. First, we removed all residues in the decoys with missing backbone atoms ($C\alpha$, N, C, and O). Second, we extracted the sequences from the decoy structure files and aligned these sequences with the native sequence of the protein of interest

Table 1. Properties of the different protein decoy sets used in this study.

Decoy set	Nprot[h]	Nres[h]	Ndecoys[h]	RMSD	MT	GDT-TS	Q
Titan-HRD [a]	142	127 (35)	854 (119)	2.4 (0.5)	2.7 (1)	0.76 (0.1)	0.85 (0.04)
Titan-HRD* [a]	1155	111 (35)	994 (138)	2.6 (0.6)	2.7 (1)	0.75 (0.1)	0.85 (0.04)
TASSER Set II [b]	55	80 (17)	438 (98)	6.3 (1.5)	9.3 (3.2)	0.54 (0.05)	0.77 (0.03)
hg Structal [c]	28	150 (7)	29 (0)	4.1 (1.2)	4.4 (1.5)	0.71 (0.07)	0.85 (0.04)
4-state [c]	7	64 (4.9)	664 (15)	5.2 (1.4)	8.3 (2.9)	0.53 (0.11)	0.75 (0.05)
fisa [c]	4	60 (10)	500 (0.4)	7.5 (1.8)	8.6 (1.7)	0.47 (0.06)	0.75 (0.06)
fisa CASP3 [c]	5	88 (15)	1437 (390)	12 (1.6)	21 (4.1)	0.3 (0.03)	0.67 (0.02)
lmds [c]	10	53 (10)	433 (79)	7.7 (1.1)	12 (2.6)	0.46 (0.04)	0.72 (0.03)
lattice ssfit [c]	8	71 (10)	1997 (1.5)	9.9 (1.0)	17 (2.4)	0.3 (0.03)	0.64 (0.02)
Rosetta-All [d]	41	82 (25)	999 (0.5)	12 (1.4)	29 (5.6)	0.27 (0.03)	0.61 (0.02)
Rosetta-Tsai [d]	29	63 (9.4)	1862 (43)	7.4 (2.1)	11 (3.9)	0.46 (0.08)	0.73 (0.04)
Rosetta-Baker [d]	57	88 (20)	100 (0)	8.5 (1.4)	15 (3.3)	0.45 (0.05)	0.76 (0.03)
CASP5 [e]	41	202 (78)	117 (41)	13 (3.7)	29 (14)	0.38 (0.12)	0.68 (0.08)
CASP6 [e]	39	172 (71)	216 (34)	13 (4.9)	27 (16)	0.39 (0.12)	0.70 (0.08)
CASP7 [e]	64	183 (80)	349 (40)	10 (3.4)	17 (10)	0.47 (0.11)	0.75 (0.07)
CASP8 [e]	77	187 (81)	334 (67)	8.8 (3.1)	13 (8.6)	0.54 (0.11)	0.79 (0.06)
CASP9 [e]	81	180 (81)	402 (95)	11 (4.9)	19 (14)	0.49 (0.12)	0.77 (0.07)
CASP-HRD [e]	109	188 (79)	192 (72)	2.8 (0.4)	2.2 (0.6)	0.76 (0.03)	0.89 (0.02)
CASP10-stage1 [f]	93	232 (102)	18 (1.9)	13 (4.3)	20 (9.4)	0.46 (0.08)	0.76 (0.05)
CASP10-stage2 [f]	93	232 (102)	132 (7.6)	11 (3.7)	17 (8.2)	0.55 (0.03)	0.80 (0.03)
TSA TM>0.5 [g]	242	179 (77)	291 (119)	6.3 (2.67)	9.4 (5.5)	0.63 (0.09)	0.82 (0.05)
TSA TM <0.5 [g]	303	110 (48)	602 (436)	12 (3.9)	23 (12)	0.34 (0.1)	0.68 (0.07)

[a]Training set (Titan HRD) and test set (Titan HRD*) from the Titan High resolution decoy set [20], available at http://titan.princeton.edu/2010-10-11/Decoys/.
[b]Tasser Set II is a structurally non-redundant set of protein structures and decoys derived with the program TASSER. It is available at http://zhanglab.ccmb.med.umich.edu/decoys/.
[c]Decoy sets from the Decoys 'R' us repository http://dd.compbio.washington.edu.
[d]Different decoy Rosetta-based decoy sets (see text for details), available at http://depts.washington.edu/bakerpg/decoys/.
[e]Collection of models from the successive CASP5 to CASP9 experiments, available from the CASP web site http://predictioncenter.org. CASP-HRD is a high resolution subset of the union of the five sets CASP5 to CASP9, which includes models that have a TM-score larger than 0.5 and a RMSD less than 4 Å to the native structures.
[f]The Stage_1 and Stage_2 decoy sets used in the CASP10 quality assessment category, available from the CASP web site http://predictioncenter.org. For details on how these sets are prepared, see [48].
[g]All high and low resolution targets (TSA TM-score>0.5)/(TSA TM-score <0.5) are listed in Files S1 and S2 respectively found in the supporting information.
[h]Nprot is the number of different proteins in the dataset, Nres is the average number of residues computed over all proteins in a dataset, and Ndecoys is the average number of decoys per proteins, averaged over the dataset. RMSD, MT, GDT-TS, and Q are the distance measures between the decoys and the corresponding native structures, averaged over all decoys and all proteins. We provide both the average values and the average mean absolute deviations (in parenthesis).

(where the native sequence is derived from the ATOM record in the corresponding PDB file). If these alignments include trailing unmatched residues either in the decoys or in the native structure, these residues are removed until all sequences are identical. We found that this procedure was necessary for some of the decoy sets described above.

Assessing the quality of decoy selection: R-score

Given a distance measure and an energy function, an ensemble of decoy protein conformations contains a "best" distance model, i.e. the conformation that is closest geometrically to the native structure, as well as a "best" energy model, i.e. the model whose energy is the lowest. Ideally, these two "best" models should be the same; in practice however, they are different due to shortcomings of the potential energy function. To quantify this difference we introduce the R-score as follows. Let $\mathcal{D}$ be the ensemble of decoys and let X_i be one of its elements. The corresponding native structure is N. We define the mapping S_d from $\mathcal{D}$ to $\mathbb{R}$ as $S_d(X_i) = d(X_i, N)$, i.e. the distance between the decoy X and N, where d can be any of the four distance measures defined above. We name X_E the decoy with the lowest energy, i.e. $E(X_E) \leq E(X) \quad \forall X \in \mathcal{D}$. In parallel, we name X_d the decoy closest to N with respect to the distance d, i.e. $S_d(X_d) \leq S_d(X) \quad \forall X \in \mathcal{D}$. The R score for d and E is defined as:

$$R(d,E) \equiv \begin{cases} \dfrac{S_d(X_E) - \langle S_d \rangle}{S_d(X_d) - \langle S_d \rangle} & \text{if } |S_d(X_E) - \langle S_d \rangle| \leq |S_d(X_d) - \langle S_d \rangle| \\ -1 & \text{otherwise} \end{cases} \quad (10)$$

where $\langle S_d \rangle$ is the average value for S_d over the decoy set $\mathcal{D}$. $R(d,E)$ is designed to assess how well E mimics S in finding the best decoy. It takes values between -1 and 1 where 1 indicates that the energy has picked the best decoy. We fix the lower limit at -1 to avoid having outliers being assigned very low negative values. Note, that if an ensemble does not contain outliers then 0 is the random expectation. If we furthermore assume that the distances $S_d(X)$ are uniformly distributed then $(1 - R(d,E))/2$ is the fraction of decoys with a distance to the native structure better than $S_d(X_E)$. The R score can also be seen as the ratio between the Z-score of the best energy model, $(S_d(X_E) - \langle S_d \rangle)/\sigma(S_d)$, and the Z-score of the best distance model, $(S_d X_d - \langle S_d \rangle)/\sigma(S_d)$, where $\sigma(S_d)$ is the standard deviation for S_d over the decoy set $\mathcal{D}$.

Assessing how well the energy functions mimic a funnel in the neighborhood of the native structure

To measure how far the energy E is from the desired linear funnel shape given by Equation 1 relative to the distance measure d we report the Pearson's correlation coefficient $Corr(d,E)$ between the energy values $E(X_i)$ and distance measures $S_d(X_i)$ over all decoys X_i in the decoy set:

$$Corr(d,E) = \frac{1}{N-1} \sum_{i=1}^{N} \frac{S_d(X_i) - \langle S_d \rangle}{\sigma(S_d)} \frac{E(X_i) - \langle E \rangle}{\sigma(E)}, \quad (11)$$

where $\langle . \rangle$ and $\sigma(.)$ stand for the mean and standard deviation over the decoy set considered.

Comparing two distance measures d_1 and d_2

In the two previous subsections, we have defined a R-score $R(d,E)$ and a correlation coefficient $Corr(d,E)$ to measure how well an energy function E mimics a distance measure d. Both

quantities can be used as is to compare two distance measures d_1 and d_2. Indeed, d_2 can be assimilated to a pseudo energy function, akin to the definition of FlexE given in equation 4. The R-score and correlation coefficient between d_1 and d_2 are then simply $R(d_1, d_2)$ and $Corr(d_1, d_2)$, respectively. $Corr(d1, d2)$ measures the dependence between $d1$ and $d2$ over a decoy set, while $R(d1, d2)$ checks the "quality" of the best decoy identified by d_2, as measured by d_1. Note that this R-score between distance measures may not be symmetric.

Results and Discussion

The diversity of the distance measures

There is no unique way to compare three dimensional shapes. When comparing protein structures, two main classes of distance measures have been proposed, those based on a Euclidean distance between the positions of the atoms of the two proteins (after proper translation and rotation of one of them), and those based on the intrinsic geometry of the structures. We have considered two examples in each class, namely RMSD and GDT-TS* for the former, and MT and Q* for the latter. A full description of these four distance metrics is given in Material and Methods. As these measures capture changes of different geometric properties of the protein structures, there is no reason to believe that they are equivalent. To test the degrees to which these distances differ, we have compared them on three different sets of decoys, namely Titan-HRD, CASP-HRD, and TSA, using two different report scores, $Corr$ and R, where $Corr$ is the Pearson's correlation coefficient that measures how well d_1 mimics d_2 over a large range of distance values while R measures how (metrically) wrong the best candidate of one distance measure (i.e. the decoy with the smallest distance to its corresponding native structure) is when measured by another distance (see Materials and Methods for details). Results for $Corr$ and R are given in tables 2 and 3, respectively.

The correlations between the distance measures are high on the Titan-HRD set of decoys, with values above 0.87 for the correlation coefficients. The corresponding R-scores are above 0.76. If we assume uniform distributions of the native-decoy distances over a decoy set, the best decoy by one distance measure on average is ranked within the top 5% and within the top 12% by another distance measure for R scores of 0.9 and 0.76, respectively. These high scores are expected, as the Titan-HRD decoys are high resolution, usually very close to their native structure counterparts (see Table 1). It is interesting however that the R score between RMSD and Q* is relatively low (0.76), even on this high resolution data set. This low value indicates that a "good" decoy defined by Q* may explore a range of RMSD values. In contrast, a decoy that is close to the native structure with respect to RMSD usually has a high percentage of native contacts, as highlighted by the R score between Q* and RMSD of 0.87. In fact, we observe that the best RMSD decoy is generally scored better by the three other distance measures.

While CASP-HRD also contains high resolution decoys that are close to their corresponding native structures (with RMSD <4 Å and TM scores above 0.5), the four distance measures we tested are less dependent on this dataset than on Titan-HRD, both globally as scored by correlation coefficients and locally (i.e. in picking a "best" decoy), as highlighted by the R scores. We see two possible reasons for these differences between the two groups of decoy sets. First, the decoys in Titan-HRD are homogeneous, as they all contain the same hydrophobic cores as the native structures. In contrast, the CASP decoys were derived with many different methods, leading to heterogeneity in their geometry.

Table 2. Correlations between the four distance measures.

Test set	Distance d_1	Distance d_2			
		RMSD	MT	GDT-TS*	Q*
Titan-HRD	RMSD	1[a]	0.92 (0.06)	0.92 (0.04)	0.87 (0.08)
	MT	0.92 (0.06)	1	0.92 (0.03)	0.94 (0.03)
	GDT-TS*	0.92 (0.04)	0.92 (0.03)	1	0.95 (0.03)
	Q*	0.87 (0.08)	0.94 (0.03)	0.95 (0.03)	1
CASP-HRD	RMSD	1	0.74 (0.16)	0.73 (0.14)	0.6 (0.19)
	MT	0.74 (0.16)	1	0.72 (0.13)	0.83 (0.07)
	GDT-TS*	0.73 (0.14)	0.72 (0.13)	1	0.74 (0.13)
	Q*	0.6 (0.19)	0.83 (0.07)	0.74 (0.13)	1
CASP10-stage1	RMSD	1	0.83 (0.16)	0.71 (0.24)	0.68 (0.24)
	MT	0.83 (0.16)	1	0.73 (0.2)	0.82 (0.14)
	GDT-TS*	0.71 (0.24)	0.73 (0.2)	1	0.86 (0.12)
	Q*	0.68 (0.24)	0.82 (0.14)	0.86 (0.12)	1
CASP10-stage2	RMSD	1	0.78 (0.16)	0.51 (0.22)	0.49 (0.19)
	MT	0.78 (0.16)	1	0.52 (0.2)	0.69 (0.14)
	GDT-TS*	0.51 (0.22)	0.52 (0.2)	1	0.64 (0.17)
	Q*	0.49 (0.19)	0.69 (0.14)	0.64 (0.17)	1
TSA	RMSD	1	0.92 (0.06)	0.8 (0.15)	0.82 (0.11)
	MT	0.92 (0.06)	1	0.78 (0.14)	0.85 (0.08)
TM-score> 0.5	GDT-TS*	0.8 (0.15)	0.78 (0.14)	1	0.89 (0.12)
	Q*	0.82 (0.11)	0.85 (0.08)	0.89 (0.12)	1
TSA	RMSD	1	0.8 (0.12)	0.59 (0.24)	0.56 (0.18)
	MT	0.8 (0.12)	1	0.54 (0.2)	0.68 (0.14)
TM-score <0.5	GDT-TS*	0.59 (0.24)	0.54 (0.2)	1	0.67 (0.22)
	Q*	0.56 (0.18)	0.68 (0.14)	0.67 (0.22)	1

[a]Pearson's correlation coefficient $Corr(d_1,d_2)$ between the two distance measures d_1 and d_2. We provide both the average value and the mean absolute deviation (in parenthesis) over the data set considered.

Second, we cannot exclude an effect of sample size, as on average the sets included in Titan-HRD contain four times more decoys and larger average mean absolute deviation of distance measures than the sets included in CASP-HRD (see Table 1).

TSA, which stands for "Test Sets All" is a large heterogeneous collection of decoy sets that were generated by many different techniques (see Materials and methods for details). Some of these decoy sets are high-resolution, i.e. contains mostly native-like structures, while others are more diverse, containing decoys that are very different from their corresponding native structures, both in terms of secondary structure content and three-dimensional organization. To assess the importance of this diversity, we selected within the TSA group of decoy sets two subgroups, those for which the decoys have average TM score larger than 0.5, and those with average TM score smaller than 0.5. This 0.5 cutoff was again chosen based on the observation made by Xu and Zhang that two decoys belong to the same fold when their TM-scores to a native structure is higher than 0.5 [47]. Table 1 shows that TSA TM-score> 0.5 generally contain longer chains with fewer decoys when compared to the TSA TM-score <0.5 set. The two sets are fully listed in File S1 and File S2. Tables 2 and 3 show that the distance measures behave on the high-resolution subgroup (TM> 0.5) as on the Titan-HRD test set, i.e. with high correlations and high R scores, meaning that they are very similar to each other. On the low-resolution subgroup (TM <0.5) however, the distance

measures are poorly correlated with each other, with most correlation coefficients in the range 0.5 to 0.7. Both results confirm that when two structures are very close to each other, different distance measures quantify their differences in a similar manner. When the two structures however are very different, different distance measures will focus on different geometric differences, leading to differences in their behaviors. We observe however one exception in Table 2, in that RMSD and MT clearly remains correlated (0.80) even for the diverse subgroup of TSA with TM <0.5. The reason for this exception is unclear.

The CASP 10 Stage_1 and Stage_2 test sets usually include longer proteins than the other sets considered here, with decoys that are far from their native counterparts. In the Stage_1 sets there are very few decoys per target (by construction, see Methods above) and relatively large average mean deviations of the distance measures. For the Stage_2 test sets there are more decoys per target; these decoys however are usually very similar to each other, leading to very low mean absolute deviations for the GDT-TS* and Q* distance measures, and consequently to low correlations and R scores between the measures. As an example, the correlation between RMSD and GDT-TS* for the Stage_2 decoy sets is only 0.51 and their non symmetric R scores are R(RMSD,GDT-TS*) = 0.71 and R(GDT-TS*,RMSD) = 0.73, respectively. These low values are good indicators of significant

Table 3. Comparing the best models picked by different distance measures.

Test set	Distance d_1	Distance d_2			
		RMSD	MT	GDT-TS*	Q*
Titan-HRD	RMSD	1[a]	0.88 (0.12)	0.91 (0.09)	0.76 (0.17)
	MT	0.94 (0.06)	1	0.92 (0.08)	0.91 (0.07)
	GDT-TS*	0.96 (0.04)	0.94 (0.07)	1	0.91 (0.08)
	Q*	0.87 (0.09)	0.92 (0.07)	0.89 (0.09)	1
CASP-HRD	RMSD	1	0.71 (0.26)	0.79 (0.22)	0.49 (0.38)
	MT	0.76 (0.22)	1	0.76 (0.22)	0.76 (0.23)
	GDT-TS*	0.8 (0.22)	0.68 (0.27)	1	0.48 (0.39)
	Q*	0.57 (0.33)	0.81 (0.16)	0.66 (0.24)	1
CASP10-stage1	RMSD	1	0.81 (0.24)	0.75 (0.31)	0.79 (0.23)
	MT	0.9 (0.13)	1	0.85 (0.19)	0.94 (0.09)
	GDT-TS*	0.79 (0.24)	0.78 (0.24)	1	0.82 (0.2)
	Q*	0.78 (0.22)	0.88 (0.14)	0.8 (0.23)	1
CASP10-stage2	RMSD	1	0.76 (0.22)	0.71 (0.3)	0.63 (0.29)
	MT	0.83 (0.18)	1	0.73 (0.24)	0.83 (0.19)
	GDT-TS*	0.73 (0.26)	0.65 (0.24)	1	0.59 (0.29)
	Q*	0.62 (0.29)	0.82 (0.18)	0.62 (0.23)	1
TSA	RMSD	1	0.9 (0.11)	0.84 (0.19)	0.81 (0.18)
	MT	0.94 (0.07)	1	0.88 (0.14)	0.92 (0.09)
TM-score> 0.5	GDT-TS*	0.85 (0.16)	0.79 (0.21)	1	0.73 (0.24)
	Q*	0.79 (0.18)	0.89 (0.11)	0.81 (0.16)	1
TSA	RMSD	1	0.83 (0.19)	0.73 (0.27)	0.71 (0.27)
	MT	0.87 (0.14)	1	0.74 (0.27)	0.88 (0.14)
TM-score <0.5	GDT-TS*	0.74 (0.27)	0.7 (0.27)	1	0.67 (0.27)
	Q*	0.68 (0.27)	0.85 (0.16)	0.68 (0.27)	1

[a]R-score $R(d_1,d_2)$ between the two distance measures d_1 and d_1. We provide both the average value and the mean absolute deviation (in parenthesis) over the data set considered.

differences between their ranking of the decoys included in CASP10 Stage_2 test sets.

Training knowledge-based potentials with different distance measures

We have derived two new smooth knowledge-based residue pair potentials, PPD and PPE. Both potentials are based on distances between the $C\alpha$ atoms of the protein structure of interest. For each of the 210 types of amino acid pairs, the two potentials are written as a weighted sum of smooth spline functions, whose weights are optimized so that the total energy of a protein model resembles the distance between the model and a reference structure (usually taken to be the native structure), as described by equation 1. The two potentials differ however on which pairs of residues are taken into account. While PPD includes all pairs of residues from the protein structure P considered, PPE only include those pairs whose inter $C\alpha$ distance is consistently below a cutoff value in an ensemble of protein models similar to P. The idea behind PPE, derived from Eickholt et al. [36], is that the various models in the ensemble contain complementary information which can be pooled together to build a contact map of consistent residue-residue contacts that are more likely to be informative. Our interest here is to assess the influence of the distance measure used to train the two potentials. We have trained PPD and PPE on the Titan-HRD* training set with the four distance measures

introduced above separately, and tested the corresponding four versions of the potentials against the Titan-HRD, CASP-HRD, and TSA test sets in their abilities to mimic any of the four distance measures. All parameters describing the amino acid pair spline potentials are listed in the file Force Field S1. The encoding used and the spline basis used is described in Readme Force Field S1. Both files are in the supporting information.

Figure 1 shows some examples of the b-spline expanded pair potentials. As expected, the pair potentials are repulsive for short inter-residue distances and have a first minimum between 4 Å and 6 Å and this preferred distance relatively independent of the training metric. For longer pair distances it is seen that most PPD pair potentials have a local minimum around 10 Å whereas the PPE pair potentials tend to have a local maximum at this distance. One plausible explanation is that as PPE does not identify new contacts for these large distances; it may then set higher energy values for remote decoys. The exact placement of the minimum as well as the depth of the potential differs for the different pair potentials. While these differences may seem small, they add up when we sum over all the interactions.

We computed both the correlations between energy and the distance measure, and the R scores that compare the best decoys picked based on energy with the decoys closest to their corresponding native structures. Results are given in Table 4 for the correlation coefficients, Table 5 for the R scores, and in

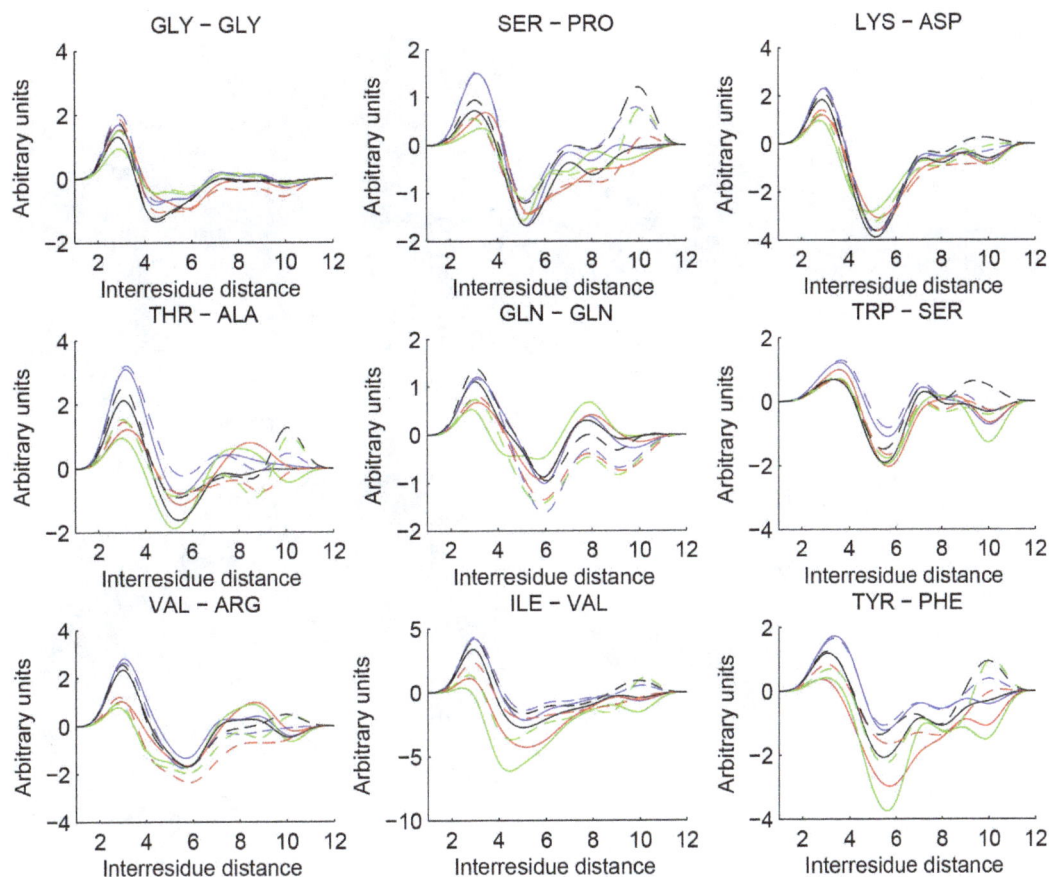

Figure 1. Showing nine different types of residue pair interactions for our single model method PPD (continuous lines) and our consensus method PPE (dotted lines) when trained on RMSD (blue), MT(red), GDT-TS(green) and Q(black).

Figures 2 and 3 for a comparison of these scores. We draw from these tables and figures the four main conclusions described below.

First, we find that both potentials PPD and PPE perform very well on the Titan-HRD test set, for all distance measures used for training and testing the potential. The corresponding mean correlation coefficients (averaged over all decoys sets in Titan-HRD) are usually above 0.8, indicating that the energy functions order the decoys in the same manner as the distance measures. In parallel, the R scores are also high, with most values well above 0.65, indicating that the decoys with the lowest energies are usually among the decoys that are close to the corresponding native structures. We should note however that PPD and PPE were trained on Titan-HRD*. While Titan-HRD and Titan-HRD* are different (see Methods), they both contain decoys that were generated with the same principles, with the significant constraint that they maintain the hydrophobic cores of the corresponding native structures. The exceptional performance of PPD and PPE may therefore not be surprising in light of this comment. Indeed, as we test these potentials on different decoy sets with more diverse populations of decoys, we observe a decrease in performance that follows the increase in diversity (in the order Titan-HRD - TSA (TM >0.5) - CASP-HRD - TSA (TM <0.5). This decrease in performance is illustrated in Figure 2.

Second, the ensemble potential PPE performs better than the single structure potential PPD, again for all the distance measures used to train and test the potentials. The differences between the two potentials are large for the high resolution decoys sets in Titan-HRD and TSA (TM>0.5), but become statistically insig-

nificant for very diverse decoy sets such as those in TSA (TM < 0.5). We believe that these differences illustrate the power of generating consensus information from an ensemble. In PPE, we only consider those contacts there are consistently below a given distance cutoff in the whole decoy set to which the protein of interest belongs. This initial filtering is clearly an advantage for Titan-HRD, as it will select the contacts in the hydrophobic cores which are native, and will ignore the contacts that fluctuate significantly due to the sampling procedure used to generate the decoys. It remains an advantage for high quality decoy but becomes less pertinent for highly diverse decoys.

Third, the performances of the two potentials PPD and PPE depend on the choice of the distance used in the training step. For example, the correlations between PPE and any of the four distance measures increase on average by 0.09 when it is trained on MT instead of RMSD (Table 4). Similar differences are observed for the R scores between PPE and the four distance measures (Table 5). More generally, it is best to train the potentials on a distance measure that is directly based on intrinsic interresidue distances, such as MT that follows the elastic network of the protein of interest, or Q* that counts the number of contacts that fall below a given distance cutoff, than on a distance measure based on extrinsic Euclidean distances, such as RMSD. Interestingly, we find that GDT-TS* behaves more like the intrinsic distance measures MT and Q* than RMSD, even though it is also based on extrinsic distances. The reason for this discrepancy is unclear.

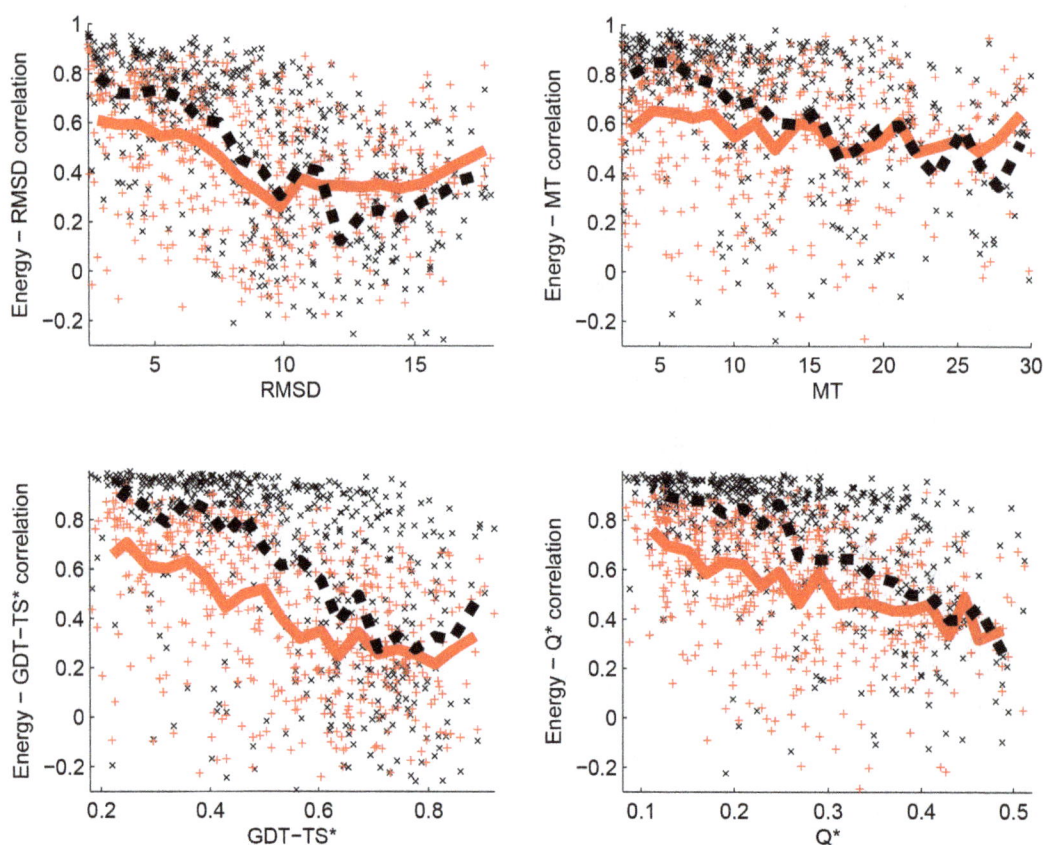

Figure 2. Energy-distance correlations as a function of the quality of the decoy set. For each decoy set in Titan-HRD, CASP-HRD, and TSA (a total of 797 sets), we plot the correlation Corr(E, d_1) as a function of the mean value of d_1 over the decoy set, where E is either the PPD energy (red, plus sign +) or the PPE energy (black, cross sign x) trained on the set Titan-HRD with the distance measure d_1, and d_1 is one of the fourth distance measures considered, namely RMSD (panel A), MT (panel B), GDT-TS* (panel C), and Q* (panel D). The corresponding running means computed over 20 equidistant intervals for PPD (red, solid line) and PPE (black, dashed line) are shown. Clearly, the quality of the correlation energy-distance decreases as the diversity of the decoy set increases.

Finally, we observe that the ability of an energy function to pick a "good" decoy (i.e. with native-like characteristics) is contingent to how well this energy function correlates with a distance measure between decoys and native structure. This is illustrated in Figure 2. This observation validates the approach of sculpting (training) a potential to mimic a distance measure.

Comparison with other energy functions

We have compared the two energy functions PPD and PPE with two well established all-atom statistical potentials RAPDF [49] and GOAP [7] and with a semi-empirical physical potential, AMBER99SB-ILDN [50], for all decoy sets in Titan-HRD, CASP-HRD, and TSA. Results for correlations between energy and distance measures and for R scores are given in Tables 4 and 5, respectively.

As intuitively expected, the performances of AMBER99SB-ILDN are very poor. This is most likely an artifact due to the presence of a few steric clashes in the decoys, and not a reflection of the quality of this potential. While it would be possible to improve this performance by applying an initial energy minimization on all decoys, this result by itself highlights that such a physical potential cannot be used directly to order a set of decoys, unless some pre-processing is applied.

RAPDF is a knowledge-based statistical potential that is based on a direct conversion of the distributions of inter-atomic distances

observed in native protein structures into energy values that are then used to assess how native-like a model is [49]. It is not based on any information from existing decoy sets, and it is not trained to mimic some differences between decoys and native structures. It is therefore not surprising that it does not perform as well as PPD and PPE, especially on the Titan-HRD as both PPD and PPE were trained on decoys resembling those included in this data set.

GOAP is an all-atom orientation-dependent knowledge-based statistical potential that includes a distance-based term and an angle-dependent contribution [7]. The distance-based term is an all-atom statistical potential that is based on the reference state that was introduced with the DFIRE potential [51]. The angle dependent component of GOAP is based on the geometric orientation of local planes. GOAP is found to perform significantly better than RAPDF on all datasets tested in this study. This is not a surprise, as GOAP includes much more information than RAPDF due to its angle term. We find however that GOAP performs only marginally better than PPD and worse than PPE. This illustrates the benefit of training a potential on a decoy set. PPD and PPE are only Ca based potentials; they have been trained however to mimic distances between non-native models and native structures of proteins.

The performances of RAPDF and GOAP depend on the distance measure used for testing. We observe that they are particularly good when the statistical potentials are tested on

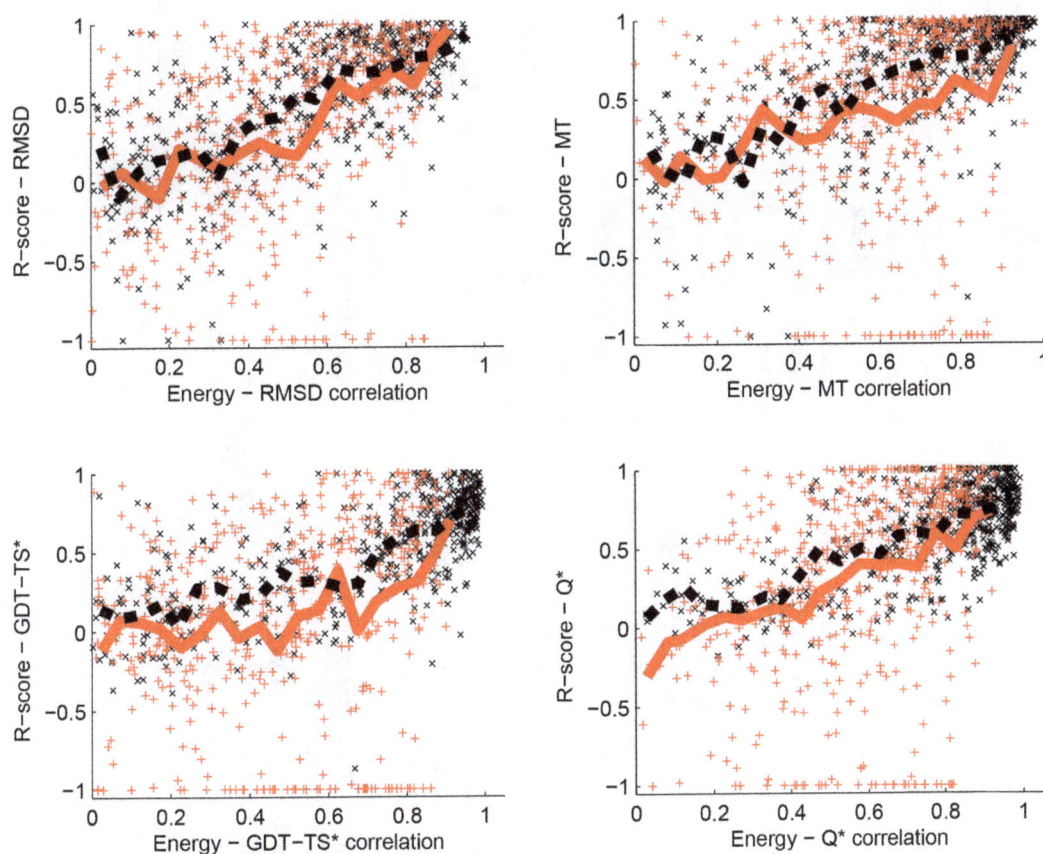

Figure 3. R scores versus Energy-distance correlations. For each decoy set in Titan-HRD, CASP-HRD, and TSA, we plot the R score $R(d_1,E)$ as a function of the correlation coefficient $Corr(d_1,E)$, where E is either the PPD energy (red, plus sign +) or the PPE energy (black, cross sign x) trained on the set Titan-HRD with the distance measure d_1, and d_1 is one of the fourth distance measures considered, namely RMSD (panel A), MT (panel B), GDT-TS* (panel C), and Q* (panel D). The corresponding running means computed over 20 equidistant intervals for PPD (red, solid line) and PPE (black, dashed line) are shown. Note that $R(d_1,E)$ compares the best decoy picked based on the energy value E with the decoy closest to the native structure according to the distance measure d_1. There is a clear correlation between these two values for all four distance measures.

GDT-TS*, reflecting the differences between these distance measures (see Table 2 and 3).

Performance in the CASP 10 quality assessment category

As part of the CASP experiment, state-of-the-art methods for protein structure assessment are judged on their ability to evaluate the quality of the predictions submitted as models for the targets considered in that specific experiment: this is the quality assessment category (QA). In 2012 as part of CASP10, 37 groups participated [48]. They were asked to evaluate the quality of sets of predictions (decoys) in two rounds designated as Stage_1 (20 decoys with a large variation in quality as measured by GDT-TS) and Stage_2 (150 decoys with homogeneous quality as measured by GDT-TS). The main reason for providing a small number of decoys in Stage_1 was to allow for judging assessment methods that rely on a single model independently from methods that rely on an ensemble of decoys (consensus methods), that would be tested extensively with the Stage_2 decoy sets. The three main conclusions drawn from these experiments were [48]: 1) The performances of the single model methods are usually worse than the the performances of consensus methods, 2) The Stage_2 sets are usually more difficult to rank than the Stage_1 sets, and 3) No methods were able to consistently pick the best decoy in an ensemble. The results for the participating groups can be seen in Figure 2 (average correlation) and Figure 3 (ability to pick the best

decoy) in [48]. We note that the single model method GOAP used in this study differs from the quasi-single model method GOAPQA used in CASP10QA. For the latter, the TM-score [46] to the top 5 ranked models is used as a measure of model quality.

The CASP 10 datasets have average native-decoy RMSDs of 11-13 Å. These differences are significantly larger than the 2.4 Å RMSD found in our training sets (see Table 1). Our analyses of the performances of PPD (single model) and PPE (ensemble of decoys) on the other datasets considered in this study have shown that for decoys that are far from their native counterparts, the two methods perform similarly, and in fact poorly (see top left panel of Figure 2 and Table Table 4). We observe the same behavior when PPD and PPE are applied on the CASP10 datasets (Tables 4 and 5). Similarly we expect and indeed find that the ensemble method PPE is ineffective in ranking the decoys of the CASP10 datasets when its performance is measured against the MT distance measure, and shows some prospects when its performance is measured against the GDT-TS* and Q* distance measures. The energy-GDT-TS correlations of 0.51(0.63) and 0.29(0.44) for PPD(resp. PPE) on Stage_1 and Stage_2 respectively are amongst the lowest reported for single model(resp. ensemble) methods in CASP10QA [48]. The low energy-distance correlations reported usually leads to a bad pick for the best decoy, see Figure 3. It is therefore surprising that the average ΔGDT-TS* of 0.07 between the GDT-TS*-closest decoy and the lowest energy decoy picked

Table 4. Energy-distance correlations.

Decoy set	Test Distance d_2	PPD Training distance d_1				PPE Training distance d_1				RAPDF[a]	GOAP[b]	AMBER[c]
		RMSD	MT	GDT-TS*	Q*	RMSD	MT	GDT-TS*	Q*			
Titan-HRD	RMSD	0.77 (0.05)[e]	0.82 (0.04)	0.81 (0.05)	0.79 (0.04)	0.81 (0.05)	0.88 (0.03)	0.85 (0.03)	0.82 (0.04)	0.5 (0.14)	0.64(0.11)	0.01 (0.02)
	MT	0.82 (0.04)	0.89 (0.03)	0.87 (0.03)	0.87 (0.03)	0.86 (0.04)	0.95 (0.01)	0.92 (0.02)	0.89 (0.02)	0.47 (0.16)	0.63(0.11)	0.01 (0.02)
	GDT-TS*	0.83 (0.06)	0.91 (0.02)	0.91 (0.02)	0.9 (0.03)	0.86 (0.04)	0.93 (0.02)	0.95 (0.01)	0.92 (0.02)	0.43 (0.18)	0.63(0.12)	0.001 (0.02)
	Q*	0.82 (0.05)	0.92 (0.2)	0.92 (0.02)	0.93 (0.02)	0.87 (0.03)	0.95 (0.01)	0.97 (0.01)	0.95 (0.01)	0.37 (0.22)	0.57(0.13)	0.002 (0.02)
CASP-HRD	RMSD	0.32 (0.16)	0.31 (0.16)	0.31 (0.16)	0.26 (0.16)	0.42 (0.16)	0.51 (0.17)	0.51 (0.18)	0.45 (0.17)	0.31(0.15)	0.3 (0.15)	0.01 (0)
	MT	0.45 (0.11)	0.49 (0.11)	0.49 (0.12)	0.43 (0.12)	0.55 (0.13)	0.69 (0.11)	0.68 (0.13)	0.61 (0.13)	0.37 (0.14)	0.41(0.13)	0.02 (0)
	GDT-TS*	0.38 (0.14)	0.39 (0.13)	0.39 (0.13)	0.33 (0.14)	0.51 (0.15)	0.6 (0.14)	0.65 (0.14)	0.58 (0.17)	0.39 (0.14)	0.43(0.11)	0.02 (0)
	Q*	0.46 (0.12)	0.6(0.11)	0.64 (0.1)	0.57 (0.1)	0.54 (0.12)	0.69 (0.12)	0.75 (0.12)	0.71 (0.1)	0.32 (0.15)	0.42(0.12)	0.02 (0)
CASP10-stage1	RMSD	0.44 (0.22)	0.53 (0.18)	0.53 (0.18)	0.48 (0.2)	0.5 (0.22)	0.54 (0.22)	0.52 (0.21)	0.5 (0.21)	0.18(0.24)	0.32 (0.26)	−0.03 (0)
	MT	0.47 (0.19)	0.61(0.17)	0.62 (0.17)	0.55 (0.18)	0.56 (0.16)	0.63 (0.13)	0.61 (0.13)	0.57 (0.13)	0.13 (0.22)	0.34(0.24)	−0.06 (0)
	GDT-TS*	0.4 (0.21)	0.49 (0.23)	0.51 (0.2)	0.43 (0.21)	0.57 (0.21)	0.63 (0.16)	0.63 (0.16)	0.59 (0.2)	0.22 (0.28)	0.4(0.2)	−0.05 (0)
	Q*	0.51 (0.22)	0.63(0.16)	0.63 (0.16)	0.56 (0.18)	0.68 (0.12)	0.75 (0.06)	0.75 (0.07)	0.72 (0.1)	0.24 (0.3)	0.41(0.2)	−0.05 (0)
CASP10-stage2	RMSD	0.33 (0.18)	0.34 (0.2)	0.34 (0.2)	0.31 (0.2)	0.31 (0.18)	0.37 (0.19)	0.34 (0.16)	0.3 (0.21)	0.15(0.13)	0.2 (0.14)	−0.005 (0)
	MT	0.42 (0.16)	0.49 (0.16)	0.49 (0.15)	0.45 (0.14)	0.4 (0.16)	0.5 (0.19)	0.48 (0.15)	0.42 (0.16)	0.19 (0.14)	0.29(0.14)	0.003 (0)
	GDT-TS*	0.31 (0.15)	0.29 (0.14)	0.29 (0.13)	0.25 (0.14)	0.37 (0.16)	0.42 (0.19)	0.44 (0.18)	0.38 (0.18)	0.29 (0.14)	0.37(0.17)	0.007 (0)
	Q*	0.45 (0.22)	0.56(0.14)	0.58 (0.12)	0.52 (0.15)	0.51 (0.18)	0.62 (0.17)	0.66 (0.14)	0.62 (0.15)	0.28(0.13)	0.41(0.13)	−0.006 (0)
TSA	RMSD	0.62 (0.12)	0.62 (0.13)	0.63 (0.13)	0.59 (0.14)	0.74 (0.09)	0.8 (0.07)	0.78 (0.08)	0.73 (0.09)	0.5 (0.14)	0.58(0.13)	0.02 (0.01)
	MT	0.65 (0.11)	0.69 (0.1)	0.7 (0.1)	0.65 (0.12)	0.75 (0.08)	0.83 (0.06)	0.8 (0.06)	0.74 (0.07)	0.5 (0.16)	0.58(0.12)	0.03 (0.01)
TM-score > 0.5	GDT-TS*	0.6 (0.15)	0.59 (0.14)	0.6 (0.13)	0.54 (0.16)	0.78 (0.06)	0.85 (0.04)	0.84 (0.04)	0.79 (0.06)	0.61 (0.11)	0.7(0.1)	0.03 (0.01)
	Q*	0.69 (0.11)	0.71 (0.09)	0.72 (0.1)	0.68 (0.1)	0.87 (0.04)	0.94 (0.02)	0.93 (0.02)	0.9 (0.03)	0.57 (0.13)	0.67(0.11)	0.03 (0.01)
TSA	RMSD	0.3 (0.16)	0.34 (0.18)	0.34 (0.18)	0.32 (0.18)	0.29 (0.23)	0.36 (0.27)	0.34 (0.27)	0.29 (0.23)	0.16 (0.13)	0.25(0.15)	0 (0.01)
	MT	0.38 (0.14)	0.47 (0.15)	0.47 (0.15)	0.45 (0.17)	0.34 (0.23)	0.45 (0.24)	0.41 (0.23)	0.35 (0.22)	0.19 (0.14)	0.29(0.13)	−0.003(0.01)
TM-score < 0.5	GDT-TS*	0.27 (0.19)	0.27 (0.2)	0.28 (0.19)	0.24 (0.18)	0.36 (0.29)	0.44 (0.33)	0.42 (0.32)	0.36 (0.29)	0.26 (0.16)	0.33(0.19)	0.004 (0.02)
	Q*	0.41 (0.17)	0.47 (0.16)	0.46 (0.17)	0.45 (0.15)	0.53 (0.19)	0.63 (0.18)	0.61 (0.18)	0.57 (0.19)	0.23 (0.17)	0.3(0.19)	−0.004 (0.02)

[a] All-atom statistical distance-based potential [49].
[b] All-atom orientation-dependent statistical potential [7].
[c] The semi-empirical physical potential AMBER99SB-ILDN[50].
[d] PPD and PPE have been trained on the distance measure d_1, and tested against the distance measure d_2.
[e] Average value, and mean absolute deviation (in parenthesis) over the data set.

Table 5. Energy-distance Rvalues.

Decoy set	Test Distance d_2[d]	PPD Training distance d_1[d]				PPE Training distance d_1[d]				RAPDF[a]	GOAP[b]	AMBER[c]
		RMSD	MT	GDT-TS*	Q*	RMSD	MT	GDT-TS*	Q*			
Titan-HRD	RMSD	0.57 (0.17)[e]	0.52 (0.17)	0.51 (0.19)	0.48 (0.2)	0.61 (0.14)	0.63 (0.17)	0.62 (0.14)	0.61 (0.15)	0.43 (0.18)	0.56 (0.15)	0.26 (0.47)
	MT	0.73 (0.1)	0.71 (0.11)	0.69 (0.13)	0.69 (0.15)	0.76 (0.11)	0.8 (0.11)	0.79 (0.12)	0.8 (0.1)	0.5 (0.18)	0.6(0.17)	0.23 (0.53)
	GDT-TS*	0.78 (0.08)	0.74 (0.09)	0.74 (0.08)	0.73 (0.09)	0.82 (0.07)	0.86 (0.06)	0.86 (0.07)	0.86 (0.07)	0.52 (0.22)	0.67(0.12)	0.18 (0.58)
	Q*	0.73 (0.11)	0.76 (0.12)	0.73 (0.1)	0.77 (0.1)	0.77 (0.09)	0.84 (0.09)	0.85 (0.09)	0.86 (0.08)	0.37 (0.19)	0.53(0.17)	0.16 (0.51)
CASP-HRD	RMSD	0.19 (0.31)	0.00 (0.42)	-0.04 (0.48)	0.03 (0.4)	0.27 (0.3)	0.33 (0.31)	0.34 (0.27)	0.3 (0.31)	0.14 (0.37)	0.22(0.37)	-0.11 (0.34)
	MT	0.31 (0.26)	0.24 (0.3)	0.14 (0.31)	0.16 (0.36)	0.38 (0.24)	0.43 (0.22)	0.43 (0.21)	0.38 (0.25)	0.24 (0.4)	0.46(0.32)	-0.09 (0.47)
	GDT-TS*	0.14 (0.3)	-0.09 (0.4)	-0.08 (0.42)	-0.06 (0.43)	0.28 (0.23)	0.31 (0.25)	0.32 (0.24)	0.28 (0.24)	0.12 (0.44)	0.34(0.33)	-0.22 (0.38)
	Q*	0.27 (0.25)	0.35 (0.33)	0.34 (0.32)	0.33 (0.31)	0.28 (0.2)	0.39 (0.24)	0.4 (0.25)	0.41 (0.26)	0.13 (0.4)	0.43(0.26)	-0.17(0.42)
CASP10-stage1	RMSD	0.55 (0.23)	0.53 (0.3)	0.57 (0.26)	0.48 (0.39)	0.5 (0.29)	0.52 (0.26)	0.53 (0.26)	0.51 (0.27)	0.32 (0.34)	0.44(0.26)	0.13 (0.6)
	MT	0.69(0.1)	0.7(0.12)	0.72(0.12)	0.64(0.14)	0.58(0.15)	0.62(0.12)	0.63(0.13)	0.6(0.15)	0.37(0.28)	0.52(0.2)	0.16(0.6)
	GDT-TS*	0.52(0.32)	0.47(0.39)	0.52(0.37)	0.42(0.43)	0.43(0.4)	0.46(0.37)	0.51(0.33)	0.46(0.36)	0.27(0.44)	0.53(0.22)	0.15(0.37)
	Q*	0.6(0.24)	0.64(0.15)	0.67(0.14)	0.57(0.19)	0.57(0.22)	0.6(0.18)	0.63(0.17)	0.61(0.18)	0.32(0.38)	0.47(0.31)	0.19(0.5)
CASP10-stage2	RMSD	0.38(0.29)	0.23(0.36)	0.29(0.3)	0.26(0.35)	0.36(0.31)	0.35(0.32)	0.32(0.32)	0.35(0.34)	0.29(0.28)	0.39(0.29)	0.11(0.42)
	MT	0.55(0.23)	0.46(0.31)	0.5(0.29)	0.49(0.33)	0.45(0.32)	0.47(0.32)	0.47(0.32)	0.48(0.3)	0.44(0.32)	0.52(0.24)	0.17(0.45)
	GDT-TS*	0.23(0.35)	0.11(0.33)	0.14(0.36)	0.14(0.34)	0.25(0.32)	0.23(0.28)	0.29(0.32)	0.25(0.32)	0.23(0.3)	0.39(0.3)	0.01(0.27)
	Q*	0.45(0.32)	0.46(0.31)	0.51(0.29)	0.48(0.3)	0.44(0.3)	0.47(0.24)	0.51(0.27)	0.53(0.27)	0.33(0.28)	0.41(0.27)	0.13(0.43)
TSA	RMSD	0.47 (0.24)	0.22 (0.41)	0.21 (0.41)	0.22 (0.41)	0.69 (0.14)	0.69(0.13)	0.69 (0.13)	0.68 (0.14)	0.44 (0.25)	0.59(0.19)	0.24 (0.38)
	MT	0.62 (0.14)	0.4 (0.24)	0.39 (0.27)	0.41 (0.24)	0.8 (0.08)	0.81 (0.07)	0.82 (0.08)	0.8 (0.08)	0.54 (0.18)	0.74(0.1)	0.32 (0.33)
TM-score> 0.5	GDT-TS*	0.29 (0.28)	0.09 (0.43)	0.08 (0.47)	0.09 (0.45)	0.58 (0.16)	0.6 (0.16)	0.61 (0.16)	0.57 (0.18)	0.37 (0.31)	0.56(0.41)	0.09 (0.26)
	Q*	0.49 (0.19)	0.32 (0.3)	0.3 (0.33)	0.33 (0.28)	0.68 (0.14)	0.72 (0.14)	0.74 (0.13)	0.74 (0.13)	0.38 (0.29)	0.59(0.43)	0.14 (0.2)
TSA	RMSD	0.16 (0.35)	0.19 (0.3)	0.19 (0.34)	0.16 (0.32)	0.26 (0.35)	0.33 (0.37)	0.32 (0.36)	0.29 (0.35)	0.19 (0.35)	0.27(0.4)	0.04 (0.41)
	MT	0.27 (0.34)	0.38 (0.3)	0.39 (0.33)	0.37 (0.29)	0.36 (0.33)	0.44 (0.31)	0.42 (0.32)	0.41 (0.31)	0.28 (0.34)	0.4(0.33)	0.04 (0.46)
TM-score <0.5	GDT-TS*	0.07 (0.3)	0.07 (0.28)	0.1 (0.3)	0.06 (0.29)	0.26 (0.3)	0.32 (0.3)	0.32 (0.3)	0.29 (0.3)	0.19 (0.3)	0.28(0.36)	0.05 (0.3)
	Q*	0.18 (0.35)	0.28 (0.34)	0.29 (0.35)	0.3 (0.31)	0.42 (0.27)	0.5 (0.27)	0.49 (0.29)	0.49 (0.27)	0.19 (0.29)	0.31(0.32)	0.01 (0.31)

[a]All-atom statistical distance-based potential [49].
[b]All-atom orientation-dependent statistical potential [7].
[c]The semi-empirical physical potential AMBER99SB-ILDN [50].
[d]PPD and PPE have been trained on the distance measure d_1 and tested against the distance measure d_2.
[e]Average value, and mean absolute deviation (in parenthesis) over the data set.

Table 6. Assessing the best decoys selected by energy functions on different decoy datasets.

		Best	PPD	PPE	RAPDF[a]	AMBER[b]	GOAP[c]
Titan-HRD	RMSD	1.1(0.21)[d]	1.7(0.29)	1.6(0.27)	1.9(0.4)	2.1(0.55)	1.7(0.3)
	MT	0.75(0.22)	1.4(0.38)	1.2(0.38)	1.8(0.62)	2.3(0.89)	1.6(0.54)
	GDT-TS	0.94(0.02)	0.89(0.03)	0.92(0.03)	0.85(0.05)	0.8(0.09)	0.88(0.03)
	Q	0.94(0.01)	0.92(0.02)	0.93(0.02)	0.88(0.03)	0.86(0.04)	0.89(0.03)
4-state	RMSD	1.1(0.1)	3.8(0.44)	2.2(0.21)	2.1(0.22)	3.6(1.5)	1.6(0.24)
	MT	0.9(0.33)	5.8(2.1)	1.2(0.31)	2.6(0.52)	6.2(3.4)	1.5(0.38)
	GDT-TS	0.91(0.03)	0.55(0.06)	0.86(0.08)	0.8(0.04)	0.67(0.1)	0.86(0.04)
	Q	0.94(0.02)	0.75(0.03)	0.92(0.02)	0.87(0.04)	0.79(0.1)	0.9(0.02)
fisa	RMSD	3.7(0.76)	5.7(0.78)	6.5(1.4)	4.4(0.72)	8.5(1.5)	4.5(0.45)
	MT	3.8(1.5)	7.9(3.7)	5.5(2)	5.4(2.5)	10(4.2)	4.9(1.9)
	GDT-TS	0.65(0.07)	0.51(0.14)	0.54(0.08)	0.6(0.06)	0.46(0.08)	0.59(0.06)
	Q	0.82(0.02)	0.79(0.03)	0.78(0.03)	0.78(0.02)	0.73(0.02)	0.79(0.03)
fisa CASP3	RMSD	6(2)	12(1.6)	12(2.4)	12(4)	12(1)	11(1.6)
	MT	8.7(4.2)	21(7.3)	17(8.2)	23(4.5)	19(2.5)	18(7.7)
	GDT-TS	0.47(0.12)	0.32(0.01)	0.34(0.02)	0.32(0.02)	0.29(0.01)	0.33(0.04)
	Q	0.76(0.06)	0.72(0.04)	0.72(0.04)	0.68(0.04)	0.67(0.07)	0.69(0.06)
hg Structal	RMSD	1.9(0.5)	2.6(1)	2.5(0.56)	2.2(0.5)	3.3(0.71)	2.4(0.6)
	MT	1.8(0.3)	2.5(0.61)	3(0.28)	2.4(0.35)	3.7(0.8)	2.7(0.28)
	GDT-TS	0.86(0.06)	0.82(0.14)	0.85(0.07)	0.84(0.07)	0.77(0.08)	0.84(0.08)
	Q	0.93(0.03)	0.92(0.03)	0.92(0.03)	0.92(0.04)	0.89(0.03)	0.92(0.04)
lmds	RMSD	5.7(0.33)	9.9(0.72)	9.8(0.89)	9.8(0.92)	10(0.61)	10(0.65)
	MT	8(0.78)	14(3.7)	17(5.5)	16(2.5)	19(1.6)	19(4.5)
	GDT-TS	0.45(0.04)	0.29(0.04)	0.32(0.05)	0.31(0.05)	0.28(0.03)	0.3(0.03)
	Q	0.74(0.02)	0.67(0.05)	0.67(0.04)	0.65(0.04)	0.63(0.03)	0.63(0.05)
lattice ssfit	RMSD	3.8(0.46)	7.6(1.3)	7.4(1.6)	7.7(1.9)	8(2.6)	8.5(1.2)
	MT	5.2(2.2)	9.8(4.5)	10(5.1)	11(4.8)	12(6.4)	12(5)
	GDT-TS	0.62(0.06)	0.45(0.07)	0.48(0.07)	0.49(0.12)	0.45(0.07)	0.44(0.04)
	Q	0.8(0.06)	0.74(0.07)	0.75(0.04)	0.74(0.05)	0.72(0.07)	0.72(0.06)
CASP5	RMSD	6.7(2.9)	13(6.2)	11(6)	10(5.2)	11(6.1)	10(5.2)
	MT	8.5(4.2)	20(11)	18(7.7)	20(12)	22(11)	20(9.9)
	GDT-TS	0.58(0.19)	0.36(0.17)	0.48(0.24)	0.44(0.19)	0.46(0.21)	0.5(0.23)
	Q	0.82(0.09)	0.72(0.13)	0.77(0.09)	0.7(0.1)2	0.72(0.13)	0.75(0.1)
CASP6	RMSD	4.8(1.5)	10(5.1)	11(4.5)	9.7(5.1)	12(5.9)	8(3.1)
	MT	5.4(1.9)	23(11)	18(4.3)	19(5.7)	24(15)	12(4)
	GDT-TS	0.64(0.14)	0.33(0.14)	0.52(0.18)	0.49(0.27)	0.38(0.17)	0.54(0.17)

Table 6. Cont.

		Best	PPD	PPE	RAPDF[a]	AMBER[b]	GOAP[c]
CASP7	Q	0.85(0.06)	0.7(0.07)	0.79(0.08)	0.75(0.11)	0.68(0.11)	0.79(0.09)
	RMSD	4.5(1.8)	8.8(4.9)	7.1(3.1)	7.9(3.9)	11(5.1)	7.8(3.4)
	MT	3.8(1.6)	9.5(3.8)	6.6(2.8)	10(4.3)	18(9.1)	8.3(3.5)
	GDT-TS	0.66(0.13)	0.49(0.21)	0.56(0.14)	0.56(0.17)	0.43(0.21)	0.58(0.13)
CASP8	Q	0.88(0.04)	0.81(0.08)	0.85(0.06)	0.81(0.06)	0.74(0.08)	0.82(0.06)
	RMSD	4.1(1.3)	7.4(2.7)	6.4(1.8)	9.8(5.5)	9.7(5.2)	7.5(3.1)
	MT	3.2(1.3)	10(4.1)	6.1(2.2)	14(6.4)	15(8)	8.7(2.7)
	GDT-TS	0.7(0.1)	0.53(0.17)	0.63(0.13)	0.51(0.22)	0.47(0.19)	0.61(0.16)
CASP9	Q	0.89(0.04)	0.81(0.07)	0.85(0.06)	0.79(0.09)	0.75(0.1)	0.83(0.07)
	RMSD	4.8(1.4)	9.7(4.9)	7.5(2.6)	9.4(4.7)	9.8(4.9)	8.2(3.3)
	MT	3.7(1.3)	14(7.8)	7.3(2.6)	12(3.9)	13(5.2)	8.5(2.6)
	GDT-TS	0.68(0.1)	0.4(0.15)	0.6(0.12)	0.52(0.19)	0.51(0.21)	0.57(0.15)
TASSER Set II	Q	0.88(0.04)	0.75(0.12)	0.85(0.04)	0.8(0.09)	0.79(0.09)	0.83(0.07)
	RMSD	3.2(1)	5.6(1.9)	5.2(1.4)	5.2(1.6)	6.4(2.1)	5.4(1.9)
	MT	3.7(1.2)	7.1(2.3)	6.6(2.5)	6.7(2.2)	11(5.4)	6.8(2.2)
	GDT-TS	0.69(0.09)	0.57(0.12)	0.59(0.12)	0.59(0.1)	0.52(0.13)	0.59(0.12)
Rosetta-All	Q	0.85(0.05)	0.81(0.06)	0.82(0.06)	0.8(0.05)	0.75(0.08)	0.79(0.05)
	RMSD	6.4(1.3)	11(2.1)	11(2.2)	12(3.2)	16(4.7)	11(2.6)
	MT	12(4.7)	22(8.9)	26(7.2)	25(9.9)	47(14)	24(7.5)
	GDT-TS	0.41(0.06)	0.29(0.04)	0.29(0.04)	0.28(0.04)	0.25(0.04)	0.28(0.04)
Rosetta-Baker	Q	0.72(0.06)	0.64(0.07)	0.64(0.07)	0.62(0.07)	0.59(0.07)	0.62(0.07)
	RMSD	4.7(2.2)	7.5(3.4)	8.4(4.3)	7.6(4.1)	8.2(2.7)	6.9(3.6)
	MT	6.9(3)	13(7.5)	15(9.2)	13(7.8)	13(6.8)	11(5.4)
	GDT-TS	0.6(0.21)	0.47(0.15)	0.46(0.13)	0.48(0.15)	0.46(0.13)	0.5(0.18)
Rosetta-Tsai	Q	0.84(0.08)	0.77(0.1)	0.77(0.11)	0.77(0.01)	0.76(0.07)	0.79(0.09)
	RMSD	2.8(0.8)	6.9(2.8)	5(1.4)	7.3(1.8)	5.7(2)	6.2(2.1)
	MT	3(1.1)	9.3(4.9)	5.5(2.2)	10(5.8)	8.3(3)	7.1(2.3)
	GDT-TS	0.72(0.08)	0.47(0.08)	0.59(0.09)	0.45(0.06)	0.54(0.09)	0.52(0.12)
CASP-HRD	Q	0.86(0.05)	0.77(0.08)	0.81(0.08)	0.74(0.07)	0.77(0.07)	0.77(0.06)
	RMSD	2(0.56)	2.6(0.73)	2.6(0.71)	2.6(0.76)	2.8(0.7)	2.6(0.74)
	MT	1.1(0.5)	2(0.8)	1.7(0.61)	2(0.83)	2.3(0.11)	1.7(0.79)
	GDT-TS	0.83(0.07)	0.75(0.06)	0.78(0.07)	0.77(0.07)	0.75(0.06)	0.78(0.07)
CASP10-stage1[e]	Q	0.93(0.03)	0.9(0.03)	0.91(0.03)	0.89(0.04)	0.88(0.04)	0.91(0.03)
	RMSD	4.7(1.2)	6.6(2.3)	7.1(2.4)	8.2(3.2)	12(4.7)	7.3(2.4)
	MT	4(2)	6.6(2)	8(1.6)	10(3.5)	20(3.4)	7.9(2.2)

Table 6. Cont.

		Best	PPD	PPE	RAPDF [a]	AMBER [b]	GOAP [c]
	GDT-TS	0.71(0.1)	0.62(0.13)	0.63(0.15)	0.57(0.16)	0.55(0.19)	0.63(0.14)
	Q	0.88(0.04)	0.84(0.04)	0.85(0.04)	0.82(0.06)	0.8(0.06)	0.84(0.06)
CASP10-stage2 [e]	RMSD	4(1.1)	5.9(1.7)	5.9(1.6)	6.7(1.9)	9(2.2)	6.2(1.6)
	MT	3.1(1)	5.4(1.6)	5.8(1.6)	6.6(1.8)	14(1.9)	5.4(1.5)
	GDT-TS	0.73(0.09)	0.64(0.14)	0.66(0.11)	0.65(0.12)	0.65(0.11)	0.67(0.11)
	Q	0.89(0.04)	0.86(0.03)	0.87(0.04)	0.86(0.05)	0.85(0.04)	0.86(0.04)

[a]All-atom statistical distance-based potential [49].
[b]The semi-empirical physical potential AMBER99SB-ILDN [50].
[c]All-atom orientation-dependent statistical potential [7].
[d]Average value, and mean absolute deviation (in parenthesis) over the data set.
[e]Only ensembles who contains a decoy with a GDT-TS> 0.4 are included. Compare with Figure 2 in [48].

by PPE on the CASP10 Stage_2 data sets places PPE in the middle of the CASP10 participating methods (see [48] Figure 2(A)).

The results for PPD, PPE, AMBER99SB-ILDN, RAPDF and GOAP on CASP 10 stages 1 and 2 are given in Tables 4 - 6 where PPD and PPE were trained and tested on the same distance measure. Clearly, GOAP has a better performance than PPD when GDT-TS* is chosen as a measure of distance. It is however noteworthy that PPD performs better than GOAP when measured by RMSD and MT instead. It is encouraging that the distance dependent C-alpha potential, PPD, as a single model method has a performance that is comparable to the state-of-the-art orientation-dependent all-atom potential, GOAP. We find that PPD is good at selecting a decoy that is close to the native structure (Table 6).

Concluding Remarks

The recent literature on generating knowledge-based potentials for protein structure modeling makes no secrets of their limitations and problems. Knowledge-based potentials are energy functions derived primarily from databases of protein structures and sequences. They can be divided into two classes. Potentials from the first class are based on a direct conversion of the distributions of some geometric properties observed in native protein structures into energy values, while potentials from the second class are trained to mimic quantitatively the geometric differences between incorrectly folded models (also called decoys) and native structures. Both potentials are designed to assess how native-like a model structure is. There is no consensus however on which geometric property should be considered, on how to convert a statistical distribution into an energy for the first class, and on how energy and geometry should be related in the second class.

In this paper, we focused on the relationship between energy and geometry when training knowledge-based potentials from the second class. We assumed that the difference between the energy of a decoy and the energy of its corresponding native structure must be linearly related to the distance between the decoy and the native structure. We trained two distance-based Cα potentials accordingly, one based on all inter-residue distances (PPD), while the other had the set of all these distances filtered to reflect consistency in an ensemble of decoys (PPE). Compared to other methods that follow the same approach however, we did not assume that the distance between a decoy and the native structure is the traditional RMSD. Instead, we tested four different distance measures, two based on extrinsic geometry (RMSD and GTD-TS*), and two based on intrinsic geometry (Q* and MT). We found that it is usually better to train the potentials using the latter type of distances.

We have found that both PPD and PPE perform extremely well on the high resolution decoy set Titan-HRD, with correlation coefficients between energy and distance usually well above 0.8. PPE always performs better than PPD on this set, emphasizing the benefits of capturing consistent information in an ensemble. While we trust the general trends highlighted by these results, we tone down the importance of In extensive testing on available decoy sets and models from the Critical Assessment of protheir exceptional character as they may only reflect the specificity of the Titan-HRD data set. tein Structure Prediction (CASP) experiments we find that PPD yields better energy-distance correlations than one of the state of the art single model potentials, GOAP [7]. We note however that the sophisticated distance-based and orientation-based statistical potential GOAP is better at picking the best decoys and has a better though comparable performance for fixed energy-distance correlation. It should be noted that PPD and PPE are Cα-based, while GOAP is an all-atom potential. We believe that this demonstrates that a very efficient training of a simple distance-based pair potential can generate a very effective measure for assessing protein structure models.

There is still room for improvement in training knowledge-based potentials. We limited our study to pairwise potentials; we will test different geometric properties of protein structures in future studies. We plan to include the potentials described here into a structure minimization package, to assess their performances in improving non-native protein structure models.

References

1. Zhang Y (2009) Protein structure prediction: when is it useful? Curr Opin Struct Biol 19: 145–155.
2. Moult J, Fidelis K, Kryshtafovych A, Tramontano A (2011) Critical assessment of methods of protein structure prediction (CASP)-round IX. Proteins: Struct Func Bioinfo 79: 1–5.
3. Cozzetto D, Kryshtafovych A, Tramontano A (2009) Evaluation of CASP8 model quality predictions. Proteins: Struct Func Bioinfo 77: 157–166.
4. Kryshtafovych A, Fidelis K, Tramontano A (2011) Evaluation of model quality predictions in CASP9. Proteins: Struct Func Bioinfo 79: 91–106.
5. Anfinsen C (1973) Principles that govern the folding of protein chains. Science 181: 223–230.
6. Lazaridis T, Karplus M (2000) Effective energy functions for protein structure prediction. Curr Opin Struct Biol 10: 139–145.
7. Zhou H, Skolnick J (2011) GOAP: a generalized orientation-dependent, all-atom statistical potential for protein structure prediction. Biophys J 101: 2043–2052.
8. Skolnick J (2006) In quest of an empirical potential for protein structure prediction. Curr Opin Struct Biol 16: 166–171.
9. Summa C, Levitt M (2007) Near-native structure refinement using *in vacuo* energy minimization. Proc Natl Acad Sci (USA) 104: 3177–3182.
10. Zhu J, Fan H, Peiole X, Honig B, Mark A (2008) Refining homology models by combining replica-exchange molecular dynamics and statistical potentials. Proteins: Struct Func Bioinfo 72: 1171–1188.
11. Chopra G, Kalisman N, Levitt M (2010) Consistent refinement of submitted models at CASP using a knowledge-based potential. Proteins: Struct Func Bioinfo 78: 2668–2678.
12. Amautova Y, Scheraga H (2008) Use of decoys to optimize an all-atom forcefield including hydration. Biophys J 95: 2434–2449.
13. Bhattachary D, Cheng J (2013) 3Drefine: consistent protein structure refinement by optimizing hydrogen bonding network and atomic level refinement. Proteins: Struct Func Bioinfo 81: 119–131.
14. Rohl C, Strauss C, Misura K, Baker D (2004) Protein structure prediction using Rosetta. Methods Enzymol 383: 66–93.
15. Zhang Y, Kolinski A, Skolnick J (2003) Touchstone II: A new approach to ab initio protein structure prediction. Biophys J 85: 1145–1164.
16. Benkert P, Tosatto S, Schomburg D (2008) QMEAN: A comprehensive scoring function for model quality assessment. Proteins: Struct Func Bioinfo 71: 261–277.
17. Zhang Y, Skolnick J (2004) Automated structure prediction of weakly homologous proteins on a genomic scale. Proc Natl Acad Sci (USA) 101: 7594–7599.
18. Zemla A (2003) LGA: a method for finding 3D similarities in protein structures. Nucl Acids Res 31: 3370–3374.
19. Perez A, Yang Z, Bahar I, Dill K, MacCallum J (2012) FlexE: using elastic network models to compare models of protein structure. J Chem Theory Computat 8: 3985–3991.
20. Rajgaria R, McAllister S, Floudas C (2006) A novel high resolution Cα–Cα distance dependent force field based on a high quality decoy set. Proteins: Struct Func Bioinfo 65: 726–741.
21. Samudrala R, Levitt M (2008) Decoys 'R'Us: A database of incorrect conformations to improve protein structure prediction. Protein Science 9: 1399–1401.
22. Tsai J, Bonneau R, Morozov A, Kuhlman B, Rohl C, et al. (2003) An improved protein decoy set for testing energy functions for protein structure prediction. Proteins: Struct Func Bioinfo 53: 76–87.
23. McLachlan A (1979) Gene duplications in the structural evolution of chymotrypsin. J Mol Biol 128: 49–80.
24. Horn B (1987) Closed form solution of absolute orientation using unit quaternions. J Opt Soc Am 4: 629–642.
25. Coutsias E, Seok C, Dill K (2004) Using quaternions to calculate RMSD. J Comp Chem 25: 1849–1857.
26. Kaindl K, Steipe B (1997) Metric properties of the root-mean square deviation of vector sets. Acta Cryst A 53: 809.
27. Tirion M (1996) Large amplitude elastic motions in proteins from a single-parameter, atomic analysis. Phys Rev Lett 77: 1905–1908.
28. Tama F, Sanejouand Y (2001) Conformational change of proteins arising from normal mode calculations. Protein Eng 14: 1–6.
29. Bohr J, Bohr H, Brunak S, Cotterill R, Fredholm H, et al. (1993) Protein structures from distance inequalities. J Mol Biol 231: 861–869.
30. Summa C, Levitt M (2007) Near-native structure refinement using in vacuo energy minimization. Proc Natl Acad Sci (USA) 104: 3177–3182.
31. Bahar I, Atilgan A, Erman B (1997) Direct evaluation of thermal fluctuations in proteins using a single-parameter harmonic potential. Folding and Design 2: 173–181.
32. Atilgan A, Durell S, Jernigan R, Demirel M, Keskin O, et al. (2001) Anisotropy of fluctuation dynamics of proteins with an elastic network model. Biophys J 80: 505–515.
33. Toda M (1967) Vibration of a chain with nonlinear interaction. J Phys Soc Japan 22: 431–436.
34. Røgen P, Koehl P (2013) Extracting knowledge from protein structure geometry. Proteins: Struct Func Bioinfo 81: 841–851.
35. de Boor C (1978) A practical guide to splines. New York: Springer-verlag.
36. Eickholt J, Wang Z, Cheng J (2011) A conformation ensemble approach to protein residue-contact. BMC structural biology 11: 38.
37. Handl J, Knowles J, Lovell S (2009) Artefacts and biases affecting the evaluation of scoring functions on decoy sets for protein structure prediction. Bioinformatics 25: 1271–1279.
38. Güntert P, Mumenthaler C, Wüthrich K (1997) Torsion angle dynamics for NMR structure calculation with the new program DYANA. J Mol Biol 273: 283–298.
39. Park B, Levitt M (1996) Energy functions that discriminate x-ray and near-native folds from well-constructed decoys. J Mol Biol 258: 367–392.
40. Simons K, Kooperberg C, Huang E, Baker D (1997) Assembly of protein tertiary structures from fragments with similar local sequences using simulated annealing and bayesian scoring functions. J Mol Biol 268: 209–225.
41. Keasar C, Levitt M (2003) A novel approach to decoy set generation: designing a physical energy function having local minima with native structure characteristics. J Mol Biol 329: 159–174.
42. Huang E (1999) A combined approach for ab initio construction of low resolution protein tertiary structures from sequence. In: Pacific Symposium on Biocomputing. volume 4, pp. 505–516.
43. Xia Y, Huang E, Levitt M, Samudrala R (2000) Ab initio construction of protein tertiary structures using a hierarchical approach. J Mol Biol 300: 171–185.
44. Simons K, Ruczinski I, Kooperberg C, Fox B, Bystroff C, et al. (1999) Improved recognition of native-like protein structures using a combination of sequence-dependent and sequence-independent features of proteins. Proteins: Struct Func Bioinfo 34: 82–95.
45. Zhang J, Zhang Y (2010) A novel side-chain orientation dependent potential derived from random-walk reference state for protein fold selection and structure prediction. PLoS One 5: e15386.
46. Zhang Y, Skolnick J (2004) Scoring function for automated assessment of protein structure template quality. Proteins: Struct Func Bioinfo 57: 702–710.
47. Xu J, Zhang Y (2010) How significant is a protein structure similarity with TM-score = 0.5? Bioinformatics 26: 889–895.

Acknowledgments

The authors want to thank the anonymous reviewers for constructive criticism and careful reading of the first version of this manuscript.

Author Contributions

Conceived and designed the experiments: MC PK PR. Performed the experiments: MC. Analyzed the data: MC PK PR. Contributed reagents/materials/analysis tools: MC. Wrote the paper: MC PK PR.

48. Kryshtafovych A, Barbato A, Fidelis K, Monastyrskyy B, Schwede T, et al. (2014) Assessment of the assessment: evaluation of the model quality estimates in CASP10. Proteins: Struct Func Bioinfo 82: 112–126.

49. Samudrala R, Moult J (1998) An all-atom distance-dependent conditional probability discriminatory function for protein structure prediction. J Mol Biol 275: 895–916.

50. Lindorff-Larsen K, Piana S, Palmo K, Maragakis P, Klepeis JL, et al. (2010) Improved side-chain torsion potentials for the Amber ff99SB protein force field. Proteins: Struct Func Bioinfo 78: 1950–1958.

51. Zhou H, Zhou Y (2002) Distance-scaled, finite ideal-gas reference state improves structure-derived potentials of mean force for structure selection and stability prediction. Protein Sci 11: 2714–2726.

222222

An Odorant-Binding Protein Is Abundantly Expressed in the Nose and in the Seminal Fluid of the Rabbit

Rosa Mastrogiacomo[1]◑, Chiara D'Ambrosio[2]◑, Alberto Niccolini[3], Andrea Serra[1], Angelo Gazzano[3], Andrea Scaloni[2]*, Paolo Pelosi[1]*

1 Department of Agriculture, Food and Environment, University of Pisa, Pisa, Italy, **2** Proteomics & Mass Spectrometry Laboratory, ISPAAM, National Research Council, Napoli, Italy, **3** Department of Veterinary Sciences, University of Pisa, Pisa, Italy

Abstract

We have purified an abundant lipocalin from the seminal fluid of the rabbit, which shows significant similarity with the sub-class of pheromone carriers "urinary" and "salivary" and presents an N-terminal sequence identical with that of an odorant-binding protein (rabOBP3) expressed in the nasal tissue of the same species. This protein is synthesised in the prostate and found in the seminal fluid, but not in sperm cells. The same protein is also expressed in the nasal epithelium of both sexes, but is completely absent in female reproductive organs. It presents four cysteines, among which two are arranged to form a disulphide bridge, and is glycosylated. This is the first report of an OBP identified at the protein level in the seminal fluid of a vertebrate species. The protein purified from seminal fluid is bound to some organic chemicals whose structure is currently under investigation. We reasonably speculate that, like urinary and salivary proteins reported in other species of mammals, this lipocalin performs a dual role, as carrier of semiochemicals in the seminal fluid and as detector of chemical signals in the nose.

Editor: Sabato D'Auria, CNR, Italy

Funding: The authors have no support or funding to report.

Competing Interests: The authors have declared that no competing interests exist.

* Email: andrea.scaloni@ispaam.cnr.it (A. Scaloni); ppelosi@agr.unipi.it (PP)

◑ These authors contributed equally to this work.

Introduction

Odorant-binding proteins (OBPs) of vertebrates are a sub-class of lipocalins [1–2], a protein super-family including retinol-binding protein [3], ß-lactoglobulin [4] and many other members that differ for amino acid sequence and physiological function but share the highly conserved structure of the ß-barrel, a sort of cup made of 8 antiparallel ß-sheets enclosing a binding cavity for hydrophobic ligands [5-10]. Vertebrate OBPs are binding proteins of about 150–160 amino acids firstly identified in the nasal epithelium of mammals and classified as carriers for odorants and pheromones [11–17]. Several members of this family have been isolated from different mammals, such as bovine, pig, rabbit and others [18–25], as well as in amphibians [26]. OBPs bind to a large variety of small organic molecules, including odorants and pheromones, with a broad specificity and dissociation constants in the micromolar range [9,27–31].

Despite the detailed structural and functional information available for several OBPs, their physiological role in olfaction is still not clear [15–17,32–33]. A carrier for hydrophobic odorants across the aqueous nasal mucus seems reasonable, but a more specific function in detecting chemical messengers cannot be excluded. This idea is based on the expression of several OBPs in the same species, with different and complementary spectra of binding [30,34]. Moreover, there is clear evidence that insect

OBPs, a class of proteins structurally different from those of vertebrates, but probably with similar functions [35], are often required for a correct detection of odors and pheromones [36–37], and are also involved in the discrimination of different semiochemicals [38–39].

Whatever their role and detailed mechanism of action, it is reasonable to hypothesise that OBPs from vertebrates might be involved in the detection of pheromones, rather than general odorants. This idea is suggested by the small number of OBP sub-types reported in mammals, as compared to those from insects, and their expression in the vomeronasal organ (an organ dedicated to pheromone perception) [40–42] or in glands of the nasal respiratory epithelium [43], but not in the olfactory mucosa. The sole exception of the human OBP, which was detected in the mucus of the olfactory cleft, but not in the lower nasal regions [44], might be explained with the fact that the vomeronasal organ is absent or non-functioning in humans. However, strong evidence for the involvement of OBPs in detecting pheromones comes from their expression in organs dedicated to the synthesis and the delivery of pheromones [33]. In fact, OBPs similar or identical to those identified in the nose have also been reported as expressed in non-sensory organs and secreted in biological fluids involved in pheromonal communication. Best studied examples include the "major urinary proteins" (MUPs) of mouse and rat [7,45–48], which are synthesised in the liver and excreted in the urine at concentrations of several mg/mL, the "salivary proteins" (SALs) of the boar, abundantly produced by the submaxillary glands [10,19,34], and the so-called "aphrodisin" identified in the vaginal secretion of the hamster [49–50]. In each species, these proteins are produced in the above-mentioned organs in a sex-specific fashion, while they are expressed in the nose equally in both sexes [51]. When released in the urine, saliva or other secretions, such proteins are loaded with organic compounds known to be the species-specific pheromones, while in the nose they are void. In particular, it has been reported that murine MUPs, when excreted in the urine, are complexed with known animal pheromones, such as 2-sec-butylthiazoline and 3,4-dehydro-exo-brevicomin [47,52]. Similarly, pig SALs, when isolated from the saliva, carries the boar-specific pheromones 5α-androst-16-en-3-one and 5α-androst-16-en-3-ol [19].

Although the few cases reported above have been studied in detail, the use of OBPs as carriers of pheromones to be released in the environment might be much more common and widespread. The sweath of horses contains large amounts of an OBP-like protein complexed with putative semiochemicals [25], while the salivary lipocalins of several mammals, often reported as allergens [53–55], might perform similar functions. Chemical communication in the rabbit has not been widely studied. A single pheromone has been so far described, namely the volatile compound 2-methyl-2-butenal, which was isolated from the milk and shown to trigger a very clear and robust response in the puppies [56–57]. Information on rabbit OBPs is limited to our previous work reporting the isolation and partial characterization of three members from the nasal tissue [18,23]. The present study was aimed at further investigating the putative role of rabbit OBPs as carriers of pheromones to be released in the environment and describes an OBP expressed only in the nose of both sexes and in seminal fluid.

Experimental Procedures

Materials

Rabbit bodies were kindly provided by a local abbattoir and dissected within an hour after death or kept at −20°C for a few days. Rabbit seminal fluid was collected using an all-glass artificial vagina equipped with a jacket where warm water was circulated.

Ethics statement

All operations were carried out in strict accordance with the recommendations for handling laboratory animals of the National Research Council (CNR) of Italy. The protocol was approved by the Committee on the Ethics of Animal Experiments of the Italian CNR (Permit Number: 01-2014 of February 18, 2014). All efforts were made to minimize suffering of the animals.

RNA extraction and cDNA synthesis

Total RNA was extracted using TRI Reagent (Sigma), following the manufacturer's protocol. cDNA was prepared from total RNA by reverse transcription, using 200 units of SuperScript™ III Reverse Transcriptase (Invitrogen) and 0.5 mg of an oligo-dT primer in a 50 μL reaction volume. The mixture also contained 0.5 mM of each dNTP (GE-Healthcare), 75 mM KCl, 3 mM MgCl2, 10 mM DTT and 0.1 mg/ml BSA in 50 mM Tris-HCl, pH 8.3. The reaction mixture was incubated at 50°C for 60 min and the product was directly used for PCR amplification or stored at −20°C.

Polymerase chain reaction

Aliquots of 1 μL of crude cDNA were amplified in a Bio-Rad Gene Cycler thermocycler, using 2.5 units of Thermus aquaticus

DNA polymerase (GE-Healthcare), 1 mM of each dNTP (GE-Healthcare), 1 μM of each PCR primer, 50 mM KCl, 2.5 mM $MgCl_2$ and 0.1 mg/ml BSA in 10 mM Tris-HCl, pH 8.3, containing 0.1% v/v Triton X-100. At the 5′ end, we used a specific primer (rabOBP3-fw: 5′-CACAGCCACTCGGA-3′) corresponding to the sequence encoding the first five amino acids of the mature protein. At the 3′ end, we used an oligo-dT to first obtain the correct sequence of the gene, then a specific primer (rabOBP3-rv: 5′-TTAGGCGGCTCCGCCGTC-3′) encoding the last five residues and the stop codon, to check the presence of the gene in different tissues. After a first denaturation step at 95°C for 5 min, we performed 35 amplification cycles (1 min, at 95°C; 30 sec, at 50°C; 1 min, at 72°C) followed by a final step of 7 min, at 72°C.

Cloning and sequencing

The crude PCR products were ligated into a pGEM (Promega) vector without further purification, using a 1:5 (plasmid:insert) molar ratio and incubating the mixture overnight, at room temperature. After transformation of E. coli XL-1 Blue competent cells with the ligation products, positive colonies were selected by PCR using the plasmid's primers SP6 and T7 and grown in LB/ampicillin medium. DNA was extracted using the Plasmid MiniPrep Kit (Euroclone) and custom sequenced at Eurofins MWG (Martinsried, Germany).

Preparation of the tissue extracts

Crude extracts were prepared by homogenization of the corresponding tissues in 10 mL of 20 mM Tris-HCl pH 7.4 (Tris buffer) per gram of tissue, using a Polytron homogenizer, followed by centrifugation at $20,000 \times g$ for 20 min. The clear supernatant was immediately used for SDS-PAGE and Western blotting experiments.

Purification of the seminal protein

Lipocalins from rabbit seminal fluid were purified through a 1×30 cm Superose 12 column in 50 mM ammonium bicarbonate, as previously reported [23]. Selected fractions were then pooled, dialysed against 20 mM Tris-HCl, pH 7.4, and applied to a 1.5×25 cm Whatman DE-52 column. Elution was performed using a linear 0.1–0.4 M NaCl gradient, in 20 mM Tris-HCl, pH 7.4. Each fraction was analysed using 12% SDS-PAGE.

Protein digestion and peptide separation

Rabbit seminal fluid OBP was resolved by SDS-PAGE, excised from the gel, triturated, in-gel reduced, S-alkylated and digested with trypsin, as previously reported [56]. Gel particles were extracted with 25 mM NH_4HCO_3/acetonitrile (1:1 v/v) by sonication, and digests were concentrated. Peptide mixtures were either desalted using μZipTipC$_{18}$ pipette tips (Millipore) before MALDI-TOF-MS analysis, directly analyzed by nanoLC-ESI-LIT-MS/MS (see below) or simply resolved on an Easy C_{18} column (100×0.075 mm, 3 μm) (Proxeon) using a linear gradient of acetonitrile containing 0.1% trifluoroacetic acid in aqueous 0.1% trifluoroacetic acid, at a flow rate of 300 nL/min, for 80 min. In the latter case, collected fractions were concentrated and analyzed by MALDI-TOF-MS.

Protein alkylation under native conditions

Protein samples for disulfide assignment were alkylated with 1.1 M iodoacetamide in 0.25 M Tris-HCl, 1.25 mM EDTA, and 6 M guanidinium chloride, pH 7.0, at 25 °C for 1 min in the dark. Samples were separated from excess salts and reagents by passing the reaction mixture through a PD10 column (Amersham Biosciences), as previously reported [59]. Protein samples were finally digested and resolved by LC as mentioned above.

Glycopeptide enrichment

To isolate glycopeptides, rabbit seminal fluid OBP digest aliquots were solved in 80% acetonitrile, 2% formic acid and loaded on GELoader tips (Eppendorf, Germany), which were plugged with 3M Empore C8 extraction disk material (3M Bioanalytical Technologies, MN) and packed with ZIC-HILIC (200 Å, 10 μm, zwitterionic sulfobetaine functional groups) resin (Sequant, Sweden) [60]. Loaded microcolumns were washed twice with 15 μL of 80% acetonitrile, 2% formic acid. Glycopeptides were first eluted with 10 μL of 2% formic acid and then with 5 μL of 50% acetonitrile, 2% formic acid; pooled fractions were analyzed by MALDI-TOF-MS, as described below.

Peptide deglycosylation and disulfide reduction

Glycopeptides were directly deglycosylated on the MALDI target by treatment with 0.2 U of PNGase F (Roche) in 50 mM NH_4HCO_3, pH 8, at 37 °C, for 1 h. Then, 2 μL of 0.1% trifluoroacetic acid was added to reaction mixtures, which were desalted on μZipTipC18 pipette tips (Millipore) before MALDI-TOF-TOF-MS analysis [61].

Disulfide-containing peptides were directly reduced on the MALDI target by treatment with 10 mM mM DTT in 50 mM NH_4HCO_3, pH 8, at 37 °C, for 1 h. Then, 2 μL of 0.1% trifluoroacetic acid was added to reaction mixtures, which were desalted on μZipTipC18 pipette tips (Millipore) before MALDI-TOF-TOF-MS analysis [61].

MS analysis

Peptide mixtures were analyzed by nLC-ESI-LIT-MS/MS using a LTQ XL mass spectrometer (ThermoFinnigan, USA) equipped with a Proxeon nanospray source connected to an Easy-nLC (Proxeon, Denmark) [58]. They were resolved on an Easy C_{18} column (100×0.075 mm, 3 μm) (Proxeon) using a linear gradient of acetonitrile containing 0.1% formic acid in aqueous 0.1% formic acid, at a flow rate of 300 nL/min, for 25 min. Spectra were acquired in the range m/z 400–1800. Acquisition was controlled by a data-dependent product ion scanning procedure over the 3 most abundant ions, enabling dynamic exclusion (repeat count 1 and exclusion duration 1 min). The mass isolation window and collision energy were set to m/z 3 and 35%, respectively.

During MALDI-TOF-MS analysis, entire protein digests or selected peptide fractions were loaded on the instrument target together with 2,5-dihydroxy-benzoic acid (10 mg/mL in 70% v/v acetonitrile, 0.1% v/v trifluoroacetic acid) or α-cyano-4-hydroxycinnamic acid (saturated solution in 30% v/v acetonitrile, 0.1% v/v trifluoroacetic acid) as matrices, using the dried droplet technique; a 384-spot ground steel plate (Bruker Daltonics) was used to this purpose. Spectra were acquired in the m/z range 500–5000 on a Bruker Ultraflextreme MALDI-TOF-TOF instrument (Bruker Daltonics) operating either in reflectron mode or linear mode. Instrument settings were: pulsed ion extraction = 100 ns, laser frequency = 1000 Hz, number of shots per sample = 2500–5000 (random walk, 500 shots per raster spot). Mass spectra were calibrated externally using nearest neighbour positions loaded with Peptide Calibration Standard II (Bruker Daltonics), with quadratic calibration curves. MS/MS spectra were acquired in LIFT mode. Data were elaborated using the FlexAnalysis software (Bruker Daltonics).

nLC-ESI-LIT-MS/MS data were searched by using MASCOT (version 2.2.06) (Matrix Science, UK) against an updated rabbit EST database containing available protein sequences (NCBI 28/11/2013, 212376 sequences). As searching parameters, we used a mass tolerance value of 2 Da for precursor ion and 0.8 Da for ion fragments, trypsin trypsin and/or slymotrypsin (cleavage at Lys, Arg, Phe, Tyr, Trp and Leu) as proteolytic enzymes, a missed cleavages maximum value of 2, Cys carbamidomethylation and Met oxidation as fixed and variable modification, respectively. Protein candidates with more than 2 assigned unique peptides with an individual Mascot ion score >25 and a significant threshold ($p<0.05$) were further considered for protein identification. In the case of glycopeptides or disulfide-containing peptides, MALDI-TOF mass signals were assigned to peptides, glycopeptides or disulfide-containing peptides using the GPMAW 4.23 software (Lighthouse Data, Denmark). This software generated a mass/fragment database output based on protein sequence, protease selectivity, nature of the amino acids susceptible to eventual glycosylation/oxidation and the molecular mass of the modifying groups. Searching parameters were set as mentioned above; mass values were matched to protein regions using a 0.02% mass tolerance value. MALDI-TOF-TOF searching parameters were set with tolerances of 100 ppm and 0.5 Da for MS and MS/MS data, respectively. Glycosylation or disulfide assignments were always confirmed by additional MS experiments on deglycosylated or reduced peptides, respectively.

Ligand-binding experiments

The affinity of the fluorescent probe N-phenyl-1-naphthylamine (1-NPN) was measured by titrating a 2 μM solution of the protein with aliquots of 1 mM 1-NPN solved in methanol to reach final concentrations of 2–16 μM. The probe was excited at 337 nm and the maximum emission wavelength was 415 nm. Dissociation constant was evaluated using GraphPad Prism software. Affinities of other ligands were measured in competitive binding assays, by titrating a solution containing the protein and 1-NPN both at the concentration of 4 μM with 1 mM solutions of each competitor in methanol to reach final concentrations of 0–16 μM. Dissociation constants of the competitors were calculated from the concentrations of ligand halving the initial fluorescence value of 1-NPN (IC_{50}), using the equation:

$$K_D = IC_{50}/1 + 1 - NPN/K_{1-NPN}$$

1-NPN being the free concentration of 1-NPN and K_{1-NPN} being the dissociation constant of the complex protein/1-NPN.

Results

Identification and purification of an OBP from the rabbit seminal fluid

With the aim of identifying OBPs expressed in rabbit non-sensory organs, we verified the occurrence of a protein in the male semen that showed a cross-reactivity with a polyclonal antiserum raised against the boar salivary lipocalin (pig SAL) [19]. This protein, which migrated in SDS-PAGE as a blurred band at about 23 kDa, was very abundant in the seminal liquid but was not present in the sperm cells. Figure 1 reports the electrophoretic analysis of the supernatant and the pellet obtained by centrifugation of the crude semen. The weaker cross-reactivity of the pellet was due to a contamination with the seminal fluid and disappeared completely after washing the pellet three times with buffer. Protein concentration in the semen was estimated to be about 10–20 mg/mL. This protein was then purified by gel filtration chromatography on a Superose-12 column, followed by anion-exchange

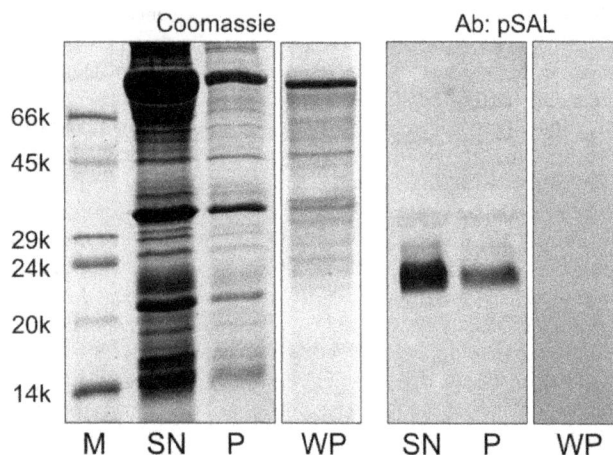

Figure 1. SDS-PAGE analysis of rabbit sperm and corresponding Western blotting. SN, soluble fraction; P, sperm cells; WP, sperm cells after washing three times with buffer. A strong cross-reactivity with a polyclonal antiserum raised against pig SAL [19] was observed for a protein migrating at about 23 kDa. Staining was much stronger in the soluble fraction; the weak reactivity observed for the sperm cells disappeared after washing the cells, thus indicating the absence of the protein in this sample.

chromatography on a DE-52 resin. Figure 2 reports the SDS-PAGE profile of selected fractions from the first purification step, together with the corresponding Western blotting, as well as of the purified protein that was used for further studies.

In order to characterize the nature of this seminal protein, we performed a MALDI-TOF peptide mass fingerprinting analysis on its tryptic digest following reduction with dithiothreitol and alkylation with iodoacetamide (data not shown). MS results matched to a sequence reported in the NCBI EST database (entry EL341998) annotated as UTE-7, which corresponded to a cDNA isolated from rabbit uterus. The sequence at the protein N-terminus of UTE-7 is identical with that of a rabbit OBP (rabOBP3) we had previously isolated from the nasal tissue [23]. Since the identity of some nucleotides in the EST entry mentioned

above was not determined and the sequence was partial, we again cloned the corresponding cDNA and sequenced it; data are reported in Supplementary Figure S1. Our analysis provided a complete nucleotide assignment, together with very few base corrections, finally ascertaining a corresponding protein sequence as made of 161 amino acids. Finally, massive peptide mapping nanoLC-ESI-LI-MS/MS experiments on a tryptic digest ascertained the nature of the protein N- and C-terminus, verifying about 93% of its amino acid sequence (Table S1).

Tissue expression

To detect the site of synthesis for this seminal protein, we performed PCR experiments on samples of cDNA prepared from different parts of male and female reproductive organs. To first identify the full sequence of the gene (Figure S1), we used a specific primer at the 5′-end encoding the first five amino acids of the sequence reported in the database as UTE-7 (acc. no: EL341998) and an oligo-dT at the 3′-end. Then, we used the same primer at the 5′-end and a second specific primer at the 3′-end encoding the last five residues and the stop codon, to check for the presence of this gene in different organs. In particular, olfactory and respiratory epithelium from both sexes, prostate, epididymis, testis, uterus, uterine tubes, ovaries, vagina and vaginal vestibule were evaluated. Amplification bands were obtained only for the prostate as well as for the respiratory epithelium of both sexes. Parallel cloning and sequencing of samples from these tissues always yielded the same sequence (Figure S1), excluding the occurrence of various protein isoforms. The specificity of protein expression in these tissues was confirmed at the protein level by Western-blotting experiments (Figure 3). On this basis, we can conclude that the protein previously named as UTE-7 is not produced in the uterus, nor in any part of the female reproductive system, but was probably found in such organ as result of a sample contamination. On the other hand, the sequence we report here very likely corresponds to the protein (rabOBP3) we had

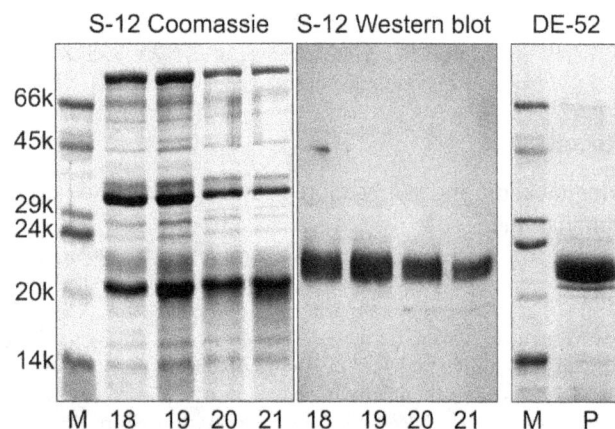

Figure 2. Purification of the rabbit seminal fluid OBP. A sample of crude seminal fluid, as obtained after sperm centrifugation, was resolved at first by gel filtration chromatography on a Superose-12 column and then by anion-exchange chromatography on a DE-52 column (see Materials and Methods section for details). The protein was eluted as a pure component, as verified by SDS-PAGE.

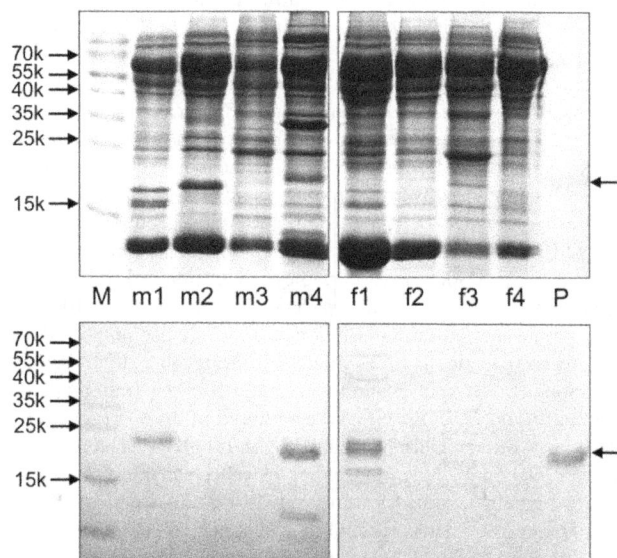

Figure 3. Expression of rabOBP3 in different tissues of male (m) and female (f) rabbit individuals. SDS-PAGE analysis of different rabbit tissues and corresponding Western blotting are shown. M: molecular weight markers; m1: nasal respiratory tissue; m2: epididymis; m3: testis; m4: prostate; f1: nasal respiratory tissue; f2: uterine tubes; f3: ovaries; f4: uterus; P: purified rabOBP3.

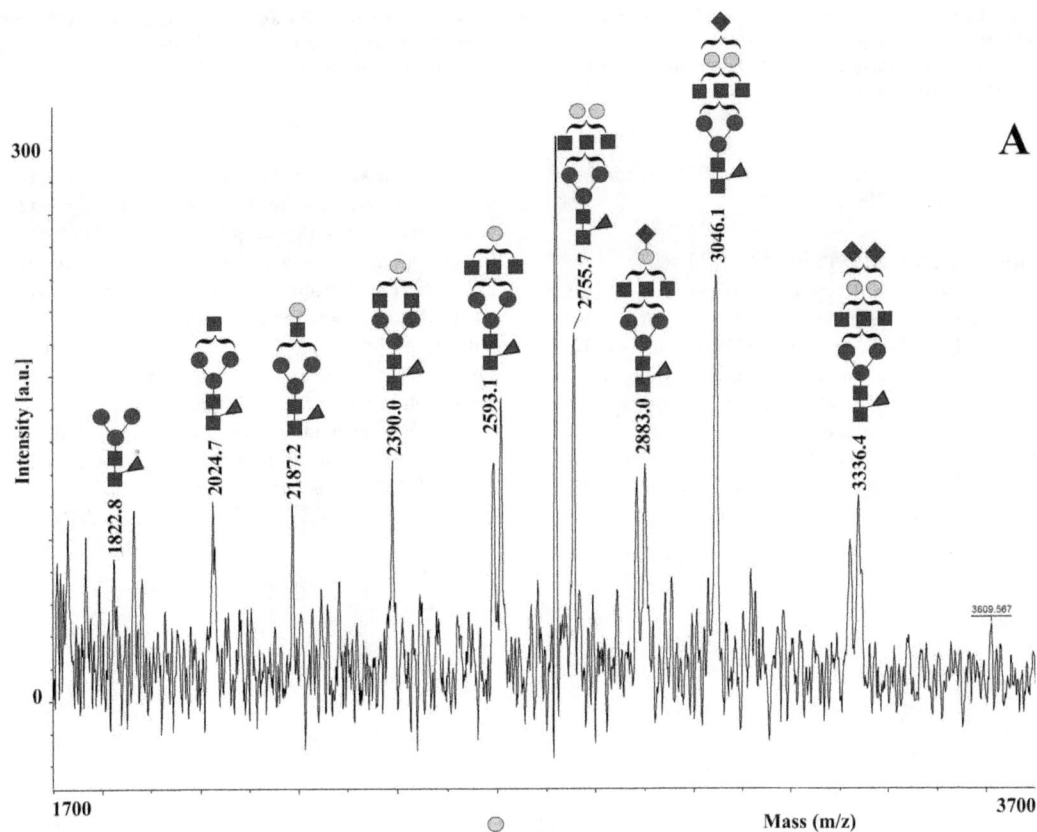

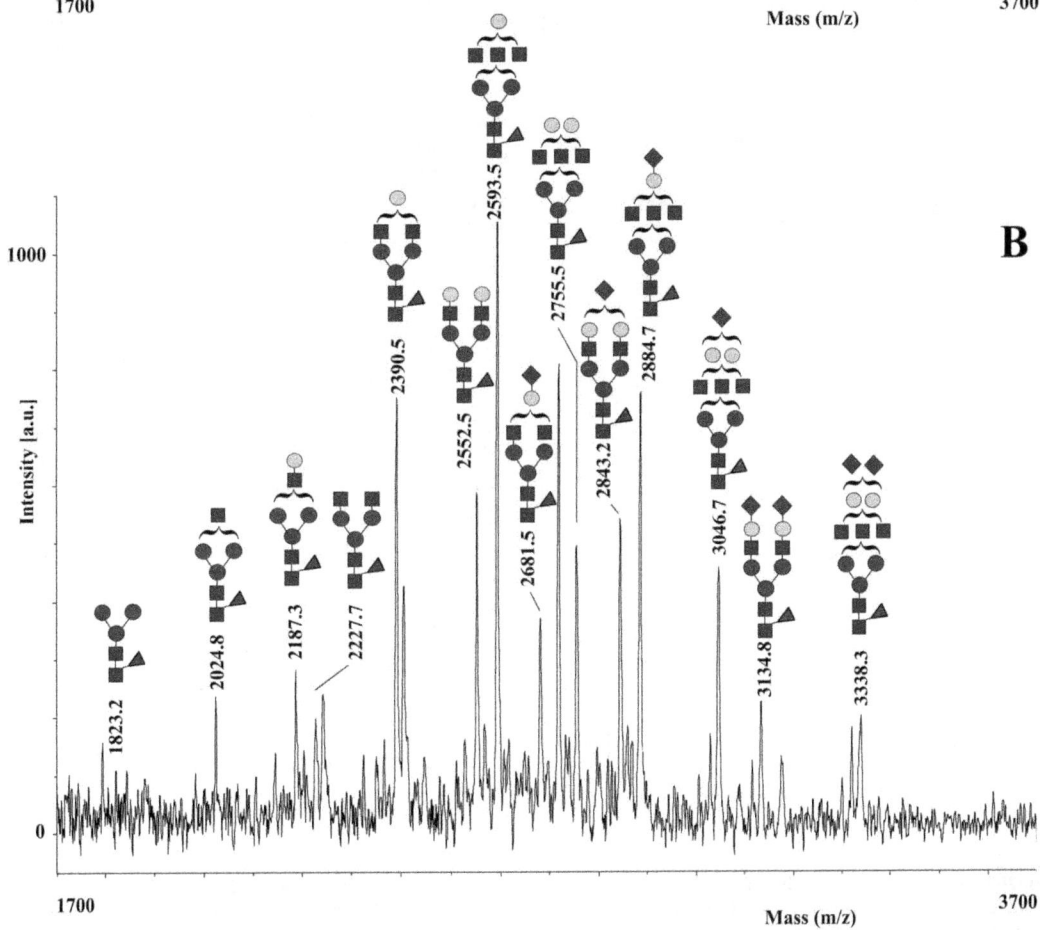

Figure 4. MALDI-TOF-MS analysis of the purified tryptic glycopeptides from rabOBP3 as obtained after HILIC enrichment and nanoLC separation. Spectra acquired in linear mode of the fractions eluting at 15 and 16 min are reported in panel A and B, respectively; shown are the mono-, bi- and tri-antennary complex-type glycan structures N-linked to Asn44 in peptide (44–50). ■, N-acetyl-glucosamine; ●, mannose; ○, galactose; ◄, fucose; ◆, N-acetyl-neuraminic acid.

previously isolated from the nasal epithelium [23]. Accordingly, we decided to rename UTE-7 as rabOBP3.

Post-translational modifications in rabOBP3

The blurred band and the discrepancy between the calculated (18 kDa) and apparent (23 kDa) molecular mass of the intact protein observed in SDS-PAGE, its broad MH$^+$ signal in MALDI-TOF-MS (data not shown) and the occurrence of two putative N-linked glycosylation sites (Asn29 and Asn44) in the corresponding amino acid sequence (as predicted by bioinformatic analysis) suggested that rabOBP3 could be a glycoprotein, similarly to what reported for pig SAL, horse EquC1 and some murine/rat MUPs [25,34,62]. To evaluate protein glycosylation and assign potential

modification site(s), a rabOBP3 sample resolved by SDS-PAGE was *in gel* reduced, alkylated with iodoacetamide and digested with trypsin. The corresponding peptide digest was then enriched for glycopeptides on a HILIC column and resolved by nanoLC into different fractions, which were then analyzed by MALDI-TOF-MS. Fractions eluting at 15 and 16 min showed a similar pattern of multiple signals in the mass spectrum (Figure 4A and B). On the basis of the measured mass values and known pathways of glycoprotein biosynthesis, all these peaks were assigned to peptide (44–50) having a pentasaccharide core N-linked to Asn44, and bearing mono-, bi- and tri-antennary complex glycan structures (theor. MH$^+$ values: m/z 1821.8, 2024.9, 2187.1, 2228.2, 2390.3, 2552.5, 2593.5, 2681.6, 2755.7, 2843.7, 2884.8, 3046.9, 3135.0

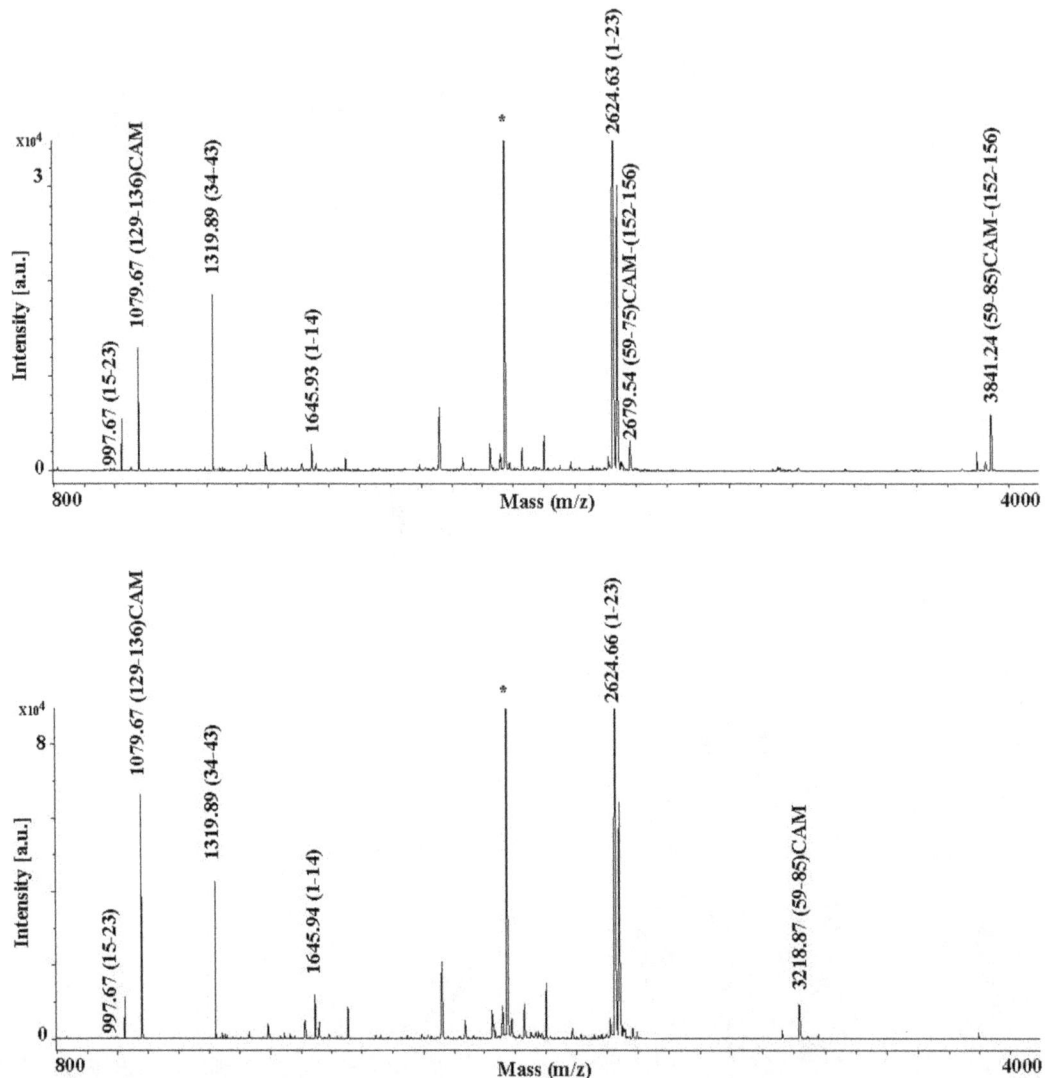

Figure 5. MALDI-TOF-MS analysis of the tryptic digest of rabOBP3 alkylated with iodoacetamide under denaturing, non-reducing conditions before (top) and following (bottom) treatment with dithiothreitol. Constant and variable signals are labelled in the spectra acquired in reflectron mode to highlight reduced and oxidized residues present under native conditions. Trypsin-derived peptides are indicated with an asterisk.

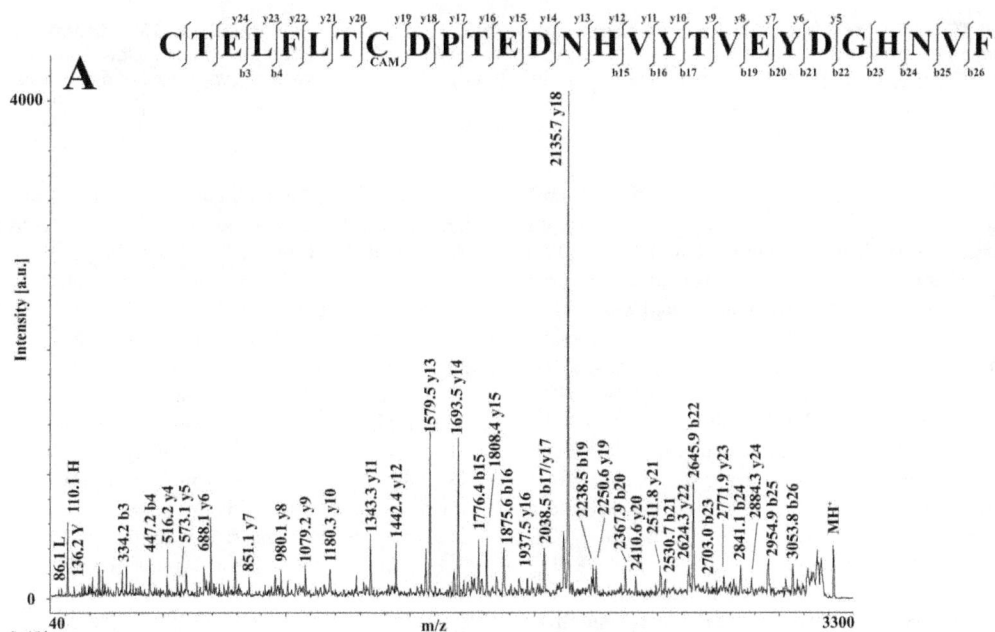

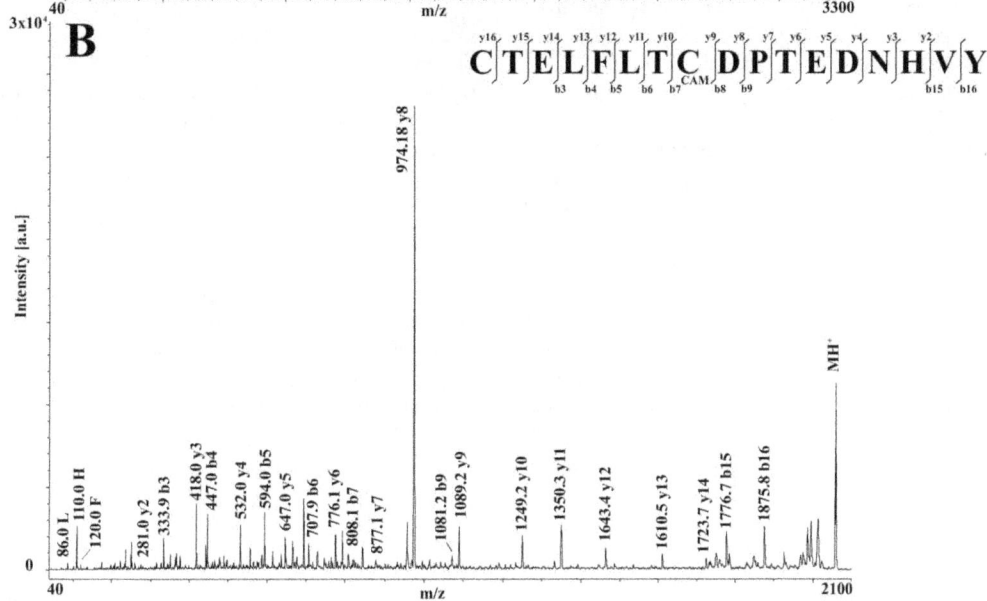

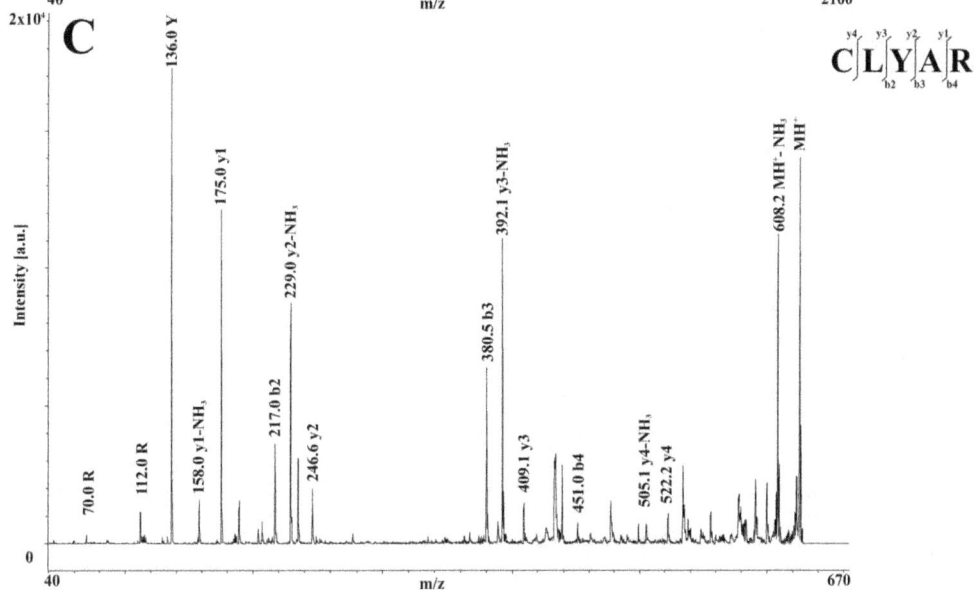

Figure 6. MALDI-TOF-TOF spectra of the disulfide-containing tryptic peptides from alkylated rabOBP3 following treatment with dithiothreitol. Fragmentation spectra of the peptides (59–85)CAM, (59–75)CAM and (152–156) are shown in panels A, B and C, respectively. In all cases, Cys residues originally involved in the S-S bond are present in a reduced status, the remaining ones occurring as carboxamidomethylated derivatives.

and 3338.2). After PNGase treatment, glycopeptides in both fractions collapsed to a unique component (peptide 44–50) having a MH$^+$ signal at m/z 784.08 (data not shown). MALDI-TOF-TOF-MS analysis of the deglycosylated peptide confirmed the expected Asn44>Asp conversion. Multiple signals associated with glyco-peptides were also detected in the mass spectrum of the fractions eluting at 21 and 22 min. On the basis of measured mass values (exp. MH$^+$ values: m/z 3124.8, 3327.9, 3490.2, 3530.9, 3693.2, 3733.5, 3855.4, 3896.6, 3983.8, 4059.1, 4146.4, 4187.7, 4350.1, 4437.0, 4641.2 and 4932.9) and the relative intensities, these peaks were associated to peptide (34–50) having the same glycan structures reported in Figure 4 as N-linked to Asn44 (theor. MH$^+$ values: m/z 3123.3, 3326.5, 3488.7, 3529.7, 3691.9, 3732.9, 3854.0, 3895.0, 3983.1, 4057.2, 4145.3, 4186.3, 4348.4, 4436.5, 4639.7 and 4931.0). No signals related to the non-glycosylated peptide counterparts were detected in any LC fractions either from the entire protein digest or its glycopeptide-enriched portion, thus suggesting that rabOBP3 was completely modified at this site. On the other hand, no glycopeptides containing the other putative N-linked glycosylation site (Asn29) were observed in the tryptic digest or its HILIC eluate either before and after nanoLC separation; conversely, the corresponding non-glycosylated counterparts were always detected in both cases, thus demonstrating that no modification occurred at this site.

To evaluate protein thiol status and assign disulfide-bridged Cys residues, if present, rabOBP3 was treated with 1.1 M iodoaceta-mide under denaturing, non-reducing conditions and purified by size-exclusion chromatography. The alkylated protein was then digested with trypsin and split in two samples that were treated or not with DTT; Figure 5 shows the MALDI-TOF mass spectrum

of each sample. In addition to a number of common signals present in both spectra, the digest deriving from the protein not treated with DTT uniquely showed the presence of a clear MH$^+$ signal at m/z 3841.24, which was associated with the disulfide-containing peptides (59–85)CAM-(152–156) resulting from an aspecific cleavage at Phe85. A faint MH$^+$ peak at m/z 2679.54 was also observed; this signal was assigned to the smaller disulfide-containing peptide homologue (59–75)CAM-(152–156) derived from an aspecific hydrolytic event at Tyr85. Conversely, the digest treated with DTT showed the absence of the signals mentioned above and the exclusive occurrence of a MH$^+$ peak at m/z 3218.87, which was associated with the peptide (59–85)CAM. Due to its reduced mass value, no signal assigned to the peptide (152–156) was observed. These result confirmed the occurrence of one cysteine (Cys59 or Cys66) involved in a disulfide bond with Cys152 in the above-mentioned peptides, the remaining one being in a reduced status. On the other hand, both samples showed the presence of a MH$^+$ signal at m/z 1079.67, which derived from the peptide (129–136)CAM; the latter result demonstrate that rabOBP3 contains Cys133 as free thiol under native conditions.

To definitively assign the Cys residues involved in the protein S-S bond, disulfide-containing peptides (59–85)CAM-(152–156) and (59–75)CAM-(152–156) were then purified by nanoLC and reduced with DTT directly on the MALDI target. Resulting products showed MH$^+$ peaks at m/z 3220.2 and 2058.6, which were associated with the expected reduced peptides (59–85)CAM and (59–75)CAM, respectively, both having the Cys residue originally involved in the S-S bond in a reduced status and the remaining one as carboxamidomethylated species. In both cases, the occurrence of the reduced peptide (152–156) was also observed

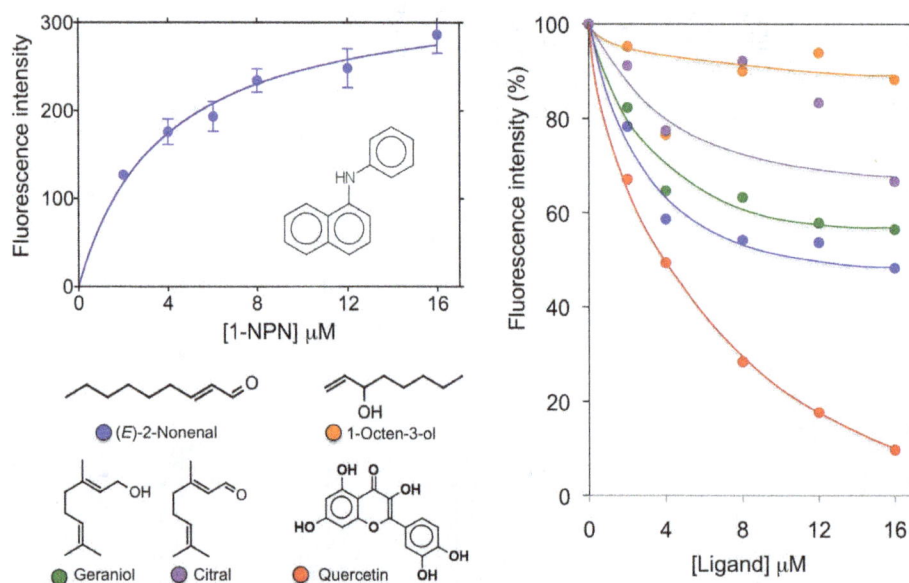

Figure 7. Binding of 1-NPN (left) and selected ligands (right) to rabOBP3 purified from seminal fluid and delipidated with dichloromethane. The protein binds the fluorescent probe 1-NPN with a dissociation constant of 3.8 μM (SD 0.9, n = 3). None of the ligands tested exhibited strong affinity to the protein, except quercetin, for which a physiological role does not seem plausible. Calculated dissociation constants are 2.2, 7.8 and 11.2 μM for quercetin, 2-nonenal and geraniol, respectively.

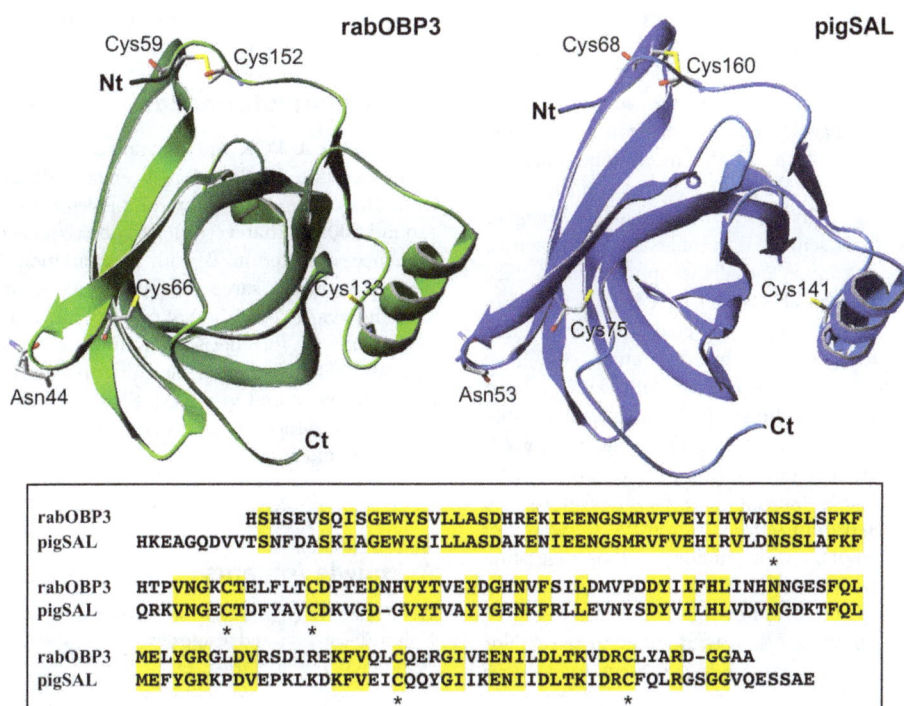

Figure 8. Three-dimensional model of rabOBP3 as built by using the crystallographic structure of pig SAL (Boar salivary lipocalin, PDB ID: 1 GM6) as a template [10]. Molecular model of rabOBP3 and pig SAL are shown in the left- and right-top panel, respectively. The corresponding sequence alignment is shown in the bottom panel, where conserved amino acids are highlighted in yellow. The conserved N-glycosylation site (Asn44), and oxidized (Cys59 and Cys152) and reduced (Cys66 and Cys133) residues are indicated by specific labelling (top) or asterisks (bottom).

in the corresponding MS spectra (exp. MH⁺ signal at m/z 625.2). MALDI-TOF-TOF-MS analysis of the reduced peptides (59–85)CAM and (59–75)CAM finally assigned the thiol group to Cys59, definitively proving the existence of a disulfide bond in rabOBP3 linking together Cys59 and Cys152 (Figure 6).

Endogenous ligands of rabOBP3

Since pig SAL and murine/rat MUPs carry species-specific pheromones as endogenous ligands, we then searched for compounds that might be complexed with rabOBP3. Gas-chromatographic separation coupled with MS (GC-MS) analysis of a dichloromethane extract of the protein from rabbit seminal liquid showed the presence of several peaks, to none of which we could confidently assign a defined chemical structure.

Ligand-binding assays showed that rabOBP3 reversibly binds to the fluorescent probe N-phenyl-1-naphthylamine (1-NPN) with a dissociation constant of 3.8 µM (SD 0.9, n = 3). Competitive binding assays, performed with some common plant volatiles indicated significant, but modest affinity to 2-nonenal and geraniol. On the other hand, quercetin efficiently displaced 1-NPN from the complex, but is difficult to propose a role as a rabbit semiochemical for this compound (Figure 7).

Three-dimensional model of rabOBP3

Based on the significant (52%) sequence identity between rabOBP3 and pig SAL (Figure 8, bottom), a three-dimensional molecular model of the first protein was built up as deriving from the crystal structure of the latter (Boar salivary lipocalin, PDB ID: 1 GM6) (Figure 8, top). The good quality of this model was assessed by ANOLEA and GROMOS evaluations, which calculated small positive energy values for very few amino acids

scattered along the sequence. Although not fixed as initial structural constrains before the modelling procedure, a *post hoc* evaluation of the rabOBP3 model was in perfect agreement with the protein post-translational modifications determined in this study. In fact, Asn44 occurred at the most external position in a loop extending its side chain into the solvent, while Cys59 and Cys152 were present in the model with their S atoms at a distance compatible with the presence of a disulfide bridge (Figure 8). The latter result was not surprising, based on the high conservation of cystine moieties in rabOBP3, pig SAL, murine/rat MUPs, and other proteins [49–63]. As expected, the remaining cysteine residues (Cys66 and Cys133) occurred too far apart to be linked together, in a condition compatible with a reduced state.

Discussion

When the first OBP of vertebrates was discovered in the nasal tissue of the cow [11–12], its sequence similarity with urinary proteins of rodents immediately suggested a function in chemical communication for these polypeptides [64], which had been described several years earlier, but whose presence in the urine had represented an unsolved puzzle until then [65–66]. Since that time, the occurrence of proteins of the same class or even identical in olfactory organs and in secretions used in chemical communication has been well documented both in vertebrates and in insects. These polypeptides can be recognised among the family of OBPs on the basis of sequence similarity. Besides the urinary proteins of mouse and rat, OBPs of vertebrates include the boar salivary lipocalin SAL [30], the horse Equc1 (abundantly secreted in sweat) [25] and the hamster aphrodisin occurring in the vaginal discharge [49]. On the other hand, the human genome contains a pseudogene for a protein of this group, which presents a mutation at the donor site of the second intron, thus disrupting the corresponding ORF [67].

Insects OBPs have been reported in the sex organs. In particular, mosquito *Aedes aegypti* and lepidopteran *Helicoverpa armigera* OBPs, which also occur in the insect antennae, are produced in the male reproductive organ and are transferred to the female during mating. It has been shown that *H. armigera* OBP, when extracted from semen, is complexed with potential pheromones for the species and eventually is found on the surface of fertilised eggs [68]. In vertebrates, OBPs have been reported in reproductive organs: aphrodisin is secreted in the vaginal discharge of the hamster [47–48], while in humans the gene encoding an OBP is expressed in the prostate [69]. Data reported in this study suggest that also in the seminal liquid of the rabbit, OBPs might act as pheromone carriers. Unfortunately, information on rabbit pheromones is limited to the suckling pheromone, which directs pups towards the nipple [56–57]. Among the volatiles we have extracted from seminal rabOBP3, we were not able to identify any compound with confidence, thus suggesting that endogenous ligands of rabOBP3 might not be among common natural chemicals. In line with this consideration, preliminary competitive binding assays with common terpenoids and fatty acids excluded these compounds as protein endogenous ligands.

In conclusion, we propose that OBPs as pheromone carriers are likely present in the seminal fluid of other mammals. The isolation of OBPs in reproductive organs and the identification of their endogenous ligands could lead to the discovery of novel pheromones mediating behaviour between sexes, such as male competition, in mammals as it has been shown in some insect species. Besides the knowledge advancement in the biology of mammals, such information might suggest strategies to improve

rearing conditions of economically important species, such as rabbit, cattle, pigs and horses.

Supporting Information

Figure S1 (A) PCR amplification of the gene encoding rabOBP3 in the prostate (P), as well as in male (mR) and female (fR) nasal respiratory tissue. All three samples gave amplification bands of around 500 bp, that were cloned and sequenced yielding the same sequence, reported in (**B**) with its translation. Similar experiments performed in the same conditions on uterus (Ut), uterine tubes (Tb) and ovaries (Ov) did not produce any amplification bands. (**C**) Alignment of the derived mature amino acid sequences of rabOBP3 cloned from nose and prostate, and compared with the sequence stored in the NCBI EST database as UTE-7 (entry EL341998). Mnose: male nasal tissue; Fnose: female nasal tissue; Prost: prostate.

Acknowledgments

We thank Ms Olga Favilli of the Department of Veterinary Sciences, University of Pisa and Pampaloni Farm, Fauglia, Pisa, for help in the collection of rabbit semen.

Author Contributions

Conceived and designed the experiments: RM AN A. Scaloni PP. Performed the experiments: RM CD AN A. Serra A. Scaloni. Analyzed the data: RM CD A. Serra A. Scaloni PP. Contributed reagents/materials/analysis tools: AG A. Scaloni A. Serra PP. Wrote the paper: A. Scaloni PP.

References

1. Flower DR (1996) The lipocalin protein family: structure and function. Biochem J 318: 1–14
2. Flower DR (2000) Experimentally determined lipocalin structures. Biochim Biophys Acta 1482: 46–56
3. Monaco HL, Rizzi M, Coda A (1995) Structure of a complex of two plasma proteins: transthyretin and retinol-binding protein. Science 268: 1039–1041
4. Sawyer L, Kontopidis G (2000) The core lipocalin ß-lactoglobulin. Biochim Biophys Acta 1482: 136–148
5. Bianchet MA, Bains G, Pelosi P, Pevsner J, Snyder SH, et al. (1996) The three dimensional structure of bovine odorant-binding protein and its mechanism of odor recognition. Nat Struct Biol 3: 934–939
6. Tegoni M, Ramoni R, Bignetti E, Spinelli S, Cambillau C (1996) Domain swapping creates a third putative combining site in bovine odorant binding protein dimer. Nat Struct Biol 3: 863–867
7. Böcskei Z, Groom CR, Flower DR, Wright CE, Phillips EV, et al. (1992) Pheromone binding to two rodent urinary proteins revealed by X-ray crystallography. Nature 360: 186–188
8. Spinelli S, Ramoni R, Grolli S, Bonicel J, Cambillau C, et al. (1998) The structure of the monomeric porcine odorant binding protein sheds light on the domain swapping mechanism. Biochemistry 37: 7913–7918
9. Vincent F, Spinelli S, Ramoni R, Grolli S, Pelosi P, et al. (2000) Complexes of porcine odorant binding protein with odorant molecules belonging to different chemical classes. J Mol Biol 300: 127–139
10. Spinelli S, Vincent F, Pelosi P, Tegoni M, Cambillau C (2002) Boar Salivary Lipocalin: Three-dimensional X-Ray Structure and Androstenol/Androstenone Docking Simulations. Eur J Biochem 269: 2449–2456
11. Pelosi P, Pisanelli AM, Baldaccini NE, Gagliardo A (1981) Binding of 3H-2-isobutyl-3-methoxypyrazine to cow olfactory mucosa. Chem Senses 6: 77–85
12. Pelosi P, Baldaccini NE, Pisanelli AM (1982) Identification of a specific olfactory receptor for 2-isobutyl-3-methoxypyrazine. Biochem J 201: 245–248
13. Bignetti E, Cavaggioni A, Pelosi P, Persaud KC, Sorbi RT, et al. (1985) Purification and characterization of an odorant binding protein from cow nasal tissue. Eur J Biochem 149: 227–231
14. Pevsner J, Trifiletti RR, Strittmatter SM, Snyder SH (1985) Isolation and characterization of an olfactory receptor protein for odorant pyrazines. Proc Natl Acad Sci USA 82: 3050–3054
15. Pelosi P (1994) Odorant-binding proteins. Crit Rev Biochem Mol Biol 29: 199–228
16. Pelosi P (1996) Perireceptor events in olfaction. J Neurobiol 30, 3–19
17. Tegoni M, Pelosi P, Vincent F, Spinelli S, Campanacci V, et al. (2000) Mammalian odorant binding proteins. Biochim Biophys Acta 1482: 229–240
18. Dal Monte M, Andreini I, Revoltella R, Pelosi P (1991) Purification and characterization of two odorant binding proteins from nasal tissue of rabbit and pig. Comp Biochem Physiol 99B: 445–451
19. Marchese S, Pes D, Scaloni A, Carbone V, Pelosi P (1998) Lipocalins of boar salivary glands binding odours and pheromones. Eur J Biochem 252: 563–568
20. Pes D, Mameli M, Andreini I, Krieger J, Weber M, et al. (1998) Cloning and expression of odorant-binding proteins Ia and Ib from mouse nasal tissue. Gene 212: 49–55
21. Paolini S, Scaloni A, Amoresano A, Marchese S, Napolitano E, et al. (1998) Amino acid sequence post-translational modifications binding and labelling of porcine odorant-binding protein. Chem Senses 23: 689–698
22. Ganni M, Garibotti M, Scaloni A, Pucci P, Pelosi P (1997) Microheterogeneity of odorant-binding proteins in the porcupine revealed by N-terminal sequencing and mass spectrometry. Comp Biochem Physiol 117B: 287–291
23. Garibotti M, Navarrini A, Pisanelli AM, Pelosi P (1997) Three odorant-binding proteins from rabbit nasal mucosa. Chem Senses 22: 383–390
24. Pes D, Pelosi P (1995) Odorant-binding proteins of the mouse. Comp Biochem Physiol 112B: 471–479.
25. D'Innocenzo B, Salzano AM, D'Ambrosio C, Gazzano A, Niccolini A, et al. (2006) Secretory proteins as potential semiochemical carriers in the horse. Biochemistry 45: 13418–13428
26. Millery J, Briand L, Bezirard V, Blon F, Fenech C, et al. (2005) Specific expression of olfactory binding protein in the aerial olfactory cavity of adult and developing *Xenopus*. Eur J Neurosci 22: 1389–1399
27. Dal Monte M, Centini M, Anselmi C, Pelosi P (1993) Binding of selected odorants to bovine and porcine odorant binding proteins. Chem Senses 18: 713–721
28. Pevsner J, Hou V, Snowman AM, Snyder SH (1990) Odorant-binding protein characterization of ligand binding. J Biol Chem 265: 6118–6125
29. Hérent MF, Collin S, Pelosi P (1995) Affinities of nutty and green-smelling compounds to odorant-binding proteins. Chem Senses 20: 601–610

30. Loebel D, Marchese S, Krieger J, Pelosi P, Breer H (1998) Subtypes of odorant binding proteins: heterologous expression and assessment of ligand binding. Eur J Biochem 254: 318–324

31. Vincent F, Ramoni R, Spinelli S, Grolli S, Tegoni M, et al. (2004) Crystal structures of bovine odorant-binding protein in complex with odorant molecules. Eur J Biochem 271: 3832–3842

32. Pelosi P (1998) Odorant-binding proteins: structural aspects. Ann NY Acad Sci 855: 281–293

33. Pelosi P (2001) The role of perireceptor events in vertebrate olfaction. Cell Mol Life Sci 58: 503–509

34. Loebel D, Scaloni A, Paolini S, Fini C, Ferrara L, et al. (2000) Cloning, post-translational modifications, heterologous expression, ligand-binding and modelling of boar salivary lipocalin. Biochem J 350: 369–379

35. Pelosi P, Zhou J-J, Ban LP, Calvello M (2006) Soluble proteins in insect chemical communication. Cell Mol Life Sci 63: 1658–1676

36. Xu P, Atkinson R, Jones DN, Smith DP (2005) Drosophila OBP LUSH is required for activity of pheromone-sensitive neurons. Neuron 45: 193–200

37. Matsuo T, Sugaya S, Yasukawa J, Aigaki T, Fuyama Y (2007) Odorant-binding proteins OBP57d and OBP57e affect taste perception and host-plant preference in Drosophila sechellia. PLoS Biol 5: e118

38. Swarup S, Williams TI, Anholt RR (2011) Functional dissection of Odorant binding protein genes in Drosophila melanogaster. Genes Brain Behav 10: 648–657

39. Sun YF, De Biasio F, Qiao HL, Iovinella I, Yang SX, et al. (2012) Two Odorant-Binding Proteins Mediate the Behavioural Response of Aphids to the Alarm Pheromone (E)-ß-farnesene and Structural Analogues. PLoS One 7: e32759

40. Pevsner J, Hwang PM, Sklar PB, Venable JC, Snyder SH (1988) Odorant-binding protein and its mRNA are localized to lateral nasal gland implying a carrier function. Proc Natl Acad Sci USA 85: 2383–2387

41. Miyawaki A, Matsushita F, Ryo Y, Mikoshiba K (1994) Possible pheromone-carrier function of two lipocalin proteins in the vomeronasal organ. EMBO J 13: 5835–5842

42. Ohno K, Kawasaki Y, Kubo T, Tohyama M (1996) Differential expression of odorant-binding protein genes in rat nasal glands: implications for odorant-binding protein II as a possible pheromone transporter. Neuroscience 71: 355–366

43. Avanzini F, Bignetti E, Bordi C, Carfagna G, Cavaggioni A, et al. (1987) Immunocytochemical localization of pyrazine-binding protein in bovine nasal mucosa. Cell Tissue Res 247, 461–464.

44. Briand L, Eloit C, Nespoulous C, Bézirard V, Huet JC, et al. (2002) Evidence of an odorant-binding protein in the human olfactory mucus: location, structural characterization, and odorant-binding properties. Biochemistry 41, 7241–7252.

45. Cavaggioni A, Mucignat-Caretta C (2000) Major urinary proteins, alpha(2U)-globulins and aphrodisin. Biochim Biophys Acta 1482: 218–228

46. Cavaggioni A, Findlay JB, Tirindelli R (1990) Ligand binding characteristics of homologous rat and mouse urinary proteins and pyrazine binding protein of calf. Comp Biochem Physiol B 96: 513–520

47. Robertson DHL, Beynon RJ, Evershed RP (1993) Extraction characterisation and binding analysis of two pheromonally active ligands associated with major urinary protein of the house mouse (Mus musculus). J Chem Ecol 19: 1405–1416

48. Hurst JL, Payne CE, Nevison CM, Marie AD, Humphries RE, et al. (2001) Individual recognition in mice mediated by major urinary proteins. Nature 414: 631–634

49. Singer AG, Macrides F, Clancy AN, Agosta WC (1986) Purification and analysis of a proteinaceous aphrodisiac pheromone from hamster vaginal discharge. J Biol Chem 261: 13323–13326

50. Vincent F, Löbel D, Brown K, Spinelli S, Grote P, et al. (2001) Crystal structure of aphrodisin, a sex pheromone from female hamster. J Mol Biol 305: 459–469

51. Scaloni A, Paolini S, Brandazza A, Fantacci M, Marchese S, et al. (2001) Purification, cloning and characterisation of novel odorant-binding proteins in the pig. Cell Mol Life Sci 58: 823–834

52. Bacchini A, Gaetani E, Cavaggioni A (1992) Pheromone binding proteins in the mouse Mus musculus. Experientia 48: 419–421

53. Rouvinen J, Rautiainen J, Virtanen T, Zeiler T, Kauppinen J, et al. (1999) Probing the molecular basis of allergy Three-dimensional structure of the bovine lipocalin allergen Bos d2. J Biol Chem 274: 2337–2343

54. Hilger C, Kuehn A, Hentges F (2012) Animal lipocalin allergens. Curr Allergy Asthma Rep 12: 438–447

55. Mechref Y, Zidek L, Ma W-D, Novotny MV (2000) Glycosilated major urinary protein of the house mouse: characterization of its N-linked oligosaccharides. Glycobiology 10: 231–235

56. Virtanen T, Kinnunen T, Rytkönen-Nissinen M (2012) Mammalian lipocalin allergens—insights into their enigmatic allergenicity. Clin Exp Allergy 42: 494–504

57. Schaal B, Coureaud G, Langlois D, Giniès C, Sémon E, et al. (2003) Chemical and behavioural characterization of the rabbit mammary pheromone. Nature 424: 68–72

58. Charra R, Datiche F, Casthano A, Gigot V, Schaal B, et al. (2012) Brain processing of the mammary pheromone in newborn rabbits. Behav Brain Res 226: 179–188

59. Salzano AM, Novi G, Arioli S, Corona S, Mora D, et al. (2013) Mono-dimensional blue native-PAGE and bi-dimensional blue native/urea-PAGE or/SDS-PAGE combined with nLC-ESI-LIT-MS/MS unveil membrane protein heteromeric and homomeric complexes in Streptococcus thermophilus. J Proteomics 94: 240–261

60. Scaloni A, Monti M, Angeli S, Pelosi P (1999) Structural analysis and disulfide-bridge pairing of two odorant-binding proteins from Bombyx mori. Biochem Biophys Res 266: 386–391

61. Picariello G, Ferranti P, Mamone G, Roepstorff P, Addeo F (2008) Identification of N-linked glycoproteins in human milk by hydrophilic interaction liquid chromatography and mass spectrometry. Proteomics 8: 3833–3847

62. Hilvo M, Baranauskiene L, Salzano AM, Scaloni A, Matulis D, et al. (2008) Biochemical characterization of CA IX, one of the most active carbonic anhydrase isozymes. J Biol Chem 283: 27799–27809

63. Perez-Miller S, Zou Q, Novotny MV, Hurley TD (2010) High resolution X-ray structures of mouse major urinary protein nasal isoform in complex with pheromones. Protein Sci 19: 1469–1479

64. Cavaggioni A, Sorbi RT, Keen JN, Pappin DJC, Findlay JBC (1987) Homology between the pyrazine-binding protein from nasal mucosa and major urinary proteins. FEBS Lett 212: 225–228

65. Finlayson JS, Asofsky R, Potter M, Runner CC (1965) Major urinary protein complex of normal mice: origin. Science 149: 981–982

66. Dinh BL, Tremblay A, Dufour D (1965) Immunochemical study of rat urinary proteins: their relation to serum and kidney proteins. J Immunol 95, 574–582

67. Zhang Z-D, Frankish A, Hunt T, Harrow J, Gerstein M (2010) Identification and analysis of unitary pseudogenes: historic and contemporary gene losses in humans and other primates. Genome Biology 11: R26

68. Sun YL, Huang LQ, Pelosi P, Wang CZ (2012) Expression in antennae and reproductive organs suggests a dual role of an odorant-binding protein in two sibling Helicoverpa species. PLoS One 7: e30040

69. Lacazette E, Gachon A-M, Pitiot G (2000) A novel human odorant-binding protein gene family resulting from genomic duplicons at 9q34: differential expression in the oral and genital spheres. Hum Mol Genetics 9: 289–301

Migratory Birds Reinforce Local Circulation of Avian Influenza Viruses

Josanne H. Verhagen[1][*][♪], Jacintha G. B. van Dijk[2][♪], Oanh Vuong[1], Theo Bestebroer[1], Pascal Lexmond[1], Marcel Klaassen[2,3], Ron A. M. Fouchier[1]

1 Department of Viroscience, Erasmus MC, Rotterdam, The Netherlands, 2 Department of Animal Ecology, Netherlands Institute of Ecology (NIOO-KNAW), Wageningen, The Netherlands, 3 Centre for Integrative Ecology, School of Life and Environmental Sciences, Deakin University, Geelong, Australia

Abstract

Migratory and resident hosts have been hypothesized to fulfil distinct roles in infectious disease dynamics. However, the contribution of resident and migratory hosts to wildlife infectious disease epidemiology, including that of low pathogenic avian influenza virus (LPAIV) in wild birds, has largely remained unstudied. During an autumn H3 LPAIV epizootic in free-living mallards (Anas platyrhynchos) — a partially migratory species — we identified resident and migratory host populations using stable hydrogen isotope analysis of flight feathers. We investigated the role of migratory and resident hosts separately in the introduction and maintenance of H3 LPAIV during the epizootic. To test this we analysed (i) H3 virus kinship, (ii) temporal patterns in H3 virus prevalence and shedding and (iii) H3-specific antibody prevalence in relation to host migratory strategy. We demonstrate that the H3 LPAIV strain causing the epizootic most likely originated from a single introduction, followed by local clonal expansion. The H3 LPAIV strain was genetically unrelated to H3 LPAIV detected both before and after the epizootic at the study site. During the LPAIV epizootic, migratory mallards were more often infected with H3 LPAIV than residents. Low titres of H3-specific antibodies were detected in only a few residents and migrants. Our results suggest that in this LPAIV epizootic, a single H3 virus was present in resident mallards prior to arrival of migratory mallards followed by a period of virus amplification, importantly associated with the influx of migratory mallards. Thus migrants are suggested to act as local amplifiers rather than the often suggested role as vectors importing novel strains from afar. Our study exemplifies that a multifaceted interdisciplinary approach offers promising opportunities to elucidate the role of migratory and resident hosts in infectious disease dynamics in wildlife.

Editor: Michael Lierz, Justus-Liebeig University Giessen, Germany

Funding: This work was sponsored by grants from the Dutch Ministry of Economic Affairs, the Netherlands Organization for Scientific Research (NWO; grant 820.01.018) and contract NIAID NIH HHSN266200700010C. The funders had no role in study design, data collection and analysis, decision to publish, or preparation of the manuscript.

* Email: j.h.verhagen@erasmusmc.nl

♪ These authors contributed equally to this work.

Introduction

Migratory and resident (i.e. sedentary) hosts are thought to fulfil different, non-mutually exclusive, roles in infectious disease dynamics in wild animal populations, although empirical evidence is largely lacking. For one, migratory hosts may transport pathogens to new areas, resulting in the exposure and potential infection of new host species, thereby contributing to the global spread of infectious diseases [1]. Resident hosts, immunologically naïve to these novel pathogens, may subsequently act as local amplifiers. For instance, the global spread of West Nile Virus (WNV) is considered to be greatly facilitated by migratory birds introducing the virus to other wildlife and humans in many parts of the world [2]. Similarly, the introduction of Ebola virus into humans in the Democratic Republic of Congo, Africa, in 2007 coincided with massive annual fruit bat migration [3].

Additionally, migratory hosts may amplify pathogens upon arrival at a staging site, either because they are immunologically

naïve to locally circulating pathogens [4] and/or as a consequence of reduced immunocompetence due to the trade-off between investment in immune defences and long-distance flight [1]. Correspondingly, pathogen prevalence or the risk of disease outbreaks may locally be reduced when migratory hosts depart [1]. Consistent with the role for migrants, residents in this scenario are suggested to act as reservoirs, permanently maintaining pathogens within their population and transmitting them to other hosts, including migrants [5,6]. Given these potentially distinct roles for migratory and resident hosts in the spatial and temporal spread of infectious diseases, it is important to differentiate between migratory and resident hosts when aiming to improve our understanding of the ecology, epidemiology, and persistence of diseases in wild animal populations.

Wild bird populations are considered the reservoir hosts of low pathogenic avian influenza A viruses (LPAIV). Predominantly birds from wetlands and aquatic environments (orders Anseriformes and Charadriiformes) are infected with LPAIV [7], causing transient and mainly intestinal infections [8,9], with no or limited signs of disease [10]. LPAIV can be classified in subtypes based on antigenic and genetic variation of the viral surface glycoproteins hemagglutinin (HA) and neuraminidase (NA). All subtypes that have been recognized to date, notably HA subtypes 1 through 16 (H1-H16) and NA subtypes 1 through 9 (N1-N9), have been found in wild birds [11]. Recently, novel influenza viruses were identified in fruit bats that are distantly related to LPAIV (H17N10, H18N11), indicating that bats, alongside wild birds, harbour influenza viruses and might play a distinct role in the dynamics of this infectious disease [12,13].

Despite a large number of studies on the ecology and epidemiology of LPAIV in wild birds, only few studies have focussed on the role of resident and migratory hosts in the dynamics of this infectious disease. Resident bird species likely facilitate LPAIV transmission, while migratory bird species harbour high LPAIV subtype diversity after arrival at the wintering grounds [14,15]. In most of these studies resident and migratory hosts belonged to different bird species, with presumably different LPAIV susceptibility. However, many bird species are composed of a mixture of resident and migratory individuals, so called partial migrants [16]. Individuals that belong to the same species but use distinct migratory strategies, may differ in morphology and behaviour (e.g. body size, dominance; [17]), immune status and pathogen exposure. As a consequence, resident and migratory individuals of a single species may respond differentially to LPAIV infection and hence their contribution to local, and consequently global, LPAIV infection dynamics may differ. Hill et al. investigated the role of migratory and resident hosts of a single bird species in LPAIV infection dynamics. In their study, no differences were detected in LPAIV prevalence between migratory and resident host populations [18]. However, migrants likely introduced LPAIV subtypes from their breeding areas to the wintering grounds and residents likely acted as LPAIV reservoirs facilitating year-round circulation of limited subtypes [18]. A similar study in the same species conducted at a local scale instead of a macro-ecological scale, showed that susceptible migratory hosts were more frequently infected with LPAIV than residents, which had probably driven the epizootic in autumn [19]. LPAIV epizootics in wild birds are likely to take place at local spatial and temporal scales, since LPAIV infections are generally short (i.e. up to a week; [20]), and most virus particles are shed within the first few days after infection [21]. Yet, the precise role of migratory and resident hosts during local LPAIV epizootics in terms of virus introduction and reinforcement, including host immunity, has remained largely unstudied.

We build on the study of van Dijk et al. [19] to investigate the role of migratory and resident hosts of a single bird species during a local LPAIV epizootic. Throughout an H3 LPAIV epizootic at the wintering grounds in autumn 2010, we sampled a partly migratory bird species, the mallard (*Anas platyrhynchos*), and connected host migratory strategy with (i) H3 virus kinship, (ii) H3 virus prevalence and shedding, and (iii) H3-specific antibody prevalence. H3 LPAIV is a dominant subtype in wild ducks in the northern hemisphere [22,23]. This study provides a detailed description of a monophyletic H3 LPAIV epizootic importantly associated with the influx of migratory mallards.

Materials and Methods

Ethics statement

Capturing free-living mallards was approved by the Dutch Ministry of Economic Affairs based on the Flora and Fauna Act (permit number FF/75A/2009/067 and FF/75A/2010/011).

Handling and sampling of free-living mallards was approved by the Animal Experiment Committee of the Erasmus MC (permit number 122-09-20 and 122-10-20) and the Royal Netherlands Academy of Arts and Sciences (KNAW) (permit number CL10.02). Free-living mallards were released into the wild after sampling. All efforts were made to minimize animal suffering throughout the studies.

Study species and site

Mallards are considered a key LPAIV host species, together with other dabbling duck species of the *Anas* genus, harbouring almost all LPAIV subtype combinations found in birds to date [11]. Mallards are partially migratory, meaning that the population exists of both migratory and resident birds. Along the East Atlantic Flyway, mallards breeding in Scandinavia, the Baltic, and northwest Russia migrate to winter at more southern latitudes in autumn, congregating with the resident populations that breed in Western Europe, including the Netherlands [24].

During the 2010 LPAIV epizootic described here, free-living mallards were caught in swim-in traps of a duck decoy [25]. The duck decoy was located near Oud Alblas (51°52′38″N, 4°43′26″E), situated in the province of Zuid-Holland in the Netherlands. This sampling site is part of the ongoing national wild bird avian influenza virus (AIV) surveillance program (dd 2014-09-20), executed by the department of Viroscience of Erasmus MC, where mallards, free-living and hunted in the near surrounding, were sampled for LPAIV from 2005 onwards.

Sampling

During the LPAIV epizootic (i.e. from August until December 2010) studied here, the duck decoy was visited, on average, seven times per month capturing approximately 11 birds per visit. Each captured mallard was marked using a metal ring with an unique code, aged (juvenile: <1 year, adult:>1 year) and sexed based on plumage characteristics [26]. For virus detection, cloacal and oropharyngeal samples were collected using sterile cotton swabs as LPAIV may replicate in both the intestinal and respiratory tract of wild birds [27]. Swabs were stored individually in virus transport medium (Hank's balanced salt solution with supplements; [28]) at 4°C, and transported to the laboratory for analysis within seven days of collection. For detection of antibodies to AIV, blood samples (<1 ml, 2% of the circulating blood volume) were collected from the brachial vein, which were allowed to clot for approximately 6 h before centrifugation to separate serum from red blood cells [29]. Serum samples were stored at −20°C until analysis. To determine a bird's migratory strategy using stable hydrogen isotope analysis, the tip (1–2 cm) of the first primary feather of the right wing was collected and stored in a sealed bag at room temperature. Of recaptured birds, both swabs and a blood sample were collected.

Migratory strategy

In the study of van Dijk et al. [19], the origin (and hence, migratory strategy) of mallards sampled during the 2010 LPAIV epizootic was determined using stable hydrogen isotope analysis in feathers. Stable isotope signatures in feathers reflect those of local food webs [30]. During the period of growth (i.e. moult), local precipitation is incorporated into these feathers [31], causing the stable hydrogen isotope (δ^2H) ratio in feathers to be correlated with δ^2H of local precipitation [32]. Across Europe, a gradient of δ^2H in feathers is found in mallards [33]. Based on feather δ^2H and additional criteria, van Dijk et al. [19] classified mallards as resident, local migrant (i.e. short distance) and distant migrant (i.e. long distance). A resident bird had grown its feathers near the duck decoy (was captured during moult) and was recaptured multiple times either before or during the LPAIV epizootic. A local and distant migratory bird was seen and sampled once, i.e. only during the LPAIV epizootic and was not captured within one year before this epizootic. Based on feather δ^2H values of local (-103.5 to $-72.6‰$) and distant migrants (-164.5 to $-103.7‰$) and using a European feather δ^2H isoscape of mallards [33], local migrants originated roughly from central Europe and distant migrants roughly from north-eastern Europe. We used similar criteria to assess the migratory strategy of mallards caught during the H3 LPAIV epizootic. For 149 individual birds in this study we were unable to assign them to either the resident or migratory population and these were excluded from analyses, except the genetic analysis.

For full details on the stable hydrogen isotope analysis, see van Dijk et al. [33]. In short, feathers were cleaned and air-dried overnight. Feather samples were placed into silver capsules, stored in 96 well trays and shipped to the Colorado Plateau Stable Isotope Laboratory (Northern Arizona University, Flagstaff, USA). Stable hydrogen isotope analyses were performed on a Delta Plus XL isotope ratio mass spectrometer equipped with a 1400 C TC/EA pyrolysis furnace. Feather δ^2H values are reported in units per mil (‰) relative to the Vienna Standard Mean Ocean Water-Standard Light Antarctic Precipitation (VSMOW-SLAP) standard scale.

Virus detection, isolation and characterization

As part of the national wild bird AIV surveillance program — including the 2010 LPAIV epizootic — LPAIV infection of free-living and hunted mallards was assessed using cloacal and oropharyngeal swab samples. RNA from these samples was isolated using the MagnaPure LC system with a MagnaPure LC total nucleic acid isolation kit (Roche Diagnostics, Almere, the Netherlands) and analysed using a real-time reverse transcriptase-PCR (RT-PCR) assay targeting the matrix gene. Matrix RT-PCR positive samples were used for the detection of H5 and H7 influenza A viruses using HA specific RT-PCR tests [28,34]. All matrix positive samples were used for virus isolation in embryonated chicken eggs and characterized as described previously [28].

Matrix RT-PCR positive samples collected during the 2010 LPAIV epizootic for which virus culture was not successful, were screened for the presence of H3 influenza A viruses using a H3 specific RT-PCR test (n = 126). Additionally, matrix RT-PCR positive samples collected half year prior to the LPAIV epizootic (November 2009-July 2010) were screened for the presence of H3 influenza A viruses to determine whether H3 LPAIV was detected in mallards prior to the epizootic (n = 20). Amplification and detection were performed on an ABI 7500 machine with the taqman Fast Virus 1 Step Master mix reagents (Applied Biosystems, Nieuwerkerk aan den IJssel, the Netherlands) and 5 µl of eluate in an end volume of 30 µl using 10 pmol Oligonucleotides RF3226 (5'-GAACAACCGGTTCCAGAT-CAA -3') and 40 pmol RF3227 (5'- TGGCAGGCCCACA-TAATGA-3') and 10 pmol of the double-dye labelled probe RF3228 (5'-FAM-TCCTRTGGATTTCCTTTGCCATAT-CATGC-BHQ-3'). Primers and probe were designed with the software package Primer Express version 3.01 (Applied Biosystems, Nieuwerkerk aan den IJssel, the Netherlands), based on avian H3 nucleotide sequences obtained from Genbank (www.ncbi.nlm.nih.gov).

The degree of virus shedding from the cloaca and the oropharynx during the LPAIV epizootic was based on the cycle threshold (C_T) value, i.e. first real-time matrix RT-PCR amplification cycle in which matrix gene amplification was detected. The C_T-value is inversely proportional to the amount of viral RNA in a sample.

Sequence analysis and phylogeny

To investigate H3 LPAIV diversity in time and space among resident and migratory mallards during the LPAIV epizootic, we performed a genetic analysis focussed on the HA segment, one of the two most variable gene segments of LPAIV. Nucleotide sequences of the HA gene segment were obtained from virus isolates that were previously characterized by hemagglutination inhibition (HI) assay as H3 LPAIV. The RT-PCR and sequencing of the HA segment was performed using HA specific primers (5'-GGATCTGCTGCTTGTCCTGT-3' and 5'- GRATAAG-CATCTATTGGAC-3'), as described previously [35].

A total of 86 HA gene segments of 1576 nt in length were included in the genetic analysis. The genetic analysis comprised H3 nucleotide sequences obtained from (i) residents and migratory mallards during the 2010 LPAIV epizootic (n = 23), (ii) additional H3 LPAIV isolates from the national wild bird surveillance program of Erasmus MC (n = 35), and (iii) a BLAST analysis using public databases available as of 29 November 2013 (www.ncbi.nlm.nih.gov, http://www.gisaid.com), from which only European virus sequences with a known isolation date were retrieved (n = 28). Duplicate and incomplete sequences were removed. Nucleotide sequences were aligned using the software MAFFT version 7 (http://mafft.cbrc.jp/alignment/software/).

H3 nucleotide sequences were labelled based on sampling site, year of virus isolation, and host migratory strategy (i.e. resident, local migrant, distant migrant). During the 2010 LPAIV epizootic, H3 nucleotide sequences were obtained from 23 viruses, isolated from residents (n = 3), from local migrants (n = 13), from distant migrants (n = 2), and from birds of which the migratory strategy could not be assessed (n = 5). This was supplemented with 12 H3 nucleotide sequences obtained from viruses isolated from mallards sampled in the duck decoy in different years, notably in 2008 (n = 11) and 2011 (n = 1). There were 31 H3 nucleotide sequences from virus samples collected at other sampling locations in the Netherlands and elsewhere in Europe between 1999 and 2011. Of these virus samples, 18 originated from locations within the province of Zuid-Holland (5–30 km from the duck decoy), i.e. from Berkenwoude (n = 13) (51°57'00"N, 4°41'36"E), Lekkerkerk (n = 2) (51°53'41"N, 439'24"E), Oudeland van Strijen (n = 2) (51°46'56"N, 4°30'56"E) and Vlist (n = 1) (51°59'13"N, 4°45'56"E). Eleven viruses were isolated from birds in coastal regions in the Netherlands (i.e. 115–200 km from the duck decoy), i.e. Schiermonnikoog (n = 1) (53°28'41"N, 6°9'24"E), Vlieland (n = 1) (53°16'42"N, 5°1'22"E), Westerland (n = 8) (52°53'39"N, 4°56'32"E) and Wieringen (n = 1) (52°54'0"N, 4°58'11"E). Outside the Netherlands, two H3 sequences were from viruses isolated in Hungary in 2009. The remaining 20 H3 nucleotide sequences originated from multiple locations throughout Europe (i.e. Belgium, Czech Republic, Germany, Iceland, Italy and Switzerland) and Russia.

A Maximum Likelihood (ML) phylogenetic tree was generated using the PhyML package version 3.1 using the GTR+I+G model of nucleotide substitution, performing a full heuristic search and subtree pruning and regrafting (SPR) searches. The best-fit model of nucleotide substitution was determined with jModelTest [36]. Tree was visualized using the Figtree program, version 1.4.0 (http://tree.bio.ed.ac.uk/software/figtree). Overall rates of evolutionary change (i.e. number of nucleotide substitutions per site per year) and time of circulation to the most recent common ancestor (TMRCA) in years was estimated using the BEAST program version 1.8.0 (http://beast.bio.ed.ac.uk/). To accommodate variation in the molecular evolutionary rate among lineages, the uncorrelated log-normal relaxed molecular clock was used. Isolation dates were used to calibrate the molecular clock. Three independent Bayesian Markov Chain Monte Carlo (MCMC) analyses were performed for 50 million states, with sampling every 2,000 states. Convergence and effective sample sizes of the estimate were checked with Tracer v1.6 (http://tree.bio.ed.ac.uk/software/tracer/). Uncertainty in parameter estimates was reported as the 95% highest posterior density (HPD) [37]. Nucleotide sequences are online available under the accession numbers as listed in Table S1 and S2.

Serology

To assess whether mallards had H3-specific antibodies during the 2010 LPAIV epizootic, all sera were first tested for the presence of AIV antibodies specific for the nucleoprotein (NP) using a multispecies blocking enzyme-linked immunosorbent assay (bELISA MultiS-Screen Avian Influenza Virus Antibody Test Kit; IDEXX Laboratories, Hoofddorp, the Netherlands), following manufacturer's instructions. Each plate contained two positive and two negative controls. Samples were tested in duplicate. An infinite M200 plate reader (Tecan Group Ltd, Männedorf, Switzerland) was used to measure the absorbance (i.e. OD-value) at 620 nm. Samples were considered positive for the presence of NP antibodies when signal-to-noise ratios (i.e. mean OD-value of the sample divided by the mean OD-value of the negative control) were <0.5. NP antibody positive serum samples were subsequently tested for the presence of H3-specific antibodies using the HI assay according to standard procedures [38]. Briefly, sera were pre-treated overnight at 37°C with receptor destroying enzyme (Vibrio cholerae neuraminidase) and incubated at 56°C for 1 h. Two-fold serial dilutions of the antisera, starting at a 1:10 dilution, were mixed with 4 hemagglutinating units of A/Mallard/Netherlands/10/2010 (H3N8) in 25 μl and were incubated at 37°C for 30 min. Subsequently, 25 μl 1% turkey erythrocytes was added and the mixture was incubated at 4°C for 1 h. Hemagglutination inhibition patterns were read and the HI titre was expressed as the reciprocal value of the highest dilution of the serum that completely inhibited agglutination of turkey erythrocytes.

Statistics

Birds were considered LPAIV positive when either cloacal or oropharyngeal swabs were positive. To exclude samples of birds that had been sampled twice within the same infectious period during the 2010 LPAIV epizootic, we used an interval of at least 30 days between the day that a bird tested LPAIV positive and the next sampling day. Mallards may shed virus up to 18 days [21].

During the LPAIV epizootic, 709 cloacal and oropharyngeal swabs were collected from 472 mallards of which 129 individuals were recaptured. Of these swabs, 84 tested positive for H3 LPAIV, 35 tested LPAIV positive but H3 negative (i.e. matrix-positive H3-negative), and 583 swabs tested LPAIV negative. Of 7 matrix-positive swabs we were unable to determine H3-positivity. To test H3 virus prevalence and shedding, we included H3-positive and H3-negative swabs (i.e. matrix-negative and matrix-positive). Swabs from birds of which the migratory strategy could not be assessed (n = 269) or with undefined age and sex (n = 13) were excluded. The exclusion of birds of which the migratory strategy could not be assessed did not affect the temporal pattern of H3 LPAIV prevalence. In total we included 420 cloacal and oropharyngeal swabs from 305 individual birds, of which 55 birds were sampled more than once (Table S3).

During the LPAIV epizootic, 428 serum samples were collected from 364 mallards of which 52 individuals were recaptured. Of

these serum samples, 9 tested positive for H3-specific antibodies, 98 tested positive for AIV antibodies but negative for H3-specific antibodies (i.e. NP-positive H3-negative), and 321 sera tested negative for AIV antibodies. To investigate H3-specific antibody prevalence, we included H3-specific antibody positive and H3-specific antibody negative sera (i.e. NP-negative and NP-positive). Sera from birds of which the migratory strategy could not be assessed (n = 96) or with undefined age and sex (n = 5) were excluded. Thus in total we included 320 sera samples from 281 individual birds, of which 30 birds were sampled more than once (Table S3).

A generalized linear mixed model (GLMM) was used in the analysis of H3 virus prevalence, with migratory strategy (i.e. resident, local migrant, distant migrant), age, sex and month as fixed factors, all two-way interactions with migratory strategy, and individual bird as random factor. The interactions between migratory strategy and age, migratory strategy and sex, and migratory strategy and month were tested to assess whether H3 virus prevalence differed per age class, sex and month for the three categories of migratory strategy. The fixed factors age and sex were merely included in the models to conduct the interactions. A general linear model (GLM) was used to test for differences in prevalence of H3-specific antibodies, with migratory strategy and month as fixed factors. Linear models (LMs) were used to determine differences in the degree of virus shedding of H3 LPAIV-particles based on viral RNA from the cloaca and the oropharynx (i.e. C_T-value) with migratory strategy and month as fixed factors. A Tukey's post hoc test was performed to detect differences in H3 LPAIV prevalence between the three categories of migratory strategy and months. All analyses were conducted using R 2.14.1 [39]. Package lme4 was used to fit the GLMM [40] and multcomp to perform a Tukey's post hoc test [41].

Results

Virus prevalence

Each year, from 2005 until 2011, LPAIV prevalence in mallards peaked between the end of summer (August) and the beginning of winter (December), with some exceptions in March 2009 and June 2011 (Figure 1A). Detection of the various HA subtypes varied per year, with most virus isolates found in autumn, notably H2 to H8, H10, and H12. H3 LPAIV was isolated from mallards every year, except in 2007 and 2009, and was the dominant HA subtype in 2006, 2008 and 2010 (Figure 1B).

During the 2010 LPAIV epizootic, mallards were infected with H3 LPAIV (84 of 709, 12%) and with other LPAIV subtypes, namely H4, H6 and H10 (35 of 709, 5%; Figure 1B). The H3 LPAIV epizootic started on the 12th of August 2010 (Figure 2A) and H3 virus prevalence differed between months (Table 1). H3 virus prevalence increased in September, peaked in October, and decreased in November and December (Figure 2A and 2C). Shortly before the 2010 LPAIV epizootic, a single mallard of unknown origin was infected with H3 LPAIV on the 10th of February 2010, followed by a period of five months where no H3 infections were detected among 536 mallards sampled.

Local and distant migrants were more often infected with H3 LPAIV (37 of 113, 33% and 22 of 98, 22% respectively) than residents (20 of 209, 10%; Figure 2C, Table 1). The peak month of the H3 LPAIV epizootic differed between the three mallard populations (Table 1): in local migrants H3 LPAIV infection peaked in September, whereas in residents and distant migrants infection peaked in October (Figure 2C). At the start of the H3 LPAIV epizootic (12th of August), three residents and one local migrant were infected with H3 LPAIV, with their populations

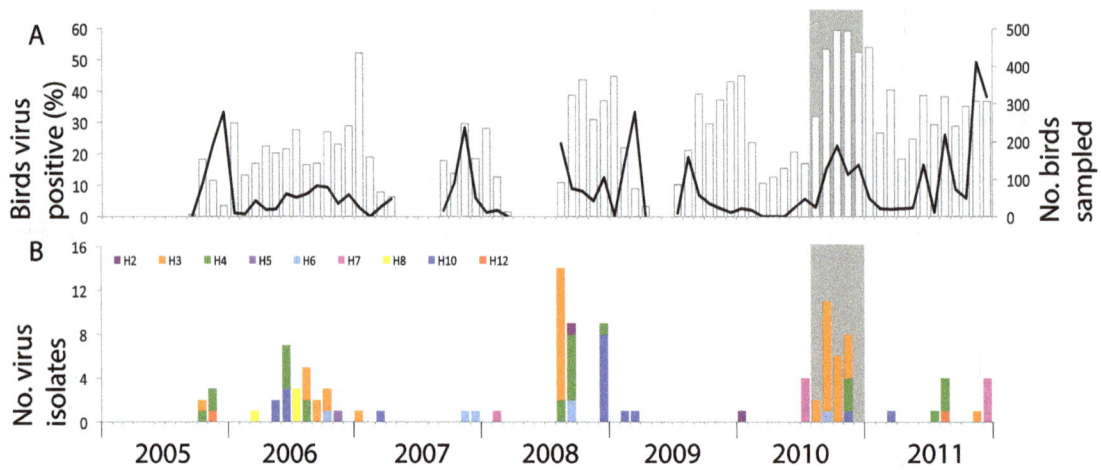

Figure 1. Prevalence and subtype diversity of low pathogenic avian influenza viruses (LPAIV) in mallards sampled at Oud Alblas, the Netherlands, 2005–2011. The grey-shaded area indicates the H3 LPAIV epizootic from August until December 2010. (A) Number of free-living and hunted birds sampled (bars, right Y-axis) and percentage of birds tested virus positive based on M RT-PCR (line, left Y-axis). (B) Number of virus isolates per HA subtype: H2 (purple), H3 (orange), H4 (green), H5 (light purple), H6 (light blue), H7 (pink), H8 (yellow), H10 (dark blue) and H12 (red).

constituting respectively 88% and 12% of the sampled mallard population. Two weeks later (26th of August), the first distant migrant infected with H3 LPAIV was detected (44% of the sampled mallard population). In September and October, most mallards infected with H3 LPAIV were local migrants (respectively 12 of 22 and 15 of 35 total H3 LPAIV positives), while local

migrants comprised respectively 24% and 40% of the sampled mallard population. In October, 11 residents and nine distant migrants were infected with H3 LPAIV, the latter constituting only 17% of the sampled mallard population. In November, only nine local and five distant migrants were infected with H3 LPAIV (comprising respectively 29% and 25% of the sampled mallard

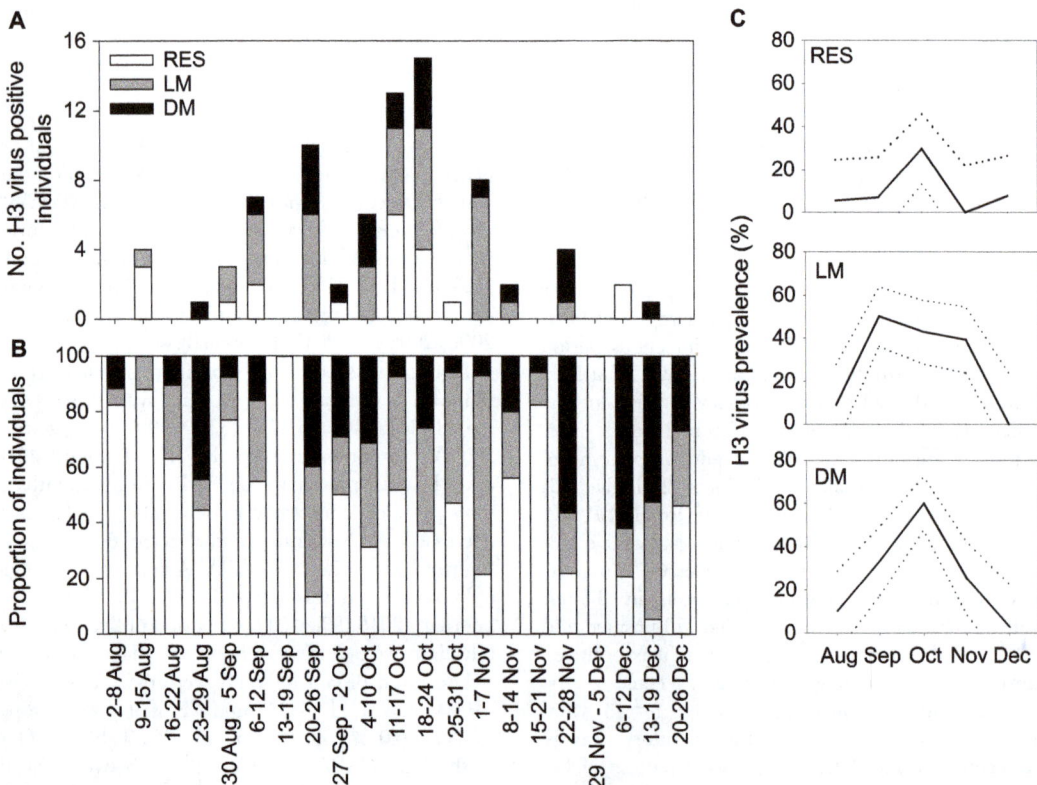

Figure 2. Prevalence of H3 low pathogenic avian influenza viruses (LPAIV) in residents, local and distant migratory mallards during the H3 LPAIV epizootic in 2010. For residents (RES), local migrants (LM) and distant migrants (DM) the (A) number of H3 virus positive individuals per week, (B) proportion of individuals sampled per week, and (C) H3 virus prevalence (±95% CI) per month are depicted.

Table 1. Linear model test results of the analysis of H3 low pathogenic avian influenza virus (LPAIV) prevalence during the LPAIV epizootic in 2010.

Variable	H3 virus prevalence	
	X^2	p-value
Age	0.144	0.705
Sex	0.659	0.417
Month	44.928	**<0.001**
Migratory strategy	23.681	**<0.001**
Migratory strategy * Age	0.777	0.678
Migratory strategy * Sex	0.558	0.757
Migratory strategy * Month	21.510	**0.006**

Besides migratory strategy, age, sex, month and two-way interactions were included. Significant values (p<0.05) are shown in bold.

population). The last month of the H3 LPAIV epizootic, only one distant migrant and two residents were infected with H3 LPAIV, although distant migrants and residents constituted respectively 43% and 32% of the sampled mallard population.

Virus shedding

H3 virus shedding from the cloaca and oropharynx did not differ between the three mallard populations ($F_{2,10} = 1.051$, p = 0.385 and $F_{2,63} = 0.025$, p = 0.976, respectively). Nor were there any differences in the monthly amount of H3 virus shed from the cloaca and oropharynx during the H3 LPAIV epizootic ($F_{3,10} = 1.945$, p = 0.186 and $F_{4,63} = 1.124$, p = 0.353, respectively).

Antibody prevalence

During the 2010 LPAIV epizootic, NP-specific LPAIV antibody prevalence increased from September onwards to 60% in December (Figure S1). During the H3 LPAIV epizootic, the proportion of local and distant migrants with H3-specific antibodies (3 of 106, 3% and 4 of 96, 4% respectively) was similar to that in residents (2 of 118, 2%; $X^2 = 0.543$, p = 0.762; Figure 3). There were no differences in H3-specific antibodies between months ($X^2 = 6.996$, p = 0.136). During the H3 LPAIV epizootic, H3-specific antibodies were detected on four sampling dates. On the 5th of August, before the start of the H3 LPAIV epizootic, one distant migrant had H3-specific antibodies (while distant migrants constituted 14% of the sampled mallard population). During the H3 LPAIV epizootic, the first resident with H3-specific antibodies was sampled on the 21st of September, with 9% of the sampled mallard population comprised of residents. After the peak of the H3 LPAIV epizootic (1st of November), two local migrants, one distant migrant and one resident had antibodies specific for H3 LPAIV. That day, local migrants constituted the largest proportion of the sampled mallard population (71%). At the end of the epizootic (21st of December), only migrants (local migrant: 1, distant migrant: 2) had specific antibodies against H3 LPAIV (constituting 38% and 44% of the sampled mallard population, respectively).

Virus kinship

The HA gene sequences of the H3 LPAIV strains isolated from free-living mallards during the H3 LPAIV epizootic were monophyletic, suggesting the outbreak resulted from a single virus introduction. Although migratory mallards kept arriving at the study site during the H3 LPAIV epizootic, the genetic analysis indicates that no other H3 LPAIVs were introduced. The

estimated time to the most recent common ancestor of the H3 LPAIV strains of the epizootic was spring 2009 (TMRCA 12 May 2009, LHPD95% 1 July 2008, UHPD95% 18 November 2009). The H3 LPAIV strain detected in a single mallard at our study site prior to the H3 LPAIV epizootic (10th of February 2010) differed from the H3 LPAIV strains of the epizootic (HA could only be sequenced partially and is not shown in the tree), and was therefore unlikely to have seeded the outbreak. Furthermore, the H3 LPAIV strains isolated during the H3 LPAIV epizootic were not closely related to isolates obtained from mallards at our study site in autumn 2008 (sequence identity 0.958–0.967), or November 2011 (sequence identity 0.954–0.957; Figure 4). However, the H3 LPAIV strains isolated from the H3 LPAIV epizootic were genetically closely related to H3 isolates from mallards at two sampling sites 8 to 12 km away from the study site one year later, in autumn 2011 (i.e. locations Berkenwoude and Vlist; Figure 4).

H3 LPAIV strains isolated from the resident, local and distant migratory population belonged to the same cluster with little variation in nucleotide sequences (sequence identity 0.995–1; detail of Figure 4). No consistent substitutions were detected in the nucleotide sequences that correlated with the migratory strategy of birds. Evolutionary divergence of the HA of H3 LPAIV was $2.5^{e\text{-}3}$ nucleotide substitutions per site per year, which is lower than reported by Hill et al. [18]: $1.38 \, (\pm 0.40)^{e\text{-}2}$.

Discussion

Studying the role of resident and migratory hosts in the spread and circulation of pathogens in animal populations is crucial for increasing our understanding of the ecology and epidemiology of infectious diseases in wildlife. We studied virus and antibody prevalence in free-living mallards during an autumn LPAIV epizootic of subtype H3 at a local scale, focussing on the distinct role that resident and migratory hosts might have played in the introduction and circulation of this virus subtype. Although alternative interpretations cannot be entirely excluded, our findings suggest that the H3 LPAIV causing the epizootic was present in resident mallards prior to the arrival of migrants, followed by virus amplification importantly associated with the arrival of migratory mallards.

H3 LPAIV isolations from residents, local and distant migrants belonged to the same genetic cluster (Figure 4). However, we cannot fully exclude the possibility that novel introductions of H3 LPAIV, or other LPAIV HA subtypes, by migratory birds occurred that were subsequently outcompeted by the dominant

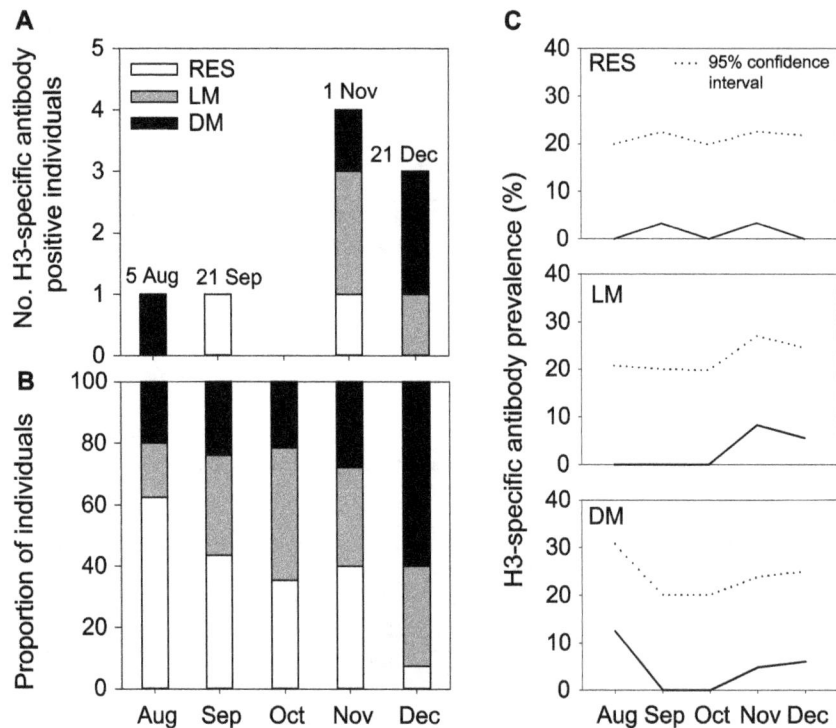

Figure 3. Prevalence of avian influenza H3-specific antibodies in residents, local and distant migratory mallards during the H3 low pathogenic avian influenza virus (LPAIV) epizootic in 2010. For residents (RES), local migrants (LM) and distant migrants (DM) the (A) number of H3-specific antibody positive individuals, (B) proportion of individuals sampled, and (C) H3-specific antibody prevalence (± 95% CI) per month are depicted.

epizootic H3 LPAIV strain and thus remained undetected during our monitoring (i.e. competitive exclusion principle; [42]). For instance, another H3 LPAIV epizootic in the area (i.e. Berkenwoude in 2008) resulted from multiple virus introductions. The H3 LPAIV that induced the 2010 epizootic was closely related to H3 LPAIV strains isolated in the near surrounding one year after the epizootic (i.e. Berkenwoude and Vlist in 2011). This suggests that after the epizootic H3 LPAIV may have overwintered and had been maintained locally. H3 virus prevalence in migratory mallards was higher (especially in distant migrants) and more prolonged (especially in local migrants) than in resident individuals. This finding corresponds with the results of van Dijk et al. [19] who found a three-fold increase in overall (i.e. non LPAIV-subtype specific) virus prevalence in migratory mallards. However, during the peak of the H3 LPAIV epizootic many residents were also infected with H3 LPAIV, which may be a consequence of the local amplification and increased viral deposition in the environment (i.e. water and sediment) at the study site. The local amplification may thus be a self-reinforcing process.

At the start of the H3 LPAIV epizootic, almost exclusively resident birds were infected with H3 LPAIV. However, it is not surprising that the majority of H3 LPAIV infections were found in residents, since the sampled mallard population consisted mainly out of resident birds (88%). What is remarkable though is that one week after detection of the first H3 LPAIV infections, no migrants were infected while a large proportion of the sampled mallard population consisted of migrants (~40%). Either migratory birds were not, or to a lesser extent, susceptible to H3 LPAIV infection, or contact rates and the amount of H3 virus particles in the surface water were still too low to infect arriving migrants. Interestingly, the peak of virus infection in October in the resident population

was mainly induced by recaptured resident birds (i.e. captured multiple times) (Figure S2). H3 virus prevalence in primary residents (i.e. captured for the first time) remained relatively low and increased in December. Potentially recaptured residents were trap-prone and had a higher probability of being exposed—and consequently becoming infected—than primary residents. In addition, in October the population of recaptured residents sampled was three-times higher than the population of primary residents sampled, increasing the probability of virus detection in recaptured residents.

During the H3 LPAIV epizootic, H3-specific antibodies were detected in both resident and migratory mallards, albeit in very few individuals and at low titres. A week before the start of the H3 LPAIV epizootic, H3-specific antibodies were found in a distant migrant (5th of August). We cannot exclude that this individual was infected with H3 LPAIV either during migration, at a stop-over site or at the breeding grounds. Hypothetically, this individual could have been infected with H3 LPAIV when transiting through southern Sweden (i.e. feather hydrogen stable isotope -129.2‰ suggest it originated from southern Scandinavia, Baltic States or Russia; [33]), introducing this virus to the wintering grounds. H3 LPAIV is detected frequently in mallards sampled in southern Sweden in early autumn [43]. Although our genetic analysis does not support this theory, it should be noted that only few H3 LPAIV originating from Sweden or other northern European countries were available and were included in the genetic analysis.

Several local and distant migrants had H3-specific antibodies after the peak of the H3 LPAIV epizootic. Since these birds were captured once during the H3 LPAIV epizootic, we cannot exclude that an H3 LPAIV infection outside the study site triggered this antibody response (i.e. genetically different H3 LPAIV were

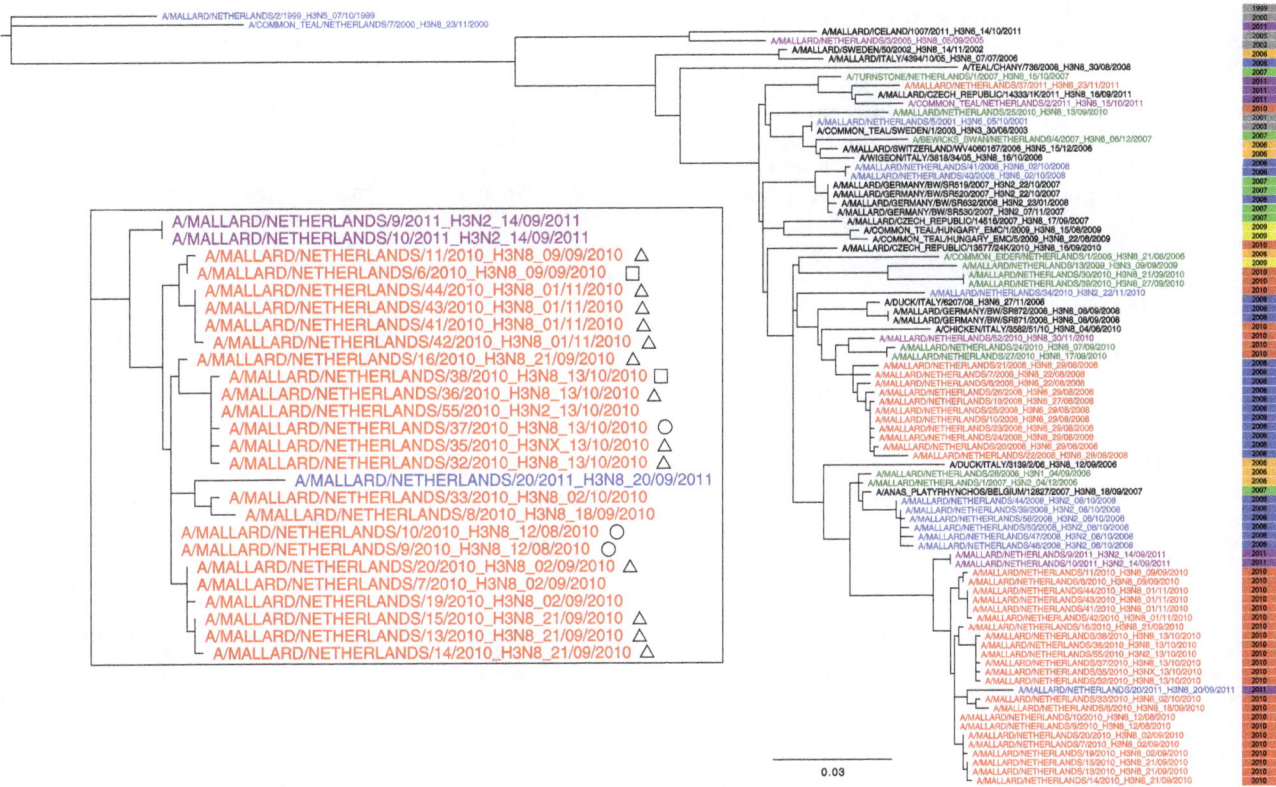

Figure 4. Phylogenetic analysis of HA gene of H3 low pathogenic avian influenza virus (LPAIV) isolated during the H3 LPAIV epizootic in 2010. The Maximum Likelihood (ML) tree contains samples of wild birds collected at various locations in and outside the Netherlands from 1999 until 2011. Each sampling location within the Netherlands is grouped by colour: Oud Alblas (red); Berkenwoude (blue); Lekkerkerk aan de IJssel, Oudeland van Strijen and Vlist (purple); Schiermonnikoog, Vlieland, Westerland and Wieringen (green). Locations are closely situated to the study site (i.e. duck decoy near Oud Alblas), except the locations shown in green, which are located at the coast. Year of virus isolation is listed next to isolate and grouped by colour. Detail of ML tree contains samples of the H3 LPAIV epizootic described in this study and migratory strategy of mallards: residents (RES; circle), local migrants (LM; triangle) and distant migrants (DM; square).

isolated at other locations in the Netherlands). Resident mallards with H3-specific antibodies most likely have been infected by the H3 LPAIV of the epizootic. Only 20% (1 of 5) of residents that had been infected with H3 LPAIV during the epizootic had H3-specific antibodies when recaptured (i.e. recaptured within 31 days since longevity of detectable HA specific antibodies is short; [44]). As result of H3 LPAIV infection an H3 specific antibody titre may have been generated, yet not detected due to antibody dynamics and timing of sampling, and/or sensitivity of the HI assay.

In conclusion, by combining virology, serology and phylogeny analyses with stable isotopes we demonstrate that a local H3 LPAIV epizootic in mallards was likely induced by a single virus introduction into susceptible residents, followed by a period of local virus amplification that was associated with the influx of migratory mallards. In addition to the study of Hill et al. [18], who showed long-distance movement of LPAIV genes by migrating mallards on a macro-ecological scale, we showed an association between local amplification of H3 LPAIV and the arrival of migratory mallards at the wintering grounds at a much smaller ecological scale. We suggest an additional role for migrating mallards as local amplifiers, based on the difference in H3 LPAIV prevalence between resident and migratory mallards upon arrival at the wintering grounds. This study exemplifies the difficulty of elucidating the role of migratory and resident hosts in infectious disease dynamics in wildlife, but provides encouraging indications

that the here presented multifaceted approach may open a window on these processes.

Supporting Information

Figure S1 Prevalence of avian influenza-specific antibodies in free-living mallards during H3 epizootic. This figure shows prevalence of avian influenza virus nucleoprotein (NP)-specific antibodies in mallards (*Anas platyrhynchos*) during the H3 low pathogenic avian influenza virus epizootic in 2010.

Figure S2 Prevalence of H3 influenza virus in resident and migratory mallards during H3 epizootic. This figure shows H3 low pathogenic avian influenza virus (LPAIV) prevalence in resident mallards (i.e. primary captured and recaptured), local and distant migratory mallards, during the H3 LPAIV epizootic in 2010.

Table S1 The H3 influenza virus strain names and accession numbers used in this study. This table includes all H3 influenza virus strain names and accession numbers used in this study.

Table S2 The sequence information of H3 influenza viruses from GISAID's EpiFlu Database. This table includes details of H3 influenza viruses downloaded from GISAID's EpiFlu Database.

Table S3 Sample collection for influenza virus and antibody detection from free-living mallards. This table includes number of samples collected for influenza virus and antibody detection from free-living mallards (*Anas platyrhynchos*) during the H3 low pathogenic avian influenza virus epizootic.

Acknowledgments

We thank Teunis de Vaal for catching mallards in the duck decoy at Oud Alblas and assisting with sampling the birds. Peter de Vries, Audrey van Mastrigt and Lennart Zwart are also thanked for their help in the field. We thank Ger van der Water, Judith Guldemeester and Kim Westgeest for logistical and technical assistance, and Richard Doucett and Melanie Caron of the Colorado Plateau Stable Isotope Laboratory for performing the stable hydrogen isotope analysis. The sequences of the H3 LPAIV used in this study are available from the Influenza Research Database (IRD) (http://www.fludb.org) and GISAID EpiFlu Database (http://www.gisaid.org) and listed in Table S1 and S2. This is publication 5677 of the NIOO-KNAW. We thank David Stallknecht and an anonymous reviewer for comments on earlier versions of this paper.

Author Contributions

Conceived and designed the experiments: JV JVD MK RF. Performed the experiments: JV JVD OV TB PL. Analyzed the data: JV JVD. Contributed reagents/materials/analysis tools: JVD RF. Wrote the paper: JH JVD MK RF.

References

1. Altizer S, Bartel R, Han BA (2011) Animal migration and infectious disease risk. Science 331: 296–302. doi:10.1126/science.1194694.
2. Rappole JH, Hubálek Z (2003) Migratory birds and West Nile virus. J Appl Microbiol 94: 47S–58S. doi:10.4269/ajtmh.2009.09-0106.
3. Leroy EM, Epelboin A, Mondonge V, Pourrut X, Gonzalez JP, et al. (2009) Human Ebola outbreak resulting from direct exposure to fruit bats in Luebo, Democratic Republic of Congo, 2007. Vector-Borne Zoonotic Dis 9: 723–728. doi:10.1089/vbz.2008.0167.
4. Leighton FA (2002) Health risk assessment of the translocation of wild animals. Rev Sci Tech Off Int Epizoot 21: 187–195.
5. Waldenström J, Bensch S, Kiboi S, Hasselquist D, Ottosson U (2002) Cross-species infection of blood parasites between resident and migratory songbirds in Africa. Mol Ecol 11: 1545–1554. doi:10.1046/j.1365-294X.2002.01523.x.
6. Haydon DT, Cleaveland S, Taylor LH, Laurenson MK (2002) Identifying reservoirs of infection: aconceptual and practical challenge. Emerg Infect Dis 8: 1468–1473.
7. Webster RG, Bean WJ, Gorman OT, Chambers TM, Kawaoka Y (1992) Evolution and ecology of influenza A viruses. Microbiol Rev 56: 152–179.
8. Daoust PY, van de Bildt M, van Riel D, van Amerongen G, Bestebroer T, et al. (2012) Replication of 2 subtypes of low-pathogenicity avian influenza virus of duck and gull origins in experimentally infected mallard ducks. Vet Pathol 50: 548–559. doi:10.1177/0300985812469633.
9. Höfle U, van de Bildt MWG, Leijten LM, van Amerongen G, Verhagen JH, et al. (2012) Tissue tropism and pathology of natural influenza virus infection in black-headed gulls (*Chroicocephalus ridibundus*). Avian Pathol 41: 547–553. doi:10.1080/03079457.2012.744447.
10. Kuiken T (2013) Is low pathogenic avian influenza virus virulent for wild waterbirds? Proc R Soc B-Biol Sci 280: 20130990. doi:10.1098/rspb.2013.0990.
11. Olsen B, Munster VJ, Wallensten A, Waldenström J, Osterhaus ADME, et al. (2006) Global patterns of influenza A virus in wild birds. Science 312: 384–388. doi:10.1126/science.1122438.
12. Tong SX, Li Y, Rivailler P, Conrardy C, Castillo DAA, et al. (2012) A distinct lineage of influenza A virus from bats. Proc Natl Acad Sci U S A 109: 4269–4274. doi:10.1073/pnas.1116200109.
13. Tong S, Zhu X, Li Y, Shi M, Zhang J, et al. (2013) New world bats harbor diverse influenza A viruses. Plos Pathog 9: e1003657. doi:10.1371/journal.ppat.1003657.
14. Stallknecht DE, Shane SM, Zwank PJ, Senne DA, Kearney MT (1990) Avian influenza viruses from migratory and resident ducks of coastal Louisiana. Avian Dis 34: 398–405.
15. Ferro PJ, Budke CM, Peterson MJ, Cox D, Roltsch E, et al. (2010) Multiyear surveillance for avian influenza virus in waterfowl from wintering grounds, Texas coast, USA. Emerg Infect Dis 16: 1224–1230. doi:10.3201/eid1608.091864.
16. Lack D (1943) The problem of partial migration. Br Birds 37: 122–130.
17. Chapman BB, Brönmark C, Nilsson JA, Hansson LA (2011) The ecology and evolution of partial migration. Oikos 120: 1764–1775. doi:10.1111/j.1600-0706.2011.20131.x.
18. Hill NJ, Takekawa JY, Ackerman JT, Hobson KA, Herring G, et al. (2012) Migration strategy affects avian influenza dynamics in mallards (*Anas platyrhynchos*). Mol Ecol 21: 5986–5999. doi:10.1111/j.1365-294X.2012.05735.x.
19. van Dijk JGB, Hoye BJ, Verhagen JH, Nolet BA, Fouchier RAM, et al. (2014) Juveniles and migrants as drivers for seasonal epizootics of avian influenza virus. J Anim Ecol 83: 266–275. doi:10.1111/1365-2656.12131.
20. Latorre-Margalef N, Gunnarsson G, Munster VJ, Fouchier RAM, Osterhaus ADME, et al. (2009) Effects of influenza A virus infection on migrating mallard ducks. Proc R Soc B-Biol Sci 276: 1029–1036. doi:10.1098/rspb.2008.1501.
21. Hénaux V, Samuel MD (2011) Avian influenza shedding patterns in waterfowl: implications for surveillance, environmental transmission, and disease spread. J Wildl Dis 47: 566–578.
22. Krauss S, Walker D, Pryor SP, Niles L, Li CH, et al. (2004) Influenza A viruses of migrating wild aquatic birds in North America. Vector-Borne Zoonotic Dis 4: 177–189. doi:10.1089/1530366042162452.
23. Munster VJ, Baas C, Lexmond P, Waldenström J, Wallensten A, et al. (2007) Spatial, temporal, and species variation in prevalence of influenza A viruses in wild migratory birds. PLoS Pathog 3: 630–638. doi:10.1371/journal.ppat.0030061.
24. Scott DA, Rose PM (1996) Atlas of Anatidae Populations in Africa and Western Eurasia, Wetlands International Publication No. 41. Wageningen: Wetlands International. 336 p.
25. Payne-Gallwey R (1886) The book of duck decoys, their construction, management, and history. London: J. van Voorst. 154 p.
26. Boyd H, Harrison J, Allison A (1975) Duck wings: a study of duck production. A WAGBI Publication. Chester: Marley Ltd., and the Harrison Zoological Museum. 112 p.
27. Fouchier RAM, Munster VJ (2009) Epidemiology of low pathogenic avian influenza viruses in wild birds. Rev Sci Tech Off Int Epizoot 28: 49–58.
28. Munster VJ, Baas C, Lexmond P, Bestebroer TM, Guldemeester J, et al. (2009) Practical considerations for high-throughput Influenza A virus surveillance studies of wild birds by use of molecular diagnostic tests. J Clin Microbiol 47: 666–673. doi:10.1128/jcm.01625-08.
29. Hoye BJ (2012) Variation in postsampling treatment of avian blood affects ecophysiological interpretations. Methods Ecol Evol 3: 162–167. doi:10.1111/j.2041-210X.2011.00135.x.
30. Peterson BJ, Fry B (1987) Stable isotopes in ecosystem studies. Annu Rev Ecol Syst 18: 293-320. doi:10.1146/annurev.ecolsys.18.1.293.
31. Hobson KA (1999) Tracing origins and migration of wildlife using stable isotopes: a review. Oecologia 120: 314–326. doi:10.1007/s004420050865.
32. Hobson KA, Wassenaar LI (1997) Linking breeding and wintering grounds of neotropical migrant songbirds using stable hydrogen isotopic analysis of feathers. Oecologia 109: 142–148. doi:10.1007/s004420050068.
33. van Dijk JGB, Meissner W, Klaassen M (2014) Improving provenance studies in migratory birds when using feather hydrogen stable isotopes. J Avian Biol 45: 103–108. doi:10.1111/j.1600-048X.2013.00232.x.
34. Fouchier RAM, Schneeberger PM, Rozendaal FW, Broekman JM, Kemink SAG, et al. (2004) Avian influenza A virus (H7N7) associated with human conjunctivitis and a fatal case of acute respiratory distress syndrome. Proc Natl Acad Sci U S A 101: 1356–1361. doi:10.1073/pnas.0308352100.
35. Hoffmann E, Stech J, Guan Y, Webster RG, Perez DR (2001) Universal primer set for the full-length amplification of all influenza A viruses. Arch Virol 146: 2275–2289. doi:10.1007/s007050170002.
36. Posada D (2008) jModelTest: Phylogenetic model averaging. Mol Biol Evol 25: 1253–1256. doi:10.1093/molbev/msn083.
37. Westgeest KB, Russell CA, Lin XD, Spronken MIJ, Bestebroer TM, et al. (2014) Genomewide analysis of reassortment and evolution of human influenza A(H3N2) viruses circulating between 1968 and 2011. J Virol 88: 2844–2857. doi:10.1128/jvi.02163-13.
38. Hirst GK (1943) Studies of antigenic differences among strains of influenza a by means of red cell agglutination. J Exp Med 78: 407–423. doi:10.1084/jem.78.5.407.
39. R Development Core Team (2012) R: A Language and Environment for Statistical Computing. Vienna, Austria: R Foundation for Statistical Computing.
40. Bates D, Maechler M, Bolker B (2012) Lme4:linear mixed-effects models using S4 classes. R package version 0.999999-0. http://CRAN.R-project.org/package=lme4.
41. Hothorn T, Bretz F, Westfall P (2008) Simultaneous inference in general parametric models. Biometrical J 50: 346–363. doi: 10.1002/bimj.200810425
42. Hardin G (1960) Competitive exclusion principle. Science 131: 1292–1297. doi:10.1126/science.131.3409.1292.

43. Latorre-Margalef N, Tolf C, Grosbois V, Avril A, Bengtsson D, et al. (2014) Long-term variation in influenza A virus prevalence and subtype diversity in migratory mallards in northern Europe. Proc R Soc B-Biol Sci 281: 20140098. doi:10.1098/rspb.2014.0098.

44. Curran JM, Robertson ID, Ellis TM, Selleck PW, O'Dea MA (2013) Variation in the responses of wild species of duck, gull, and wader to inoculation with a wild-bird-origin H6N2 low pathogenicity avian influenza virus. Avian Dis 57: 581–586.

Permissions

The contributors of this book come from diverse backgrounds, making this book a truly international effort. This book will bring forth new frontiers with its revolutionizing research information and detailed analysis of the nascent developments around the world.

We would like to thank all the contributing authors for lending their expertise to make the book truly unique. They have played a crucial role in the development of this book. Without their invaluable contributions this book wouldn't have been possible. They have made vital efforts to compile up to date information on the varied aspects of this subject to make this book a valuable addition to the collection of many professionals and students.

This book was conceptualized with the vision of imparting up-to-date information and advanced data in this field. To ensure the same, a matchless editorial board was set up. Every individual on the board went through rigorous rounds of assessment to prove their worth. After which they invested a large part of their time researching and compiling the most relevant data for our readers.

The editorial board has been involved in producing this book since its inception. They have spent rigorous hours researching and exploring the diverse topics which have resulted in the successful publishing of this book. They have passed on their knowledge of decades through this book. To expedite this challenging task, the publisher supported the team at every step. A small team of assistant editors was also appointed to further simplify the editing procedure and attain best results for the readers.

Apart from the editorial board, the designing team has also invested a significant amount of their time in understanding the subject and creating the most relevant covers. They scrutinized every image to scout for the most suitable representation of the subject and create an appropriate cover for the book.

The publishing team has been an ardent support to the editorial, designing and production team. Their endless efforts to recruit the best for this project, has resulted in the accomplishment of this book. They are a veteran in the field of academics and their pool of knowledge is as vast as their experience in printing. Their expertise and guidance has proved useful at every step. Their uncompromising quality standards have made this book an exceptional effort. Their encouragement from time to time has been an inspiration for everyone.

The publisher and the editorial board hope that this book will prove to be a valuable piece of knowledge for researchers, students, practitioners and scholars across the globe.

List of Contributors

Chris A. Dejong and Joanna Y. Wilson
Department of Biology, McMaster University, Hamilton, Ontario, Canada

Amie J. Radenbaugh, Singer Ma, Adam Ewing, Joshua M. Stuart and Jingchun Zhu
University of California Santa Cruz Genomics Institute, Department of Biomolecular Engineering, University of California Santa Cruz, Santa Cruz, California, United States of America

Eric A. Collisson
Division of Hematology/Oncology, University of California San Francisco, San Francisco, California, United States of America

David Haussler
University of California Santa Cruz Genomics Institute, Department of Biomolecular Engineering, University of California Santa Cruz, Santa Cruz, California, United States of America
Howard Hughes Medical Institute, Chevy Chase, Maryland, United States of America

Vinícius S. Nunes, Maria Isabel N. Cano and Marcela Segatto
Departamento de Genética, Instituto de Biociências, Universidade Estadual Paulista (UNESP), Botucatu, São Paulo, Brazil

Marcelo S. da Silva
Departamento de Genética, Instituto de Biociências, Universidade Estadual Paulista (UNESP), Botucatu, São Paulo, Brazil
Universidade Estadual deCampinas (UNICAMP), Campinas, São Paulo, Brazil

Peter J. Myler
Seattle Biomedical Research Institute, Seattle, Washington, United States of America
Department of Global Health, University of Washington, Seattle, Washington, United States of America
Department of Biomedical Informatics and Medical Education, University of Washington, Seattle, Washington, United States of America

Elton J. R. Vasconcelos
Seattle Biomedical Research Institute, Seattle, Washington, United States of America

Zalman Vaksman, Natalie C. Fonville and Hongseok Tae
Virginia Bioinformatics Institute, Virginia Tech, Blacksburg, Virginia, 24061, United States of America

Harold R. Garner
Virginia Bioinformatics Institute, Virginia Tech, Blacksburg, Virginia, 24061, United States of America
Genomeon LLC, Floyd, Virginia, 24091, United States of America

Nicolas Tchitchek and Angela L. Rasmussen
Department of Microbiology, University of Washington, Seattle, Washington, United States of America

David Safronetz, Heinz Feldmann and Hideki Ebihara
Laboratory of Virology, Division of Intramural Research, National Institute of Allergy and Infectious Diseases, National Institutes of Health, Rocky Mountain Laboratories, Hamilton, Montana, United States of America

Craig Martens, Kimmo Virtaneva and Stephen F. Porcella
Genomics Unit, Research Technologies Section, National Institute of Allergy and Infectious Diseases, National Institutes of Health, Rocky Mountain Laboratories, Hamilton, Montana, United States of America

Michael G. Katze
Department of Microbiology, University of Washington, Seattle, Washington, United States of America
Washington National Primate Research Center, University of Washington, Seattle, Washington, United States of America

Jun Uetake
Transdisciplinary Research Integration Center, Minato-ku, Tokyo, Japan
National Institute of Polar Research, Tachikawa, Tokyo, Japan

Sota Tanaka, Hideaki Motoyama and Satoshi Imura
Faculty of Science, Chiba University, Chiba, Chiba, Japan

Kosuke Hara
Graduate School of Science, Kyoto University, Kyoto, Japan

Yukiko Tanabe
Institute for Advanced Study, Waseda University, Shinjuku-ku, Tokyo, Japan

Denis Samyn
Department of Mechanical Engineering, Nagaoka University of Technology, Nagaoka, Nigata, Japan

Shiro Kohshima
Wildlife Research Center, Kyoto University, Kyoto, Kyoto, Japan

Hinco J. Gierman, Kristen Fortney, Glenn J. Markov, Justin D. Smith and Stuart K. Kim
Depts. of Developmental Biology and Genetics, Stanford University, Stanford, CA, United States of America

Jared C. Roach, Hong Li, Gustavo Glusman and Leroy Hood
Institute for Systems Biology, Seattle, WA, United States of America

Natalie S. Coles and L. Stephen Coles
Gerontology Research Group, Los Angeles, CA, United States of America
David Geffen School of Medicine, University of California Los Angeles, Los Angeles, CA, United States of America

Bhanupratap Singh Chouhan, Konstantin Denessiouk, Alexander Denesyuk and Mark S. Johnson
Structural Bioinformatics Laboratory, Biochemistry, Department of Biosciences, Åbo Akademi University, Turku, Finland

Jarmo Käpylä and Jyrki Heino
Department of Biochemistry, University of Turku, Turku, Finland

Rodrigo Pessôa, Jaqueline Tomoko Watanabe, Paula Calabria and Alvina Clara Felix
Virology Department, São Paulo Institute of Tropical Medicine, University of São Paulo, São Paulo, Brazil

Ester C. Sabino
Department of Infectious Disease/Institute of Tropical Medicine, University of São Paulo, São Paulo, Brazil

Michael P. Busch
Blood Systems Research Institute, San Francisco, California, United States of America

Paula Loureiro
Pernambuco State Center of Hematology and Hemotherapy, Recife, Pernambuco, Brazil

Sabri S. Sanabani
Clinical Laboratory, Department of Pathology, Hospital das Clínicas, School of Medicine, University of São Paulo, São Paulo, Brazil

Arnfinn Lodden Økland, Are Nylund, Steffen Blindheim, Kuninori Watanabe and Heidrun Plarre
Department of Biology, University of Bergen, 5020 Bergen, Norway

Aina-Cathrine Øvergård
SLRC-Sea Lice Research Center, Institute of Marine Research, 5817 Bergen, Norway

Sindre Grotmol
Department of Biology, University of Bergen, 5020 Bergen, Norway 3 SLRC-Sea Lice
Research Center, Department of Biology, University of Bergen, 5020 Bergen, Norway

Carl-Erik Arnesen
Firda Sjøfarmer AS, 5966 Eivindvik, Norway

John B. Hogenesch
Institute for Biomedical Informatics, University of Pennsylvania School of Medicine, Philadelphia, PA, United States of America
Institute for Translational Medicine and Therapeutics, University of Pennsylvania School of Medicine, Philadelphia, PA, United States of America
Department of Pharmacology, University of Pennsylvania School of Medicine, Philadelphia, PA, United States of America

Jonathan M. Toung and Nicholas Lahens
Genomics and Computational Biology Graduate Program, University of Pennsylvania School of Medicine, Philadelphia, PA, United States of America

Gregory Grant
Institute for Biomedical Informatics, University of Pennsylvania School of Medicine, Philadelphia, PA, United States of America
Institute for Translational Medicine and Therapeutics, University of Pennsylvania School of Medicine, Philadelphia, PA, United States of America
Department of Genetics, University of Pennsylvania School of Medicine, Philadelphia, PA, United States of America

Fan Fan, Mallikarjuna R. Pabbidi, Stanley V. Smith and Richard J. Roman
Department of Pharmacology and Toxicology, University of Mississippi Medical Center, Jackson, Mississippi, United States of America

Aron M. Geurts and Howard Jacob
Human and Molecular Genetics Center, Medical College of Wisconsin, Milwaukee, Wisconsin, United States of America

David R. Harder
Department of Physiology and Cardiovascular Research Center, Medical College of Wisconsin, Milwaukee, Wisconsin, United States of America

Xiaoyan Xu, Yubang Shen and Jianjun Fu
Key Laboratory of Exploration and Utilization of Aquatic Genetic Resources, Shanghai Ocean University, Ministry of Education, Shanghai 201306, PR China

Liqun Lu
National Pathogen Collection Center for Aquatic Animals, College of Fisheries and Life Science, Shanghai Ocean University, 999 Huchenghuan Road, 201306 Shanghai, PR China

Jiale Li
Key Laboratory of Exploration and Utilization of Aquatic Genetic Resources, Shanghai Ocean University, Ministry of Education, Shanghai 201306, PR China
E-Institute of Shanghai Universities, Shanghai Ocean University, 999 Huchenghuan Road, 201306 Shanghai, PR China

Zehra Agha
Department of Biosciences, Faculty of Science, COMSATS Institute of Information Technology, Islamabad, Pakistan
Department of Human Genetics, Nijmegen Centre for Molecular Life Sciences, Radboud University Medical Centre, Nijmegen, the Netherlands
Department of Bioinformatics and Biotechnology, International Islamic University, Islamabad, Pakistan

Zafar Iqbal, Lisenka E. L. M. Vissers, Christian Gilissen, Joris A. Veltman and Rolph Pfundt
Department of Human Genetics, Nijmegen Centre for Molecular Life Sciences, Radboud University Medical Centre, Nijmegen, the Netherlands

Maleeha Azam, Humaira Ayub, Syeda Hafiza Benish Ali and Moeen Riaz
Department of Biosciences, Faculty of Science, COMSATS Institute of Information Technology, Islamabad, Pakistan

Hans van Bokhoven
Department of Human Genetics, Nijmegen Centre for Molecular Life Sciences, Radboud University Medical Centre, Nijmegen, the Netherlands

Department of Cognitive Neurosciences, Donders Institute for Brain, Cognition and Behaviour, Nijmegen, The Netherlands

Raheel Qamar
Department of Biosciences, Faculty of Science, COMSATS Institute of Information Technology, Islamabad, Pakistan
Department of Biochemistry, Al-Nafees Medical College & Hospital, Isra University, Islamabad, Pakistan

Maximilian Nepel, Jürg Schönenberger and Veronika E. Mayer
Division of Structural and Functional Botany, Department of Botany and Biodiversity Research, University of Vienna, Vienna, Austria

Hermann Voglmayr
Division of Systematic and Evolutionary Botany, Department of Botany and Biodiversity Research, University of Vienna, Vienna, Austria
Institute of Forest Entomology, Forest Pathology and Forest Protection, Department of Forest and Soil Sciences, BOKU-University of Natural Resources and Life Sciences, Vienna, Austria

Martin Carlsen and Peter Røgen
Department of Applied Mathematics and Computer Science, Technical University of Denmark, Kongens Lyngby, Denmark

Patrice Koehl
Department of Computer Science and Genome Center, University of California Davis, Davis, CA, United States of America

Rosa Mastrogiacomo, Paolo Pelosi and Andrea Serra
Department of Agriculture, Food and Environment, University of Pisa, Pisa, Italy

Chiara D9Ambrosio and Andrea Scaloni
Proteomics & Mass Spectrometry Laboratory, ISPAAM, National Research Council, Napoli, Italy

Alberto Niccolini and Angelo Gazzano
Department of Veterinary Sciences, University of Pisa, Pisa, Italy

Josanne H. Verhagen, Oanh Vuong, Theo Bestebroer, Pascal Lexmond and Ron A. M. Fouchier
Department of Viroscience, Erasmus MC, Rotterdam, The Netherlands

Jacintha G. B. van Dijk
Department of Animal Ecology, Netherlands Institute of Ecology (NIOO-KNAW), Wageningen, The Netherlands

Marcel Klaassen
Department of Animal Ecology, Netherlands Institute of Ecology (NIOO-KNAW), Wageningen The Netherlands
Centre for Integrative Ecology, School of Life and Environmental Sciences, Deakin University, Geelong, Australia

Index

A

American College Of Medical
Genetics And Genomics (acmg), 82
Annelid Capitella Teleta, 1
Annotation And Analysis, 54
Autoregulation Of Cerebral Blood Flow, 142
Avian Influenza Viruses, 212, 216, 220

B

Bf1 Recombinants, 101, 107, 109-110, 112
Blood Donors In Pernambuco, 101, 112
Blood Feeding Copepod Parasite, 114

C

Carton Galleries, 173, 179-181
Chaetothyrialean Fungi, 173, 179-180
Complement (cypome), 1
Conserved Template Sequence, 26
Cypome Annotation, 5

D

De Novo Assembly, 64, 116, 152-153, 160-161
Deep Sequencing Of Hiv-1, 101
Detection Theory, 130, 136
Disappearing African Glacier, 65
Disseminating 71_bf1, 101
Distance Measures, 183-184, 186-193, 198
Dna Repair Defects, 40, 42
Dna Repair Deficiencies, 40
Dual-specificity Protein Phosphatase-5, 142, 150

E

Early Chordate Origin, 85
Exome Sequencing Analysis, 40
Exome Sequencing Identifies, 162
Exome-wide Somatic Microsatellite Variation, 40

F

Fawn Hooded Hypertensive (fhh) Rats, 142
Fhh.1bn Rats, 142-150
Full-length Proviral Genomes, 101
Fungal Hyphae, 173, 178

G

Generalized Linear Mixed Model (glmm), 215
Genomic Characterization, 24, 114, 129

Glacier Biology, 65
Grass Carp Ctenopharyngodon Idella Transcriptome, 152

H

Heterozygotic Loci, 41, 48, 50-52
High Diversity, 173, 176
Homologue Detection, 97
Human Gene Mutation Database (hgmd), 78

I

Identification Of Rna-dna Sequence, 130
Identify Mirna Targets, 152
Immunoprecipitation Assays, 31
Indirect Immunofluorescence (iif), 33
Intellectual Disability, 162, 165, 171-172

L

Local Circulation, 212
Low Dna Allelic Frequencies, 15, 19
Low Specificity, 173

M

Microsatellites (mst), 40
Migratory Birds, 212, 217-218, 220
Moss Gemmae, 65, 68-69, 71-72
Motile Aeromonad Septicemia, 152-153
Mst Minor Allele Caller, 40
Myogenic Response, 142-143, 146-151

N

Neotropical Ant-plant Association, 173
Neurocognitive Disorder, 162
Next-generation Sequencing (ngs), 152
Novel Biogenic Aggregation, 65

O

Odorant-binding Protein, 201, 210-211
Optimal Growing Temperatures, 65

P

Parasitic Copepod, 114, 127-128
Pathogenic Variants, 78
Phylogenetic Position, 74, 114, 124-125
Phylogenetic Tree Construction, 63, 97
Proteins, 1, 13-14, 17, 22, 27, 31, 39, 61, 64, 85-86, 91, 93-95,
99, 114, 117, 119-122, 127-128, 140, 144, 146, 152-153, 155-
156, 161-162, 170-171, 183, 186-189, 192, 199-202, 209-211

Putative Leishmania Telomerase Rna (leishter), 26

R
Rabbit, 30-31, 33, 57-58, 143, 201-204, 209-211
Radia, 15-24
Reduced Immunocompetence, 212
Ribonucleoprotein Complex, 27, 31, 33
Rna And Dna Integrated Analysis, 15
Rna-sequencing (rna-seq), 130

S
Salmon Lice, 114-117, 119, 127-128
Salmon Louse (lepeophtheirus Salmonis), 114
Salmon Louse Chalimi Stages, 114
Seminal Fluid, 201-204, 208, 210
Sequencing, 15, 18-20, 22, 24-25, 31, 37, 39-43, 45-46, 51-55, 61-64, 73, 75-76, 79-84, 86, 99, 101-103, 106-114, 116-117, 128, 130-131, 133, 135-138, 140-141, 143, 145, 152-155, 157, 159-172, 176, 181, 202, 204, 210, 214
Sigmavirus/dimarhabdoviruses Cluster, 114
Somatic Mutation Detection, 15
Somatic Single Nucleotide, 15-16
Sperm Cells, 201, 203-204
Steroidogenic Cyps, 11-12

Supercentenarians (110 Years Or Older), 75
Syrian Hamster (mesocricetus Auratus) Transcriptome, 54

T
The Cancer Genome Atlas (tcga), 15
Three Novel Candidate Genes, 162
Train And Test Knowledge-based Potentials, 183
Trans-splicing, 26-27, 32, 37-38
Transmission Electron Microscopy (tem), 115, 117
Tropical Regions, 65
Two New Species In Rhabdoviridae, 114
Two Novel Circulating Recombinant Forms (crf) 70_bf1, 101

U
Unique Ecosystem, 65

V
Vertebrate Integrin Ai Domains, 85

W
Whole-genome Sequencing, 75-76, 83-84
World's Oldest People, 75

Z
Zinc-finger Nuclease Knockout, 142

www.ingramcontent.com/pod-product-compliance
Lightning Source LLC
Chambersburg PA
CBHW080531200326
41458CB00012B/4399